BOSTON STUDIES IN THE PHILOSOPHY OF SCIENCE

VOLUME IV

BOSTON STUDIES IN THE PHILOSOPHY OF SCIENCE

VOLUME IV

PROCEEDINGS OF THE BOSTON COLLOQUIUM FOR THE PHILOSOPHY OF SCIENCE 1966/1968

EDITED BY ROBERT S. COHEN AND MARX W. WARTOFSKY

D. REIDEL PUBLISHING COMPANY / DORDRECHT-HOLLAND

ISBN-13: 978-94-010-3380-0 e-ISBN-13: 978-94-010-3378-7
DOI: 10.1007/978-94-010-3378-7

PREFACE

The fourth volume of *Boston Studies in the Philosophy of Science* consists mainly of papers which were contributed to our Colloquium during the past few years. The volume represents a wide range of interests in contemporary philosophy of science: issues in the philosophy of mind and of language, the neurophysiology of perceptual and linguistic behavior, philosophy of history and of the social sciences, and studies in the fundamental categories and methods of philosophy and the inter-relationships of the sciences with ethics and metaphysics. Papers on the logic and methods of the natural sciences, including biological, physical and mathematical topics appear in the fifth volume of our series. We have included in the present volume the first English translation of the classic and fundamental work on aphasia by Carl Wernicke, together with a lucid and appreciative guide to his work by Dr. Norman Geschwind.

The papers were not written to form a coherent volume, nor have they been edited with such a purpose. They represent current work-in-progress, both in the United States and in Europe. Although most of the authors are philosophers, it is worth noting that we have essays of philosophical significance here written by a sociologist, an anthropologist, a political scientist, and by three neurophysiologists. We hope that collaboration among working scientists and working philosophers may develop further.

The fourth and fifth volumes of *Boston Studies in the Philosophy of Science* complete the publication of papers which were presented during the period of support of our Colloquium by the U.S. National Science Foundation. It is a pleasure to acknowledge once again our thanks to the Foundation for such generous support over many years.

R. S. COHEN
M. W. WARTOFSKY

TABLE OF CONTENTS

NORMAN GESCHWIND*

THE WORK AND INFLUENCE OF WERNICKE

The following paper consists primarily of sections of two previous publications dealing with Carl Wernicke and his influence. The first section is based on the paper 'Carl Wernicke, The Breslau School and The History of Aphasia' (Geschwind, 1966), while the second is based on 'Wernicke's Contribution to the Study of Aphasia' (Geschwind, 1967).[1]

I. THE BRESLAU SCHOOL

The political difficulties of Germany at the close of the 18th century were reflected in the fortunes of the German universities, a large number of which either disappeared or merged with other institutions. The list of the universities affected included some of the most distinguished of the medieval foundations. Wittenberg, so closely associated with Luther and the origins of the Reformation, dissolved to be united with Halle in 1815. It was not alone. The two decades from 1798 to 1818 also witnessed the total or partial dissolution of the universities of Mainz, Cologne, Bamberg, Dillingen, Duisburg, Rinteln, Helmstedt, Salzburg, Erfurt, Altdorf, Frankfurt an der Oder, Ingolstadt and Münster.

It might easily have been expected that this massive disappearance of ancient seats of learning would have seriously weakened the position of German scholarship. The political conditions of the time might appear at first view to have been equally unfavorable to the flowering of learning. After the Congress of Vienna, Germany consisted of 35 principalities and four free cities, each in total control of its own internal affairs.

As is so often the case, however, the obvious rational expectation turns out to be incorrect and the 19th century was, in fact, a period of great success for German scholarship. There were several factors which played a role in this upsurge of learning. For one thing periods of confusion are sometimes more conducive to that passion and freedom which is so essential to great scholarship than are periods of undisturbed tranquility. There were perhaps even more important reasons for the growth of the

German universities despite the buffetings of the time. Reforms are often easier in periods of upheaval when ancient practices lose their emotional force. Many of the universities which did disappear in the early 1800's were, in fact, despite their illustrious histories, only medieval relics, which deserved to vanish. Their dissolution in some cases strengthened other institutions which absorbed their faculties. The place of these extinct institutions was taken by newer and more vigorous institutions more suited to the new times. Two of the greatest universities, Berlin and Bonn, were founded only in the second decade of the 19th century, while Münich, another great seat of learning, was reorganized.

The crazy-quilt pattern of independent principalities probably was favorable rather than detrimental to the growth of learning. In such countries as France and England, in which the medieval fractionation of power among independent duchies and baronies had long since been abolished, there had been a powerful trend toward centralization of cultural life in or near the capital, to the detriment of provincial cities and universities. In Germany, by contrast, almost any professor could feel satisfied that he was in or near a capital. This situation persisted for most of the 19th century in Germany in contrast to the situation in France and England. The existence of many small independent states with a high level of sophistication must have had an effect on scholarship similar to that which the existence of independent Scandinavian countries has on performance in the Olympic games.

Another benefit of the decentralization of power was that many of these states took pride in their universities and supported them more handsomely than would a central government anxious to concentrate its efforts on the aggrandizement of the capital. Many of the German princes had in fact prided themselves on their roles as patrons of learning.

There was still one further element in German university life which was beneficial to the widespread development of scholarship. One of the most useful features of medieval German university life had been carried over into the 19th century despite the decline of so many of the outworn traditions. The German student (often the able and ambitious one, although all too frequently the one whose main qualification was wealth), was likely to spend time in several universities, wandering like his medieval forbear to sit at the feet of great masters of learning. The great scholar or scientist could thus build his own prestige and not have to depend on that of his institution. This system also favored an interchange of ideas that kept intellectual life in ferment. Even if the German professor was perhaps too often the supremely remote *Geheimrat* who brooked no disagreement, the tendency for the student to go elsewhere insured the infusion of new ideas into the system. At a higher level this interchange continued and there was a strong tradition of appointment of professors from the outside which tended to prevent departments from freezing into a uniform pattern. This tendency to appoint outsiders extended even beyond national boundaries to include all of the German-speaking world and even some non-German speaking countries where German scholarship was dominant. The orbit thus included not only the German states but also much of the Austrian empire and Switzerland, and even Holland.

German neurology and psychiatry, like the rest of German medicine, shared in the benefits of this system. It received tremendous impetus from the growth of nationalism and yet manifested the psychological and material advantages of decentralization and participated in the cross-fertilization resulting from the free movement of scholars. While in France and England the growth of neurology and psychiatry was very closely tied to London and Paris, these fields of knowledge flowered widely throughout the German speaking world. The independence of the many great centers was so well established that it survived even the unification of Germany and the increased emphasis on Berlin as the capital of the amalgamated empire. This impetus carried well into the 20th century. Permanent major contributions came from such widely dispersed centers as Berlin, Breslau, Frankfurt am Main, Graz, Halle, Heidelberg, Jena, Königsberg, Leipzig, Münich, Prague, Vienna and Zürich and from such workers as Westphal, Jolly, Bonhoeffer, Wernicke, Alzheimer, Liepmann, Oppenheim, Edinger, Goldstein, Kleist, Krafft-Ebing, Wagner von Jauregg, Anton, Freud, Hitzig, Kraepelin, Nissl, Berger, Flechsig, Meynert, von Gudden, Grashey, Pick, Pötzl, Forel, Bleuler, Foerster, von Monakow, Vogt, Brodmann and Spielmeyer, to name only some of the most distinguished of those working at the turn of the century or immediately afterwards.

Aphasia was one of the most vigorously pursued areas of study in neurology in the forty years from 1874 to 1914 and its fountainhead was to be Carl Wernicke, working in one of the most peripheral of the German universities, Breslau. This city was the capital of the Prussian province of lower Silesia, which bordered on Poland. It was the eighth largest of the German cities and was a major industrial center. It must have been very different from the great universities of the German heartland because of its nearness to and long historical associations with Slavic Europe. It bordered on the Polish province of Galicia, which itself had a very heavy admixture of Jews on a background of Poles and other Slavs. Today, Breslau itself is no longer a German city but with a much reduced population it is once again part of Poland under the name Wroclaw.

Carl Wernicke was himself born in Silesia, in Tarnowitz (now Tarnowskie Gory) not far from the Galician border and now also part of Poland. He grew up in a family of very constrained circumstances which were only worsened by the death of his father when Wernicke was 17. Despite these difficulties, his mother's ambitions for her son led her somehow to provide for his further education in the hope that he would become a minister. Just before his final examination she too died, never fully to appreciate the fruits of her efforts. He was graduated in medicine in 1870 at the age of 22. After a brief period in the university eye clinic and as assistant to a surgeon in the Franco-Prussian War, he joined the psychiatric service at Breslau.

In 1874, at the age of only 26, he published his epoch-making work on aphasia, *Der aphasische Symptomencomplex* subtitled *Eine psychologische Studie auf anatomischer Basis.* This 72-page monograph was to set the tone for research in aphasia over the next 40 years. To a considerable extent this work must be regarded as the real beginning of modern research in aphasia. Aphasic disorders had been recognized sporadically over many centuries, and there had indeed even been some impressive clinical studies. Benton (1964) has recently reviewed the prehistory of aphasia and given an impressive account of the forerunners.

The modern history of aphasia begins, however, clearly with Paul Broca. Critchley (1961) and Joynt (1964) have recently reviewed extensively and put into proper perspective the achievements of Broca. There has been considerable argument over the difficult question of priority between the supporters of Broca and those of Dax. As far as the modern history of aphasia is concerned, however, Dax, even if the claims for him are correct, can still belong only to the period of prehistory. His work like that of the other forerunners was at best an isolated flash of insight which set no new activities in motion. While it was Broca who first succeeded in arousing widespread interest in aphasia, to Wernicke goes the credit for the establishment of the theoretical framework which guided most of the significant investigations over the next 50 years. In the second portion of this paper we will discuss in some detail Wernicke's contribution and the reasons for its predominant influence.

Wernicke's influence was the result not only on his work but also on his ability, not common to all distinguished figures, of developing great students. The roster of those who came under his influence at Breslau is an impressive one. It was not merely that his students were distinguished. They learned to develop to full advantage that combination of talents which had served Wernicke so well, a profound respect for careful observation and a feeling for the elucidation of mechanisms.

Wernicke's own activities extended into many fields and most of his contributions are now parts of the foundations of neurology. He elucidated (with Mann) the pattern of weakness in spastic hemiplegias, described what is now known as Wernicke's disease and its pathology, proved the origin of cortical deafness in bilateral temporal lesions, traced the pyramidal tract to the spinal cord and established much of the picture of cortical sensory loss. He contributed heavily to the delineation of the symptom pictures of organic psychoses. He brought to recognition the important distinction between remote memory and the ability to learn new material, or recent memory, as it is now conventionally called in neurology.

This great production had its profound effect on his pupils. The first of these students in one sense was Karl Bonhoeffer since he was first of Wernicke's assistants to be offered the 'Habilitation' (the first step in being formally permitted to teach publicly at a university). This followed Bonhoeffer's demonstration that a patient with unilateral chorea (until then always regarded as a functional psychiatric disorder) had a lesion in the mid-brain. Bonhoeffer succeeded to the professorship in Breslau after Wernicke's departure for Halle, and subsequently was to hold the chair of psychiatry and neurology in Berlin. He devoted himself to many problems including aphasia and cerebellar speech disturbances; he wrote the classical descriptions of alcoholic delirium and other alcoholic psychoses and of the memory disturbances in Korsakow's syndrome. He delineated the generally accepted symptom picture of the toxic psychoses.

If the leading academic psychiatrist of Germany after World War I was a pupil of Wernicke's so was the leading neurosurgeon of Germany,

Otfrid Foerster, who eventually became chief of the neurological clinic in Breslau. In a series of papers published over nearly 40 years he recorded an impressive list of contributions. He pioneered in the field of rehabilitation, made extensive studies of disturbances of movement and co-ordination and of the surgical treatment of these disorders. His work on sensation is classical and included studies on the anatomy of the sensory pathways and on the neurosurgical treatment of pain. Some of his greatest contributions came from his studies of stimulation of the human brain and associated studies on epilepsy and its surgical treatment. He wrote little on aphasia but even this little reflected the influence of Wernicke's teaching. The first description of the speech disturbances resulting from frontal parasagittal excision as presented by his assistant Schwab (1927), is an example.

While Bonhoeffer and Foerster were the most prominent, other students of Wernicke also made great advances. What is especially remarkable is the influence that he and his assistants had on the study of the aphasias and related disorders. Wernicke himself had established the foundations of the study of aphasia. His assistant Lissauer (1889) was to apply the same principles to an analysis of the problems of the agnosias, in a paper opening with a masterful classical clinical description. Still another assistant, Hugo Liepmann, produced almost single-handedly in a few years the classical clinical pictures and analyses of the apraxias. Of all his students Liepmann was to bring Wernicke's method of analysis of the higher functions to its most brilliant development and to prove more than anyone else its ultimate worth. His name belongs in this field as one of the creators of the foundations, along with Broca and Wernicke.

Hugo Liepmann's early background contrasted sharply with that of Wernicke. While the master had grown up in difficult circumstances, Liepmann was born in 1863 to wealth and remained a man of independent means throughout his life. He had begun his career in philosophy with a Ph.D. thesis on 'The Atoms of Leucippus and Democritus' and he produced a later work on Schopenhauer. There is a story told by his pupil and friend Hildebrandt that, displeased with philosophy, he hurled a volume by a prominent neo-Kantian into a corner and turned to medicine. Perhaps the story of this dramatic gesture is apocryphal, but Liepmann did turn to a new career.

Liepmann received his second degree, that in medicine, in 1894, at the age of 31. He was already five years older than Wernicke had been when he produced his early masterwork, but Liepmann was not long in making up the lost time. In that same year he produced his first neurological work, a pioneer study on the experimental production of hallucinations. It was his medical dissertation and was entitled, 'The Deliria of Alcoholics and Experimentally Induced Visual Hallucinations', a neglected path-finding work in a field in which interest has revived only since the last war.

In 1895 he went to Breslau to spend his four most important formative years. In this period he produced his study of pure word-deafness and confirmed at postmortem Lichtheim's anatomical predictions. In 1900 he returned to Berlin and over the next decade produced his series of papers on apraxia. It was he who first used this term in its modern sense. Liepmann's first great paper in this area was published in 1900, the study of the famous *Regierungsrat*. His accomplishment was two-fold. He began by demonstrating that a patient who at first appeared to be globally demented had in fact a combination of remarkable never previously observed disturbances. He then proceeded to predict, on the basis of the known anatomy of the cortex and its fiber pathways, the localization of the lesions. Two years after the publication of this paper and its predictions as to the underlying pathology, the patient died and the autopsy confirmed Liepmann's expectations. Wernicke was especially delighted with this work as well he might be. It remains to this day the most brilliant example of the 'psychological analysis on an anatomical basis' which the master had first developed a quarter of a century earlier.

Liepmann continued this work in the following years. In 1907 he and Otto Maas presented the clinical picture of disconnection of the hemispheres by a lesion in the corpus callosum which was demonstrated at postmortem. Kurt Goldstein (1908) also a former student of Wernicke's, was to confirm this a year later with another proven case. Other case reports followed, including one by Bonhoeffer (1914). This work was generally neglected in the English-speaking world and only today are we rediscovering the callosal syndromes which were common knowledge to Liepmann's contemporaries.

If the Breslau school was to find its scientific activities so richly awarded, it was not to find the outside world quite so satisfying. Inadequate formal recognition, indeed even active administrative persecution and violent events harassed the careers of the major figures of this school.

Wernicke was to find himself involved in difficulties from the start. I have already mentioned the financial difficulties of his beginnings intensified by the loss of his father and the death of his mother before he had even received his degree. Following the publication of his early master work in 1874 he left Breslau to work in Berlin at the Charité Hospital under Westphal. But within two years he was dismissed because as Liepmann put it, "He came into conflict with the direction of the Charité in a non-scientific and non-official situation, which was intensified by Wernicke's obstinacy...". What the situation was I have not found recorded, but one rumor at least is that it concerned a woman. Wernicke went into private practice with no academic connection. A possible appointment at Heidelberg was lost when he became ill for six months. He applied for an appointment at the municipal mental hospital Dalldorf (whose name had the association to Berliners that Bedlam had to Londoners) but was turned down, it is said at the instigation of the great pathologist Virchow who was such a powerful force in medicine in Berlin and who would not forgive him the battle at the Charité. The Charité incident was never to be forgotten by Wernicke. Not only was Virchow angered, but also Althoff, the administrative official who controlled appointments in the Prussian universities harassed Wernicke for the rest of his life.

Finally in 1885 he received an appointment at Breslau. Wernicke's old friend, Förster, the head of the ophthalmic clinic, had succeeded in obtaining this post for his brilliant former pupil despite the opposition of Althoff who had blocked all other routes of advance successfully. The years here were, however, to have a bitter end. Wernicke had used the municipal psychiatric hospital as the source of most of his clinical material. He had tried fruitlessly for some years to get a university clinic built but had repeatedly been refused on economic grounds. Liepmann records his repeated complaint, only too familiar to those in academic life that "12 million (marks) are granted without any further question for a warship, but the measly million for a clinic is not to be had". The idea of competition with a warship may sound quaint to our ears, but the complaint is still well known. The city fathers of Breslau who had objected to the arrangement finally withdrew their hospital connection. The psyciatric clinic now existed only on paper! For the next two years Wernicke was permitted to use patients for demonstration in clinical lectures to students but even this privilege was then withdrawn. Wernicke still had a neurological clinic available to him and a private clinic. The marked decline in the workload of the department, however, had the advantage that it gave him considerably more free time to finish his *Foundations of Psychiatry*. It seems clear, however, that Wernicke lacked the ability to convince himself of the benefits of adversity.

This incredible slap in the face was not all he had to suffer in these years. Three major professorships fell vacant. According to some he refused the chair vacated by Krafft-Ebing in Vienna in 1902. Niessl von Mayendorf claims, however, that Wernicke would have been delighted to take over the chair once held by Meynert but that, before an approach was made to him, several Viennese newspapers carried the statement that Wernicke would refuse if offered the position. According to Mayendorf the rumor became so strongly established that Wernicke never actually received the offer.

A vacancy was created at Munich by the retirement of Bumm but was filled by Kraepelin, Wernicke's great opponent in purely psychiatric matters. Finally, with the death of Jolly, there became available the directorship of the University Nerve Clinic at the Charité in Berlin which Wernicke had left so unceremoniously 28 years before. The medical faculty of the university nominated Wernicke as their first choice for the chair. But the education ministry, under the direction of Wernicke's life-long antagonist Althoff, who had maintained his grudge over all these years, turned him down, and Ziehen, the professor at Halle, was chosen instead.

In 1903 the direction of the Breslau psychiatric hospital was returned to Wernicke, who for five years now had been the director of a clinic which existed only on paper. The five years had left their mark, however, and in 1904 when Wernicke received an offer from Halle (the chair having been vacated by Ziehen's move to Berlin) he accepted it. He thus took over the chair founded by Hitzig, the founder of the experimental study of localization in the brain of man. The brief period of slightly more than a year was to be perhaps the happiest in Wernicke's life. Here he was now in charge of a beautiful university clinic. He seemed to have relished leaving Breslau in many ways. He had never participated in sports because of his small, weakly build but now took enthusiastically to athletic activities such as mountain-climbing, skating, and cycling. This was to be his undoing.

On June 13, 1905 Wernicke and his assistant, Berthold Pfeifer, went cycling in the Thuringian forest. They met a wagon loaded with logs and as they cycled past it, Wernicke fell. A rear wheel of the wagon ran over his breast-bone. He apparently surmised that his injury was fatal and told Pfeifer that he probably had a hemorrhage into the pericardium. He was taken to a nearby country inn. Wernicke forbade his wife to be brought so that she would be spared his miserable death, but nevertheless she came. He was concerned that the wagon driver should not be held responsible and insisted that he alone was to blame. It is recorded that in a quiet moment during this period, he said, in the nomenclature of his own psychiatric textbook, "I am dying as the result of autopsychic disorientation"; the latter phrase Kleist interprets as meaning incorrect assessment of his own ability (to ride a bicycle). His friend von Koschützky recorded that after four miserable days of conscious suffering, he raised his voice for the first time to denounce the clumsiness of Fate. He died of his injuries on June 17, 1905 at the age of 57. The advances in thoracic surgery to be made a few years later might have saved him.

Of his important pupils Bonhoeffer was perhaps the most unlike him personally. He appears from the statements of his biographers and the accounts of those who worked with him to have had the classical Greek quality of measure in all things. Hiw own career had none of the upsetting character of Wernicke's. He left Breslau in 1903 to assume the professorship in Königsberg but left there within a year to take over the chair at Heidelberg which Kraepelin had vacated in his move to Munich. Althoff, Wernicke's old nemesis, was angered by this change and threatened that Bonhoeffer would never again receive an appointment at a Prussian university. But with Wernicke's departure from Breslau Bonhoeffer was now called to take over his chair in that city. Althoff could not interfere since he was no longer in office. It is worth noting here parenthetically the vindictiveness of the fate which had pursued Wernicke. The advantages he might eventually have garnered from the final departure of his old opponent from office were made illusory by his own death a few months later. In 1912 Bonhoeffer went to Berlin where he remained for the remainder of his life.

This early history of uninterrupted success however did not test Bonhoeffer's mettle and it was not until the tragic rise of Nazism in Germany that Bonhoeffer's tremendous personal strength was demonstrated. The rise of Hitler affected everything including the usually well-ordered life of a professorial neuropsychiatric unit. Bonhoeffer handled these problems, as he had all others, with Apollonian equanimity. One of his assistants, Fred Quadfasel, who, despite an impeccable East Prussian Protestant ancestry, had shown his dislike of the new regime and was put in prison. Bonhoeffer told Quadfasel that being in prison was no disgrace since members of his own family had been imprisoned during the revolution of 1848. Subsequently it was Bonhoeffer who arranged with a cousin of his, a nobleman in the diplomatic corps to recover the passport confiscated from Quadfasel and thereby enabled him with the help of others to get the immunity of a diplomatic railroad car and thus leave the country.

Bonhoeffer's difficulties as the leading psychiatrist of the capital of the Reich must have been prodigious but as recorded by Stertz he dealt with these problems with sovereign calm. He never had a picture of Hitler on his wall, and after the passage of the sterilization laws by the Nazis he never reported a patient as hereditarily defective. Yet to the remark by an assistant that he hated the Nazis he replied, "A psychiatrist does not hate, he understands". His moral fiber seems to have been conveyed to others of his family circle. Two of his sons and two sons-in-law died at the hands of the Nazis. Of these the most distinguished was his son, Pastor Dietrich Bonhoeffer who despite his youth was one of the leading figures in German protestant theology (his books have been widely republished in recent years) and who was one of the leading conspirators along with Dohnyani, one of the sons-in-law, in one of the plots to overthrow Hitler. Arrested and kept in prison, he refused under torture to give the names of his conspirators. He was killed as an act of spite on April 9, 1945 and his brother Klaus on April 22 – both died within a month of the end of the war.

Somehow the elder Bonhoeffer survived it all. A son who had also come through this period became one of Germany's leading physical chemists. Karl Bonhoeffer lived, as Stertz tells us, to enjoy an 80th birthday and golden wedding anniversary surrounded by a large family circle. The *Monatsschrift für Psychiatrie und Neurologie*, which had been founded by Wernicke and Ziehen, and which Bonhoeffer had edited for many years, honored him with a special Jubilee volume, edited by his former assistant Paul Jossmann who had come to the United States during the

Nazi regime. The American Psychiatric Association elected him an honorary member. Bonhoeffer regarded this, as Jossmann pointed out, as a token of the revived internationality of science rather than as a personal honor. He had himself noticed the warning signs of the stroke of which he was to die on December 4, 1948. In the announcement of his death in the newspapers was the phrase, "Despite severe suffering a happy life".

Of all the pupils of Wernicke Foerster was to have the most brilliant worldly success. He refused to leave Breslau but the world beat a path to his door. The list of those who came to study with him is a roll-call of the modern greats of neurosurgery and honors fell on him one after another. Sherrington even expressed the view that he was worthy of a Nobel prize. Personal tragedies, however, clouded his life. His first-born daughter died shortly after birth. In the early 1930's his talented assistant Schwab lost his wife in childbirth and shortly thereafter killed the child and committed suicide. The coming of the Nazis added further burdens. His wife was of Jewish extraction and some of his outstanding assistants had been Jewish. In addition he was himself probably suspect for his political views and for having spent a considerable period in Russia as the physician of the hemiplegic and aphasic Lenin. Foerster lacked the Olympian detachment of Bonhoeffer in the face of the Nazis and personally suffered more from the political changes. He died on June 15, 1941 of tuberculosis. His wife in caring for him had contracted the same disease and died two days later.

Of all of Wernicke's Breslau assistants, Kleist records that Liepmann was closest to him, and it was certainly he who was to repeat most closely the first brilliant achievement of the master. His career was to be much like Wernicke's. After leaving Breslau, he returned to Berlin to work at Dalldorf where Wernicke had been refused a post so many years earlier. For nearly the remainder of his career he was to stay here, never to be offered a professorial chair despite the wide recognition of his work and his acceptance in Berlin as one of the outstanding lights of German neuropsychiatry. His disability was not his personality, but another even less changeable one. Liepmann was Jewish and as the great sociologist Max Weber had pointed out, Jews could not expect to advance in German academic life. Perhaps one of the reasons for the popularity of medicine among German Jews was that it was one of the few fields in which a man could make a significant scientific accomplishment without an academic appointment. Kurt Goldstein records that Liepmann was told that he would be offered an academic position if he converted. Although not religious, he refused on principle. Liepmann lost 60 pounds during World War I because he would eat no more than the ration allotted to his patients. It is not likely that such a man would have been tempted by the suggestion that he compromise with prejudice.

His later years were darkened by the onset of Parkinson's disease. On his 60th birthday the *Zeitschrift für die gesamte Neurologie und Psychiatrie* honored him. Two years later, faced with the relentless progression of his disease, he committed suicide. Niessl von Mayendorf (1925–1926) began his obituary with a dramatically apt quotation from King Lear,

> Oh you mighty Gods!
> This world do I renounce; and in your sights
> Shake patiently my great affliction off....

and went on to write in an equally dramatic vein, "Hopelessly languishing away he tore the thread of his life from the ever-spinning Norns and cut it with manly decisiveness...".

His death was to be like Wernicke's in his long hovering between life and death. Hildebrandt recorded that long after he had poisoned himself and become severely clouded, his mind cleared remarkably and it was thought for a time that he might survive. During this lucid interval he told his wife, "Read the Phaedo", Plato's account of the death of Socrates, a curious reminder of Liepmann's early interest in Greek philosophy. So died Wernicke's most direct heir, "a worthy representative", as Kurt Goldstein (1953) was to put it, "of a great creative epoch in neurology".

II. WERNICKE'S INFLUENCE ON THE HISTORY OF APHASIA

In this part I will treat in detail Wernicke's contribution to the study of aphasia, in particular his classic work, *Der aphasische Symptomencomplex,* published in 1874.

In order to put this publication in perspective, let us briefly review its background. Wernicke had received his medical degree in 1870 at the age of 22 from the University of Breslau. He worked after this for six months in the ophthalmology clinic under Förster, who was probably the first to recognize the young man's exceptional talent, and then went off as an assistant to a surgeon in the Franco-Prussian war. On his return to Breslau he took up a post as assistant in the psychiatric clinic. Carl Weigert, the developer of the stain for myelin that led to such important advances in neuroanatomy, was a fellow assistant. During this time Fritsch and Hitzig reported their important discoveries on the localized excitability of the cerebral cortex of man. Wernicke spent six months during this period with the great neuroanatomist Meynert in Vienna and this was probably the most important influence of his life. Having returned from Vienna he soon afterwards published *Der aphasische Symptomencomplex* which was completed in May, 1874, less than four years after his graduation and after only three years of training in neurology. He was only 26 years old in the month that he completed the manuscript.

The major influence on this work was the short period with Meynert who influenced not only Wernicke but also many others, including Flechsig, the pioneer of developmental studies of the brain, the great anatomist Forel and Freud. Meynert was one of the very first to be concerned with the relationship of anatomy to function, an interest that he transmitted to Wernicke. Thus, in his early work Meynert correlated the increased size of certain portions of the motor pathways in the brain stem and spinal cord of the kangaroo with the great preponderance of the hind-limbs in that animal, and showed the contrasting pattern in the bat with its predominance of the forelimbs, thus also establishing himself as one of the pioneers in comparative neuroanatomy. He correlated development with order of myelination thus laying the ground for Flechsig's later work. He brought order into the fiber systems of the brain by distinguishing the projection pathways connecting the cortex to the lower centers, the association pathways connecting different regions of the same hemisphere, and the comissural systems joining corresponding points in the two hemispheres. He pioneered the study of cortical cytoarchitecture by being the first to point out regional differences of cellular architecture in the cortex. He was the first to point out that the anterior parts of the brain had motor functions while the posterior parts were sensory in nature. In reviewing these accomplishments it is easy to see why it was written of him that "Erst seit Meynert ist das Gehirn beseelt" ["Only since the work of Meynert has the brain been given life"]. As I will point out later, Meynert made a specific contribution to the study of aphasia that has generally been overlooked and which influenced Wernicke's thoughts on aphasia.

THE WORK AND INFLUENCE OF WERNICKE

In the short space of 13 years between the publications of Broca in 1861 and Wernicke in 1874 interest in aphasia had grown rapidly. Bastian and Hughlings Jackson were already active in England, as were also a large number of French and German writers, and the literature of the subject was already sizeable. Wernicke's contribution to this growing literature was entitled *Der aphasische Symptomencomplex: eine psychologische Studie auf anatomischer Basis* [The Symptom-complex of Aphasia: a Psychological Study on an Anatomical Basis]. In the very opening of the paper he points out that he will apply Meynert's teachings to the problem of language and aphasia. Following this brief introduction we come to the first section in which Wernicke immediately makes clear his theoretical position.

He begins this section by pointing out that the doctrine of the equopotentiality of the brain which had been advocated by the great French physiologist Flourens had been recognized as untenable for some time but that it was equally incorrect to substitute for it the view of the phrenologists who arbitrarily assigned individual territories of the cortex to mental functions (as arbitrarily defined by the terminology of ordinary speech). Wernicke thus rejected the approach to cerebral function which denies all localization, but he equally strongly rejected the approach which was to appear in more sophisticated form in other writings on aphasia, i.e., the one which assumed that for every specifiable language function there was a particular piece of cortex assigned to that function, the approach of such authors as Henschen and above all Kleist (who, although a student of Wernicke's, eventually adopted a theoretical view of quite different nature from his).

In rejecting these extreme approaches which we may label as the holistic and mosacist respectively, Wernicke offers his alternative: "...only the most elementary psychic functions can be assigned to defined areas of the cortex..." and as examples of such elementary functions he cites visual, olfactory, and tactile perceptions in the central projection areas of the respective sensory nerves. "Everything which goes beyond these simplest functions, the association of different impressions into a concept, thinking, consciousness, is an achievement of the fiber tracts which connect the different regions of the cortex to each other, the so-called association system of Meynert. We will see later to what extent these more complicated achievements are even now amenable to examination." This then is the underlying thesis: The higher mental functions are not properties of localized regions of the brain, but are constructed out of the association of different regions, and it is the anatomical connections of these regions which make possible these associations.

Wernicke now turns to developing his ideas in detail. It is a characteristic of all nervous tissue, he writes, to preserve the effects of stimulation beyond the duration of the stimulus itself. In peripheral nerves this effect is brief and accounts for some of the phenomena of after-images, but in the cortex these residua of previous excitations may remain for long periods and it is these residua for which the name 'memory-images' is suggested.

Basing himself on what then were recent physiological and anatomical observations, Wernicke points out that the parts of the brain lying anterior to the central fissure of Rolando are motor regions, while the remaining portions of the hemispheres are sensory in function. Thus the memory images of movements must lie in the frontal lobe, those of sensations in the appropriate regions of the other lobes.

He now turns to the problem of volitional movement. This is distinguished from reflex movement in two ways: (1) It does not occur in immediate response to an outside stimulus but can result from the memory images of former perceptions which need not be aroused in all instances by an external stimulus. Hence, although volitional movements differ from reflex movements, they have an essential unity with them. (2) The volitional movement is adapted more closely to its goal. It is, so to speak, 'preformed' being built up of memory images of previous movements.

Volitional movements are thus also reflex movements but are distinguished by their dependence on remembered stimuli arising within the cortex which arouse remembered movements stored elsewhere in the cortex. As the number of sensory and motor images increases, the more associational possibilities exist, the greater becomes the variety of movements, and the greater becomes the choice of movements which may be available in any given situation, the exact selection depending on all the factors involved at any one time. Thus, writes Wernicke in closing Section I, "The only correct scientific definition of free will agrees most closely with this mechanistic concept of the origin of spontaneous movement".

In Section II Wernicke turns to the specific problem of aphasia. He comments that although Broca's views as to the importance of the posterior end of the third frontal convolution[2] for language had encountered considerable opposition, the publication of so many cases in agreement with his ideas, and the lack of effective counter views had led to continuing acceptance of Broca's position. In the interim, however, Meynert had published a most important work in which he demonstrated that the central projection of the auditory nerve lay in the cortex of the Sylvian fissure, and to this cortical region he assigned the designation of *Klangfeld* (sound-feld). He supported his view by numerous autopsy findings in aphasics in which either the insular region or the adjoining parts of the Sylvian fissure were altered. Many academic physicians had adopted this view since it corresponded to the vaguely felt need to bring hearing into association with the process of speech and because it helped to account for the fact that most cases of aphasia in which Broca's area was intact had changes in the region specified by Meynert, i.e. in the temporal operculum and adjoining regions.

I might comment here that the significance of Meynert's contribution to the study of aphasia has generally been overlooked. In the discussions of the history of aphasia that I have so far consulted, he is mentioned only as Wernicke's teacher. Nielsen (1946) in his bibliography cites two works by Meynert on aphasia published in 1866 but does not refer to these in the text. It is important to appreciate that it was apparently he who first convincingly showed that aphasia could occur in temporal lobe lesions. It was Wernicke's function to complete the cycle started by Meynert by showing how this fact achieved meaning in terms of Meynert's own studies of the brain, and to point out the important fact that this new localization corresponded to a clinical picture different from that seen in lesions of Broca's area.

Wernicke now develops his central argument. The frontal lobe was already known at that time to be motor in function. Hitzig had shown that stimulation of the lower end of the Rolandic cortex in monkeys led to bilateral movements of mouth and tongue. Broca's area lying immediately anterior to this region must be the center for the representation (i.e., memory-images) of the movements elicited from this region while the first temporal gyrus lying at the central projection of the auditory nerve must be the center for the images of sounds. The fibers between these regions, which Wernicke thought to run in the insula, must be the mediators of the psychic reflex arc between the heard and spoken word. It should be pointed out that Wernicke (1908) was later to modify his view on the path between the two cortical speech areas and to accept von Monakow's view that it ran not in the insula but in the arcuate fasciculus, arching from the posterior temporal region around the back end of the Sylvian fissure and then running forward in the white matter of the parietal lobe until it reached the posterior inferior frontal region.

Wernicke now gives his analysis of the types of aphasia to be expected from various lesions. Peripheral lesions of the auditory pathway, not involving the first temporal gyrus, lead to deafness but not to disturbances in speech production in the adult, who already has a rich store of earlier acquired sound images which he can reproduce at will. But if peripheral deafness occurs before there is a consolidated store of sound images, the child remains mute since he has acquired no store of sound images with which to arouse movements of speech. Sound thus has a special relationship to speech which is shared by no other sense organ. As Wernicke notes, mutism occurring in a deaf child is a common and easily understood disorder, but the rare situation of a child endowed with normal hearing who cannot learn to speak rightfully deserves the name of congenital aphasia.

Wernicke enlarges his discussion of deaf-mutism to include an intriguing suggestion whose specultative character he is quick to admit. Of what significance, he asks, is unilateral deafness in childhood? This would lead to deaf-mutism only if the right ear were affected and only the left hemisphere capable of acquiring language, which seems unlikely. But perhaps right-sided deafness would lead the child to develop the right hemisphere more for language than the left since he is now depending on his left ear. Could, he asks, such a process account for the existence of people with speech in the right hemisphere? This argument of Wernicke's would have seemed totally untenable a few years ago since the representation of each ear in both cerebral hemispheres (unknown at the time Wernicke was writing his paper) would seem to invalidate it. However, the recent emphasis on the Broadbent technique (Kimura, 1961) has again highlighted the more intimate relationship of each ear to the opposite hemisphere and perhaps warrants re-consideration of this speculation of Wernicke's.

Wernicke continued his analysis of aphasia to discuss the effect of a lesion of the first temporal gyrus, which has come to be called Wernicke's area in his honor by later authors.[3] The sound images of the names of objects will be lost. From this region arise the fibers which run from the auditory to the motor region, but these are now destroyed. In addition the association fibers from this region by which other sensory images are aroused will be destroyed. "The patient is thus neither capable of repeating nor understanding the spoken word." The representations of speech movements are retained since these were learned earlier and thus "the capacity to speak is preserved but with some limitation. For in usual act of speaking the sound image always appears unconsciously to be innervated simultaneously (as is indeed readily understandable from the process of development of speech), to be so to speak simultaneously hallucinated, and thereby to exercise a continuing correction on the course of the movement representations. The deaf person ... still makes full use of this corrective mechanism." This stress on internal correction has a very contemporary ring to it. The patient is thus also aphasic in speech as a result of the loss of this unconscious correction exerted by the sound image. "The errors consist in the ready confusion of words ... He uses the same words now correctly, now incorrectly ...". In the milder cases it is names of objects and actions which are most affected but the conjunctions and other words necessary for sentence structure are preserved although these can be lost in more severe cases. But in all cases two cardinal features persist – the patient has many words, and he lacks comprehension of spoken language.

In order to understand Wernicke's next comment we must try to place ourselves in the position of his readers. Wernicke was describing what he thought to be a new clinical syndrome. Although he was not cognizant of the fact, Bastian (1869) and Schmidt (1871) had already described aphasias with comprehension disturbances, but these papers had gone unnoticed and it was Wernicke who was the first to succeed in calling receptive aphasias to the attention of the medical world. Wernicke had, however, gone further than these authors in linking incomprehension of language to a type of speech disturbance different from that of Broca's aphasia, i.e., to what we would now speak of as a fluent paraphasic disturbance. Finally he had linked this new syndrome to a localization different from that of the classical Broca's aphasia.

Wernicke writes, "That cases with such manifestations have not been observed, or at least not published, until now, rests not only on the rarity of such cases but also on the fact that thoroughly experienced and intelligent physicians regard this condition as a confusional state – as I myself have had the opportunity to observe. For the psychiatrically trained man who knows the clinical forms of confusional states, the diagnosis presents not the least difficulty." It is perhaps unfortunate that nearly a hundred years after the publication of Wernicke's paper the error he pointed out is still a common one, and patients with fluent paraphasic syndromes are still often misdiagnosed as confused or psychotic.

Wernicke continues to list some further features of his new syndrome. A large potential vocabulary is a cardinal feature of this form, i.e., although the patient may fail to find words or uses them incorrectly, he will at times produce almost any word, in contrast to the situation in Broca's aphasia. There is no hemiplegia, in contrast again to Broca's aphasia. "Agraphia is present. Writing is a conscious movement which is learned with intimate dependence on the sound and is always executed under its direction. Self-observation teaches us (and clinical experience agrees with this) that between the movement of writing and the concept there is no direct connection such as must be assumed between the speech movement and the concept." Wernicke now argues, however, that the educated reader will not be alexic since he has learned to understand the written word without a rendering into spoken form although he will read aloud aphasically. The uneducated by contrast who always reads aloud and who therefore must turn the written word into its auditory form will be alexic. This supposition would probably not be accepted today since clinical experience does not support it. Both the educated and uneducated develop alexia with this syndrome. Wernicke goes on to point out that with other lesions in the visual regions, agraphia and alexia alone could occur.

A comment is appropriate here on this syndrome of fluent paraphasic speech and disturbance in writing and comprehension difficulty for spoken and written language which came to be known as Wernicke's aphasia. It should be noted that Wernicke did not use the term 'word-deafness' which was introduced by later writers. Furthermore a certain number of authors state incorrectly that Wernicke claimed that a lesion of the first temporal gyrus produced the syndrome of pure word-deafness, i.e., difficulty in comprehension of speech out of all proportion to disturbances in the other modalities. Wernicke did not describe this syndrome nor did he even envisage its existence in this paper. It was not until Lichtheim (1885) that the theoretical possibility of this syndrome was suggested, and only later was it described clinically by Liepmann who was one of Wernicke's assistants in Breslau (Liepmann, 1898). Furthermore, Wernicke did not conceive of the syndrome he described as being a composite one, i.e., he did not argue that four separate lesions were necessary to produce the full syndrome, but asserted instead that as a result of the way in which language is learned and used all modalities would be affected by a disturbance of the auditory memory images.

Wernicke now considered the effect of a lesion involving the fibers connecting the sound field to the area of representations of movement but leaving each one of these regions intact. Wernicke called the resulting disturbance conduction aphasia (Leitungsaphasie). The patient understands everything, and can speak freely, but his speech is altered in the same way as in the form of aphasia just described resulting from a lesion of the auditory field. For, although the sound images are preserved in conduction aphasia they cannot exert an influence on the choice of words. The patient can judge, however, whether his own productions are correct or incorrect and shows concern over his errors. In pure form, the patient will show in speech a great similarity with the sensory aphasic, but his comprehension is intact. This form of aphasia will show disturbances in reading and writing depending on how these were learned in childhood.

The problem of conduction aphasia continued to trouble Wernicke over the next thirty years, since it seemed to him that the clinical pictures and anatomical findings he had expected did not seem to coincide. In fact he saw many patients with paraphasic speech and good comprehension but did not always find the expected lesions. He had on the other hand seen lesions of the insula but without this clinical picture.

This confusing situation became clarified in two ways. In the first place Wernicke had placed the pathway between the posterior and anterior speech regions in the insula. But Wernicke (1908) was later to be convinced by von Monakow's researches that the major pathway between the two speech areas lay in the arcuate (or superior longitudinal) fasciculus in the lower parietal region. Most later students of this problem have agreed with this localization.

There was, however, an even more important source of error which resulted from the fact that Wernicke had failed to analyze his own diagram correctly. He had omitted the deduction that there should be a disturbance in repetition in conduction aphasia. Lichtheim (1885) correctly added this deduction and cited a case showing this picture. Freud (1953) returned to this problem in 1891. Freud argued as Lichtheim had that the disconnecting lesion between the motor and sensory area should produce a loss of repetition in the face of intact comprehension. Freud goes on to remark that this situation is highly unlikely. Yet within the next 20 years the triad of paraphasia, intact comprehension and impaired repetition was to become well known. In 1904, Karl Kleist, then an assistant of Wernicke's at Halle demonstrated to Wernicke himself a case which finally convinced him of the existence of this entity. Kleist was to describe several such cases. In addition the syndrome was described by others such as Liepmann and Pappenheim and Pötzl and Stengel. Kurt Goldstein (who had also been a student under Wernicke) was to call the complex of paraphasia with intact comprehension and poor repetition 'central aphasia.[4]' In recent years interest has again revived in this entity and its existence has been reconfirmed.

Wernicke lists finally the effects of destruction of Broca's area. The patient understands everything but is mute or employs only a few simple words, which he uses mostly for the designation of objects. Most previously described cases of aphasia belonged to this group. Wernicke thought that agraphia would not result unless the patient had been accustomed to speak to himself while writing. The patient will understand written language unless he has always read aloud.

Finally he turns to lesions in the descending pathways from the frontal lobe. These he assumes could produce aphasia, but may not do so since the descending fibers may be rearranged so that individual parts of the speech act might be impaired, thus leading to dysarthria. Wernicke shows no knowledge in this analysis of the role of callosal connections. His failure to appreciate these facts made him believe that a *complete* lesion of one cerebral peduncle could lead to aphasia. His pupil Bonhoeffer (1914) was to add refinement to this analysis by considering the role of callosal connections and innervation of the speech organs by the right hemisphere in providing an alternative route from Broca's area in low lesions of the pyramidal tract.

Wernicke closes this analysis by assuming that in many cases the patient improves because of eventual substitution by the other hemisphere, a view that has remained a source of unsettled controversy even today.

Wernicke now discusses the problem of those deaf-mutes who by means of very highly complicated special teaching techniques learn to produce speech. He argues that in such patients the auditory soundfield must play much less of a role than it does in the normal acquisition of speech. Such patients could therefore become aphasic but their lesions would involve different parts of the brain from those which would have to be damaged in normals. In particular, lesions of the first temporal gyrus might well be ineffective in producing aphasia. He urges the importance of studying post-mortem material from deaf-mute aphasics. We see here again Wernicke's emphasis on the fact that aphasia is produced not by lesions of fixed language areas but rather by damage to those areas which in any individual have been used in acquiring speech.

Wernicke now turns to a discussion of 'Asymbolie', a term introduced by Finkelnburg. He argues that loss of recognition of an object results from the loss of the optic, tactile, and auditory components which go to make up the concept. The patient so affected would differ from the aphasic in that he would indeed fail to understand the significance of objects he has perceived. Wernicke admits that current anatomical knowledge is insufficient to specify the locus of the lesions. We see here a discussion of what later was to be called agnosia, to the study of which Lissauer (1889), one of Wernicke's assistants, made significant contributions.

Section III of the paper gives histories and clinical findings of ten patients. The first case was that of Suzanne Adam, a 59-year old lady who showed very poor comprehension and repetition, fluent, mildly paraphasic speech, alexia, and agraphia, without gross neurological findings otherwise. She showed marked improvement, except for her agraphia. She did not come to post-mortem examination. However, writes Wernicke, this case recalled to him another earlier case, who had not been so well studied, because, he says, at the time he saw her he had not yet worked out a correct analysis of the symptom complex of aphasia. This was the case of Susanne Rother, a 75-year old lady who had been regarded when first admitted as suffering from aphasia with confusion. Her speech, while not so plentiful as that of the first cited case, was fluent and paraphasic. (It should be noted that the word paraphasia was never used by Wernicke at this time and came into literature later. He speaks instead of misused and distorted words.)

She had severe comprehension deficit which made some people think she was deaf. At post-mortem the Sylvian branch of the left middle cerebral artery was occluded. There was softening of the 1st temporal gyrus, and this was the only focal lesion found in the hemispheres.

It is on this case 2, admittedly studied inadequately, that the Wernicke argument rests since the remainder of the cases discussed did not show the cardinal syndrome described in the paper, that of fluent paraphasic speech with comprehension difficulty. Case 8 (Louise Funke), however, came to post-mortem after the paper was completed and her findings were given in a brief appendix to the paper. In life she had shown the use of only one word 'ja' and understood nothing. At post-mortem a major branch of the Sylvian branch of the middle cerebral artery was thrombosed. There was a large Sylvian infarct which included both Broca's area and the 1st temporal gyrus. Wernicke writes, "...in both cases [this one and case 2] the first temporal gyrus and its anastomosis with the second were involved. Both suffered from sensory aphasia. Could this agreement be the result of a coincidence?"

On this meager anatomical evidence his argument rested. Like Broca's first case it was a slim reed on which to found so thorough-going a theory of aphasia. Other authors have claimed that none of Wernicke's cases supported his assertions, but this seems difficult to accept since the clinical and post-mortem findings of case 2 and case 8 were clearly consistent with his thesis although they were not sufficient to prove it. Henry Head (1926) dismissed Wernicke's conclusions, but without taking into account the results in cases 2 and 8. It is perhaps ironic that Head himself cites cases of his own which he could have advanced in support of Wernicke's localizations. Thus his case 14 was a case of what in his nomenclature was called syntactical aphasia, the lesion of which Head places "in and around the upper temporal gyri and the parts beneath them". He says of the case that "...his speech was jargon...his words poured out in phrases which...were in most cases incomprehensible. He could not repeat a sentence said to him... He was unable to find names for common objects... Comprehension of spoken words was obviously defective...". It is almost as if he were quoting directly from Wernicke. Why Head so summarily dismissed Wernicke when his own evidence supported him so strongly must remain a mystery.

Wernicke closes his paper with some further comments. Baginsky, he notes, had spoken of 'centrifugal' and 'centripetal' forms of aphasia. But, he says, it is one thing to create hypothetical centers and quite another to construct a theory by converting known anatomical data into psychological form. He stresses the necessity of further anatomical knowledge. In the appendix he credits Hitzig with recently published views on aphasia similar to his own independently developed ideas.

At this point it is reasonable to consider why this paper has so great an influence. One might entertain various theories. A consideration of some features of Wernicke's career may help to clarify some of the possibilities. He was only 26 when the paper was written, without previous publications, unknown, without academic rank, of an undistinguished family, in a provincial university, without the support of major well-known figures. One can effectively dismiss the view that it was sheer prestige, academic authority or influence which were the sources of the effect of this paper. The growing prestige of German science in general would not account for the selection of his work over that of other better known German authors. A general tendency to accept diagram-making would again not be an adequate explanation since many writers of the day were producing their own diagrams.

Nor could the effect of the paper be attributed solely to Wernicke's later career. In the first place it had achieved a wide influence long before he himself had attained advancement. His academic progress was disappointingly slow and for several years in Berlin he was without academic connections. Having once appointed to the professorship in Breslau, he failed to move on from there to any of the more distinguished chairs. Hence, he never achieved great political influence within the German academic world. Furthermore he did very little further work on aphasia, leaving this area primarily to others. He advanced many other areas of neurology in later years but the aphasias and related disturbances were left to his students and others to prosecute. In fact, it is important to realize that he considered his major activity to be his psychiatric investigations eventually to be summarized in his text *Grundriss der Psychiatrie in klinischen Vorlesungen*. But while his work on aphasia was so successful, his psychiatry was to have little influence, although his psychiatric writing appeared during his period of greatest prestige. The same distinguished students who carried on his ideas in aphasia made the gesture of supporting their teacher's psychiatric teachings, but without enthusiasm. Yet

Wernicke's psychiatry was also based on a simple diagram which failed of acceptance. The accusation of 'brain-mythology' was made about his psychiatric teaching, yet his anatomical ideas on aphasia succeeded despite the unsuccessful attempts to apply the same epithet to them. In fact, the most successful psychiatric investigator of the period was Emil Kraepelin, who differed fundamentally from Wernicke. His psychiatry was not based on anatomy, was without diagrams, and was thoroughgoingly based on clinical descriptions.

Since this was a period in which psychiatry and neurology were still intimately linked in Germany, the very same people who accepted Wernicke's diagrammatic anatomical approach to aphasia were simultaneously adopting a very different psychiatric teaching. This should make us wary of accepting any simple conclusion that there was a general tendency to accept anatomical explanations or simple diagrams without question. Furthermore, scientific controversy raged in the German academic life of that period and disagreements were aired freely and at great length.

Perhaps only a more thorough investigation will unearth the full reasons for the acceptance of Wernicke's first paper, and I can here only suggest some possibilities. In the first place his paper brought new clinical facts to attention. As I have already mentioned, Bastian (1869) and Schmidt (1871) were the first to call attention to the sensory varieties of aphasia (and Wernicke, 1908, who had been unfamiliar with these works was later to give them full credit), but their papers had gone unnoticed and it was Wernicke's work which succeeded in disseminating knowledge of these forms. He stressed the existence of fluent aphasias in contrast to the previously well-known forms, and called attention to the error of diagnosing these as confusional or psychiatric syndromes. He brought all of these elements together into a syndrome described for the first time. Furthermore, he had resolved the problem of the existence of cases of aphasia without damage to Broca's area by showing that Meynert's demonstration of cases of aphasia with temporal lobe lesions was clarified by the existence of a syndrome of temporal aphasia different from that of frontal aphasia. He established the localization of his syndrome of sensory aphasia, to be confirmed by later authors. It is important in viewing the paper to see it in the context of its period. Neuroanatomy was only beginning. The exact course of the auditory pathways was still unknown and Wernicke pointed out that the somesthetic pathways were still a mystery. Standards of clinical examination were hardly developed, and pathological examination was based on gross findings. This paper came at the beginnings but was to be one of the forerunners of the great revolution in neurology.

Yet the most important contribution of this paper lay at a still deeper level. It is reflected in Wernicke's subtitle to this work, *A Psychological Study on an Anatomical Basis.* It is worthwhile to consider the previous history of aphasia in order to understand this point more fully.

At first glance one might assume that Broca's success should have greatly strengthened the cause of phrenology since his discovery was the result of an attempt to verify or disprove a prediction made by workers in the phrenological tradition. Success in prediction, however, as Alfred North Whitehead once indicated, is a criterion which despite its practical appeal is often lost on scientists who demand something further. Successful predictions arising from certain kinds of theories, e.g., from astrology, are likely to do little good for their promulgators. At the other extreme, even the most successful prediction based on curve-fitting by the addition of further arbitrary constants will, as Whitehead asserts, leave the scientist dissatisfied despite the practical utility of such results. The scientist demands that a prediction arise from a hypothesis which itself is integrated into a wider body of knowledge. It is precisely this bridging of gaps which made so dramatic such work as Newton's on color or Hodgkin and Huxley's on the nature of ionic movements in the nerve impulse. It is not merely that the demand for elegance and parsimony is so abruptly met, that Occam's injunction is heeded beyond expectation and the number of basic entities is not only not increased but actually reduced, but that this linkage of areas not previously seen to be related is among the most fruitful of all stimuli to research. Phrenology even in its more sophisticated forms simply could not supply a theory which linked to the rest of science. Its very success in Broca's original discovery may have only hastened its own downfall in increasing the appetite for a theory that could link meaningfully to other knowledge.

It was precisely this demand that Wernicke satisfied. He supplied the localization of aphasia not merely with a further new set of facts, he provided a theory which made it appear possible to tie these phenomena to knowledge already acquired about the nervous system and thus made possible the development of a scientific approach to aphasia. On the basis of this theory one was able to predict the existence of syndromes not previously seen and to devise experimental means of testing hypotheses. No one else had supplied this. Little wonder that the appearance of this paper by an unknown 26-year old physician with less than four years of clinical experience and only six months study of anatomy under Meynert in Vienna was considered so astonishing.

It was this characteristic of Wernicke's approach that made it so fruitful over the next forty years and that has given it such viability. However brilliant are some of the clinical insights of a Jackson or a Head, they suffer from not being readily derivable from anatomical facts or being fruitful in deriving new experiments. The Wernicke approach has been forced to adapt itself repeatedly to criticism but in the end the core of Wernicke's method, while certainly by no means the only useful technique, has remained the most consistently useful one over the years.

The predictive and explanatory power of this theory appears in the work of Lichtheim (1885) who extended Wernicke's analysis and was thus able to predict correctly the lesion responsible for pure word-deafness. Dejerine's analyses of isolated acquired reading deficits was developed on a further extension of this analysis as were the contributions of Liepmann and Bonhoeffer to the clinical pictures resulting from callosal lesions.

It may seem to some that perhaps this account of the Breslau school and its approach to aphasia neglects the existence of other influences in the history of the study of the higher functions of the nervous system. In most histories of aphasia the view is advanced that there were two separate streams, the localizationist initiated by Wernicke and the anti-localizationist school to which such names as Marie, Head, von Monakow, Goldstein, and Pick are often ascribed.

It would take us too far afield from our discussion to treat this problem thoroughly. I have dealt with this issue elsewhere (Geschwind, 1964) particularly as concerns the position of Kurt Goldstein, and will only briefly recapitulate here. Any careful investigation of the original literature will in fact reveal that none of these classic critics brought evidence to discredit the major conclusions of Wernicke or his followers. Henry Head's syntactic aphasia, for example, is given the same clinical description and the same localization that half a century earlier Wernicke had given to that form of aphasia which now bears his name. Head's verbal aphasia obviously corresponds to the classical Broca's aphasia and indeed was given the same localization by Head. Despite Marie's objections, his localizations of aphasia were highly classical. As Liepmann was to point out, despite von Monakow's criticism of him the list of the localizations of apraxic lesions advanced by the great Russian-Swiss neurologist was identical with his own much older list. Finally Goldstein, as I have shown, supported in his writings over forty years (including his book on language in 1948), the main conclusions of Wernicke's approach and consistently praised Liepmann's epoch-making work.

Why is there this discrepancy between the common view of the existence of two opposed streams and the views expressed above? I believe that the answer will be found in a closer examination of the literature of aphasia in pre-World War I Europe. On reading this literature one soon realizes that there was no wholesale rejection of the Wernicke approach by any of the major figures. But if the Wernicke approach remained, as it were, the central theme of nearly all the approaches, it was to be elaborated in different ways. Wernicke had placed the balance in his subtitle: *A Psychological Study on an Anatomical Basis*. On the one side the theory could be twisted into being the basis of a multiplication of specific centers, a neophrenological approach as it were. But such multiplication of ad hoc centers had been rejected, as we have noted, by Wernicke. The extreme mosaicist view suffers from the same difficulties as the phrenological view: mechanisms become unimportant according to such a view; the whole point of linking psychology and anatomy vanishes. This pole of extreme mosaicism was supported not by Wernicke or his followers, such as Liepmann but by others such as Jendrassik, Henschen, and Kleist.

If the theory of aphasia ran into danger from a tendency by some towards excessive parcellation, it also ran into difficulties from the tendency of others to overpsychologize the study of aphasia, an approach which in its extreme form denied the importance of anatomy, or argued that aphasia must be considered completely separately on the psychological and anatomical sides. Even this view, however, was generally built on the core of the Wernicke approach. Thus, as I have pointed out elsewhere, Kurt Goldstein, who is often regarded as the spearhead of the psychological approach actually supported the main anatomical tenets of the Wernicke school; indeed some of Goldstein's writings, for example, on callosal syndromes, were in explicit support of Liepmann's analysis which was obviously in the classical Wernicke tradition. It is striking that the localizationists were often criticized for speaking of a concept 'center'. Yet this notion reappeared extensively in Goldstein as the 'concept field' (Begriffsfeld). Goldstein hastened to add that by this he meant no localized region but an extensive area of the brain outside the speech area (which he divided classically into sensory and motor zones). Yet Wernicke himself had disavowed a concept center as a localized entity and used this term in essentially the same sense as Goldstein's later use of Begriffsfeld.

The Wernicke view thus corresponded to neither of the views usually stressed in most histories. He was neither a mosaicist nor a holist, but placed the major emphasis on the development of the higher functions by means of the pattern of connections between regions representing elementary cerebral functions.

It is perhaps not surprising that the differences between views become smaller when they are looked at from our current vantage point. In the heat of scientific dispute it may be easily overlooked that investigators, in fact, agree far more than they disagree. Indeed if it were not for this fundamental agreement many disputes would be futile. The central theme of Wernicke's approach survived the extremist pressures on it. But it should not be presumed that aphasia stopped with Wernicke, or even with Liepmann any more than neurophysiology stopped with the establishment of its foundations by their distinguished contemporary Sherrington. The study of aphasia gained from the critiques. The attacks by the holistically oriented forced a constant reappraisal of clinical criteria so that non-focal phenomena could be distinguished from focal disturbances. Although such localizationists as Charcot had stressed the effects of early experience, the holists had more than anyone else beaten the drum for the importance of this factor. The localizer Bastian was praised by Freud and Goldstein for emphasizing that differences in clinical picture might reflect not anatomical distinctions but intensity of damage – the holists however emphasized this point more persistently. Writers like Goldstein, agreeing with such highly anatomically oriented workers as Niessl von Mayendorf, stressed the equipotentiality not of the whole brain, which was clearly untenable in any strict sense, but of the individual more or less localizable 'center'; certainly it was the holist Goldstein with his emphasis on the flexibility of the brain who made the greatest attack on the problem of re-education in aphasia; yet it was the localizer Foerster, one of

Wernicke's star pupils who had been the first to establish rehabilitation as a major neurological activity. The holistically oriented were more active in linking linguistics to aphasia, yet Pick, one of the pioneers in this approach, had been of the first to support Wernicke's views on the localization of sensory aphasias. The holists were thus in no sense unimportant and they left their permanent seal on thinking about aphasia; they studied new problems, fostered therapy and forced refinement of clinical observation and anatomy. In the end, they had, so to speak, a profound effect on the course of the stream but they did not establish a new channel.

The mosaicists were also useful. If their anatomy had no link to psychology their researches could be used by others who were more oriented in this direction. They made fundamental advances in our knowledge of the gross and fine anatomy of the cortex. If the holistically oriented had an aversion to discovering new syndromes, the mosaicists performed a valuable service in discovering remarkably isolated disturbances which cried for explanation. It is perhaps conforting to consider that the activities of highly skilled and intelligent men, even when they are wrong, are rarely totally wasted. The history of aphasia would have been poorer if the field had been diverted into either channel; its future history will be the richer if the many channels explored by those on either bank of the main stream are investigated further.

We can perhaps close this essay most appropriately with a quotation from one of those who are usually not regarded as being in the direct Wernicke tradition, although in fact, as we have noted he accepted a major portion of Wernicke's teachings and indeed even employed the connectionist approach in his analysis of several problems. Kurt Goldstein first learned his neurology from Wernicke in Breslau. Over half a century later he was to write the following words of this early experience, "...his way of examining patients and his demonstrations were so elucidating and stimulating that we who had the good fortune to attend his clinics were deeply influenced in our further consideration of neurological and psychiatric problems. We could never forget him. His influence can be seen in the work of each of his many pupils, not a small number of whom became men of stature in their own right in the profession..." (Goldstein, 1953).

Harvard Medical School, and Neurological Unit,
Boston City Hospital, Boston

BIBLIOGRAPHY

In writing this paper the author has drawn heavily on many personal conversations with Drs. Fred A. Quadfasel and Paul Jossmann, both former assistants of Bonhoeffer in Berlin, and Professor Ernst Jokl, a former assistant of Foerster's, all widely cognizant of the history of German neurology and psychiatry. He has also consulted extensively the original scientific writings of the members of the Breslau school and their contemporaries. Other information sources are included in the following list.

Bastian, H. C.: 1869, 'On the Various Forms of Loss of Speech in Cerebral Disease', *Brit. For. Med.-Chir. Rev.* 209–236; 470–492.

Benton, A. L.: 1964, 'Contributions to Aphasia before Broca', *Cortex* **1**, 314–327.

Bonhoeffer, K.: 1914, 'Klinischer und anatomischer Befund zur Lehre von der Apraxie und der "motorischen Sprachbahn"', *Mschr. Psychiat. Neurol.* **35**, 113–128.

Critchley, M.: 1961, 'Broca's Contribution to Aphasia Reviewed a Century Later', in *Scientific Aspects of Neurology* (ed. by H. Garland), Williams & Wilkins, pp. 131–141.

Dejerine, J.: 1892, 'Contribution à l'étude anatomo-pathologique et clinique des différentes variétés de cécité verbale', *C. R. Soc. Biol. (Paris)* **44**, 61–90.

Encyclopaedia Britannica, 1960.

Freud, S.: 1953, *On Aphasia; a Critical Study*, International Universities Press, New York.

Geschwind, N.: 1964, 'The Paradoxical Position of Kurt Goldstein in the History of Aphasia', *Cortex* **1**, 214–224.

Geschwind, N.: 1966, 'Carl Wernicke, The Breslau School and the History of Aphasia', in *Brain Function*, vol. III: *Speech, Language and Communication*, UCLA Forum in Medical Sciences, No. 4 (ed. by E. C. Carterette), University of California Press, Berkeley and Los Angeles, pp. 1–16.

Geschwind, N.: 1967, 'Wernicke's Contribution to the Study of Aphasia', *Cortex* **3**, 449–463.

Goldstein, K.: 1908, Zur Lehre von der motorischen Apraxie', *J. Psychol. u. Neurol.* **11**, 169–187; 270–283.

Goldstein, K.: 1927, 'Die Lokalisation in der Grosshirnrinde; nach den Erfahrungen am kranken Menschen', in *Handbuch der Normalen und Pathologischen Physiologie*, vol. X (ed. by A. Bethe *et al.*), Springer, Berlin, pp. 600–842.

Goldstein, K.: 1948, *Language and Language Disturbances; Aphasic Symptom Complexes and Their Significance for Medicine and Theory of Language*, Grune & Stratton, New York.

Goldstein, K.: 1953, 'Carl Wernicke (1848–1904)', in *The Founders of Neurology* (ed. by W. Haymaker), Thomas, Springfield, pp. 406–409.

Goldstein, K.: 1953, 'Hugo Karl Liepmann (1863–1925)', in *The Founders of Neurology* (ed. by W. Haymaker), Thomas, Springfield, pp. 326–329.

Haymaker, W. (ed.): 1953, *The Founders of Neurology*, Thomas, Springfield.

Head, H.: 1926, *Aphasia and Kindred Disorders of Speech*, Cambridge University Press, Cambridge.

Isserlin, M.: 1925, 'Hugo Liepmann zum Gedächtnis', *Z. Neurol. Psychiat.* **99**, 635–650.

Jossmann, P. B.: 1949, 'Professor Karl Bonhoeffer, 1868–1948', *Am. J. Psychiat.* **106**, 159–160.

Joynt, R.: 1964, 'Paul-Pierre Broca: His Contribution to the Knowledge of Aphasia', *Cortex* **1**, 206–213.
Kimura, D.: 1961, 'Some Effects of Temporal Lobe Damage on Auditory Perception', *Canadian J. Psychol.* **15**, 156–165.
Kirchoff, T. (ed.): 1921, 1922, *Deutsche Irrenärzte* (2 vols.), Springer, Berlin.
Kleist, K.: 1959, 'Carl Wernicke (1848–1905)', in *Grosse Nervenärzte*, vol. II (ed. by K. Kolle), Thieme, Stuttgart, pp. 106–128.
Kolle, K. (ed.): 1956, 1959, 1963, *Grosse Nervenärzte* (3 vols.), Thieme, Stuttgart.
Lichtheim, L.: 1885, 'On Aphasia', *Brain* **7**, 433–484.
Liepmann, H.: 1895, 'Ueber die Delirien der Alkoholisten und über künstlich bei ihnen hervorgerufene Visionen', *Arch. Psychiat. Nervenkr.* **27**, 172–232.
Liepmann, H.: 1898, 'Ein Fall von reiner Sprachtaubheit', *Psychiatrische Abhandlungen*, Breslau.
Lissauer, H.: 1889, 'Ein Fall von Seelenblindheit nebst einem Beitrage zur Theorie derselben', *Arch. Psychiat. Nervenkr.* **21**, 222–271.
Nielsen, J. M.: 1946, *Agnosia, Apraxia, Aphasia*, Hoeber, New York.
Penfield, W. and Roberts, L.: 1959, *Speech and Brain Mechanisms*, Princeton University Press, Princeton.
Schwab, O.: 1927, 'Über vorübergehende aphasische Störungen nach Rindenexzision aus dem linken Stirnhirn bei Epileptikern', *Deutsche Z. Nervenheilk.* **94**, 177–184.
Wernicke, C.: 1908, 'The symptom-complex of aphasia', in *Diseases of the Nervous System* (ed. by A. Church. Appleton), New York, pp. 265–324.

REFERENCES

* Supported in parts by grants NB-06209 and NB-08472 from the National Institutes of Health to the Boston University School of Medicine.

[1] Ed. Note: We acknowledge with thanks permission of the author and publisher to reprint these materials.

[2] Wernicke usually employs the nomenclature of Leuret which called this the first frontal convolution.

[3] Some recent authors, e.g., Penfield and Roberts (1959), have used the term to designate an area far larger than that referred to by Wernicke, but this is not in keeping with classical usage. The larger area specified by these authors is not cytoarchitectonically homogeneous. Furthermore, different types of aphasic syndrome occur with lesions of different portions of this large area. It would seem reasonable on all grounds to retain the classical usage.

[4] Some authors have recently used Goldstein's term central aphasia in a much broader sense than Goldstein himself did. He specifically equated his 'central aphasia' to conduction aphasia.

CARL WERNICKE*

THE SYMPTOM COMPLEX OF APHASIA

A Psychological Study on an Anatomical Basis

Meynert's theory of the connections of the brain contains the foundations of a precise physiology of the central nervous system. It provides only a broad, general outline, to be sure, but one of such brilliant inner truth that it can even now be applied without hesitation to individual cases.

The present study is an attempt to make practical use of Meynert's neuroanatomical teachings in an area in which fundamental principles of this kind are badly needed but up to now have scarcely been applied – the normal speech process and the disturbances of this process, which are known as aphasia.

Students of Meynert's work will readily recognize to what extent I am either merely repeating Meynert's views or presenting my own contributions. Because of the strict dependence on neuroanatomy which the treatment of this subject demands, it has been impossible to give Meynert credit explicitly in every instance in which it is due to him. In any case, everything of value in the present study is ultimately based on Meynert's work, since the theory put forward here follows almost automatically from the study of Meynert's writings and dissections.

I

The doctrine of the unity of consciousness as Flourens formulated it, in which all regions of the cerebral mantle are considered to be equipotential, has long been untenable.[1] Not, however, in the sense in which the phrenologists considered it untenable: they assigned psychic functions, arbitrarily differentiated on the basis of linguistic usage – magnanimity, dependence, parental love, sexual drive, etc., to areas of the cerebral cortex designated in an equally arbitrary way. Only the most elementary psychic functions can be assigned to specific areas of the cerebral cortex – e.g., visual perception, to the central zone of projection of the optic nerve, sensations of smell to that of the olfactory nerve, tactile sensations in the little finger to the central projection of the ulnar nerve, etc. The surface of the brain is a mosaic of very simple elements of this kind, which are characterized by their anatomical connections with the periphery of the body. Everything beyond these elementary functions, such as the linking of different sense impressions to form a concept, thought, and consciousness, is a function of the fiber tracts that connect different cortical regions with each other, i.e., a function of the association systems, to use Meynert's term. We will see later to what extent these complicated functions are even now accessible to our comprehension.

The *sense impressions* projected onto the cerebral cortex from the outside world *last longer than the external stimulus affecting the sense organ; they can reappear in the form of memory images (Erinnerungsbilder)* independently of the stimulus that produced them, *although in less vivid form.* This seems at first to constitute a specific difference between the discharge of nervous excitation in the cerebral cortex and in other pathways and way-stations in the nervous system. In the latter it would seem that when the excitation process has been completed there is no further trace of it.

It can, however, easily be shown that memory of some sort is characteristic of the entire nervous system. On the basis of experience, we can advance the following formulation: the resistance of a particular pathway to nervous stimulation is reduced with repeated use. Thus reflex movements take place most easily over pathways that have already been used frequently, and in the form that has occurred most often. The same principle applies to the practice of particular movements, as in playing the piano, and to the learning of distinctive acoustic impressions, as in listening to a symphony. When epilepsy is acquired as the result of a blow or a fall on the head, the resulting anemia of the pons, produced mechanically by a single incident, leads to generalized convulsions. From this moment on the pathway is affected by much more trivial influences, and even the physiological variations in circulation associated with emotions, etc. suffice to elicit a new epileptic attack. It is generally acknowledged that the more often the same process takes place the more difficult a cure becomes. Finally, the phenomenon of after-images proves, as Meynert has correctly emphasized, that the sensitive elements of the retina also have the capacity to retain excitement longer than the duration of the stimulus. If the sensation of color does eventually die out, it is primarily because, given the small number of sensitive elements, those which were just functioning are taken over by a new stimulus so that the residual sensation is drowned out. The cerebral cortex, however, with its approximately 600 million nerve cells, according to Meynert's estimate, provides a sufficient number of storage areas in which the countless sense impressions delivered by the outside world can be stored in succession. Thus the fact that cortical neurons undergo lasting molecular changes after stimuli lasting only a brief time cannot be considered unusual and specific to them. This property of the cortical neurons should instead be considered as the analogue of what takes place in peripheral nerves, heightened in degree by the special anatomical conditions in the cortex.

From here on we will call these residues of past excitation which are present in the cerebral cortex memory images, in distinction to the sense impressions themselves. Now, for reasons to be discussed later, we must assume that bodily movements and changes in the state of the musculature, give rise to sensations, memory images of which also remain in the cerebral cortex. For the sake of brevity we shall call these memory images of sensations of movements either representations of movement (Bewegungsvorstellungen) or movement images, and we will consider them equivalent to other memory images.

The entire surface of the brain can be divided into two large areas of functionally different significance: the frontal lobe, i.e., the entire area in each hemisphere anterior to the fissure of Rolando, and *the temporo-occipital lobe* (regarded as a single unit). The first is a motor area and contains representations of movement; the second is sensory, containing memory images of past sense impressions. The parietal lobe proper, which lies between them, is a transitional area. The nature of its functions is still under discussion.[2]

This assertion, one of the foundations of the new cerebral physiology, is based on the following considerations. According to Meynert, the portion of the cerebral peduncle that brings about voluntary movements has its central origin primarily in the cortex of the frontal lobes. Focal lesions in the cortex of the frontal lobes produce convulsions and paralyses on the opposite side. Among all the diseases of the surface of the brain, general paresis, which is clearly a motor disorder, is associated with atrophy predominantly of the frontal lobes, according to Meynert's measurements of brain weight.[3] Experimental proofs of the motor function of the frontal lobes were advanced by Fritsch and Hitzig, and later by

Nothnagel. All the sites in the cerebral cortex the stimulation of which produced movement and the removal of which produced paralytic phenomena in circumscribed muscle groups were located in the frontal portions of the brain; the rest of the brain surface was unresponsive to direct stimulation.[4]

The paralytic phenomena produced by extirpation of these centers of movement deserve special notice. They were defined by Fritsch and Hitzig as a disturbance of muscular sensation, and Nothnagel designated them directly as ataxia; in both cases the notion of an actual paralysis was rejected. Experience shows, however, that in humans paralysis is observed when these cortical areas are destroyed. Further proof of this is furnished by one case published by Hitzig, where true paralyses appeared in general paresis. In answering Nothnagel's objections one should also consider how easily bilateral movements are produced by *one* hemisphere when body parts are involved that have previously always been innervated together. A voluntary impulse from the left speech center suffices for speech movements carried out equally by the cranial muscles of both sides.[5] A beginning piano player trying to play with the right hand also moves the left fingers, and independence of the two hands is reached only after years of practice. In this case the act of innervation is produced only in the left hemisphere, but the cells of the anterior horn of the spinal cord, which have previously always been innervated together and have thus become firmly associated, now operate in unison by means of their connections. There is some justification for the paradoxical sounding assertion that the presence of both hemispheres is an unconditional prerequisite not for the simultaneous action of bilateral muscle groups but for their separate action. But if innervation from one hemisphere suffices to set both extremities in motion in a human who has practiced an isolated movement of the upper extremities a thousand times, how much more effective will it be in a dog or a rabbit, animals which only rarely practice consciously isolated innervation of an extremity. However, their incorrect interpretations of the observed motor disturbances does not lessen the significance of the achievement of the above-mentioned authors. That achievement consists in proving experimentally *that the same areas of the surface of the brain, the stimulation of which produces movement – that is, those which are motor centers in the strict sense – are also the seat of muscle sensation, of representations of the degree and type of muscle innervation, in short, of the movement images we postulated above.* The clumsiness that one observes in the movements of general paretics, such as their awkward gait, lacking all elasticity and precision, is an example of such disturbances of sensations of movement in man, and these signs have always been interpreted as paralyses.

There are thus motor areas in the brain that are connected with a centrifugal path and are also the seat of representations of movement. Since we are excluding innate ideas from the outset, the question arises, how these representations of movement are laid down in the cerebral cortex. In answer to this question the results of Bruecke and Lotze's physiological and philosophical deductions agree with the conclusions Meynert draws from anatomical facts. According to Bruecke and Lotze, movements must be divided into primary and secondary movements. The primary are the reflex movements, which long precede voluntary movements in the child. From them the child acquires his representations of movement, which he then puts to use in the secondary, conscious movements produced by the will. Now, complicated preformed reflex apparatuses are found in the thalamus and the corpora quadrigemina, and it is well known that after removal of the hemispheres their presence still suffices to produce all types of movement reflexly. Since the function of these ganglia is not dependent on the presence of the hemispheres, the important group of fibers that reaches the frontal lobes from the thalamus partly as the anterior thalamic peduncle and partly as a component of the internal capsule can be only a centripetal path through which sensations of innervation of reflex movements are transmitted to the frontal lobes. This is how Meynert conceives the situation, and his views thus provide an anatomical basis for the ideas presented here.

The anatomical facts make it indisputably clear that the occipito-temporal lobe is a sensory structure. All the sensory nerves whose central course to the cerebral cortex is known, terminate in the cortex of the occipito-temporal lobe. These are the olfactory nerve, the optic nerve, and that part of the posterior column of the spinal cord that does not have connections with the cerebellum. Until now we have been able to trace the central pathway of the acoustic nerve only into the cerebellum, but on physiological grounds, there is no doubt that it goes on further, to the cerebrum. Analogy with the other sensory nerves, then, requires us to place its central termination also in the occipito-temporal lobe.

This also must hold for the central pathway of the gustatory nerve, which is still completely unknown. The abundance of granule cell layers in a limited area of the occipito-temporal lobe, i.e., primarily in the region of the fissura hippocampi, also supports the view that this lobe has a sensory function. This sort of granule cell layer is found elsewhere primarily in sensory regions – the olfactory bulb, the retina, the substantia gelatinosa of the fifth nerve, and the substantia gelatinosa of the head of the posterior horn of the spinal cord.

As we have said, memory images, of sensations on the one hand, of the forms of movement of one's own body on the other hand, are the elements acquired from the outside world which make up the content of consciousness. Even a completely elementary act of consciousness, e.g., a voluntary movement, can be completely understood in terms of these units. Since voluntary movement characterizes even the lowest animals, its use as an example satisfies the requirement of maximum simplicity. Furthermore, it is *the* characteristic of the animal organism, and its preeminent importance thus justifies a thorough discussion.

Voluntary movement is distinguished from reflex movement by two properties:

(1) *It does not follow the stimulus instantaneously but rather owes its elicitation to memory images of earlier sensations* which are reawakened only in response to an external stimulus. Griesinger emphasized this comparison with reflex movement in his excellent treatise 'Über psychische Reflexactionen' (On Psychic Reflex Actions) by deriving both spontaneous movement ('psychic reflex action') as well as reflex movement from stimuli reaching the cerebral cortex along centripetal paths. The spontaneous movements, for example that are derived from so-called instincts (such as the mating instinct, the instinct to feed oneself, etc.), are certainly evoked, as is obvious, by residues of sensations. But the same is true of all other spontaneous movements, even though a greater number of sensations and more complicated ones may be involved in their production.

(2) Spontaneous movement is also distinguished from reflex movement *by its pre-formed pattern – polished, specific and adapted to the attainment of a goal.* It is thus characterized, by the presence of a representation of the movement to be carried out, and, as was shown above, this representation also must be conceived as a residue of sensation (a memory image).

If we apply the schema of reflex movement to spontaneous movement, taking these distinguishing characteristics into consideration, we get the schematic representation shown in Figure 1.

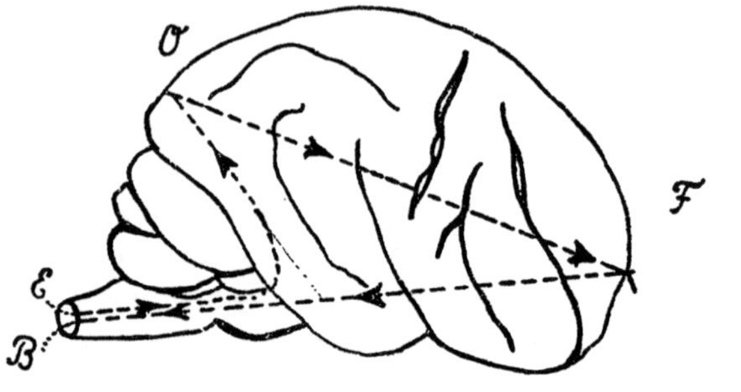

Fig. 1.

The sensation *E* reaches a point in the occipito-temporal lobe along the path *EO* and leaves a memory image there. On the occurrence of a new external impression, the stimulus, which has been latent, as it were, becomes effective again and is transmitted to a motor point in the frontal lobe *F* through some portion of the great fiber tracts connecting the occipito-temporal lobe with the frontal lobe. Once it has reached this point, the excitation in the centrifugal path *FB* gives rise to a movement.

This movement is endowed with the second characteristic of spontaneous movement, i.e., purposeful organization, as a result of the fact that, as was shown above, the locations of the representations of movement and of the release of movement in the cerebral cortex coincide (cf. p. 37). *Thus the path EOFB suffices for a complete explanation of voluntary movement on the pattern of a reflex process.*

There are numerous anatomical pathways which could mediate such psychic reflex movements – most of the cerebral white matter consists of such association bundles, some with simple trajectories, some with more complicated ones. The courses of these tracts have been described with great exactitude by Burdach (1825), on the basis of fiber preparations.

On the other hand, the use of these pathways, i.e., the ready transmission of excitation along them, is no accident; rather, it can easily be demonstrated that two places in the brain that have once functioned together regularly remain associated (Figure 2).

Let *x* be a cell in the corpora quadrigemina, which as a reflex center mediating the connection of the optic pathway *E* with the motor cell *y*, brings about the turning in of the eye *B*. This process does not take place without the participation of consciousness, and in fact, by means of the

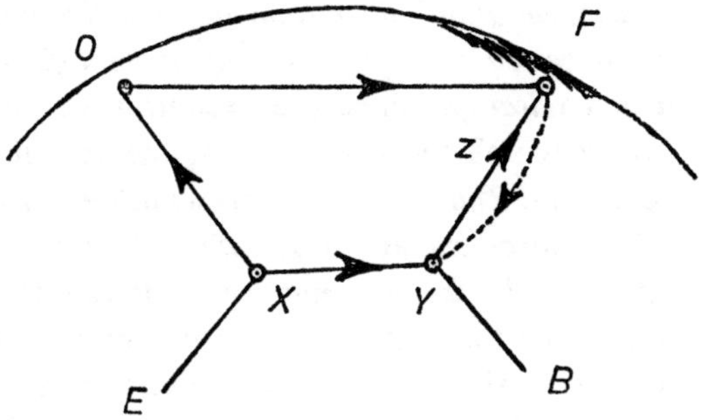

Fig. 2.

paths connecting the corpora quadrigemina with the cerebral cortex, both a sensation *O* and a representation of movement *F* are laid down in the cortex. If we assume a certain quantity of vital energy to be necessary for a stimulus to act as such, then a part of that energy is claimed for the paths *xO* and *yZF*, which must be considered secondary connections for the transmission of the excitation.[6] At the same time the excitatory process must follow the path *OF* as well, both ends of which are excited. If we compare the process that takes place in any nervous pathway during excitation with the course of a wave beginning at *E* in a system of tubes filled with fluid – for we can certainly conjecture that the process is an electrical one, although we may by no means assume it as certain – then the participation of the path *OF* in the movement becomes clear. Since the resistance of a nervous pathway is weakened by use, from now on the memory image *O* will easily elicit the representation of movement *F* and the voluntary movement itself over the centrifugal path *FyB*.

The genesis of every voluntary movement can be understood in terms of this simple schema, although of course there will usually be a multiple number of memory images functioning as the initial members of the reflex arc outlined above. This will be true even of the most elementary spontaneous movements, the instinctive ones.

However, just as memory images which have become associated with one another will sometimes be activated together in response to an external stimulus, so too, at the other end of the reflex arc, many representations of movement, associated by simultaneity or succession, will be called into consciousness at the same time. *The calling into consciousness of a representation of movement and the movement itself, however, can be conceived only as resulting from different degrees of intensity of cellular excitation*, since the excitation of the cell must be strong enough to overcome the resistance of the centrifugal path if it is to produce movement. *Thus the only representations of movement that will actually lead to movement* are those with already well established fiber connections with *the largest number* of initial members of the reflex arc – presupposing that the excitation transmitted from all initial members is of equal intensity – or those associated with *the most intense* memory images, or, finally, those offering the paths of least resistance to the incoming excitation. In the last two cases the strength of the single excitation compensates for the small number of excitations.

This scheme provides for the possibility of a choice of movements; the more memory images the individual has at his disposal, and the more practiced he is in associating memory images, the greater is his choice. The only correct scientific definition of free will is one which fits in most closely with this mechanistic conception of the genesis of spontaneous movement.

II

The purpose of the preceding discussion was to expound the general laws which can be shown at the present time, on the basis of anatomical and physiological facts, to be operative in the genesis of spontaneous movements. That speech movements must be included among the consciously performed voluntary movements needs no explanation. This follows simply from the fact that they must be learned by the child like every other spontaneous movement. Their development follows the development of consciousness so closely that they can actually be regarded as a measure of the latter. They are performed well only after the child has already attained a certain degree of skill in many other conscious movements.

The primary speech movements, i.e., those performed by the child before the development of consciousness, are of a reflex-mimetic[7] nature and are elicited from the same regions of the tegmentum of the pons and the medulla in which the acoustic nerve originates. In this region are located large nerve cells with many processes which must be regarded as extensions of the motor nuclei of the facial nerve, the vagus, and the hypoglossal nerve which, according to Meynert, are anatomically connected through cell processes with the fibrae arcuatae, coming from the site of origin of the acoustic nerve. All experimental results indicate that the respiratory center is also located in the very extensive area in which the acoustic nerve originates. In the case of the newborn child it has been satisfactorily proven that the presence of the medulla suffices for the unarticulated scream, a muscle action which despite its elementariness presupposes the combined actions of the expiratory muscles and the constrictors of the glottis.[8] Probably even more complicated patterns of muscular activity, adequate even for the formation of one-syllable words, take place reflexly, but perhaps in a more complex apparatus, the cerebellum. Because its functions are unknown, however, the cerebellum serves only too well to explain every gap in our knowledge; let us therefore disregard it and admit that it is difficult to conceive of the formation of whole words by means of the reflex apparatus of the medulla. Now, whether whole words or only fragments of them are produced reflexly in the medulla, in any case a sound image of the word or syllable is transmitted to some sensory portion of the brain from the site of the reflex action; the sensation of innervation of the movement performed is laid down in the frontal motor areas as a representation of a speech movement (cf. p. 37). The sound image and the representation of movement remain associated through some group of fibers in the white matter (cf. Figure 2). When, later, the spontaneous movement, the consciously uttered word, takes place, the associated representation of movement is innervated by the memory image of the sound.

The innervation is now transmitted along the pyramidal tract, as is the case in all other voluntary movements, to the muscles that function in the speech act.

Before we can provide an anatomical foundation for the conception of the speech process that has just been developed, we must review previous attempts to localize a center for speech in certain anatomical regions of the brain.

I will pass over the older works on this subject, which have already been summarized frequently and in detail, and turn to Broca, the author who first abandoned the consideration of more extensive areas of the surface of the brain and dared to postulate an anatomically distinct circumscribed speech center. As is well known, he located the faculty of speech in the posterior end of the so-called third frontal convolution. According to Leuret, who counts from the fissure of Sylvius, it is actually the first and thus he locates it in that portion of the lowest (and outermost) convolution of the operculum which lies anterior to the junction of the operculum with the anterior central convolution. Valid objections were raised from the outset against this view. However, many cases of speech disturbance that corroborated it were published, while the cases which were not in accordance with it were very far from being able to provide a new positive theory. As a result, this view has continued to win ground for itself and still has a very great number of adherents.

In the meantime, Meynert published a study that attracted a great deal of attention. In this work he demonstrated the connection of the acoustic nerve with the cortex of the fissure of Sylvius by means of a fiber bundle to which he gave the name *Acusticusstrang* (longitudinal bundle of the auditory nerve). Meynert called the whole region in which this bundle originated, a region coinciding approximately with the extent of the claustrum, a sound field. He supported his view with numerous post-mortem findings in aphasics in whom either the region of the insula itself or the adjacent part of the fissure of Sylvius was pathologically altered. A large number of academic physicians who had taken part in the controversy now tended to Meynert's view, especially since it met the half-felt need to relate the acoustic nerve to the process of speech. Remarkably enough, many of them held to the notion that all of the cortex around the fissure of Sylvius constituted a speech organ, even when Meynert himself disavowed the existence of the longitudinal bundle of the auditory nerve as such and expressed the view that the acoustic nerve had no direct connection with the cerebrum but reached it only indirectly by way of the cerebellum.[9] Their continued support of this notion is explained by the post-mortem findings. In most cases of aphasia where Broca's area was not found to be affected, there were changes in the area on which Meynert had laid stress. Recently Sander and Finkelnburg have again strongly emphasized this fact. Meynert also still holds to his earlier view of the significance of this area and advances in support of it, not only the post-mortem results, but also the fact that the claustrum is composed of association cells, i.e., spindle-shaped cells, and the fact that the claustrum has intimate connections with the other association systems of the cerebrum.

Cases such as the one reported by Simon, which is really equivalent to an experiment, seem to me to have shown beyond a doubt that destruction of Broca's area causes aphasia. There is, however, equally little reason to doubt the consistent conclusion from the experience of other skilled and conscientious observers that focal lesions in the region of the fossa Sylvii and its immediate vicinity produce aphasia, and that Broca's area is therefore not the only one to function as a speech center.

Let us now specify the immediate vicinity of the Sylvian fossa. We will direct our attention to that gyrus of the cerebral convexity which circles around the Sylvian fissure in an arc directed superiorly and posteriorly. Anterior to the central fissure it runs longitudinally (as the first frontal gyrus, in Leuret's nomenclature), while its posterior limb is in the longitudinally coursing first temporal gyrus. That this entire structure is to be regarded as a single gyrus is evident from comparison with the brains of animals, e.g., those of dogs. Comparative anatomy has shown it to be a general law of the formation of the convolutions of the brain that they describe an arc around the fossa Sylvii, the peak of which is directed toward the occipital pole and both limbs of which run more or less parallel to the fossa Sylvii in the frontal and temporal portions of the brain. This law holds for humans as well.

In the first frontal convolution (in the nomenclature of Leuret), described earlier, we must differentiate the crest of the gyrus and the two banks. Only the crest of the convolution lies exposed on the convexity of the brain; of the two banks, one is turned medially toward the second frontal convolution, while the other forms a direct continuation of the insular cortex. It protrudes over the insula and lies flat against it on all sides. Just as consideration of comparative anatomy shows the whole first primitive convolution to be a unitary structure, so also in its inner structure over its entire extent it shows a peculiarity already sufficiently emphasized by Meynert – the claustrum is wrapped around the insula to form a surface facing the insula and can be followed quite a way toward the crest of the convolution. The entire region of the first primitive convolution thus shares in the characteristic of the insular cortex: the innermost layer of the cortex, containing spindle cells, condenses to form a separate gray mass.

Furthermore, I have succeeded, by means of fiber dissections, in showing a characteristic arrangement of the bundles of fibers which lie just under the cortex on the external surface of the first primitive convolution, facing the fossa Sylvii. They form an uninterrupted sheet of radial fibers which with the myelinated core of the first primitive convolution, extend to the base of the deep cleft which everywhere separates the first primitive convolution from the insula (Burdach's anterior cleft, superior cleft, and inferior cleft), then bridge the cleft and end in the insular cortex. The insula thus resembles a large spider gathering into itself the fibre bundles which stream into it radially from all parts of the first primitive convolution. We thus get here, as nowhere else in the entire brain, the impression of a true center for some group of functions.

The existence of fibrae propriae or laminae arcuatae (Arnold), between the insular cortex and the convolutions of the convexity has also been demonstrated. Since to my knowledge these have not previously been described and since they constitute a major proof of the unitary character of the entire first primitive convolution and the insular cortex, the reader will permit me to speak further of them.

Fibrae propriae can be demonstrated without difficulty between all adjacent convolutions of the convexity and of the medial surface. It seems questionable however whether the sulcus centralis Rolando, which cuts into the primitive convolution system vertically, is bridged by such fibres; the results of fiber dissections in this region are completely unclear. The situation, however, in respect to the deep cleft which appears to demarcate the region of the insula from the first convolutional arc surrounding it is completely different. It is known that embryologically the insula forms the organizational midpoint of the entire convexity of the hemisphere. Its cortex forms the part of the wall of the cerebral vesicle that is fused with the basal ganglia internally and is therefore so restricted in growth externally that a depression, the fossa Sylvii, arises on the convexity of the cerebral vesicle.

In this region, local conditions pose great difficulties for dissection of fiber bundles. It is possible to carry it out successfully only in brains which have been well hardened in alcohol, but in just such brains it is difficult to bend back the overhanging portions of the hemisphere which cover the insula itself. But if this is not done, then one is left working completely in the dark and is forced to assume extremely uncomfortable and unnatural positions in order to avoid blocking the only source of light. The dissection is easiest in the tissue, which, running vertically first upward and then downward, bridges the cleft between the operculum and the insula (Burdach's superior cleft). The fiber layer is also of considerable thickness in this area and on the average exceeds $\frac{1}{2}$ cm. Afterwards, the dissection is most easily carried out in the anterior cleft, between the frontal lobe and the insula. The deep cleft which runs forwards and upwards between this part of the frontal lobe and the operculum is completely covered with fibrae propriae. The dissection is most difficult in the temporal lobe, first because of the deep location of the inferior cleft, which is bridged there, but also because the laminae arcuatae are thinner. The situation is still more difficult in the deep hollow where the posterior ends of the superior and inferior clefts come together.

In carrying out this preparation it is best first to apply the scalpel about halfway up the inner surface of the operculum and from there to dissect off the cortex both upwards and towards the insula. As is known, at the crest of the convolution the fibrae propriae are dispersed and can no longer be dissected. In the insular cortex the fibers can also not be traced very far, because they become intertwined to a great extent and descend into the depths on the side of the claustrum.

After considering the above exposition of the speech process as a spontaneous movement, it seemed thoroughly improbable *a priori* that Broca's area, located in the frontal part of the first convolutional arc, described above, could be the only speech center. The consideration of the anatomical circumstances just described, of the numerous supporting post-mortem studies, and finally of the variety in the clinical picture of aphasia thus leads compellingly to the following interpretation of the situation. The entire region of the first convolution, which circles around the fossa Sylvii serves in conjunction with the insular cortex as a speech center. The first frontal convolution, which is a motor area, is the center of representations of movement; the first temporal convolution, a sensory area, is the center for sound images. The fibrae propriae which come together in the insular cortex form the mediating psychic reflex arcs. The first temporal convolution should thus be considered the central termination of the acoustic nerve, and the first frontal convolution (which includes

Broca's area) the central end of the nerves of the muscles of speech (Figure 3).

Let *F* be the frontal, *O* the occipital, *T* the temporal lobe of a schematically drawn brain. Let *C* designate the central fissure. The arc of the first

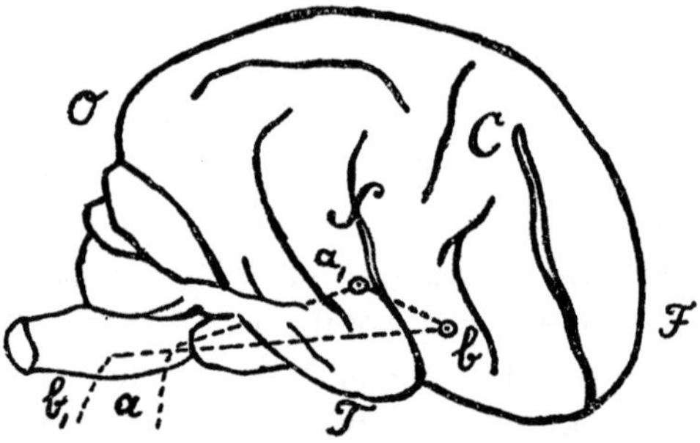

Fig. 3.

primitive convolution extends around the fossa Sylvii *S*. Within it let a_1 be the central termination of the acoustic nerve (*a* being the place where it enters the medulla). Let *b* stand for the representations of movement in the cerebral cortex which are necessary for the production of sound, connected with a_1 by association fibers a_1b running in the insular cortex. From *b* let the centrifugal path of the sound-producing motor nerves run to the medulla, where for the most part they leave the brain (the accessory and phrenic nerves extend still further downward).

Any interruption of the path aa_1 bb_1 can cause aphasia. The clinical picture will however vary according to the portion of the path affected by the interruption.

I. Let us suppose that the path aa_1 is interrupted, i.e., a pathological process has destroyed the acoustic nerve at some point in its central course. As everyday experience tells us, this produces deafness without any trace of aphasia – but only in adults who are already in firm possession of an extensive stock of sound images acquired earlier and capable of reproducing them at will. *If, however, the interruption occurs at a period in childhood when a consolidated store of sound images has not yet been formed in the cerebral cortex, mutism inevitably results.* This is the usual origin of deaf-muteness; the child has not acquired any representations of sound from which representations of movement can be called into consciousness.

The inevitable appearance of mutism in cases of congenital or early acquired deafness casts a very interesting sidelight on the significance of acoustic impressions for the development of language. It is a widespread error, held especially by philosophers and linguists (Steinthal) to consider that the most important aspect of the development of speech is the formation of concepts, a concept being the sum of the various sense impressions of an object. The concept then supposedly creates language out of an inner need. The development of language would follow the same laws in present-day man that it followed in its development among the first men. Thus language would not be the result of imitation but would instead be a spontaneous phenomenon for which the conditions – i.e., the formation of concepts – would be provided equally by all sensory spheres, there being no particularly intimate relationship with one sensory sphere, namely that of hearing. Were this the case, the person born blind would have much more reason to be mute than one born deaf, for of all the senses, vision is doubtless the one which provides us with by far the largest number of attributes of objects, and is thus by far the most important for concept formation. Although the senses of vision, touch, smell, etc. – all those essential for the concept – must be conceived as being connected by means of a portion of the association system[10] with the representations of movement which are innervated in speaking, nevertheless only the path aa_1b has such immense importance for the development of language, because it is the path over which the child learns to speak. The main task of the child learning to speak is to imitate the word he hears; he learns to connect it with a particular concept only after he has long been in possession of the word. The word is an imitation of the sound image, not of the visual or the tactile image; those born deaf no more learn to speak than a blind man learns to draw.

A situation that occurs frequently deserves special mention here. What consequences has deafness on one side for the child affected during the course of development? Mutism can result only if we proceed from the most unlikely assumption that only the speech region of the left hemisphere[11] is capable of development, the right one having atrophied at least functionally over the generations. Even if we admit that in deafness on the right-hand side the child, who is now dependent on the right hemisphere, will have more difficulty learning to speak than one deaf on the left side, nevertheless it can certainly be assumed that the presence of the acoustic path leading to the right temporal lobe will be sufficient to allow the child to learn to speak. On the other hand, this situation is of great importance in evaluating post-mortem findings, since it could account for many cases in which the seat of the process which causes aphasia is on the right side. Incidentally, the great gap which results from our ignorance of the path travelled by the acoustic nerve to the cerebrum is nowhere more noticeable than here. For the present it is still an arbitrary assumption that the left acoustic nerve ends in the right hemisphere rather than anywhere else.

Deaf-mutism is a symptom complex with which we are so familiar that the coincidence of the two symptoms has always seemed self-evident, and they have not been considered in connection with aphasia. In contrast, the rare circumstance in which children endowed with good hearing do not learn to speak has been correctly described as congenital aphasia.

II. Let us suppose the location of the sound image itself, a_1, to be affected by the lesion. This point is not identical with the entire central projection of the acoustic nerve, for there have been cases of aphasia observed in which complete loss of sound images accompanied completely preserved hearing (on both sides!). Probably the organization of the central terminations of the sensory nerves is similar to that of the motor nerves, which we will discuss later: they spread out and terminate, in accordance with the different functions of the individual fibers, at different places in the cerebral cortex. Thus that part of the central projection of the acoustic nerve which contains word-sounds may be destroyed while all noises or all musical tones may still be perceived.

If locus a_1, the cortex of the first temporal convolution, is destroyed, the sound images of the names of all possible objects will be extinguished from memory, although the concepts may still remain in their full clarity. For in most cases the sound image is of secondary importance for the concept of the object, whereas the somesthetic sense images are of critical importance for it. Disturbances which affect the visual and tactile sense images which constitute the concept (Finkelnburg's asymbolia) should thus be included not among speech disturbances but among disturbances of concepts, that is, of intelligence, even though they do form a separate, clearly defined group among those disturbances.

It is clear that the preservation of the path a_1b has no value once the sound images have been lost. The words are no longer innervated by the sound images. Furthermore the path which connected the sound heard with the other sense images of the object (association fibers going from the first temporal convolution to the other sensory areas of the temporo-occipital lobe) has also been interrupted. *The patient is thus capable neither of repeating a spoken word – since this is the characteristic function* of the path aa_1bb_1 – *nor of understanding the spoken word.* In spoken words he hears either only a confused noise that has no meaning for him or at best a language completely foreign to him, the individual sounds of which he perceives and gradually learns to understand again.

One pathway still remains by means of which representations of speech movements can be innervated. The patient with the characteristics just described has no disturbance of intelligence. Through his behavior and his comprehension of signs and gestures he makes it clear beyond a doubt that he is familiar with the sensory images of the concrete objects around him and thus with their concepts as well. Now the association of sound images with representations of movement essentially has its value in the fact that it makes it possible to learn language. Very soon after we have learned to speak a word, the intention of merely reproducing the sound disappears, making way for the intention of reproducing a specific meaning. The actual sensory images of an object, that is, are now able directly to innervate the representation of movement of a word directly. The capacity to speak is thus retained, but with a certain limitation. In normal speech the sound image always seems to be unconsciously innervated at the same time, as is easily understandable in terms of the genesis of language. The sound image seems simultaneously to be hallucinated, as it were, and thus exercises a continuous corrective function on the course of the representations of movement. The deaf person in whom only the course of the acoustic nerve is interrupted still has this corrective mechanism completely at his disposal.

The situation can be represented schematically as follows. In Figure 4, which should be compared with the ones given above, let c be a tactile image associated with a_1, and d the corresponding optical memory image. The concept is nothing more than the path cd. The child first learned to speak over the path a_1b, which, having been used thousands of times, retains a significant influence on the choice of the correct representation of movement. Later on, however, this path is no longer the one primarily used. Rather, the shorter paths *cb* and *db* are chosen, and the mere existence of the path a_1b, without its being intentionally innervated, is sufficient to insure the choice of the correct representation of movement. Thus the conjunction $d+c+a_1$, each of approximately equal intensity,

Fig. 4.

acts to produce the correct word choice. If, however, a_1 is lacking, only the sum $d+c$ has an innervative effect, and the powerful influence of a_1b is lacking.

Aside from his deficient comprehension, the patient has aphasic manifestations in speaking because of the absence of the corrective function exercised unconsciously by the sound images. These manifestations consist in the tendency readily to confuse words. The patient still possesses potentially the capacity to name everything correctly, but the degree to which he is master of that capacity varies depending on his mood and affect. He uses the same words correctly at times and incorrectly at other times, without any regularity. He has no definite vocabulary, if by vocabulary is understood those words used only correctly. Furthermore he is not aware of whether he is using the words correctly or not. In states of intense emotion, which favor explosion into movement without the intervention of associated memory images, the absence of the corrective function will be least noticeable.

The symptoms will be more or less marked depending on the intensity and extent of the pathological process that has affected the first temporal convolution. In the more serious forms, where the sound images not only of concrete objects and actions but also of the connecting words necessary for sentence formation have been lost, the diagnosis will rest on only two factors, *the quantity of words spoken and the degree of lack of comprehension of what is heard.* The fact that cases of this kind have until now not been observed or at least not yet been published is the result, not only of the infrequency of such cases but also of the fact that even thoroughly experienced and intelligent physicians interpret this condition as a confusional state, as I myself have had the opportunity to observe. For those who are psychiatrically trained and who know the clinical pictures of confusional states, the diagnosis presents no difficulties whatsoever.

By means of suggestive questioning one can establish the presence of lesser degrees of this disorder, in which the form elements important for sentence structure are still preserved, and in which the sense of a question is, in general, correctly grasped except for single words. If, when the patient is shown a glass and asked 'Is that a glass?' he does not know immediately but hesitates, reflects for a while and finally utters a strained, doubtful, 'Yes' or 'No', then the case certainly belongs in this category.

The following points are worthy of special emphasis:

(1) In partial lesions of the sensory speech center there will probably be a certain number of words which will be both correctly used in speaking and correctly understood when heard. This vocabulary can be determined through suggestive questioning. Extensive and laborious observation will thus be required to diagnose this partial form of sensory aphasia.

(2) A large potential vocabulary is a major criterion for this form of aphasia. Conditions in which only a few simple words are preserved always belong to the motor form of aphasia, to be discussed in Section IV.

(3) There is no trace of hemiplegia to be observed.

(4) Agraphia is present. Writing is a voluntary movement learned in close dependence on sound and always executed under the guidance of sound. Self-observation demonstrates – and clinical experience is in agreement with it on this point – that there is no direct connection between the writing movement and the concept such as we must assume there to be between the speech movement and the concept. In cases of partial sensory aphasia partial agraphia is also to be expected.

(5) The ability to understand written or printed symbols is a completely different matter. This will either be dependent on or independent of the existence of sound images according to the patient's educational level. The uneducated man with little practice in reading understands the written word only when he hears himself say it. The scholar, practiced since

childhood, skims over a page and understands its meaning without becoming conscious of individual words. The former will show the symptoms of alexia as well as aphasia, while the latter, in striking contrast to his inability to understand speech, will be able to understand all written material. On the other hand, in reading aloud he will be as aphasic as in spontaneous speech.

Both conditions, agraphia as well as alexia, can also be caused by a lesion in a completely different area, the visual cortical region, since the optical memory image of written symbols is indispensable for writing as well as for reading. The possibility cannot, of course, be denied, that the written symbols occupy a special cortical zone within this region, delimited precisely by means of its intimate connection with the entire speech area and that thus a very circumscribed cortical lesion might give rise to agraphia and alexia alone. It is, however, much more likely that the whole visual sensory region will be affected. The symptom complex would then contain the additional manifestation of inability to recognize seen objects. That disturbance would, however, not properly have anything to do with aphasia.

III. Let us suppose that the path a_1b, i.e., the association fibers which connect the sound image with the accompanying representation of movement, is affected, while the sound image a_1 and the representation of movement b are themselves preserved.

In this case, in contrast to the form of aphasia just described, *the patient understands everything. He can always express himself, but his ability to choose the correct word is disturbed in a manner similar to that in the form just described.* The sound image is preserved in this case, and it can also be innervated by the other sense images that form the concept. Because, however, the path a_1b is interrupted, the sound image cannot exercise its influence on the correct selection of representations of movement, or at least it does so only with greatly decreased intensity. Whereas in the previous form only the sum $c+d$ was present for the innervation of b, here the sum $c+d+a_1$ is present; a_1, however, is much less important than is normally the case because it must now be routed along the path cb. Words will thus be confused, not as badly as in the preceding form, but nevertheless very noticeable. Another corrective mechanism nevertheless can come into play here, a conscious one which, although it is little utilized in the normal speech process, can gradually completely replace the unconscious one. The acoustic nerve is intact and transmits the sound of the spoken word to the locus of sound images, which is also intact. Thus the spoken word is heard and judged correct or incorrect as the case may be. With some concentration the patient will know when he has spoken falsely and will be disturbed by it. If he is given a selection of words spoken aloud from which he is to choose the correct one, he will regularly choose correctly. He will also infallibly answer questions correctly. Such a patient will be able to practice what he wants to say by saying it quietly to himself beforehand, and if he is strong-willed and capable of intense concentration, he will in time be able to overcome his defect by means of a conscious, although laborious and time-consuming, corrective mechanism.

In pure and complete cases of this type the patient manifests a great similarity in speech to the sensory aphasic. However, he understands everything correctly and answers questions correctly. He thus gives the impression of far greater intelligence. Hemiplegia of the opposite half of the body will almost always be an accompanying manifestation.

In milder cases of this type one must be content with assuming the presence of disturbances in transmission over this path caused by some kind of circulatory impairment. In these cases there is no hemiplegia at all, and confusion of words is not as striking as the subjectively experienced difficulty in finding them. Speech is halting and filled with long pauses in which the patient visibly struggles to express himself. After long effort the patient often gives up and begins a new sentence. Again the beginning will come out smoothly, followed by the same hesitation and the same signs of effort. A similar disorder can occur even in completely healthy people who, in the middle of a conversation or a lecture cannot find a word, labor to replace it by another, and must finally make do with a poor substitute. I have observed this even among able people who were fluent and could produce unusual combinations of ideas, in short, people who while speaking, carried on an intensive intellectual activity. Steinthal, in his *Abriss der Sprachwissenschaften* (Outline of Linguistics) considered that thought alone is difficult. It is certainly much more difficult to think and speak at the same time. The functional hyperemia that facilitates communication between the sensory images that are continuously appearing de novo will necessarily lead to an anemia in the speech area.[12]

The disturbances in reading and writing which occur in the form of aphasia produced by lesions of the insular region are comprehensible only on the basis of a more detailed examination of the way both abilities are learned by the child (Figure 5).

The child comes to read by learning to bring the optical sense image of

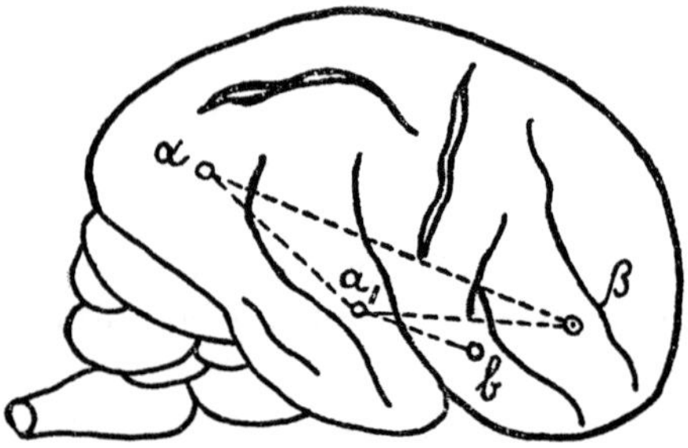

Fig. 5.

the letter, α, (one portion of the visual sensory area), into relationship with its sound image, i.e., by learning to associate them. He learns to read aloud when the sum $\alpha + a_1$ can innervate the representation of movement b over the path a_1b. The combination of sound image and optical image constitutes, however, the entire concept of the letter; it has no other attributes. Thus if the path *ab* is interrupted, what is true of every other perceived object, namely, that its concept can innervate the representation of the speech movement directly, will not be true of the letter. A patient of this type will thus not be able to read an individual letter shown to him. To this extent alexia is also always present in this form of aphasia. Whether or not the alexia extends to whole words depends upon the patient's level of education. If he has never advanced beyond reading individual letters and constructing a word out of them, then his aphasia will cause him to lose the ability to read. But if he has achieved the skill that is the norm among the educated classes, then the written word will evoke a particular concept in him, he will understand what is written and will probably find the right word for it, if he is well motivated (see above). He can, however, never read aloud an individual letter presented to him, because the existence of the path *ab* is indispensable for reading a letter aloud. The patient, however, indicates his complete comprehension of the sense of the letter in that he rejects the name of every other letter and immediately accepts the correct name when it is offered to him. In the aphasia resulting from lesions of the insular region the educated person thus suffers no disturbance of any kind in the comprehension of written language. However, this statement must unfortunately be qualified by the fact that in just this type of aphasia right hemianopia seems to be a frequent complication.[13]

Writing is learned by copying the optical sense image of the letter over the path $\alpha\beta$ (β designating the center of writing movements). Only through much practice do we achieve a close association α, β between the sound image and the representation of movement, by means of which we attain the capacity of writing without a model. However, because we first learned to write over the path $\alpha\beta$ it retains the same influence on the course of writing movements that the path a_1b exercises on the course of speech movements. Now since β, a motor center, must be located in the frontal lobe, the path $a_1\beta$ must be close to the path a_1b, and they will thus usually be affected simultaneously. Now, under no circumstances can a direct path be available from the sense images that form the concept to the motor center, over which the writing movements could be innervated while the sound images were circumvented. Thus the aphasia of the insular region usually also has agraphia as a consequence. The copying of letters and words written out for the patient will, however, always be possible, because the path $\alpha\beta$ is intact.

IV. Aphasia following lesions of the frontal lobe, caused by destruction of the speech movement images b, presents an entirely different picture. *The patient understands everything but has either suddenly become mute or has at most only a few simple words at his disposal.* Those few words are usually used to designate all possible objects. This is not, however, because he does not know the meaning of these words but rather because the patient feels the need to respond to questioning by making some sort of sound. It is the same as the mechanism by which a dog responds to everything by barking; he tries to endow the one sound at his command with every possible meaning. The powerful excitation process arriving at b from a_1 and cd can be discharged only in this one movement. *In pure cases of this type, it can easily be determined through suggestive questioning that no confusion of words is present.* Questions are always answered correctly by gestures. Every command give the patient will be carried out. Almost all the cases of aphasia described until now, i.e., those which correspond to the form described by Broca, are of this type.

If there is some extension of the pathological process, the center for the representations of the movements necessary for writing, which is probably adjacent, will also be affected and agraphia will result. There is of course a question whether it is legitimate to assume the existence of such a center. The movements used in writing are learned at a time when the child has already attained control of the use of his limbs and of the course and extent of each single movement. He thus needs only to learn new combinations of the representations of movement already present and to learn moderation in the force with which he is accustomed to performing his muscular actions. For this reason also, it is not only the left hemisphere that possesses representations of writing movements; every normal person can also write with the left hand. The left hand is no more awkward in writing, in comparison with the right hand, than it is in any other movement.

But if the pathological process extends over a larger part of the frontal lobe, the majority of the representations of movement for the opposite side of the body will be eradicated and paralysis of that side will then appear. Agraphia will in any case then appear for purely mechanical reasons. If such a patient still succeeds with the help of his left hand in holding the pen in the right hand and in guiding the right hand purposefully, then it will actually be the left hand and not the right that is writing, and this cannot be used as a proof that the left center for writing movements is still in existence.

Anyone accustomed to speaking aloud to himself as he writes, to dictating to himself, as it were, will become agraphic as a result of the close association between speech movements and writing movements, even if there is no actual compelling reason for agraphia in the localization of the lesion.

If the lesion has interrupted the fiber bundles that come from the first temporal convolution and connect the center of sound images with the representations of writing movements, then the patient will be able to copy by means of the path $\alpha\beta$, but will not be able to write spontaneously, i.e., he will be partially agraphic.

Except in the situation mentioned on p. 53, aphasia following lesions of the frontal lobe can never destroy the capacity to understand written and printed words.

V. Interruption of the path $b\ b_1$ (see Figure 3), i.e., of the fiber tract which runs toward the basal ganglia from the first frontal convolution, must have exactly the same effect as destruction of the corresponding cortical areas themselves. *This interruption will produce the same kind of motor aphasia* as has just been described. It is hardly to be expected that a lesion would ever destroy just this portion of the tract as it enters the lenticular nucleus and the corpus striatum and leave the remaining portions completely intact, i.e., it is unlikely that the lesion would simply produce aphasia without any other paralysis. This is even more the case when there is destruction of this tract within the lenticular nucleus and corpus striatum. Lesions of lesser extent within the lenticular nucleus will, however, be affecting a completely different grouping of fibers. It is very probably the case that the fibers belonging to different representations of movement reach the lenticular nucleus from different portions of the frontal lobes and are now grouped in such a way that the territory of one peripheral nerve is now represented by specific groups of nerve cells of more or less circumscribed extent. To choose one example, it can be assumed with certainty that the representation of the facial nerve in the lenticular nucleus consists of two circumscribed groups of cells situated far apart. One of these unites all the fibers for the muscles of the mouth, whether they come from the first frontal convolution and serve for speech movements or from some other convolutional area of the frontal lobe which contains perhaps representations of mimetic movement. The other circumscribed group of nerve cells unites all the fibers for the peri-orbital muscles, without regard to the representations of movement to which they correspond. Similarly, the tongue musculature will have its own nucleus where, again, fibers come together from different cortical areas representing respectively either speech movements, or chewing, or swallowing, or any other voluntary movements. From this we may conclude that as a result of circumscribed destruction within the lenticular nucleus it would never be possible for all speech movements to be affected at once. Instead, partial aphasias are produced, which are manifested by paralysis of certain muscles innervated in the course of the speech act, e.g., the muscles of the mouth supplied by the facial nerve. At the same time the other movements of the tongue and the larynx involved in speech can take place normally, so that speech is still intelligible.

There is thus no doubt that insofar as the lenticular nucleus is concerned there is a distinction only of degree, not of quality, between aphasia and alalia, and *that complete destruction of the left lenticular nucleus must produce aphasia in addition to paralysis.* This is thus an aphasia of the lenticular nucleus, and I must emphasize it to counter the habitual view, which so naively neglects the anatomical relationships of the brain.

What has been said makes it clear that interruption of the left crus cerebri leads to aphasia, and in fact interruption of the lower portion of the crus cerebri, the cerebral peduncle, is sufficient, because speech movements, as was explained above, are voluntary, learned movements.

Because pathological processes are usually rather extensive, it is obvious that cases in which two or three of the symptom complexes described appear together will be more frequent than ones showing these pure clinical pictures, which depend on more or less arbitrary anatomical boundaries. It is also true that the typical pictures described here certainly do occur (see the case histories given below), and in themselves suffice to justify our clinical classification. I believe that having directed attention to them will soon lead to the publication of a good many pertinent case-histories and post-mortem protocols. The combined forms can easily be deduced from what has already been presented. When the psychic symptoms of aphasia of the temporal lobe variety are accompanied by right-sided hemiplegia, one may conclude that the lesion affects simultaneously the temporal lobe and the insular region. Combined involvement of the frontal lobe and the insular region can be distinguished only with difficulty from that of the frontal lobe alone. Damage to the entire primitive convolutional arc causes absolutely speechlessness together with inability to comprehend speech, accompanied by agraphia and alexia.

Only a particular period in the course of the disease should be considered if one is to diagnose aphasia correctly. On the one hand, the general phenomena which accompany the onset of aphasia as they do that of most localized lesions of the brain must have disappeared. On the other hand, however, the condition ought not to have lasted so long that the possibility of compensation by the other hemisphere is already present. Fortunately these two sources of error are to some extent mutually exclusive, in that aphasia resulting from frontal damage, which is accompanied by the most severe general manifestations, is compensated for by the other hemisphere only rather late, whereas sensory aphasia, which is quite soon compensated for by the other hemisphere, causes only minor general symptoms at its onset. Thus only the practiced diagnostician will succeed in correctly judging the best time for diagnosis, and he will succeed only when he has been observing the case since its onset. I do not need to emphasize how important this point is for the evaluation of post-mortem findings.

The theory of aphasia that I have just developed requires some supplementation, without which it could be criticized for not having sufficiently considered the available useful data.

At this point I must return to the proposition which in the course of my study of this subject came to stand out more and more clearly as the basis for a fruitful interpretation of the symptoms of aphasia: the learning of speech consists in imitation of what is heard and speech is not identical with a certain level of intellectual development; thought and language are two completely independent processes, which can in fact interfere with each other. Everyday observation supports this view. All children in learning to speak pass through a stage in which they show very strikingly Romberg's echolalia; they repeat the questions directed to them instead of answering. They play with words and practice distorting them in order to be able to evaluate with confidence the sounds produced. During this period the meaning of words is of secondary importance to them. Without instruction by other humans, Caspar Hauser did not learn to speak despite his completely developed brain and despite the cunning and sensory acuity he had acquired in his struggles with animals. He learned however to speak very quickly when he came into contact with other men. A further striking proof of this is offered by deaf-mutes. People normally learn to speak only by means of hearing. If this is absent, a person remains mute, no matter how far he may develop intellectually. When considering pathological changes in speech, one must take into consideration its normal course of development.

But how is it that some deaf-mutes do manage to achieve articulated and vocal speech by means of extremely complicated special instruction? They only prove that direct connections between the visual and tactile sensory regions and the locus of representations of speech movements do exist and that in certain circumstances they could suffice to make normal speech possible. The fact that someone who has completely lost sound images can nevertheless say many things correctly, including whole sentences, cannot be explained in any other way. Nevertheless the case of deaf-mutes who learn to speak passably through suitable instruction is completely different. In them the optical and tactile sense images instead of the acoustic ones function as the initial members of the psychic reflex arc. Once such a deaf-mute is in secure possession of the capacity to speak, he too can develop aphasia. Dependening on the site of the lesion he can suffer from motor, sensory, or conduction aphasia. It is, however, only in motor aphasia among deaf mutes that the lesion is located in the same portion of the frontal lobe that has been shown to be involved in normals (Figure 6). In sensory aphasia, the optical and tactile images, *c* and *d*,

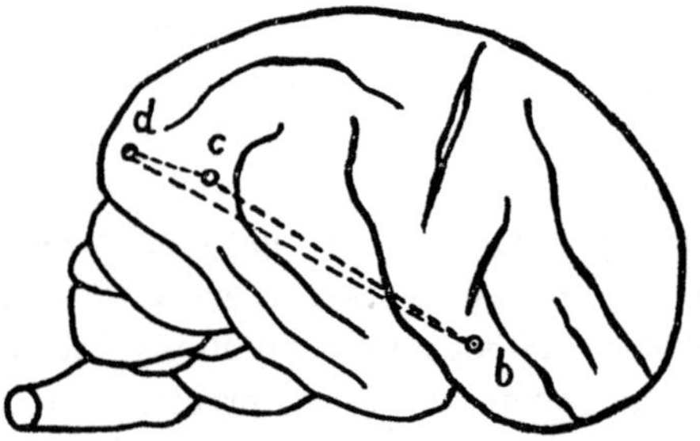

Fig. 6.

would have to destroyed. Their anatomical localization is not sufficiently well known, but it is certainly not identical with the first temporal convolution. Finally, in conduction aphasia, the fiber bundles which connect the frontal lobe with the occipito-temporal lobe, especially the superior longitudinal fasciculus of Burdach, would have to be interrupted. It would be pointless to describe the symptoms of these three forms of aphasia in deaf-mutes in further detail. It should, however, be emphasized that deaf-mutes with sensory aphasia will always suffer from asymbolia (Finkelnburg) and that the picture will thus be much more complicated and more difficult to recognize. Post-mortem studies of deaf-mutes in these cases would be of the greatest importance for the theory of aphasia, just as they are in other cases.

Let us take a closer look at the concept of asymbolia, not in our own precise sense but rather as Finkelnburg has presented it. One of Finkelnburg's patients did not recognize persons and places that used to be known to him; another patient showed a lack of understanding of spoken words, no longer made the sign of the cross when saying grace, and confused movements that she attempted to execute. The third patient confused the keys and could no longer play the piano from sheet music. The fourth was different from other aphasics in her loss of understanding of coins; the fifth lost familiarity with the symbols of religion and of the government services, and with the forms of expression of social conventions. In accordance with these examples asymbolia would be defined more or less as the misunderstanding of the meaning of an object or an action. Cases 4 and 5 especially would not be unusual at all, for half of the patients in mental institutions exhibit the same symptoms. These very symptoms accompany intellectual deficit. There could be nothing worse for the theory of aphasia than to interpret the disturbances of intelligence that accompany it – and that can occasionally be observed in every other localized disorder of the brain – as essential to the clinical picture. In doing so one would be making the same mistake as if one tried to explain the loss of consciousness that accompanies an attack of apoplexy as the result of destruction of the lenticular nucleus.

If we restrict ourselves to Finkelnburg's first examples, it is possible to formulate a better definition. Asymbolia would then be equivalent to the loss of the optical memory images of an object[14] or to the loss of any memory images essential to the concept of an object. We would still be dealing with a disturbance of intelligence, but we would at least have thus clarified a form of circumscribed intellectual deficit not previously observed, one which, like aphasia, would permit the diagnosis of a localized lesion of the brain. In order to express this more clearly I must discuss in greater detail a topic mentioned earlier.

The spoken and written name of an object is not a new attribute of the object. It is thus clearly different from the actual sensory memory images of the object. Only the latter make up the concept of the object.

The concept of a bell, for example, consists of memory images of visual, tactile, and auditory perceptions, connected (associated) with each other. These memory images are essential attributes of the bell. The spoken word 'bell', however, has nothing in common with the acoustic impression that a bell makes on us, and there is likewise not the slightest similarity between the written word 'bell' and the image of a bell. Only in a very poorly developed form of writing, e.g., hieroglyphics, could there be an exception to this. This makes it clear that it is necessary to regard the concept and the word as different. Disturbances of the concepts of the things with which we deal in the process of thinking are always disturbances of intelligence; disturbances of speech, on the contrary, cause difficulties only in the use of the conventional means of representation of the concepts.

We have only a few clues to the anatomical site of the cortical zones that function as the seat of optical and tactile sensory images. We can say only that it is probable that the arrangement of the fibers in the brain stem is preserved in their later course through the white matter of the hemispheres. Accordingly the optic radiations would run in the innermost, i.e., most medial portion of the occipito-temporal lobe. The sensory pathways from the posterior columns of the spinal cord would run just lateral to these en route to their cortical terminations. The acoustic memory images that do not participate in speech should be located in close proximity to the outermost portion of the first temporal convolution. What effect would isolated lesions in these areas have on language?

Let us first imagine that these conceptual regions were damaged in their entirety in both hemispheres; the result would be profound dementia. Language in itself would not need to suffer, although the content of what was said would be extremely improverished intellectually: the telegraph is functioning, but the telegram to be transmitted is nonsensical. The formal ability to speak is often maintained in microcephalics; they acquire speech, but they do not learn to grasp concepts, and they therefore exhibit echolalia.

If only the visual sensory region were damaged at identical points in both hemispheres[15], then the first symptom of interest would be a kind of alexia essentially different from the one previously discussed. Written symbols would then be completely lost and could not be called back into consciousness even by production of the sound; in contrast to relative alexia, which is only a consequence of aphasia, this would be absolute alexia. Agraphia would also be a consequence of this lesion. The written symbols would, however, not be specially affected compared with other optical memory images. Thus the ordinary visual impressions of the environment, i.e., familiar objects, persons, and places, would no longer be recognized. The fact that this failure of recognition has been observed only in partial form up to now can be explained by the fact that destructive lesions in both occipito-temporal lobes erase only those memory images which are affected on both sides, whereas all those destroyed in only one hemisphere are preserved. Such a case is reported in Gogol's excellent dissertation.[16] The patient bit into the soap and urinated into the wash-basin; a circle, a thermometer, a pitcher, a sand-holder, a tower across the street – all were equally foreign to him. In an earlier stage of the illness, which was more suitable for study, he showed the inability not only to express himself through speech but also to understand spoken words. The autopsy (Ebstein) showed "up to 4 cm. posterior to the fossa Sylvii an ochre-yellow colored region where the brain tissue seemed considerably softened and it was difficult to distinguish the borders of the gyri. Similar involvement of the inferior layer of the third left frontal convolution. As a result of the softening just described the convolutions that lie over the insula (the operculum) are shrivelled and the insula is exposed. On the external surface of the right occipital lobe, just at the pole, is another region, also ochre yellow in color, about the size of eight groschen, scarred and sunken, in which the surface of the cortex is softened."

III

Despite its extent, the clinical literature of aphasia is useful only to a small extent in supporting any theory based on anatomy. This is due on the one hand to the subjective attitude with which most observers have approached their cases. Thus we find that in the different case histories published by one observer the emphasis is always placed on a symptom that has perhaps not been described previously while the remainder of the psychological findings are neglected. Alternatively, in order to avoid this error, others give descriptions which are verbose because they are objective. Nevertheless even in these, what is most important is omitted, for the reason that an accurate investigation, especially in the case of psychological symptom complexes, cannot dispense with a previously formulated theory to provide it with guidelines. The second problem, however, is the unsatisfactory nature of the autopsy results. There is no doubt that the more important writers on aphasia (such as Broca, Ogle and Hughlings Jackson) were so familiar with the convolutions and fiber tracts of the brain that they could themselves give precise descriptions of the localization of lesions. But the majority of the observers who provided them with material were undoubtedly incapable of carrying out genuine post-mortem studies of the brain. Almost everyone was thus forced to use questionable material, i.e., to extrapolate so far that only rather general statements could be made.

For these reasons I realized that I was compelled to refrain from any detailed use of the clinical literature. Fortunately the few cases I had observed myself offered me sufficient material to provide evidence of another kind. For all their diversity, the clinical pictures of aphasia vary between two extremes, those of pure motor aphasia and pure sensory aphasia. The existence of these two forms must be considered an irrefutable proof of the presence of two anatomically different centers for language.

Cases of pure motor aphasia are so frequently to be found in the literature that there can no longer be any doubt of its existence or of the presence in these cases of lesions of the first frontal convolution. There is, however, to my knowledge, not a single clear case of the pure sensory form of aphasia to be found in the literature. Two such cases have come under my observation, one of whom is still in the psychiatric service of the Allerheiligen Hospital.

(1) Susanne Adam, née Sommer, the widow of a working-man, 59 years old, fell ill suddenly, with no known cause, on March 1, 1874. She complained about dizziness and headaches, but had suffered no loss of consciousness. Although she spoke in a confused manner she still expressed many things correctly, but answered questions in a completely confused way. She was able correctly to express her complaints of headache and feelings of weakness, but she mixed the meaningless word 'begräben' into everything she said. After eating lunch in her usual manner she went to bed. She was admitted to a ward in the Allerheiligen Hospital the next day. There her condition was considered to be simply a confusional state and, since no physical illness could be found, she was transferred to the psychiatric ward. There, on March 7, 1874, the results of examination were as follows. A slight, moderately aged woman, with a senile cataract on the right and surgical coloboma of the left iris, with an intelligent, receptive facial expression. No disturbance in gait, grip weak bilaterally, somewhat weaker on the left than on the right. Sensation as tested by pin-prick seems generally somewhat dulled, since light pin-prick evoked signs of pain only on the fingers, toes and face. The blood-vessels, including the temporal arteries, are very tortuous and can be felt as hard cords. Nothing abnormal in heart or lungs. Hearing, tested with a watch,

equally well preserved on both sides. Glaucomatous excavation of the optic nerve-head on the left observed with the ophthalmoscope.

She understands absolutely nothing that is said to her; but care must be taken not to betray one's meaning through gestures. When called, she answers 'yes' to her own as well as to any other name and turns around. To the superficial observer she gives the impression of confusion, because not only do her answers fail to correspond to the meaning of the questions put to her, but also because the sentences she utters are often themselves faulty in that they contain nonsensical or distorted words. Nevertheless the meaning of any of her sentences that one manages to understand at all is always reasonable. There is no trace of flight of ideas. She also behaves calmly and properly, whereas confusion of this degree would have to be accompanied by serious deterioration of behavior. Very often, especially when emotionally aroused, she succeeds in forming whole sentences which are completely correct. She often names correctly objects held before her, for example a hat, a pencil, a watch, a thaler, a 2½ groschen coin, a handkerchief, etc. At other times, however, she will not be able to find the same name. Examples of completely correct sentences, used appropriately: "The food tasted very good today", "I hope I will get well again", "The doctor gave me two groschen." There are many others of the same type. She introduces her son, who happens to be visiting, to the doctor and says, "This is my Richard, my handsome son. Isn't that right, a very handsome son." In the morning and evening she says her prayers faultlessly, just as she does after having been asked several times, about fourteen verses in all. During the day, when associating with the other patients and in a relaxed mood, she names most objects correctly, so that we must assume that under certain circumstances she would have an unlimited vocabulary at her disposal. She sings, but without words, the Tyrolean song 'Wenn ich zu meinem Kinde geh', after another patient happens to have sung it.

Thus the patient can under certain conditions say everything correctly, but she understands absolutely nothing.

Great care and strict control of the examiner's own facial expressions and gestures are necessary in order to establish this conclusion. When on rounds one comes up to her bed and asks her to show her tongue, she does so correctly, but only because she guesses the meaning of the question and imitates the behavior of the other patients, since if she is requested first to close her eyes, she sticks out her tongue. If she is given further commands, without accompanying gestures, e.g., to pick up the glass from the chair, she gives evidence of the greatest difficulty, tries sticking out her tongue, closes her eyes, shows her teeth, etc. In short, she does the things she has had frequent opportunity to see the other patients doing. While doing this she says "What should I show you now", or "What should I smear now" (schmieren), etc. "What good is that, when I don't hear it." Finally she starts to cry and exclaims, with no difficulty in speaking, "Will I ever get well again?" Finally she allows herself to be calmed down as we emphatically indicate that she will – by means of gestures, of course.

She knows how to use all objects, puts on her glasses correctly, etc. Seated at the table in order to write, she takes a pencil, which has been handed to her pointing in the wrong direction, looks at it, then turns it around and holds it correctly, but writes only up and down strokes. She turns a pen around which has also been handed to her pointing in the wrong direction, dips it correctly into the inkwell and then puts the steel point in the right position, but with no better results. After a few lines have been freshly written, the wet writing is spread before her and the sand container put in her hand. She asks the doctor questioningly "Shall I pou (pour) some on?" ("schuet" for "schuetten"). At the affirmative answer she pours the sand, puts the container away, then folds the page properly and pours the extra sand back into the container.

After a successful attempt to tell the doctor something that seems important to her, she looks at him and says ,"Do you hear that?" and is happy when she sees an affirmative gesture. She has complete alexia, nor does she understand numbers correctly, although she often uses them correctly in speaking.

Her general condition being good, she improved rapidly. On March 15, 1874, when I presented her to some colleagues of mine, she understood many things that were said to her repeatedly and forcibly. She responded correctly to her name and did not respond when she was called by a strange name. The picture as a whole was nevertheless still a typical one: she spontaneously used many words correctly but in contrast she understood extremely few words, and these only with great difficulty.

On March 18, 1874, the following conversation, taken down word for word, took place. It already shows significant progress.

"Good morning, how are you?"

"Thank you, I'm quite well."

"How old are you?"

"Thank you, I'm fine."

"How old are you?"

"Do you mean, what I'm cal, how I hear?" (Meinen Sie, wie ich hei, wie ich höre?)

"How old you are is what I wanted to know."

"Well, I don't know exactly what I'm called shear" (... wie ich so heissen schwiere), (corrected) "What I am called hear" (wie ich so heissen höre).

"Would you like perhaps to give me your hand?"

"I really don't know, what I, etc." (no sign of comprehension).

"Where is Richard?"

(Thinks for a long time), "My Sodam, my Richard." (Mein Sodam, mein Richard).

"Is there anything you would like to have?"

"Now I don't know what it's called, I know it quite well. I've swollen with middle (with it)" (ich habe ja schon geschwollt mitte) (damit). "I know well what that actually comes to be called, but it doesn't occur to me."

(She is shown the watch).

"A watch" (softly). "A pocket" (louder), "a pocket-watch, a pretty one."

She is given her glasses, she puts them on, then looks at the watch and says, "But I can't hear with this. A very pretty watch, I like it" (with visible pleasure).

The watch is held up to her ear, taken away and brought close again: "Yes, I hear that, that too, now I don't hear it, now it's coming a ways thinly."

A button is shown her: "That I know, a buron, a butto" (eine Knure, eine Knoppe – for Knopf).

When one of her braids is held she says (with a smile): "That's my..."

"Is that your hair?"

"Yes, my watch, my hore, my hairhour" (meine Uhr, meine Hore, meine Haaruhre).

A dreier (3-groschen piece) is given her: "Oh yes, that's a drekter, a dreier." A two-groschen coin: "That's a two doctor, two droschen", (finally) "two groschen."

She is given the coin: "That I'll let you many many times every possible thing that you have seen to me. I thank you will many times that you have said everything to me. Many thanks that you have been so good, that you have been so nice" (Sa lasse ich mir viel viel Mal alles Mögliche, was Sie mir haben gesehen. Ich danke halt will viel liebes Mal, dass Sie mir das Alles gesagt. Na, da dank ich viel Mal, dass Sie sind so gut gewesen, dass Sie sind so gütig gewesen).

"What is Richard doing?"

"If he will come and see on Sundays."

"How big is Richard?" (with a gesture).

"O very big, taller than you, he is 19 tahr sohr."

After a while she cries, "Oh, dear God, if I were only well again."

On March 25, 1874, the recovery had made further progress. It was noticeable that she could repeat words spoken to her, at first incorrectly but eventually correctly, and that she often corrected herself spontaneously. Her general condition was very satisfactory.

April 20, 1874. She has made further progress and now understands almost everything that is repeated to her several times. She still speaks somewhat haltingly, but for the most part correctly, and reads without stumbling. When she is asked to write about some topic of her own choice, only a few words come to her, and she is also not capable of writing words dictated to her. On the other hand she can copy fairly well words that she is given in written form and she copies all the single letters correctly. Agraphia is thus her most striking language disturbance at this time.

The case just described reminded me very much of one I had observed earlier and which presented certain similarities. My notes on it are lacking in the necessary thoroughness and precision because at the time I had not yet formed a correct analysis of the symptom complex of aphasia.

(2) Susanne Rother, 75 years old, wife of a porter, was admitted to the Allerheiligen Hospital on October 7, 1873. She showed all the signs of extreme senescence. Very advanced atheroma of all accessible vessels, and an expression of suffering on her face. Continual feeling of cold but no fever. She could walk only when supported, evidently on account of general weakness and dizziness, but she seemed to drag primarily the left leg. She lay in bed, usually moaning, and deeply wrapped up in the blankets. She urinated and defecated in bed.

Her mental condition was regarded at the time as a confusional state associated with aphasia. She answered all questions directed at her in a completely confused way, and carried out commands either not at all or in a completely confused manner, which at the time gave the impression of apraxia. The attendants thought that she was deaf because of her lack of understanding of what was said to her. Furthermore, she paid little attention to her surroundings, and in keeping with her severe malaise showed little urge to communicate. Her (spontaneously used) vocabulary thus seemed small in contrast to that of the case described above but was nevertheless large enough that motor aphasia (see above) could not be considered. The presence of aphasia could be recognized by her substitutions and distortions of words. Thus she often said correctly, "Thank you very much" (ich danke recht herzlich), but at other times, "I thank you very giving" (ich danke recht geblich). "I am very sick", "Oh, I'm so cold", "You are a good man", are expressions she used frequently. The doctor she had just called "a good man" she soon afterwards called "my little daughter" or "my little son", without distinction.

An ophthalmoscopic examination done on November 5, 1873, showed gray atrophy of the right optic nerve head.

Sensation seemed intact. Grip was equally weak on both sides. No more precise examination of sensation or motility could be undertaken.

The area of cardiac dulness was increased to the left, and there were no abnormal heart sounds.

No improvement in either mental or physical symptoms took place.

The patient died on Dec. 1, 1873, after a prolonged intestinal infection, accompanied in the last 2 days by vomiting and deep prostration.

The patient's history as given by her family showed that she had been suffering from a weakness of the left leg for 10 years. It had come on gradually; the confused speech was said to have appeared suddenly on Nov. 2, 1873. Information about her physical and mental status otherwise was lacking.

The autopsy showed edema of the pia and a mild internal hydrocephalus. The convolutions of both hemispheres and both insular regions were completely shrivelled and atrophic. Furthermore all the arteries of the brain showed extreme atheromatous degeneration. The lower branch of the artery of the left fossa Sylvii, which runs in Burdach's inferior cleft, was occluded by a thrombus adhering tightly to the wall. The entire first temporal convolution from its origin in the anastomosis with the second temporal convolution, and furthermore the entire region of origin of the latter from the first convolution (Bischof's lower parietal lobule) as well as the external portion of its longitudinal course were now a yellowish pulp to which the pia, which was cloudy, was tightly adherent. The radiation of the temporal lobe into the insula was for the most part interrupted by this softening. The insula itself and the basal ganglia showed no change. The locus of softening was not bounded by any area of increased resistance due to inflammation but passed directly over into an area of normal consistency.

In the dorsal region of the spinal cord a part of the left anterior lateral column was gray and degenerated, and in the lumbar region this was true of the posterior columns.

A few critical comments are appropriate concerning these two cases. Both patients show the distinctive symptoms of sensory aphasia. They have a relatively large vocabulary at their disposal and have completely lost the ability to comprehend speech. The course of the illness, however, differs greatly in the two cases: the first will presumably recover completely, whereas in the second case there are present serious general disturbances of the brain that finally lead to death. Data about other criteria such as agraphia and alexia are lacking in the second case.

In my opinion, and experienced neuropathologists will certainly concur, the concept of localized illness cannot be distinguished too clearly from that of general illness of the brain. In the numerous examinations of the brains of patients suffering from senile or alcoholic cerebral atrophy which I have had the opportunity to carry out in the course of 3 years at the Allerheiligen Hospital, I have always found all the convolutions, and especially those of the insula and the first primitive convolutional arc, affected without there having been any particular focal manifestations during the course of the illness. I have made the same observation in extensive meningitis, both traumatic and tuberculous, where because of the pressure exerted by the abundant exudate all the hemispheric convolutions without exception showed to a pronounced degree the shrivelling and puckering emphasized by Rokitansky, without distinctive localized symptoms having been present during the patient's life. *Simple atrophy which has attacked one convolution as a component of generalized atrophy never brings about loss of its functions and cases no focal manifestations.* We can regard this empirically based conclusion which will have to be explained in terms of the nature of the pathological process, as one of the well-established rules of neuropathology. Disregard for it in post-mortem studies and in clinical observation has already caused much confusion.

We can thus assert with certainty that the softening of the left first temporal convolution in the case of Rother was the only brain lesion which could have caused the localized symptom of aphasia that persisted throughout the whole course of the illness, and that the generalized convolutional atrophy was either the result of senescence or, which is considerably more probable, a consequence of the presence of a circumscribed focal lesion. Griesinger has rightly emphasized the deleterious influence that a focal lesion exercises on the nutrition of the brain as a whole. Here as in most cases we find a corresponding anatomical alteration, as seen in the atrophy of the convolutions.

Thus, in the case of Rother the most striking localized symptom, aphasia, is explained by the softening of the left first temporal convolution. In the case of Adam we are also justified in assuming a focal lesion in the left first temporal convolution.

A second point that deserves to be emphasized is the paresis of the left leg; it seemed to point to a lesion on the right side of the brain. But the autopsy showed the anatomical basis of it to be a gray degeneration in the dorsal part of the left lateral column. Important evidence for the independence of the two phenomena was the patient's history, according to which the paresis of the left leg had already been present 10 years before.

Finally, the crossed atrophy of the optic nerve deserves mention. If it were the rule in focal disease of the temporal lobe, it would support the theory of a complete crossing in the chiasm.[17] Neither the thalamus nor the corpora quadrigemina showed any difference in color or size.

(3) The following is a clear case of conduction aphasia. I am indebted to Professor Foerster for referring it to me.

Beckmann, a pharmacist, 64 years old, having had a few glasses of beer the night before and having slept well during the night, noticed on the morning of March 15, 1874, that he could no longer read and was even less capable of writing, although the outlines of everything were clear. He therefore travelled to Breslau on March 18 to consult an ophthalmologist. In the course of that day the first disturbances in speech appeared. There was no trace whatsoever of disturbance in his general condition.

On March 20, 1874 examination showed him to be in the following condition: A powerfully built man, in excellent nutrition, with a ruddy complexion. Physically completely robust, with no objective signs of advanced senescence. The radial artery was only moderately rigid, the pulse strong, of normal rate, skipping after every 6–10 beats. No trace of paralysis in either the face or the extremities.

Area of cardiac dulness enlarged somewhat to the left. Heart sounds weak. No emphysema.

He understands everything correctly and always answers questions correctly. He knows exactly how to use all objects. He shows no trace of motor aphasia, since his vocabulary is not reduced. He cannot, however, find words for many objects he wishes to designate. He makes an effort to find them, becoming agitated in the process, and if one names them for him he repeats the name without hesitation. This is thus the same condition that appears as a normal physiological manifestation in many healthy people and that was characterized above as conduction aphasia. He can say many things fluently, especially familiar expressions. He then comes to a word on which he stumbles, remains caught on it, exerts himself and becomes irritated. After that every word that he utters, haltingly, is nonsensical; he corrects himself over and over again, and the harder he tries the worse the situation becomes. If one guesses the intended meaning and says the word to him, he breathes a sigh of relief, "Yes, that's what I wanted to say." On the other hand, when he is relaxed, whole sentences are sometimes produced, grammatically completely correct, with a meaning different from that intended. Then he again becomes irritated and rejects the whole sentence. He often asks if what he has just said is correct.

The relations of his aphasia to his alexia and agraphia are interesting. Both conditions are present, but in very different degrees. A large printed letter is shown to him, which he is to name. Despite all his efforts he cannot recognize it; he looks around for help; his eye lights on the title of a book, printed in gold letters, and he points to the title and says, "I recognize that, that's Goethe." Right next to it, bound in the same way, is Schiller. Asked to read the title, he does not recognize it. Despite all his efforts he can discover only by comparing the two titles that it is not the same as Goethe. Thus on the street he reads the signs without paying special attention to them, but if he is directed to a specific word or a specific letter he never succeeds in finding it. From a group of numbers and letters written out for him he finds, slowly but always correctly, the one asked for. When the number or letter asked for is not among those shown him he is regularly aware of it. He also recognizes every individual letter if one says the name to him; he rejects the other names and stops at the correct one. With numbers he helps himself by fixing the number with his eye and counting off on his fingers until he gets to the number that corresponds to the visual image. That no visual disturbance is responsible for his not recognizing the letters follows first of all from the fact that he recognizes all other objects, photographs and so forth, correctly, as well as from his statement that he sees the outlines of the letters clearly. The letters still look familiar, and do not look like Chinese or cuneiform script, as would be the case with complete loss of the visual images. He also gives a direct proof of this when, with some concentration, he succeeds in copying the letters that he cannot name, as well as whole words.

He does, however, suffer from agraphia; he can copy everything but he cannot write independently. He succeeds in writing almost no letters; despite all his efforts only up and down strokes come out. He does better with simple numbers, but even two-digit numbers are a difficult task for him.

An exact determination of the visual field of both eyes, undertaken on March 25, 1874, by Prof. Foerster, showed a clear-cut hemianopia on the right. Nothing unusual was observed with the ophthalmoscope.

Further observation showed that the aphasia varied considerably in degree, sometimes being hardly noticeable but emerging very markedly at other times. Furthermore, only substantives, and among them especially place names and proper names, caused him difficulty. He was much less aphasic with his next of kin than with strangers. His aphasia was always most striking when he was being examined by a doctor. With every new failing of which he became aware during an examination, the aphasia worsened.

The patient is still under treatment. His physical state is excellent, the aphasia seems to have improved somewhat and in any case has not become worse. On the other hand, a second determination of the visual field by Prof. Foerster with the perimeter, on April 9, 1874, showed that the visual field had undergone a still further reduction: in both eyes the portion of the right visual field between the macula lutea and the periphery, where vision had previously been unclear, was now completely lost. Furthermore the left boundary of the visual field had shifted considerably to the right. The disease process was thus shown to be progressive.

An examination performed in May of this year showed significant improvement. The patient now showed an interesting symptom: he read whole words, such as his name or the names of members of his family, correctly, but he could not read the individual letters of the names correctly.

(4) Kunschkel, 50 years old, a goldsmith, had been suffering from incontinence of urine and feces for 2 years and had therefore been admitted as an inpatient in the Allerheiligen Hospital on Jan. 27, 1874. There his unmotivated and excessive coarseness towards attendants and doctors attracted attention and he was transferred to the psychiatric service on Jan. 29, 1874. His behavior there was sullen and unfriendly; he was extremely irritable and little things angered him. He had complete incontinence of urine and stool, which often seemed deliberate. He had an extremely unclear notion of why he was in the hospital. There were no signs of paraplegia or tabes. His entire skeleton was extremely distorted as the result of rickets.

On March 26, 1874, during an altercation that developed in the men's lounge and was not stopped quickly enough, he received a blow on the right ear. After that he spoke confusedly and reeled as he walked. The right half of his body drooped. The next morning a hematoma of the right ear and pronounced aphasia with paresis of the right leg were observed.

In walking he staggered and dragged the right leg. Grip was weak on both sides, somewhat more so on the left than on the right. No visible paralysis in the face. Sensation generally somewhat dulled. Pulse 90. Respirations rapid, about 22. Frequent coughing without expectoration. Rales at the bases of both lungs. Hearing was intact.

The aphasia manifested itself in the following way. He would speak quite a while without any hesitation and then, at the end of the sentence, an incorrect word would come out, without any special accentuation. If he was asked about that very word, he would try to improve it, and would then produce incorrect words and syllables irregularly mixed together, a gibberish that could be taken down only with difficulty. If, however, questions were interjected as he was doing this, he often replied in correct whole sentences. To the extent to which he was capable of speech his answers were always correct and in exact accordance with the intended meaning; questioning also established that he understood and readily interpreted everything that was said.

Alexia is present. If, however, he is given the task of chosing a particular letter from a row of letters written out for him, he can do it correctly. Similarly, when all the letters and numbers are recited to him, he recognizes every individual letter and every number.

Agraphia is also present. He can write his first and last names, but his hand shakes. He writes the word 'Goldarbeiter' (goldsmith) correctly up to the 'd'; then he stops and writes a new word that looks like 'well' in the middle of what he has already written. He is no longer capable of anything beyond this, and the result of further exertion is only irregular strokes.

The results of an investigation of the visual field were doubtful because of the patient's lack of attentiveness. Nevertheless, there is much to indicate the presence of right hemianopia.

On March 30 I was able to take down the following conversation:

"What year is this?"

"44."

"Is your father still alive?"

"No, he died in 46".

"How many years ago is that?"

"About 29 zaten daten diten."

"Do you have money?"

"Yes, I have cashed in the Luten and my things."

"How much do you have?"

"Almost 100 thaler. If I had had 100, I would have alread..." (hesitates).

"What would you have then?"

"Graetliche hubel."

"What kind of a place is this?"

"It's an imperial castle."

"What else?"

"Now there are kaisernes kis katen lebeln. There you have for example a royal of the Elizabeth casette, that is, of general casette."

"Would you like to go home?"

"It's already been four weeks and I haven't gotten away from here."

"What is this?" (a handkerchief is shown to him).

"That's a fine netet." (Filil).

"Net?" (Filet).

"No, it's not a net, it's not that strong."

"What is it then?"

"We call it a handkerchief."

"What else can we call it?"

"As we understand in nobles branches I moves in aristocrats" (nu n, wie man's in noblen Zweigen begreif ich in Adeln sich bewegt).

(A pencil is shown to him).

"I thought that was a little kolinomitz, but it isn't".

"What is it then?"

"I have it under the neumman–neu–pencil."

(A pair of glasses is shown to him).

"That's called glottes, glattes..." (Bruecke, Brikke, for Brille).

"Do I use them just for playing around?"

"Oh no, to develop the mile gold, to develop one's appearance" (um eben das meilige golden, um eben sein Ansehen zu entwickeln).

His gait has become significantly more steady. Ophthalmoscopy shows completely normal papillae.

On April 4 no sign of paralysis was any longer visible in his gait. The aphasia could be detected only in that here and there the beginning of a word came out uncertainly or there was an incorrect vowel in an unaccented syllable at the end of a word, etc. The alexia has also disappeared; in difficult words like 'Zugrundelegung' only the last syllable is read unclearly. On the other hand, foreign words or words unfamiliar to him can be repeated only in distorted form. He retains several syllables of these words but completes the word in an arbitrary way. He reads complicated numbers such as 25394 correctly. The disturbance is most striking in writing. He has to think for a long time before beginning to write. Then the beginning of a sentence dictated to him will be produced correctly, but not the end. Asked to write the word 'Goldarbeiter' (goldsmith) just after he has been speaking of a pince-nez, he begins "paen...". After a while he is completely exhausted and can no longer pull himself together enough to begin a new word.

The 2 cases just described clearly belong together. Both patients have complete comprehension of what they hear. Each still has a complete vocabulary at his command, but the degree of control varies with the mood and circumstances. One, Beckmann, is still mentally completely intact; the other, Kunschkel, is talkative, feeble-minded, lacking all self-control and any critical faculty. Thus Beckmann will get stuck in the middle of a sentence and struggle to express himself while Kunschkel will replace what is missing with meaningless words or new combinations of syllables and will feel satisfied at having completed the sentence. He also makes efforts to correct himself, but only poor ones. Both have the same kind of alexia, and agraphia is much more prominent in both than aphasia or alexia. I also strongly suspect that Kunschkel also has a right hemianopia. That would strengthen further the analogy between the two cases, which must have its basis in anatomical findings.

It was stated above that conduction aphasia cannot be the cause of alexia (except for the letters), at least not among educated people. Beckmann seems to constitute an exception to this statement. However, his condition, and probably Kunschkel's as well, is associated with right hemianopia; without this Beckmann would be able to read everything without any difficulty. The proof of this is the fact that he is actually capable of reading many things, but only in passing, when he does not focus on the words. The word he focusses on is only half there for him; it can thus awaken no specific concept in him. He lacks, however, the ability to put together the word from its individual letters. Another very striking example of his potential ability to read, in addition to the one narrated above about Goethe, was provided to me by the way he scanned a prescription. He had hardly glanced at it when he said, "This word here is aloe." The word 'coloquinten' written immediately below it he could not decipher despite all his efforts. He had perhaps involuntarily read the word with a backward glance or with peripheral portions of the retina in which there were still some preserved areas on the right.

The narrowing of the visual field, which rapidly became more pronounced as the aphasia seemed to improve, suggested at first that a progressive disease process had attacked not the insular region itself but some structure near it (probably the left optic tract[18]) which represents the left halves of both retinas. The fact that the external portion of the right halves of both retinas was also affected can be explained only through crossing of the midline toward the right[19], by the desease process. At the same time there was no paralysis or disturbance of his general condition. Everything points to a localized disease process at the tuber cinerum which extends to the left into the point of origin of the fossa Sylvii (the lamina perforata anterior) and which has now arrested its progression.

The aphasia can be attributed only to disturbances of circulation affecting the area around the actual focal lesion.

Kunschkel's case is clearly to be regarded as having been caused by trauma. The fact that the trauma affected the right side while the cerebral disturbances were left-sided is in agreement with an observation frequently made – that a blow on one side of the head produces hemorrhage and softening in the opposite hemisphere.

There are many cases of pure motor aphasia in the literature.

The existence of the pure motor form of aphasia has in any case been adequately established and is the least in need of confirmation through further cases. The following two case histories are at least of predominantly motor character.

(5) Rosina Peter, 78 years old, the widow of a coachman, suffered a stroke with paralysis of the left side, 2 years before being admitted to the Allerheiligen Hospital. Improvement was rapid, and after two weeks she was able to get out of bed and walk around. She dragged the left leg, but the left arm was soon completely usable again. Sensation on the left, however, remained diminished and the left limbs were usually colder. Furthermore the stroke resulted in generalized chorea which persisted. On account of the violence of her movements she fell from a chair in October 1873, and fractured the neck of the left femur. On her admission to the psychiatric ward, which took place on Feb. 27, 1874, her condition was found to be the following: a very aged woman, in constant restless motion, which is most marked in the right limbs and the face. She makes noises continually and the speech is babbling and incomprehensible. At the same time she opens and shuts her mouth and her eyes, rolls her tongue around in her mouth, sticking it out and to the side, waves her hands in the air, accompanying her continuous talk with equally continuous gestures. In short, she presents the picture of someone possessed. She is usually in the sitting position, and dyspneic. There is a recent fracture of the sternum right under the angle of Ludovicus, as indicated by abnormal mobility and crepitation. There is effusion of blood into the mediastinum, an irregular pericardial friction rub, cardiac arrhythmia and a radial pulse of varying fulness. Vessels are extremely atheromatous. The skin of the left breast is suffused with blood up to the sternum.

Her behavior is decidely maniacal. She hits people and throws away dishes offered to her. When taken from her bed and forced to walk, she resists violently and pushes the attendants away from her with considerable force. No conclusion is possible concerning her gait, since walking is possible only if she is supported, for the reason that she avoids as much as she can stepping on the left leg.

During the next few days she gradually became calmer and the dyspnea receded. Sleep was induced by small subcutaneous doses of morphine. Temperature high in the evening (up to 39.5 °C.), normal early in the day. Subjective feeling of well-being. She still becomes a little violent but is generally in a calm, agreeable mood. Her flinging movements are less often directed at particular targets. In the attempts to walk, which is very difficult for her, the choreic movements increase. The choreic movements occur almost uninterruptedly while she is awake but stop completely in sleep and can be overcome temporarily by exertion of the will.

No fever since March 19, 1874. General condition excellent. The dulness near the mediastinum has decreased in intensity and extent.

March 28, 1874. Vomiting yesterday evening, then a quiet night. More vomiting today. She has however not complained of other difficulties, nor has there been loss of consciousness. She is completely speechless, with preserved consciousness. She points to her mouth to indicate her inability to speak. She is capable of sticking her tongue out. She understands what is said to her.

March 29. A little improvement. She produces some indistinct sounds on attempting to speak.

March 30. Patient had a good night. This morning completely mute, cannot swallow any longer. The mandible droops, the tongue hangs out, and saliva drools from her mouth. Right hemiplegia which also involves the face.

The extensors of the right leg are contracted, and the leg can be flexed only against considerable resistance. The right arm severely paralyzed.

Sensitivity to pain generally diminished (as tested by pin-prick), most conspicuously in the right hand. Ophthalmoscopy shows a reddish gray nerve head without signs of swelling. She understands simple instructions and uses the left hand to carry them out.

On intubation of the esophagus for artificial feeding she screams and resists, but afterwards presses the doctor's hand. Death took place on April 10, 1874. The autopsy showed pachymeningitic adhesions to the vault of the cranium, osteophytes of the inner cranial wall, and atrophied convolutions on both sides. On the left there is an area of softening, extending in most places to the surface, involving the anastomosis between the first and second temporal convolutions and over which the pia is adherent. From this region the area of softening extends posteriorly only a short distance in the depths of the white matter. On the other hand it extends anteriorly in the white matter to a point beyond Broca's area. Broca's area itself is not affected, but is sunken into the outer layer of the white matter which is involved. The white matter of the central convolutions is affected in the same way.

A short distance posterior to this lesion, in the area of the convolutions at the junction with the occipital lobe, the white matter of the hemisphere is similarly softened. On the right a large part of the surface of the convexity is also affected. A large softening occupies almost the entire width of the hemisphere beyond the central fissure. The area around this lesion is hardened and the gray matter of the convolutions shows intense red striations.

Both cingulate gyri are intact, as are the crura of the fornix. In the head of the right caudate nucleus, on its ventricular surface, there is an old cyst about the size of a pea, containing serous fluid.

(6) Isidor Itzigsohn, a travelling salesman, 26 years old, was admitted to the medical clinic on March 25, 1874. I report the case with the kind permission of 'Geheimrat' Lebert. The patient had taken sick that morning, suffering first from speechlessness which was followed in a few hours by the onset of paralysis of the right side of the body. At first only the presence of motor aphasia could be established with certainty, since the patient's generally dull and apthetic condition was not suitable for a more precise delineation of his symptoms. The right arm was completely paralyzed; the right leg dragged in walking, which was, however, still possible. The right side of the face lagged in movement. Sensation, tested by pin-prick, was intact. There were diminished reflexes in the paralyzed limbs. No abnormal findings in blood vessels or heart. In the next few days the paralysis of the right leg increased and the mental condition remained unchanged. Since the patient's history gave evidence of previous syphilitic infection, an inunction treatment was ordered.

His condition improved rapidly. By April 26, 1874, the patient could walk and use his right hand reasonably well. An examination carried out at this time gave the following results. Despite complete mobility of the tongue and lips, the patient cannot speak a single word. He can articulate only syllables which contain the letter 'a' and an almost silent ending, mostly 'm'. In the attempt to repeat a given word the number and cadence of the syllables often correspond to that of the word spoken.

He understands all instructions and questions that do not require a great application of intelligence, but his behavior is insolent and capricious which makes the examination more difficult. The meaning of prepositions seems to escape him. When instructed to put the book first *on*, then *under* the paper, he does not understand that these are two different commands. When the meaning is unmistakable, he handles the prepositions correctly, as when he is asked to climb onto a chair. Similarly, he does not understand the various names of colors, although he finds the color of any object correctly on a spectrum. A number of sound images thus seem to have been destroyed.

He understands written symbols and correctly carries out simple commands given him in writing. He also knows the numbers and the individual letters.

The capacity to write is retained only insofar as he can copy everything correctly. Independently or from dictation he cannot even write his name or most of the individual letters. On the other hand he can write the alphabet, although with many errors. He writes the numbers correctly in order. He also copies printed words correctly, and does this independently of the form of the letter. Thus in one word he wrote a cursive 'r' instead of the printed 'r' in the original.

It can be clearly shown that he has no hemianopia. Hearing and vision are intact; the ophthalmoscopic examination gives normal results.

The case of Peter is interesting in many respects. The chorea appeared at first primarily in the right limbs. This is readily explained by the left hemiplegia produced by the first apoplectic attack. This shows the correctness of Meynert's theory according to which chorea is a form of hyperkinesia which involves the pyramidal tract. The hemiplegia was obviously caused by the old softening in the right hemisphere, not however, directly by it, but by the collateral edema that accompanied its onset. When such an extensive portion of the brain is suddenly cut off from its circulation (which leads to the apoplectic attack) it is not too hazardous to assume that this leads to the development of collateral edema in the immediate vicinity and hyperemia in the other hemisphere. While the former explains the hemiplegia, the latter helps us to understand the chorea. The chorea would thus be caused by hyperemia of the pyramidal tract in some part of its course, in the same way as hyperesthesia can be experimentally produced by opening the spinal pia (Meynert).

This focus of softening, which had been present for years, was now joined by a second in the left hemisphere, which produced aphasia and right-sided hemiplegia, and in a short time led to the death of the patient. The fact that the mandible drooped, that the tongue could not be moved, and that swallowing was impossible should not be interpreted simply as a sign of approaching death; these symptoms appeared when full consciousness and understanding of the situation were still present. *Like the aphasia, they must be considered to have been caused by the destruction of the white matter*, since the basal ganglia showed nothing abnormal except the little cyst in the corpus striatum. Finally, the atrophy of all the convolutions, which in this case too accompanied localized lesions of the brain, should be emphasized.

The case of Itzigsohn, the result of a focal syphilitic lesion of the brain, is certainly not one of pure motor aphasia, for the latter could not have agraphia as a consequence. The agraphia is, however, not absolute but is present only in spontaneous writing or writing from dictation. Since the ability to copy is retained, the path $\alpha\beta$ (see p. 57) is intact. Thus the writing disturbance results from the interruption of the path $a_1\beta$, which connects the sound images with the psychomotor center of writing movements. From all of these findings, it is probable that the deep recess where the anterior cleft and the superior cleft meet, an area rich in blood vessels, is the point of origin of the pathological process. Since aphasia only very infrequently results from tumors, the process might be a syphilitic encephalitis rather than a gumma. Furthermore, the partial involvement of the sensory speech center suggests that the lesion is of considerable extent.

The following case is a natural sequel to the two just described.

(7) Karl Seidel, a day laborer, 60 years old, fell ill on May 7, 1874. His attack came one as he was standing on a ladder; he was able to hold on until he was carried down. He was unsteady and could not speak, but was capable of walking home without being helped. At home paralysis of the right side was noticed. He lay there in a drowsy state for 24 hours and was admitted to the medical clinic the next day, where Geheimrat Lebert kindly allowed me to observe him.

The May 15th examination showed him to be in the following condition: a powerfully built man, with few signs of aging, lying in bed fully conscious. Appears ill, and his face is yellowish in color. Arteries tortuous but completely elastic. Dulness of the heart increased to the left, with a systolic murmur at the apex and an irregular rhythm. At rest the wrinkles on the right side of the face are obliterated, and the lips are open on the left. In speaking the right half of the mouth lags in movement. The right arm is little used spontaneously, and its movements are slow and executed with marked effort. The grip is extremely weak on the right. The right arm is cooler, and the hand is somewhat cyanotic. The disturbance of movement in the right leg is not very striking in walking; he merely shows slight dragging of the leg. When however he tries to get into bed, he cannot get the leg up and has to use his hands to help it. Sensation, as tested by pin-prick, is not markedly different on the right.

Mentally he is somewhat apathetic and unconcerned. His appetite is good. He says nothing spontaneously all day long and answers questions only with 'Yes' or 'No'. He has a hard time finding his way around the ward, cannot remember where the toilet is, and confuses objects. Thus he may urinate first into a water glass and then, when it turns out not to be large enough, into the urinal next to it.

He understands all relatively simple statements dealing with things within his visual field, answers 'Yes' or 'No' correctly and carries out all commands correctly. Spontaneously however he can say only 'Yes' or 'No'; the other things he says in answer to questions are for the most part incomprehensible, although some articulation is evident in them. He can however repeat everything correctly, although with some dysarthria, despite which the words are clearly recognizable in respect to vowels, cadence of syllables, etc. Closer examination shows that only the consonants 'c', 'd', 't', 'z', and 'k' give him trouble. When asked to stick out his tongue he only opens his mouth. He does no better even when he is shown what to do. He can move his tongue to the right or left only with difficulty and jerkily, and he cannot move it around. The mobility of the tongue is thus significantly impaired.

The following conversation is characteristic of his speech disturbance:

"Good evening."

"Good evening."

"How are you?"

"Yes."

"What is your name?"

"Yes."

"Is your name Thomas? Berthold? Schulze? Mueller?"

"No. No" (to each question).

"Is your name Seidel?"

"Yes."

"Say Seidel."

"Seidel."

"Say Berthold, Thomas, Hospital, mental hospital, masonry work." (He repeats each word correctly.)

When asked whether he can read and write he replies in the negative. He can however write his name. He writes Seidel correctly. He is asked to add his first name, Karl; instead, he writes Seidel. Karl is written out for him, but he again writes Seidel. On a new piece of paper 'garden' is written out for him to copy. The word that he now writes is a mixture of Seidel and 'garden'. It begins with 's' and contains a clear 'a' and 'd'. In the meantime on another sheet of paper the number 1874 has been written and he is asked to write 1874. He writes 1844. Now he is asked to write the number 60 and he writes 1848. When his error is pointed out to him, he writes 18 (i.e., a two-digit number) and acts as if it is now correct. When 60 is written for him he copies 66. In any case we see that he is influenced by what is written out for him. Anything in which he eventually succeeds seems to persist for a long time and interferes with any new task. He also becomes tired very quickly and then makes no further attempts.

He cannot read aloud '1874' although he copied it reasonably well. In reading his ready fatigability and tendency to perseveration cause him even greater difficulties. Out of a row of letters he reads the first correctly but not the others. He reads the number 5 correctly but not the number 9 that follows. Five fingers are held up before him and after repeated questioning he says the word "Five." Two fingers are then held up, but he can find no word for that. "Five fingers?", he is asked. "Yes", he replies. This is contested: "That's not really five fingers!" "Oh, yes."

In discussing this case, the presence of mitral insufficiency clearly suggests embolization and softening. The softening must have spared the region of the first primitive convolution and the insula, since the whole psychic reflex arc aa_1 bb_1 is intact. On the other hand, the fiber tracts *cb* and *db* have been interrupted, since the concept is no longer capable of innervating the representations of speech movement. The softening thus affects the white matter of the hemispheres. At the same time the paralysis of the right side, above all the paralysis of the tongue, points to involvement of the lenticular nucleus. Thus we must assume a relatively large area of softening which has destroyed part of the lenticular nucleus and which extends into the white matter of the left hemisphere above the lateral ventricle.

Given the patient's poor education, little can be concluded from the disturbances in reading and writing.

This case illustrates a new type of aphasia which, despite its outward similarity to the motor form, is nevertheless most closely related to the type of aphasia resulting from lesions of the insular region. The latter is the result of the interruption of the connecting paths between the motor and sensory speech centers when these are themselves intact. The designation of Seidel's condition as aphasia requires no further proof, although traces of asymbolia are also present. *The distinctive speech disturbances can, however, be explained only by interruption of the paths cb and db.* The asymbolia of which there are in any case only some traces must be interpreted as an accidental complication, i.e., as an effect of the extensiveness of the lesion and cannot explain the peculiar disturbance of speech.

This case, which is certainly very rare, provides a necessary supplement to the description of conduction aphasia in Section III.

(8) Louise Funke, a lady's companion, 59 years old, suffered an apoplectic attack on Dec. 6, 1873, and was admitted to the Allerheiligen Hospital on Dec. 8. At that time she presented a clear-cut picture of cerebral hemorrhage. There was hypertrophy of the left ventricle and some albuminuria. Paralysis of the right half of the body and speechlessness were observed.

At the beginning of March, 1874, examination showed her to be in the following condition: Complete paralysis of the right limbs. The right corner of the mouth droops, the naso-labial fold is pulled to the left, and the wrinkles are obliterated on the right. Only the left side of the mouth participates in movement. She sticks her tongue out only in a trembling and uncertain manner, without deviation to either side. The palpebral fissures are equally wide, and the eyelids can be closed on the right side as well as on the left. The right limbs can easily be moved passively, but there is pain with rapid movements. The left pupil is contracted and does not react to light. It does however react sympathetically to illumination of the other eye. The pulse is slow – 48. Sensitivity to pain, as far as it could be determined, is retained on the right, or only a little diminished.

She has only the word 'Yes' at her command, with which she responds to all questions. Trying to express herself spontaneously, she repeats "Yes, yes", ad infinitum. She understands nothing that is said to her. Simple questions and commands never produce any reaction other than the often repeated "Yes". One can however make onself understood to her reasonably well by means of gestures; at least she shakes hands when a hand is extended to her, and she sticks out her tongue when one points to her mouth or sticks one's tongue out to show her, etc. Any further examination is impossible because of her lack of comprehension of what is said to her. It can thus also not be determined whether she is deaf on one or both sides or not at all.

There is present in this case extensive destruction of the lenticular nucleus and of the first primitive convolutional arc or of its white matter.

(9) Siegmund Zwettels, a merchant, 43 years old, suffered a stroke three years ago which left his right arm paralyzed (the leg was reported to be completely intact) and left him speechless. He could say only his name, S. Zwettels. He could also write it, but otherwise he had completely lost the ability to read and write. Since then the capacity to use the right arm and speech have returned almost completely. However, he often finds words only after long deliberation and cannot immediately name objects correctly. Since his condition is accompanied by right-sided hemianopia and his intelligence is completely intact, he is in a position to give an account of the effect of the hemianopia on reading. He indicates that in reading he has to put the individual letters together and thus can read only slowly. Thus on the street he reads signs very easily when walking past them, and only has difficulty when focussing on a word. In reading individual letters he succeeds with most, but once in a while he stops and finds the letter only after he has recited the alphabet up to this point (a residue of his earlier alexia). He can read a four-digit number only after a long period of concentration and trial, during which he has to distinguish the tens from the ones. These findings are in agreement with

those we drew upon above in explaining the symptoms in the case of Beckmann.

(10) Withold von Salmonsky, a clerk, a little over 20 years old, was admitted to the Allerheiligen Hospital on March 19, 1874. There he had an epileptic attack with loss of consciousness lasting 12 minutes, and he complained of pains in the back and arms, especially when moved. Lying on the left side was painful for him. He developed a suppurative parotitis on the left, which was incised and continued to drain for a long time. His state of nutrition deteriorated. 10–12 days before his death a speech disturbance was noticed. Its course unfortunately was not adequately followed. On May 14, 1874, examination showed him to be in the following condition: a very anemic, poorly nourished individual, with atrophic musculature and a blank expression on his face. He complains of pains in the back and on the left side of the head. His pulse is full, strong, soft, 104. Respirations not accelerated. Lungs clear. The right half of the face is flattened, the naso-labial fold obliterated. The right corner of the mouth is lower than the left. The disturbances of facial movement are especially striking when he laughs. The forehead is wrinkled evenly, lid closing is intact, the right pupil is dilated and reacts to light. No paralysis of the limbs. Tapping on the left temporal region is extremely painful. Extreme hyperesthesia, limited to the lower limbs. He walks laboriously and on a broad base. Ophthalmoscopic examination shows both papillae to be grayish with quite indistinct borders, although without marked signs of swelling. It is clearly recognizable that this is an early stage of papilledema. He understands most of what is said to him and has an unlimited vocabulary at his command, but he confuses words without being aware of it. His speech is well articulated and free of mechanical impairment. He names objects shown to him correctly at times and incorrectly at other times. Thus he calls a knife 'a knife', but he also calls a clock 'a knife'. When the knife is opened and shown to him, he calls it 'an open clock'.

His reading aloud is very distinctive and displays the same disturbance of language as his spontaneous speech. He reads line after line smoothly but inserts other words arbitrarily, although he believes that he is reading correctly; in this way a nonsensical mixture of correctly and incorrectly read words is produced.

His writing was not examined.

On the basis of the symptom complex described, Dr. Friedländer, to whom I am indebted for the use of this case, diagnosed a brain abscess in the left hemisphere. The diagnosis was also supported by mild irregular variations in temperature.

On the 18th of May, left-sided pneumonia was observed. The patient became stuporous and died on May 21, 1874.

The findings of the autopsy were as follows: no unusual findings in the dura, and no injury to the skull. The pia is dry, anemic and slightly cloudy. The sulci are obliterated and the convolutions compressed against one another, especially in the left hemisphere. In taking out the brain, a fluctuant mass could be felt in the left temporal lobe. On the basal surface of the hippocampal gyrus was a yellowish-green area about as big as a groschen. The pia, however, was not essentially involved. The left Sylvian fossa was exposed carefully by removal of the pia and the blood vessels. It thus became evident that the surfaces of the first temporal convolution and the operculum which face each other have pressed against each other so that the temporal lobe juts sharply outwards and the operculum sharply inwards. The gyri of the insula are wedged between the temporal lobe and the operculum, so that the insula protrudes as a sharp ridge. The entire brain stem on the left is compressed from above downwards and is therefore broadened. The lower surface of the left cerebral peduncle has a saddle-shaped indentation. The whole Sylvian fossa can be laid bare without opening the abscess.

The abscess occupies most of the temporal lobe and indeed reaches the outer and inferior wall of the temporal horn of the lateral ventricle, the ependyma of which is not broken. It contains foul-smelling green pus and is covered by a capsule. The surrounding region is softened (macerated) to such a degree that the borderline between diseased and healthy tissue can no longer be determined. The fibers of the deepest part of the first temporal convolution, in contact with the underlying white matter, are also softened, but the white matter in the center of the convolution is intact although edematous. The abscess extends through the region where the temporal lobe meets the insula; in this region the brain substance (i.e., the junction of the caudate and lenticular nuclei) is itself infiltrated with pus and there is no capsule. The outer fibers of the crus cerebri are also involved in this region. The left optic tract has suffered from white softening and has been flattened.

Aside from being compressed, all the other parts of the brain, especially the entire first left frontal convolution and the basal ganglia, are normal.

In the lumbar part of the spinal cord, there is a circumscribed region of purulent meningitis. The spinal cord is compressed in this region by large collections of purulent material. The rest of the spinal cord shows no abnormality.

The case just described is so puzzling that it seems hardly possible to formulate a satisfactory notion of its course. Since there was no right-sided hemiplegia, the basal ganglia on the left must have been functioning despite the compression, or the compression must not have appeared until very late. If these ganglia were functioning, however, then the aphasia cannot be explained by the compression of the first frontal convolution. This negative result is the only thing which can be deduced from this case for the theory of aphasia. It is impossible to determine whether or not the white matter of the first temporal convolution, which is of importance in this case, was already softened during the patient's life; it was certainly better preserved than that of the other temporal convolutions. The clinical symptoms pointed to an aphasia of the insular region, but this can hardly be offered as an explanation if the basal ganglia were functioning normally, since the basal ganglia and the insular cortex suffered from the same compression.

On the other hand, the aphasia was not sufficiently well followed from its onset. It could have been of much earlier origin and have presented the symptoms of sensory aphasia at the time. In the course of the illness the other hemisphere could have taken over the function of the left temporal lobe, as indeed takes place very rapidly in sensory aphasia. During recovery from sensory aphasia there is, moreover, a stage in which the patient understands most of what is said and nevertheless still makes errors in the use of words.

In no circumstance can this case be used as proof for or against the localization of speech centers which I have advanced. A brain abscess, which, like a tumor, exerts pressure on various portions of the intracranial contents in accordance with laws still completely unknown to us, cannot help to decide a question of this kind. *Only a focal softening can give us information about the localized functions of the brain.*

The paralysis of the right side of the mouth is also unexplained. If it is to be related to the partial destruction of the left motor ganglia, then the part destroyed would have to contain the center for mouth movements, which was described more fully earlier.

The hyperesthesia of the lower extremities made diagnosis more difficult during life. A simple explanation for this was to be found in the circumscribed spinal meningitis.

Here I would like to take the opportunity to express my gratitude to my colleagues in the hospital, Drs. Friedländer and Weigert, for their friendly support and for willingly making their material available to me.

I am far from thinking that in the present study I have expressed completely new views on the nature of aphasia. Similar psychological and philosophical deductions appear in most of the more important writers on aphasia. Thus Baginsky has already expressed very similar views in his classification of aphasia into centrifugal and centripetal forms. My conception differs from the earlier ones in its consistently maintained anatomical foundation. Previous theories postulated theoretically different centers (a coordination center, a concept center, etc.), but paid no attention to anatomy in doing so, for the reason that the functions of the brain, completely unknown at the time, did not yet justify anatomical conclusions. It is a significantly different approach to undertake a thorough-going study of neuroanatomy and, making use of the now almost universally accepted principles of experimental psychology, to transform the anatomical data into psychological form and to construct a theory out of such material. My explanation of the speech process is only a special application of the mechanism of spontaneous movement the basic features of which are already well-established, to the movements necessary for speech. The sensory function of the occipito-temporal lobe and the motor function of the frontal lobe constitute the basic elements of the theory. Even the part that still requires anatomical proof, i.e., the localization of the sound images in the first temporal convolution, receives indisputable anatomical support (aside from that provided by post-mortem findings) in the pattern of connections, as demonstrated both in the claustrum and in the fibrae propriae.

There is a question whether it is justifiable to apply results of anatomical and physiological research which have not yet been definitively established to the formulation of a new theory. The reader may answer the questions as he thinks best; I have tried in this publication to provide him directly with the necessary information, and have not concealed or glossed over any gaps. If we consider, however, how boldly and fruitfully the available but still incomplete physiological information in some of the older texts, such as Romberg's, was applied in practice, we will not condemn the effort to attempt at long last to do the same in the field of neuropathology. The justification for my theory also depends, however, on other more essential factors.

Despite the temptation to do so in the problems considered here, I have never gone beyond the very simple hypothesis, which can hardly be challenged now, that the central termination of each nerve filament functions as a psychic element (see p. 34). Thus I had to omit from consideration several remarkable phenomena in the field of aphasia, e.g., the forgetting only of substantives or verbs. *I have deliberately avoided dealing with such phenomena, which cannot be explained in terms of these very simple elements.* I would like to state clearly here *that entire areas of cerebral physiology and the disorders of these areas, which include most psychic disturbances, are temporarily still inaccessible to scientific treatment* (in the medical sense). A clear insight into those areas which must be temporarily left aside is the most urgent desideratum for any future progress in psychiatry.

The theory of aphasia presented here is capable of subsuming under itself the so varied clinical pictures of aphasia. The very multiplicity of these pictures, which until now has provided every new observer with new riddles to solve, will no longer be so striking; we will even be able to calculate the number of possible syndromes by means of the laws of combination. *What is characteristic of all of them is an interruption of the psychic reflex arc used in the normal speech process.* This provides the reader with a clear definition of the concept of aphasia.

The analysis of the normal speech process into different centers cannot only encompass the multiplicity of the forms of aphasia within a broad framework but can also explain most of the contradictions that have hitherto appeared either within the observations of one investigator or between the views of different observers. Thus Sander states the opinion shared by him with Griesinger that the connection from the visual image to the sound image is destroyed in aphasics whereas that from the sound image to the visual image is intact. This logical contradiction is easily resolved by dividing Griesinger's sound image between a motor and a sensory center. The errors in movement that have been observed in aphasics can also be explained for the most part as the result of defective comprehension of commands (in sensory aphasia).

Aphasia falls under the more general rubric of *focal psychic disorders.* One can most easily understand what is meant by this term by considering some examples. Meynert conceives maniacal movements to be the result of excitations of the motor regions of the surface of the brain produced by stimuli originating within the brain. If such maniacal movements are limited to a circumscribed group of muscles and are constant over a fairly long period of time, then one is justified in assuming a circumscribed cortical lesion as the cause of this localized psychic symptom.[20] Asymbolia is a focal disturbance of this type, which can be diagnosed only on the basis of psychic symptoms. Those psychoses (usually classified as forms of insanity) which are accompanied by auditory or visual hallucinations are, like maniacal movements, probably caused by a stimuli originating within the brain. The fact that nevertheless these sensations are thought to originate in the sense organs and thus to have their source in the outside world is no more surprising than the pain which the amputee localizes in the absent little toe. *Focal psychic disorders are thus characterized by symptoms either of stimulation or of loss of circumscribed groups of psychic elements.*

Focal psychic symptoms can also be caused by the interruption of the nerve fibers that associate psychic elements with one another, that is, by lesions of the white matter, as has been amply demonstrated in the case of aphasia. Although lesions of the white matter often remain undiagnosed, it is a great error to regard the functions of the white matter as unimportant. All the higher mental process probably depend on the fiber tracts for their development. For obvious reasons, further research in focal psychic disorders is primarily the domain of the psychiatrist. Nevertheless, correct use of autopsy findings will be possible only when the anatomy of the white matter can be sufficiently clarified by means of histological sections (since dissections of fiber tracts have proved to be inadequate for this purpose). The author is engaged in this investigation and, having now almost completed the preparation of the material, consisting of serial sections of brains of dogs and monkeys, hopes to publish it in the near future.

Breslau, May, 1874

ADDENDUM

In the preface to Hitzig's collected writings, which appeared while the present study was in press, I find to my surprise views expressed on the significance of the cerebral cortex and even on aphasia which are almost completely identical with my own. It may thus not appear superfluous for me to state that I had already presented my theory of aphasia in November of last year to a number of colleagues to whom I was demonstrating some neuroanatomical sections. This concordance of our views gives me even greater satisfaction because we arrived at them in completely different ways and because it shows that anatomy and physiological experimentation are of equal importance for our knowledge of the brain.

During this interval the patient Funke (case 8) has died, and an autopsy was performed on June 23, 1874. Edema of the pia and generalized atrophy of the convolutions were found, but aside from this the right hemisphere was completely intact. On the left, visible even on examination of the convexity, was an extensive yellow softening which occupied almost the entire first primitive convolutional arc, thus affecting both banks of the Sylvian fissure. The pia over the lesion was adherent and cloudy. The opposing surfaces of the first primitive convolution were tightly adherent both to each other and to the external wall of the insula, so that it was possible to expose and dissect out the insula only with very great care. A major branch of the artery of the Sylvian fossa was thrombosed throughout its entire course and had been transformed into a firm, yellow and whitish cord into which there ran similarly affected smaller thread-like vessels. No trace of the lumen of the vessel was to be found.

Anterior to the central fissure the softening was limited to the first frontal convolution with only its anterior third spared. The portion of the first primitive convolution which bridges over the lower end of the central fissure, was also affected, as was the lower part of both central convolutions. Posteriorly the lesion widened and involved the entire lobule formed by the anastomosis of the first and second temporal convolutions. The pole and the more medial parts of the occipital lobe retained their normal consistency; the temporal lobe, in contrast, was for the most part softened and only the hippocampal gyrus showed normal consistency.

The insula was reduced in all its dimensions and was for the most part softened and edematous.

These findings agree with those of the patient Rother (case 2). In both cases the first temporal convolution and its anastomosis with the second were involved. Both suffered from sensory aphasia. Could this agreement be only a coincidence?

Finally, let me point out some unusual features of the courses of some of the cases presented earlier. In the patient Adam (case 1), agraphia is now almost the only remaining symptom. Beckmann (case 3) on the other hand can now write fluently, but still suffers from marked alexia. If, as is to be expected, the improvement continues in both, they will manifest isolated agraphia and alexia, conditions which would thus be shown to be normal stages in the evolution of sensory and conduction aphasia.

NOTES

* Ed. note: At the time of writing this study Carl Wernicke was Assistant Physician, in the Psychiatric Service of the Allerheiligen Hospital at Breslau. The study was first published by Max Cohn and Weigert, 1874.

[1] Nothnagel's reversion to Flourens' view after quoting experiments that definitely support the theory of localized function seems to me to contain an internal contradiction.

[2] Hitzig has recently also elicited movements from the posterior central gyrus.

[3] In the winter semester of 1872–73 Meynert gave the following figures: In delirious patients, whose brains are closest to those of normal people, in 46 men and 70 women there was an average of 41.52 and 41.90% of the cerebral cortex in the frontal lobes. In general paretics, on the other hand, in 173 men and 30 women the figures were 40.35 and 39.90%.

[4] Another motor region in the core of the brain which Nothnagel found at the posterior end of the hemisphere, very probably belongs, to judge from his description, to the cornu Ammonis, a structure already claimed by Meynert to be a motor area.

[5] In his recent experiments on apes, Hitzig finds that when the center for mouth and tongue movements, located directly behind Broca's area, is stimulated on the left side, it produces movements in the muscle groups on both sides.

[6] This may be the only basis for the action of the cerebral hemispheres in inhibiting reflexes, which has recently been emphasized again by Golz.

[7] Is it not possible that imitation is originally a reflex activity? Man's mastery of it may reflect only the heightening in the course of generations of this same imitative reflex ability in all sensory spheres which characterizes the non-human primates, our nearest relations, more than any other animal. It is a fact that even in adults many mimetic movements are of an involuntary, reflex nature. It must be even more so the case in

children, in whom consciousness does not yet exert its inhibiting influence on reflex activity.

[8] Two such cases, in which perforation (of intra-uterine hydrocephalus) was performed on a living child, have been reported to me by Dr. Grossmann, an assistant in the obstetrical clinic here. In one of the cases the autopsy was performed by Prof. Waldeyer.

[9] Meynert's anatomical preparations have convinced me that his error is quite excusable. Many who begrudge him this error would hardly have admitted it as freely as he did.

[10] As will be shown later.

[11] I will not go further here into the question of the unilaterality of the speech center. On this question see Simon, Wilks, Broadbent (the last two are reported in *Canstadts Jahresbericht*, 1872).

[12] Cf. Meynert's brilliant theory of the circulation of the brain in his *Bau der Grosshirnrinde* (Structure of the Cerebral Cortex).

[13] See the cases Beckmann and Kunschkel reported below. Of all the areas relevant to our discussion, the insula is closest to the left optic tract.

[14] In order to adopt what is correct in Finkelnburg's theory, we have taken the liberty of also applying the term asymbolia to the essential tactile and auditory images.

[15] All the available information supports the view that the sensory nerves, most of which are bilateral in function, deliver memory images to identical points in both hemispheres. The locus of sound images must thus be present on the right as well as the left, for we hear with both ears all through life. But only the left sound center is effectively connected with the motor speech center, and thus probably only the left sound center has established well-worn connections with the conceptual regions. But the right sound center can completely replace the left one very quickly, as is shown by the case of Adam (see below).

[16] *Ein Beitrag zur Lehre von der Aphasie* (A Contribution to the Theory of Aphasia), Breslau 1873.

[17] In the autopsy of a 73-year old man who suffered from senile cataracts and died of senile atrophy of the brain, I recently found the hippocampal and lingual gyri atrophied. The fornix on the same side was gray in color, and the thalamus on the same side likewise gray and very small. The opposite optic nerve was gray and degenerated.

[18] If there is a complete crossing in the chiasm, the conclusion that it is the left optic tract is not permissible. In that case, involvement of the posterior angle of the chiasm would have to be considered.

[19] Assuming a complete crossing to the left of the chiasm.

[20] I have observed such a case and published it in the recent dissertation of Czarnowsky.

NORMAN GESCHWIND

ANATOMY AND THE HIGHER FUNCTIONS OF THE BRAIN*

I. INTRODUCTION

The following paper is composed of sections of three earlier publications, all dealing with the relationship of the anatomy of the nervous system to the higher mental functions. The first section is taken from my paper 'The Development of the Brain and the Evolution of Language'. It discusses the changes in the course of evolution of the brain that made it possible to develop the understanding of names, an essential feature of human language, which is either not present or exists only in rudimentary form in other animals. The second section, taken from 'The Neural Basis of Language' discusses a presumably later step in the evolution of human language: The ability to repeat, and its anatomical substrate. In both of these papers the importance of anatomical connections between different portions of the brain is stressed. The third section from 'Disconnexion Syndromes in Animals and Man' applies these anatomical principles to certain clinical situations. It is shown here that certain syndromes resulting from localized brain damage and widely interpreted as perceptual in nature are in fact the result of destruction of associative connections between different portions of the brain.

Most students of the problem are agreed that there is something special about human language. The major alternative views may be summed up in two simple categories: (1) There is nothing *qualitatively* distinctive about human language; man only possesses *more* language than lower animals. The opposing group of views state: (2) Only man has language. Both approaches leave us with difficulties. If man has more of something than a monkey then just what particular function of the monkey's is it that is being multiplied in man? If only man has language then what is the essence of this new addition to the armamentarium of the activities of animals?

The first group of theorists are likely not to worry about the details of what it is in the animal that is multiplied in man. They tend to express

Boston Studies in the Philosophy of Science, IV.

their ideas as *Gedanken* experiments: Tell me what you want the animal to do and I can devise a training procedure to do it. The second group tend to phrase the differences in terms sometimes as ethereal as those of the first group are uncompromisingly untheoretical and practical.

Regardless of their particular position the adherents of each of these views have shared in common either an inability or a lack of interest in specifying what change in the design of the human brain permits of these quantitative or qualitative distinctions. The first part of this paper represents an attempt to bridge this gap in some small way. The linguist, even if he accepts the views presented here, may well find them disappointingly meager. Much of the forefront of pure linguistic theory places an overwhelming stress on aspects of syntax. The linguist would certainly wish to know in what way the human brain has become adapted to the production of syntactic speech. Instead he will have to be satisfied – if indeed he is satisfied at all – by a theory of object-naming. Let me therefore hasten to state that I make no pretensions for what I say except insofar as it applies to object-naming. On the other hand an interest in object-naming is not ill-placed since much of the practical use of language rests on the basis of this rather unglamorous activity. Even object-naming in itself has its more interesting sides. Thus, workers in childhood development have tended to neglect the acquisition of names since this has been regarded as the simple enlargement of the lexicon, and have concentrated on the evolution of grammar. Yet the problem of the lexicon goes beyond the mere filling of the dictionary randomly. Thus, the naming of colors, which to the adult appears like a remarkably simple task, is often acquired late, long after the child can name remarkably complicated objects. The ramifications of this extend still further into language performance. Thus, there are reasons for regarding it as not unlikely that the time of acquisition of color-naming and of reading might be highly correlated.

There is perhaps one more reason to stress object-naming. While the mechanism suggested here for the naming of things throws no direct light on the acquisition of grammar, it does throw into relief some different ways of looking at the acquisition of grammar which may well be experimentally useful.

It is perhaps appropriate to put in one final comment on this type of theorizing. The linguist may wonder whether such a theory, even if correct, is any concern of his. Let me suggest some reasons which justify interest by linguists in this type of theory. In the first place it is reasonable that anything which throws some light on the nature of language should be of interest to the linguist. A second reason is that the linguist has the ability to provide help to the student of the nervous system. The theory of object-naming to be propounded here has certain implications which might conceivably be more effectively tested in a non-Indo-European language.[1]

There is a third reason which is perhaps the most cogent one for the linguist to take some interest in these matters. I gather from various sources that the status of linguistic theories continues to be a difficult problem. In a sense the difficulty reduces to that of deciding whether the best theory is simply the most economical set of axioms from which the language behavior can be deduced, i.e., in essence a condensed description. If this is true then theoretical linguistics might appear to be only a more efficient form of description. I would wish, cautiously, to make the suggestion, that perhaps a further touchstone may be added: to what extent does the theory tie in with other, non-linguistic information, for example, the anatomical aspects of language? In the end such bridges link a theory to the broader body of scientific knowledge. I would personally not see much virtue in the views of those theoreticians who feel that language and its disturbances must be viewed *separately* on the linguistic and biological levels. Let us now return to the main topic of this paper.

II. THE CONNECTIONS OF THE CEREBRAL CORTEX

In order to specify what the differences are between man and sub-human primates let me first discuss some of the features of learning in these non-human forms. I hope I may be forgiven for some preliminary remarks about some elementary anatomical and physological aspects which may not be familiar to all of you. If the discussion *appears* very remote from our topic please rest assured that it is all really germane.

Lying along the inner surface of the temporal lobes of the brain are a group of structures of extremely complex anatomy. For the sake of simplicity I will loosely use the term 'limbic system' to refer to these and to their major connections to the hypothalamus and other structures lower in the brain. Included in the anatomical organization of this region are those parts of the brain which when stimulated give rise to sensations of taste or smell, and to certain emotions such as fear and their corresponding motor activities. For simplicity we may divide those aspects of the limbic system of interest to us into two groups: (1) Limbic motor responses – these are highly organized, essentially inborn motor sequences which mediate the motor responses involved in fear, rage, and sexual activities. (2) Limbic sensory responses – these are the subjective feelings of smell, taste, hunger and thirst and the corresponding feelings of satiation, sexual sensations, anger, and fear. It is not at all inaccurate to say that the limbic system mediates both the inborn motor sequences involved in those elementary activities intimately related to the survival of the organism or the species and the subjective experiences related to these activities.

Let us now consider by contrast the remaining 'non-limbic' portions of the surface of the brain which, in fact, make up the bulk of the exposed surface of the hemispheres in the primates, including man. In particular we will concentrate on the sensory portions of this non-limbic brain. There are three sensory regions which are to be specified here: the visual cortex, the auditory cortex, and the somesthetic cortex (which eventually receives the sensory information from the skin, muscles, bone, joints, and tendons – the term 'touch' can loosely be used to designate this group of sensations). These three regions may be said to mediate the 'non-limbic' modalities of sensation.

The zones we have so far considered, that is the 'limbic' regions on the medial surface of the temporal lobe, and the primary centers for the 'non-limbic' sensations of vision, audition and somesthesis, constitute only a small fraction of the surface of the hemisphere. These four regions, together with the classic motor cortex, are all, in the terminology of the great German neuroanatomist Flechsig, 'primordial zones', that is, regions which mature early in the development of the individual. The most commonly used criterion for such maturation is the degree of myelination of the nerve fibers in these regions. These primordial zones also differ quite markedly from each other in cellular architecture.

If we contemplate the brain of a subprimate mammal, such as a rabbit or a cat, we find that these primordial zones make up the majority of the cerebral mantle. As we move up the phylogenetic scale we find the striking fact that these zones become increasingly separated by new areas of cortex. The cortex separating the primordial zones occupies a greater percentage of the surface in the primates and achieves its greatest extent in man where it clearly occupies most of the surface of the hemispheres. It is indeed the development of this association cortex which is responsible for the higher functions of the nervous system. The association cortex matures later in the life of the individual. Flechsig called these regions 'intermediary zones' to indicate their later myelination. The different parts of the association cortex are more like each other in structure than are the different primordial zones.

Let us now turn to the problem of the connections between these different regions of the cerebral mantle. Flechsig (1901) stated an important principle for the brain of man which has in recent years been reemphasized for the brain of the subhuman primate by Bailey and Bonin (1951). The principle asserts that in man the primordial zones, i.e., the regions of early maturation, do not have any significant number of direct connections between them. Thus there are no significant direct interconnections between the limbic regions, the motor cortex, the auditory, somesthetic, and visual cortexes.[2]

Each primordial zone has a significant number of connections only to the immediately adjacent association cortex. The association cortex itself may have long connections to other regions. Let us now consider the connections of the non-limbic sensory modalities. In particular let us consider the connections of the visual cortex which are the best known and may serve as a paradigm for the others.

The primary visual cortex makes connections to the immediately adjacent visual association cortex. The visual association cortex has three important sets of connections, the first two of which I mention only in passing, and the third of which is immediately germane to our argument: (1) Connections to the visual association cortex of the opposite side via the corpus callosum. (2) Connections to the association cortex anterior to the classic motor cortex. (3) The largest connection and the one of most concern to us runs from the visual association cortex to the outer and inferior surfaces of the temporal lobe.

III. CONNECTIONS TO THE LIMBIC SYSTEM AND LEARNING

Why should the largest connection of the visual cortex eventually reach this portion of the temporal lobe? If we consider in turn this region of the temporal lobe we find that it makes connections with the limbic structures on the inner surface of the temporal lobe. In fact the limbic structures themselves constitute one of the primordial zones and therefore do not receive long connections directly from any other primordial zone but receive connections only from the nearby association cortex. The lateral and inferior temporal lobe is thus the association cortex for the limbic structures.

Let us now sum up briefly the point to which these considerations have led us. The visual cortex has its main connections to the limbic system. These are not direct connections but go by way of the association cortex adjacent to the visual cortex and the association cortex on the lateral and basal surfaces of the temporal lobe. We are now prepared – at last – to consider the functional significance of these anatomical connections.

In order to approach this problem let us ask ourselves how a monkey would respond with a rage response to the sight of an object (such as a net) which he has learned to dislike. The impulse must travel over the pathway I have just specified from the visual cortex to the limbic structures and there arouse the inborn rage response. Suppose this pathway is broken at some point? The monkey will no longer respond with rage to this visual stimulus. We will have produced a visual-limbic disconnection.

Consider now an example which is far more important for our purpose. How do we teach a monkey presented visually with, let us say, a circle and a cross to press the lever under the circle rather than that under the cross? We do this by what the psychologists call *reinforcing* one or another response. We may reinforce positively, i.e., reward the response we wish the animal to make, or we may reinforce negatively, i.e., punish the response to be rejected. I do not wish to go into the complexities of the question of reinforcement. For the sake of simplicity we may say that the primary reinforcers are limbic sensations, i.e., the rewards are such things as the taste of food, the satisfaction of hunger or thirst; the punishments are such things as the arousal of fear.

In effect then the visual learning paradigm in the monkey may be reduced to a simple formula: the animal learns to form associations of a visual stimulus to a 'limbic' stimulus. In more general terms learning in the monkey consists in forming associations between a 'non-limbic' stimulus and a 'limbic' stimulus. This is true for vision, audition, and somesthesis. From the point of view of evolution this is reasonable:

a stimulus in a non-limbic sensory modality is learned readily only to the extent that it arouses a sensation important to the survival of the individual animal or the species.

IV. ASSOCIATIONS BETWEEN NON-LIMBIC MODALITIES AND NAMING

In order to look a bit more deeply at learning in the monkey let us now turn to a somewhat different test situation, the one reported by Ettlinger (1960). Here the monkey is taught to choose one of two patterns presented visually, let us say, a circle rather than a cross, by the technique already mentioned of rewarding the choice of one item. After the monkey has learned this to a high degree so that he chooses the rewarded circle on nearly all occasions, another experiment is performed. The monkey is now permitted to palpate two solid figures, a circle and a cross respectively, which are concealed from vision. The animal is again rewarded for choosing the circle rather than the cross. Curiously enough he shows no evidence of carry-over from the identical task done visually. The tactile task is treated by the monkey as a totally new one.

How are we to interpret this experiment and others which have essentially the same result? Ettlinger and others have argued that cross-modal transfers do not tend to occur in animals. But our previous discussion has already stressed that *all* learning in the monkey is cross-modal since the process of learning depends on the formation of associations between *non-limbic* and limbic sensations which is certainly a cross-modal task. The essence indeed of these experiments is that the animal has failed to form an association between two non-limbic modalities.[3]

The same task is, of course, remarkably easy in man who makes the transfer immediately in at least this particular experiment. What is the reason for the ease with which humans do this task which is so difficult for the monkey? Some authors say that man does it by means of verbal mediation. In other words, when one learns to choose a circle over a cross, one learns to choose the 'circle'. When the task is done we again choose the object which we have named 'circle'. The monkey's failure at these particular cross-modal tasks is thus attributed to his failure to use language.

A moment's reflection, however, shows that this argument which appears so plausible at first glance, has in fact been stated backwards. In order to call a seen object a circle we must have learned to associate the visual stimulus with the auditory stimulus 'circle'. In other words, to say that the ability to perform cross-modal associations depends on the ability to name misses the real point which is, as it were, the reverse: *The ability to name depends on the ability to form non-limbic cross-modal associations,* particularly *visual-auditory and tactile-auditory.*[4] One may go further: *The ability to develop language in man probably depends on his ability to form cross-modal associations* between two non-limbic modalities. One might raise the objection, however, that a deaf man might learn language without using cross-modal associations at all. Could he not learn to read by being shown the picture of the object which corresponded to each new written word? Indeed he probably could. In order, however, for him to be able to name (in writing) things with which he palpated he would at least have to be able to form tactile-visual associations. There is, however, a further consideration. Even a visual-visual association is a link between two non-limbic stimuli, and it is questionable that even this would readily be acquired by the monkey.

We have asserted then the thesis that man develops language because he can form associations between two non-limbic stimuli. Could not limbic stimuli be useful as language? The answer of course is no: a monkey would have a great deal of difficulty arousing in another animal a smell or taste, or a feeling of hunger or thirst. We communicate by producing non-limbic stimuli – auditory, visual, or tactile.

What is the basis for the difference in the ability of man and the monkey to form pure non-limbic associations? Let us return to our anatomical considerations. As we have pointed out there are three main connections from the visual association areas of a monkey – one to the symmetrical region of the opposite side, one to the motor association area, and the largest to the lateral and basal temporal lobe which is the association cortex for the limbic system. Thus none of the *main* connections are designed to mediate visual-auditory associations. There are, however, definite connections between some of the visual and auditory association areas, as shown for example in the experiments of Sugar *et al.* (1948). These experimenters found clear evidence of fibers running from the auditory association cortex to the visual association cortex but none in the reverse direction; in other words they found no evidence for fibers which could mediate the arousal of auditory memories from a visual stimulus. Even the auditory to visual fibers they found constitute a group far smaller than the fiber systems mediating visual-limbic associations. The monkey brain thus probably contains visual, auditory, and somesthetic regions which, operating on the whole independently, feed into the limbic system stimuli which are used only the extent that they more or less immediately affect survival.[5]

Even if a small number of fibers exist which have not been discovered between the visual and auditory regions, this would not affect the argument. The more information that must be carried between two cortical regions, the more fibers necessary to carry them. The more associations formed, the more cells are needed to store the memories. The monkey does not appear to have a sufficient basis anatomically to form extensive non-limbic intermodal associations. That a limited number of such associations could be performed perhaps with extensive training is probably correct, but there is probably an inherent limit to the speed of formation or the variety of such associations. One might say that the monkey lacks adequate 'memory space' and 'channel capacity' for this type of association.

It might be worth digressing for a moment to consider the great strain put on the facilities of a brain by a demand for non-limbic intermodal associations. Consider the situation that visual, auditory, and somesthetic association areas discharge independently into the limbic system via association cortex. In order to add visual-auditory and visual-tactile associations to the visual-limbic associations one would probably have to add at least twice as many fibers and cells as are now used for the visual-limbic connections alone. The probable minimum net effect of such an extensive addition of non-limbic intermodal fibers is therefore to triple the array of cells and fibers needed in the visual association system. It is no surprise that this uneconomical situation does not appear even in the subhuman primates.

What then of the human brain? What correlates with the newly acquired or at least greatly increased ability to form non-limbic intermodal associations and thus to develop language?

It is obvious that there has been marked development of the human brain in comparison to that of even the highest apes (Connolly, 1950). Thus the gorilla has the largest brain of any of the anthropoids, but it never even reaches half the weight of an average adult human brain.

In addition the cerebellum makes up a higher percentage of the brain in the anthropoid apes so that the cerebral hemispheres of man (which constitute the part which is of concern to us) are relatively even larger than the discrepancy in whole-brain weights would suggest. Furthermore, a fully adult gorilla is a considerably larger animal than a human. An orangutan which when fully grown may weigh as much as a small human adult has a brain which is about one quarter as heavy as the human brain at maturity.

The human brain at birth is perhaps 40% of its adult size while that of the ape's is nearly 70% of full-grown size. This suggests that much of the human brain is evolutionarily late since it matures so late in the life of an individual. This supposition is amply borne out by closer inspection. The great increase in the size of man's brain is primarily in the association areas, rather than in the primary receptive areas. This growth is reflected in the frontal pole, the temporal lobe, and in the inferior parietal region. The greatest relative growth of the human brain compared to that of the subhuman primates is in the inferior parietal region, and it behooves us to look at this zone more carefully.

The impressive development of the inferior posterior parietal region (the region of the angular gyrus) is so great that some authors (e.g., Goldstein, 1927) even assert that this region is almost unique to man. Even those who deny this, such as Bonin and Bailey (1961), still stress that it is this region of the brain which has expanded most markedly in man compared to any of the apes. Its advanced state is testified to by the fact that it is one of the last regions to myelinate and was called by Flechsig one of the 'terminal zones' in contrast to the earliest myelinating 'primordial zones' and the 'intermediary zones' which myelinate at an intermediate period. According to Dr. Paul Yakovlev this region does not mature in its cellular structure until three or four years or even much later in childhood.

This region is ideally located for the purposes of our theory. It is placed between the association cortexes of the three non-limbic modalities: vision, audition, and somesthesis. It is thus admirably suited to play the role of acting as the way-station by which associations may be formed between these non-limbic modalities. This area may well be termed 'the association cortex' of the association cortexes. By providing the basis for the formation of non-limbic associations, it provides the anatomical basis for language – or at least for object-naming. Other authors have stressed the importance of the posterior parietal region for language,[6] but I hope that I have been able to give a physiological basis to this localization.

There are many problems associated with the discussion I have just presented. Indeed, rather than use the word 'problems', I would perhaps have been more correct in saying that it raises many possibilities for further investigation. Much of the data now available does support the view advanced here. I will cite a few examples of such data.

Very large lesions of the angular gyrus region produce an aphasic picture in which the patient speaks volubly but communicates little information. The speech is in particular remarkably poor in the names of objects. The reason for this clinical picture is readily deducible. The connections from the rest of the brain to the auditory association area are cut off. On the other hand since the auditory association area and the motor speech region and their connections are intact, the patient can still produce the automatisms of language freely – these are those aspects of language not demanding intermodal transfer for their use at a given instant. The comprehension of language is seriously impaired since the arousal of associations in other modalities by the auditory speech patterns is also cut off by the lesion. These patients, despite severe comprehension defect, may show perfect repetition because they have preserved the connections between the auditory association cortex and the motor speech region.

Even more revealing as to the importance of intermodal associations are certain more limited lesions. Thus in certain partial lesions of the angular gyrus (first described by Dejerine, 1891) there may be an isolated loss of the ability to comprehend written language and to write (the so-called 'pure alexia with agraphia'), while all other aspects of language are intact. The functions of the angular gyrus in auditory-visual associations may be quite dramatically brought out by considering an even more restricted lesion, that of the so-called 'pure word-blindness without agraphia' (Dejerine, 1892; Geschwind, 1962). Here the left primary visual cortex is destroyed. The patient can thus see only with his right visual cortex. In addition, however, there is usually a destruction of the posterior end of the corpus callosum. This acts to cut off the right visual region from the left angular gyrus region, and thus the patient cannot understand written language. All other aspects of language are intact.

The disturbances in these patients can be summarized as follows: they lose the ability to comprehend written language, to read music, and to name colors (but they can still match colors correctly and show other evidences of correct color perception by *non-verbal* means – see discussion in Geschwind, 1965). They tend to preserve the ability to name objects and to read numbers. How does one make sense of this constellation of impaired and preserved visual tasks? Reading is learned nearly always after speaking. While one learns in speaking to name objects with which one has had both tactile and visual experience, learning to read is almost exclusively dependent on pure visual-auditory connections (particularly after the earliest stages). When these pure connections are cut off the patient can no longer read. He can still copy the words he cannot read (which effectively rules out almost any perceptual theory of this disturbance) because visuo-motor connections in the right hemisphere are still intact. He can name objects because these arouse tactile associations further forward in the right hemisphere and a connection can be made to the speech regions across an intact part of the corpus callosum lying forward of the damaged part.

Color-naming and music-reading are lost with word-reading because they are very pure visuo-auditory tasks – we can identify an apple by its form or texture, but we cannot identify a color by these. Finally, number-reading is preserved because numbers are usually associated with a powerful somesthetic reinforcement (counting on the fingers) through a considerable part of childhood, a form of reinforcement not available for words.

The thesis presented here is that the development of object-naming depends on the presence of anatomical structures in man which are absent or poorly developed in the monkey. The assertion that object-naming depends on visual-auditory transfers is of course an old one. Wittgenstein (1953) opens his *Philosophical Investigations* with a quotation from St. Augustine's *Confessions*: "When they [my elders] named some object and accordingly moved towards something, I saw this and grasped that the thing was called by the sound they uttered when they meant to point it out....'

V. OTHER ASPECTS OF LANGUAGE

Can we expand the theory presented here to more than object-naming and can we somehow use the same approach to help us with grammar? We may appropriately quote Wittgenstein's comment on the above passage, "Augustine does not speak of there being any difference between kinds of word. If you describe the learning of language in this way you are, I believe, thinking primarily of nouns like 'table', 'chair', 'bread', and of people's names, and only secondarily of the names of certain actions and properties; and [*sc.* you are thinking] of the remaining kinds of word as something that will take care of itself". But, of course, Wittgenstein implies, and we must agree, that the other words will not take care of themselves. How would you teach a child the use of the word 'if' by the method described by Augustine?

I do not have a solution to offer at this moment. It would be difficult to conceive of grammar as depending on intermodal transfers in the straight-forward manner that object-naming does. Grammar appears at first glance to be something that grows up within the auditory association system itself, to depend on intramodal, auditory-auditory associations (which can then be transferred *en bloc* to written language by use of the cross-modal association system we have discussed).

This answer is, needless to say, hardly satisfactory, but our earlier discussion may give a clue as to some useful future approaches. We began this paper by stressing the exact method by which a monkey learns to choose a circle over a cross. Similarly, we have discussed the exact mechanism of object-naming in man. We can answer questions as to how we would teach certain things to a monkey and how we could teach a child the name of an object. Let us consider the possible applications of the method to grammar. We have some important facts, gleaned in important studies in recent years, on the child's acquisition of grammar. These facts, however, are still concerned mostly with the age of acquisition of certain grammatical features, their order of acquisition, and the learning of distinctions between features that might at first be accepted as equivalent. To some extent we can specify what rule the child follows at each age. I would suggest, as a supplement and by no means a substitute to these very important studies, that some attention be paid to the actual act of learning by the child and that some experiments be devised to study the elementary situations in which children learn to use certain words. At its origin grammar may well be acquired by mechanisms as naive as those of object-naming. Once these elementary associative steps are passed the acquisition of the more complex structures presents less difficulty. It is the very first steps which are the vital ones. Once speech is developed, it can develop an autonomy from the outside world that is all too familiar. If our studies on learning by monkeys and object-naming in man are to be of use, they should teach us to study not what grammatical rule is acquired by the child at a given age but *how* he has acquired this rule and how we could teach it to him.

Let us now turn to a different aspect of language, the ability to repeat the words that one has heard. Here we have a task of matching a motor response to an auditory stimulus, 'shadowing' speech, if I may borrow a phrase from Dr. Richard Chase.

We have very good evidence from various sources as to the anatomy of this activity. It appears to depend on a pathway running from the auditory association cortex in the posterior superior temporal region (the classical Wernicke's area) via the arcuate fasciculus, a system of fibers running in the lower parietal lobe and eventually reaching Broca's area, i.e., the association cortex lying anterior to the region of the classical motor cortex involved in movements of the speech musculature.

It should be understood clearly that the process of learning to say the name of a seen object is thus essentially a two-step operation. The child is shown an object and told, 'This is a banana'. He thus learns to associate the visual stimulus to this auditory pattern, i.e., to link the visual association cortex to the auditory association cortex by way of the angular gyrus. He is also learning to repeat the sounds of his language, i.e., to match an articulatory pattern to an auditory pattern. When he sees a banana, it now arouses an auditory stimulus in the auditory association cortex. The child repeats, as it were, this inwardly heard stimulus and thus completes the process of saying the name of the seen object.

Let me cite some of the evidence which favors this model. The independence of comprehension and repetition may readily be demonstrated. Thus, it is well known that certain children who remain mute throughout childhood may yet comprehend the spoken word normally. Conversely, one finds that certain mentally defective children may learn to repeat the sounds of their language normally and indeed even to learn long sequences such as poems or prayers, and yet understand spoken language extremely poorly. Furthermore, as will be pointed out below, the two functions of comprehension and repetition are clearly separable in the aphasias. Thus, one group of patients who lose the comprehension of the spoken word show normal repetition while another group in whom repetition fails understand language normally. Furthermore, as I will now discuss, the lesions found in these patients correspond to what would be demanded by this model.

The lesions of the aphasias are in fact among the most useful sets of information we have for studying the neurological foundations of language. Let us consider the model we have and see how it fits in with the results obtained from the study of the aphasias.

I will not say much in detail about reading and writing, but will only comment here without expansion that almost invariably one learns to read only after having learned spoken language, and that in this process we learn to associate the written word to its auditory equivalent. Reading aloud is of course essentially an inner repetition process such as we have described above for saying the names of objects. In what follows we will use the word 'read' only in the sense of comprehending written language.

As we have noted before, Broca's area is a region of motor association cortex which is most likely involved in establishing patterns of articulation to match auditory stimuli. With lesions of this region we find intact comprehension. By contrast we find that speech is slow, halting, effortful, and badly articulated.

In lesions of the speech area that leave Broca's area intact we find that speech is often fluent although abnormal. I have already pointed out that fluency is probably the result of the running-on of the intact Broca's area uncontrolled by the other parts of the speech area. But this free running-on of Broca's area can occur only if it has overlearned many language patterns, and thus we fail to see fluent aphasias in children even with lesions sparing Broca's area. The degree of fluency of those aphasias which result from posterior lesions appears to increase with age, as would be expected on the basis of this explanation. We can also understand from this explanation why fluent agraphias do not occur (or if they do are extremely rare) at any age, since the amount of practice in writing achieved by the overwhelming majority of humans in the course of a lifetime is probably less than the amount of practice in speaking achieved by most children before arrival at adolescence.

These fluent aphasias differ, however, among themselves in ways to be expected by the model. Thus, a lesion of Wernicke's area, which is a region of auditory association cortex produces the kind of fluently abnormal speech which is often found when Broca's area is left intact. Repetition is seriously impaired since this lesion involving the auditory association cortex prevents auditory stimulation from being conveyed to Broca's area. Comprehension of spoken language is impaired because auditory stimuli cannot be conveyed to the rest of the brain to arouse associations. In addition, reading comprehension is impaired, since, as we have noted earlier (and indeed as was pointed out by Wernicke himself), the process of learning to comprehend written language in most people involves learning to associate a written stimulus to an auditory one.

If we now look at a somewhat different lesion we see a somewhat different syndrome. Here the lesion is in the arcuate fasciculus which connects Wernicke's to Broca's area. The patient often has a fluent aphasia because Broca's area is preserved. Repetition is severely impaired because the connection from Wernicke's area to Broca's area is not intact. By contrast, auditory comprehension is retained because Wernicke's area is preserved as well as its connections to other parts of the brain (except for the one to Broca's area).

The last condition I will mention in this rapid survey is in many ways the most interesting. Here Broca's and Wernicke's areas and the arcuate fasciculus are unaffected but a large lesion running through the posterior parietal region effectively cuts off connections running to and from Wernicke's area. Again we often have a fluent aphasia, since the entire speech area can now run on without receiving information conveyed from the rest of the cortex. Repetition is intact, however, and indeed these patients may even be echolalic. Comprehension is, however, severely impaired since the stimulation which reaches Wernicke's area cannot arouse associations in other sensory regions.

I have rapidly run through some of the aspects of the evolution of language and of its disintegration in the aphasias. Two central processes separable from each other developmentally, with different anatomical localizations and separable from each other by disease, have been stressed. On the one hand, we have what we may loosely call 'comprehension', i.e., the process by which we learn to associate an auditory stimulus to a visual or other sensory stimulus, and on the other, we have 'repetition' by which we learn to reproduce an auditory stimulus. Much – but by no means all – of the complexity of human language is developed on the basis of these two processes.

Let us turn now to a discussion of some of the disturbances of the higher functions in man, known as the agnosias, to see what relation they may have to naming defects.

VI. THE AGNOSIAS

The agnosias have always represented an unusually difficult group of disturbances. Classically they were defined as disturbances of 'recognition' without disturbances of elementary sensation. Some observers have simply denied the existence of disturbances of this type. There has been the attempt on one side to reduce them to perceptual defects of complex nature, such as continuously shifting thresholds. On the other hand there has been the tendency to deal with such disturbances in terms of rather complex psychological malfunctions such as loss of ability to perceive Gestalts. A more classical approach was that of Lissauer (1889) who divided agnosias into apperceptive and associative, or disorders of primary and secondary recognition. The theory that I will present here will propose that the agnosias are indeed associative disorders; what will be added to the classical notion will be an analysis of just what type of associative disorder is involved and an explanation of certain clinical features that have in the past acted as stumbling blocks to analysis of these disturbances. I have already discussed in detail the agnosias of animals. The human cases are also based on disconnexion but I hope to show that the mechanisms are different in important ways from those operative in animals.

Workers in this field have tended to use a certain set of criteria for calling a disturbance an agnosia. First it is necessary to show either (1) that elementary sensation is intact or (2) that if an impairment of elementary sensation is present it is not sufficiently severe to explain the disturbance of recognition. Assuming this to have been demonstrated, how has it been possible to demonstrate that the failure of recognition was not simply a failure of naming? There have been several criteria generally employed. I will list them first and discuss their usefulness afterwards:

(1) *The 'agnosic' disturbance is a circumscribed one.* – There is a common view that aphasic misnaming cannot be confined to a single modality; even more strongly it is assumed that it cannot be confined to a class within a modality. From this point of view an aphasic misnames every-

thing; by contrast there can be a visual agnosia or even an agnosia for colours alone.

(2) *'Agnosic' errors differ in character from 'aphasic' errors.* – It is often assumed that aphasic errors are of one type, agnosic of another; the aphasic error is a 'linguistic' one, the agnosic a 'perceptual' one. According to this view, 'chair' for 'table' is an aphasic error but 'a small package' for 'a book' is an agnosic error; 'something to eat with' for 'spoon' is an aphasic error but 'pencil' for 'knife' is an agnosic error.

(3) *The 'agnosic' cannot choose the correct response.* – The aphasic who has failed to name is usually described as being readily able to accept the correct name when it is offered. The 'agnosic' by contrast is said to be incapable of selecting the correct name.

(4) *The 'agnosic' describes his perceptual difficulty.* – The aphasic is usually pictured as saying, 'I know what it is but I can't find the name'. When asked about the appearances of objects he asserts that they appear normal to him. The 'agnosic', by contrast, informs the examiner that things look different to him or that he has difficulty in seeing clearly. Alternatively he may offer dramatic descriptions of bizarre perceptions, thus apparently providing direct evidence that he is experiencing distortions of his perceptions.

(5) *The 'agnosic' cannot show the use of the object.* – This criterion is actually the most complex. The aphasic is said to be able to show the use of the object. The 'visual agnosic' cannot show the use of the object but can manipulate it correctly if it is actually placed in his hand. The 'apraxic' according to this classical analysis cannot handle the object correctly even when it is placed in his hand.

As I hope to demonstrate, all of these criteria are actually inadequate for the separation of patients into two classes of distinct physiological disturbance. The fundamental difficulty has been in the acceptance of a special class of defects of 'recognition', lying somewhere between defects of 'perception' and of 'naming'. What indeed are the criteria for 'recognition' and is it a single function? I believe in fact that there is no single faculty of 'recognition' but that the term covers the totality of all the associations aroused by any object. Phrased in another way, we 'manifest recognition' by responding appropriately; to the extent that any appropriate response occurs, we have shown 'recognition'. But this view abolishes the notion of a unitary step of 'recognition'; instead, there are multiple parallel processes of appropriate response to a stimulus. To describe the behaviour correctly we must describe the pattern of loss and preservation of responses to each particular type of stimulus.

Let us make this more concrete by referring to a particular patient's behaviour. I will turn for my example to the case of Geschwind and Kaplan (1962), the first modern case of callosal disconnexion. Let us consider his difficulty in identifying objects placed in the left hand. He incorrectly named objects placed in the left hand. He could, however, draw the object afterwards with his left hand or select it from a group tactilely or visually. It is obviously correct to describe this patient as showing correct identification by nonverbal means and incorrect identification verbally of objects held in the left hand. It is difficult to see how this could in any way be described as a 'perceptual' disorder since perception must have been intact to insure correct selection from a group or the drawing of the object.

We have interpreted this disturbance as a result of disconnexion from the speech area. In addition the patient showed other disturbances of identification not due to disconnexion from the speech area. Thus, he could not select or draw with the right hand an object held with the left, concealed from vision, or select or draw with the left hand an object held with the right. But again 'perceptual' disturbance in either hand is excluded by the patient's ability to select or draw with a given hand an object that had been held in the same hand. We can only regard the defect as associative – what was perceived by the right hemisphere could not be relayed to the left and vice versa.

Let us return for a moment to the failure to identify verbally with the left hand objects held in that hand which we have attributed to disconnexion from the speech area. It may be argued that we have evaded the fundamental issue, that although we have shown that the patient could not name the object placed in his left hand but yet could draw it or select it afterwards from a group, we have not really shown whether the patient had recognized the object. This question is based on two assumptions which I believe are incorrect. The first assumption is that 'the patient' is an entity. But in someone whose two hemispheres operate as independently as this patient's the word loses its ordinary meaning. There is no answer to 'Did the patient recognize?' There are, however, clear answers to 'Under what conditions did the right or the left hemisphere recognize?'

The second assumption is that recognition is more than the sum of the individual acts of naming, drawing, handling, selecting, etc. If it is more than these, how would one test for this faculty and how would one show that this test established some superordinate function of recognition? I do not know of an answer to this question.

I have argued that the patient's failure to identify verbally objects held in the left hand must be regarded as a result of separation of the right somesthetic region from the speech area and can in no sense be regarded as a perceptual defect. In the light of this analysis, let us now consider the patient's responses from the point of view of the classical criteria listed above which were used to distinguish 'agnosic' from 'aphasic' disturbances. This analysis will demonstrate the inapplicability of these criteria.

(1) *Our patient's disturbance was highly circumscribed* to one modality and indeed one-half of the body. Classically such circumscription was often regarded as being incompatible with an aphasia. Yet as soon as one admits the substantial restriction of speech to one hemisphere, the possibility of cutting off one-half of the brain from speech must be considered. There is no basis therefore for this classical distinction.

(2) *The errors made by our patient were not 'errors of naming'* in the sense of the commonly used criteria. He could call a 'screw-driver' a 'rubber band', a 'coin' a 'cigarette lighter'. These are not (without meaningless extension of the term) errors in sphere nor are they errors based on similarity of sound. They are, therefore, not 'aphasic' errors according to the usual criterion. But it is easy to see that if a particular part of the brain is fully disconnected from the speech area, there is no reason that an erroneous name should be related in any way to the correct term. The more complete the disconnexion, the more random the errors must be. The less complete the disconnexion, i.e. if some information gets through to the speech area, the more the errors may be related to the correct response. It should be added that in fact the classically anomic patient often makes naming errors not related by sphere or sound.

(3) *Our patient showed inability to choose the correct verbal response when it was offered to him.* – When he misnamed an object we would offer him several choices. The patient would generally not select the correct answer from the group offered. If, however, the speech area is fully disconnected from the right hemisphere, there is no reason why the

speech area should select the correct term. It is obvious that the ability to select the correct word depends on there being some connexion between the site of perception and the speech area. The ability to select the correct word from a group when the patient cannot find it spontaneously indicates a lesser degree of disconnexion. Phrased in another, diagrammatic way, one could conceive that when one offers a word to a patient he in some way compares the images or memories aroused by this word with the sensations he is receiving from the object. If there is no site for such comparison, the patient will not recognize the correct word when it is offered. I would like to point out that in fact the obvious nominal aphasic often fails to accept the correct word when it is offered. Furthermore, even when the patient insists that 'I know what it is but I can't tell you the name' he may fail to accept the correct name from a group.

(4) *The patient gave descriptions of his 'perceptions'.* – Thus, when holding a half-dollar, the patient said that it was a cigarette lighter. When asked if it had corners he said, 'Oh, yes, there's one, there's another'. He would describe incorrectly the object held in his hand in most instances. The same objection applies here as to the previous categories. If the patient's speech area is disconnected from a site of primary perception why should his speech area be able to describe what is going on at the site of the primary perception? Clearly, it should not. The fact that 'the patient' (i.e. the speech area) gives a description does not mean that we are getting an actual description of the perceptions going on in another disconnected part of the brain. We must again remember that we are dealing with more than one 'patient' here. The 'patient' who speaks to you is not the 'patient' who is perceiving – they are, in fact, separate.

(5) Our patient lacked only one of the classical criteria of 'agnosia'. According to these classical criteria, a patient with 'tactile agnosia' should mishandle objects which he has not seen but should respond correctly to the sight of the object (i.e. his behaviour should be opposite to that of the visual agnosic). Our patient in fact handled objects perfectly correctly when blindfolded. I will return in a later part of my discussion of the apraxias to this question of the use of objects.

VII. THE PROBLEM OF CONFABULATORY RESPONSE

I have discussed in some detail a disturbance which might have been called 'agnosic' and tried to show that this disturbance is much more meaningfully described as a naming defect resulting from disconnexion from the speech area. There are certain further implications of this interpretation which I would like to present now. One most important implication is that the 'introspections' of the patient as to his disability may be of little or no use to the examiner. The patient cannot 'introspect' about the activities of a piece of brain which has no connexion to the speech area. What he tells you is of little value in elucidating the mechanism and may indeed be actively misleading. Indeed, it becomes clear that many of the patient's responses can only be described as confabulatory, i.e. they are attempts to fill gaps in the information available to his speech area; phrased in more conventional terms they are attempts to explain what the patient cannot understand.[7]

These confabulatory responses have been a major source of confusion for many years, particularly in the assessment of difficulties of sensory identification. It is curious that sophistication concerning confabulatory behaviour has been much more widely developed in certain areas than others. When the patient with a severe recent memory defect asserts that the examiner spent the previous evening with him in a bar, we neither ask ourselves why the patient had hallucinations of seeing the examiner the night before nor why he is now having false memories. We see this instead as the verbal filling in of a gap. We discount the attribution by the same patient of his lack of knowledge of current events to the fact that he has never been interested in politics. Similarly, if an aphasic patient tells us that his trouble in speaking is the result of his ill-fitting dentures or a sore throat, we do not take the attitude that he is giving us any very useful insight into his illness but rather that he too is attempting to explain a gap in his performance. An even more extreme example is that of patients with denial of blindness who may be prepared to describe in detail many of the objects in the room; once we know that the patient is blind, we do not attempt to make use of his 'introspections' about his visual perceptions. We do not take seriously this patient's protestations that he has failed in some task because the light is poor or he is not wearing his glasses. Furthermore, we do not assert as an alternative that the patient is really hallucinating.

The 'higher' the function impaired, the more readily the patient may succeed in fooling others with confabulatory responses. I recall one patient with pure alexia without agraphia who asserted that his trouble in reading was due to his 'blindness in his right eye'. The resident staff correctly realized that he meant a right visual field defect when he spoke of a blind right eye. However, they incorrectly accepted the patient's attribution of his reading disturbance to this visual impairment. Only when the discrepancy between reading and object-naming was pointed out to them did they realize that they had too readily accepted the patient's 'explanation' of his disability.

There are many occasions in which patients with aphasia, dementia, blindness, or recent memory defects produce confabulatory responses which are recognized. There are probably equally numerous situations in which such confabulatory responses are not recognized as such by the examiner. Most of the conditions called 'agnosias' fall into this category. Much time has been spent in the vain attempt to analyse the supposed disorder of perception which has been regarded as underlying the patient's responses.

In summary, the theory here proposed is that most of the classical agnosias are highly isolated disturbances of naming which are the result of disconnexions from the speech area. The prominence of confabulatory responses may incorrectly lead the examiner to believe that he is dealing with a perceptual disorder.

VIII. INABILITY TO IDENTIFY COLOURS

The patient of Geschwind and Fusillo (1966) showed pure alexia without agraphia and a failure to identify colours. This disturbance could readily have been called an 'agnosia for colours'. Yet, analysis of his responses makes it clear that his difficulty could be correctly described as a disturbance in the naming of colours. By all non-verbal criteria he identified colours correctly – he matched and sorted correctly, did the Ishihara test perfectly and even properly matched colours to pictures of objects, yet he failed consistently in verbal naming. It is obviously appropriate to describe his disturbance as a colour-naming defect since by no criterion could he be shown to have a perceptual defect. His pattern of responses also illustrates the lack of usefulness of the classical criteria. Thus he exhibited a very circumscribed naming defect. Only one in many testing sessions did he misname an object, yet his colour-naming remained consistently extremely poor. He generally failed to accept the correct name of the colour when it was offered. Indeed, when we told him, for example, that something was not grey but red, he said, 'well, it looks grey to me – maybe it's a reddish-grey'. On another occasion he said, 'You say it's white but it looks tan to me'. Again the most parsimonious assumption is that he had a colour-naming defect with confabulatory response. Whenever we asked the patient to give a *verbal* account of the colour experiences of his right hemisphere, he produced a confabulatory response; if we tested the right hemisphere *non-verbally*, we got evidence of perfectly normal colour perception.

I would like to discuss the one instance of poor matching by this patient since the careful analysis of this failure was most illuminating. The patient was given a pile of colour chips which contained chips of two different shades of each of the primary colours. He was shown a chip of one of the two shades of red and told, 'This is red. Now pick up all the red chips'. He picked each chip up slowly and examined it. He finally separated out all of the chips of the selected colour as well as all the chips of one of the two shades of green. When asked what colour the chips were which he had selected he said, 'These are red and these others (indicating the green ones) are a different shade of red'. He was then able to separate these chips into two piles. When shown a chip of the *other* shade of green and asked which of the two piles it was most like, he unhesitatingly classified it with the pile of green chips.

There was clearly a sorting error here but full consideration of the patient's behaviour seems to exclude the possibility that we are dealing with a perceptual error. I believe the sequence of events to have been as follows. We had asked the patient to 'pick up all the red chips', rather than to 'pick up all the chips like this one'. He had done the task slowly probably because he did it by picking up each chip and naming it. Having misnamed a chip of one of the shades of green as red, he had then selected the remaining chips of this shade. We may consider this as a sorting error, resulting not from failure of discrimination or perception (which his subsequent behaviour in this task excluded, i.e. the rapid separation of the two colours) but as a 'secondary sorting error' which resulted from doing the task verbally. Thus, even failures in sorting should be analysed carefully to determine whether they are secondary to misnaming or are truly failures of discrimination.[8]

IX. CLASSICAL VISUAL AGNOSIA

There are many remarkable accounts of this disturbance in the literature (e.g. Lissauer, 1889; Lange, 1936; Brain, 1941; Macrae and Trolle, 1956; Ettlinger and Wyke, 1961; Hécaen and Angelergues, 1963). In going through these accounts one is immediately struck by the frequent evidence of preservation of non-verbal visual identification while verbal identification is impaired. One striking feature of many of these cases is the fact that the patient's ordinary behaviour is in marked contrast to the supposed perceptual disturbance. Thus, one may read an account of a patient who cannot 'identify' a glass of water and yet a few minutes later picks it up and drinks from it. Much stress has been laid on the fact that a patient may show normal behaviour in a 'natural' situation but not in an 'artificial' one. Once one abandons the notion of a unitary process of 'recognition', it is easy to see that the patient may fail to 'identify' an object, i.e. fail to give a verbal account of what is going on in his visual receptive regions and yet be able to respond nonverbally to a non-verbal stimulus. Most of the pathways involved in the non-verbal response may be quite different from those involved in naming. For a thirsty man to respond appropriately to the sight of a glass of water by grasping it and drinking from it is *anatomically* a different task from that of responding to a verbal request for identification. It is certainly reasonable that a naming defect should not produce gross functional impairment; by contrast it would be very difficult to understand how a 'higher-order perceptual defect' could leave ordinary perceptual function unimpaired. A similar preservation of function appears in those cases who can draw the object which they cannot recognize (Lange, 1936) a behaviour obviously compatible with a naming difficulty.

The patient of Ettlinger and Wyke (1961) made many errors in visual naming (e.g. 'a box' for 'a glove', 'for cooking' for 'a spoon') and many fewer in tactile naming. This case can be used as further evidence against the idea that an aphasic naming disorder must cut across modalities. The data of these authors show that while 'agnosic' errors (e.g. 'box' for 'glove') were more common visually than tactilely, the same was true for 'aphasic' errors (e.g. 'for cooking' for 'a spoon'). A striking feature of this case was the matching of outline shapes without error even though the patient made nine errors in naming outline pictures of ten objects.

Certainly a naming defect is the simplest explanation of this disorder. This case was similar in this respect to the earlier case of Brain (1941).

The remarkable case of Macrae and Trolle (1956) presents another example of one of these striking discrepancies. The authors point out that when the patient was pressed with regard to his method of finding his way to and from work, 'he could not visualize the street plan or the route'. Despite this he did in fact drive two miles a day to work without difficulty. I would suggest that 'visualization' here means either developing or describing an internal picture in response to a verbal stimulus – but this will not occur if the visual system is separated from the speech area. Yet, the subject may respond nonverbally to the actual route.

Patients with classical visual agnosias have often been presumed to be suffering from 'perceptual' disorder on the basis of their 'introspections' concerning their illness. The case of Lissauer (1889) illustrates how little such 'introspections' are to be trusted. Thus, Lissauer comments that when the patient misidentified an object, he would, after learning its true nature, excuse himself by saying that 'his eyes had deceived him at first'. As Lissauer comments, this statement, coupled with the fact that some of the misidentifications were so curious, might have led the examiner to believe that the patient had been having sensory illusions of some type. But if the examiner then asked the patient whether the object, now that he knew what is was, looked any different from the way it had before, he received a vigorous denial. The patient in fact now asserted that the object had looked the same before but that 'his poor memory was responsible for the fact that it had not occurred to him what the thing was called and what it was'. Had Lissauer pushed his analysis somewhat further here he might well have been able to see that the 'misidentification' was actually only misnaming. Lissauer himself comments elsewhere in the paper, 'The patient succeeded... in drawing simple objects which he didn't recognize, a striking proof that he was well able to perceive their forms'. Similarly, although the patient failed to name colours, he succeeded without any difficulty in matching tasks.

X. THE LESIONS OF CLASSICAL VISUAL AGNOSIA

At first it would be tempting to ascribe classical visual agnosia simply to disconnexion of the visual region from the speech area. This interpretation, however, runs into obvious difficulties. Thus, we have already explained in an earlier section that pure alexia without agraphia (with its frequent accompaniment of colour-naming difficulty) is the result of disconnexion of the visual cortex from the speech area. In these cases there is usually destruction of the left calcarine cortex. In addition there is a lesion of the splenium of the corpus callosum which isolates the right visual region from the speech area. It was particularly noted that these patients did not show difficulties in object-naming and often showed retained ability to read numbers. I said in the earlier discussion that it seemed likely that the explanation for the preservation of object naming was that objects could arouse somesthetic and other associations more anteriorly in the right hemisphere; these would then traverse the callosum in the preserved portions anterior to the splenium and thus reach the speech area. For the development of a naming defect for objects, failure to arouse these tactile associations (a visual-tactile disconnexion) is a necessary addition to the disconnexion of visual regions from the speech area.

The lesions found in the majority of cases of visual agnosia support this notion. Most commonly observed are very large bilateral parieto-occipital lesions. There are probably no direct connexions between the visual and the somesthetic cortex so that visual-somesthetic connexions must be mediated via the intervening association cortex. A large parieto-occipital lesion would effectively cut off such connexions. One can also see the reason for the rarity of classical visual 'agnosia'. If the lesion is too small, a large enough extent of association cortex is left to make the defect a minor one. If the lesions are too large, the patient may show gross aphasic impairment as well as marked disturbances of visual-motor connexions. In this case his impairment of speech function may make analysis exceedingly difficult. The cases with very extensive lesions might with some justification be regarded as manifesting disturbances of the general ability to manifest visual 'recognition' since all outflow channels may be cut off except possibly some primitive and not very extensive pathways descending subcortically.

There is another but much smaller group of visual 'agnosias' who have unilateral lesions. This group shows destruction of the left visual cortex. In addition there are always lesions of other structures. One group of these patients shows involvement of the splenium. Others have such extensive involvement of the white matter or of the association cortex in the left occipital and parietal lobes as to have destroyed either the fibres coming from the splenium or their terminations in the left hemisphere (Lange, 1936; Nielsen, 1937).[9] The beautifully described case of Lissauer (1889) turned out to have such a lesion (Hahn, cited by Meyer, 1895–1896) although he himself had expected bilateral lesions.

Why do these cases with unilateral lesions manifest more than pure alexia? It is difficult to be certain and I would hope that more careful study of cases in the future may provide the answer. One possibility is that in some people visual stimuli in the right hemisphere fail to arouse tactile associations or do so weakly. These patients should therefore develop a more extensive naming defect when the right visual regions are cut off from the speech area. Another possibility is that in these cases the lesion in the splenium extends further forward than it does in the cases of pure alexia without agraphia. The lesion would thus cut off not only fibres from the visual association cortex but also fibres from the somesthetic association cortex. In the case of Trescher and Ford (1937) the surgeon is said to have divided the posterior half of the corpus callosum; it seems likely that the lesion must have involved more than the splenium. The description of the case would suggest that the patient in addition to being alexic in the left half-field, also could not name objects in that field. This case therefore provides suggestive evidence for the hypothesis that lesions extending somewhat anterior to the splenium may cut off impulses from the somesthetic association cortex and thus prevent visual stimuli from reaching the speech area via this detour.

XI. THE HANDLING OF OBJECTS

It was pointed out earlier that one of the classical criteria for a visual agnosia was failure of the patient to demonstrate the use of objects shown to him. It was pointed out that according to this scheme the 'aphasic' could show the use of an object shown to him; the 'visual agnosic' could not do this but could manipulate the object correctly if it was placed in his hand; the 'apraxic' could not perform even with the object in his hand. I have reserved this criterion for special discussion since it appears at first glance to provide a very reasonable basis for separating disturbances of recognition.

This criterion, however, also fails as the mark of a separate category of 'recognition' disturbances. In the first place, as I have already commented, some patients who are severely 'agnosic' by the other classical criteria listed earlier can in fact respond correctly to the sight of objects. If one simply interpreted agnosias as deficits of 'recognition' there should be no reason why correct demonstration of the use of a seen object should be preserved in the presence of a severe disturbance of this type. Furthermore, there are severely apraxic patients who fail to respond to the sight of an object yet handle the actual object correctly when it is placed in the hand. By the classical criteria these patients should be called agnosic. Yet these same patients may be able to name the object and describe its use, behaviour certainly incompatible with a loss of 'recognition'.

As I point out in my discussion of the apraxias (Geschwind, 1965, Section V) they are also the result of disconnexions. The lesions producing these disturbances separate sensory from motor regions. It is obvious that the kind of lesion which causes classical visual agnosia will frequently although not invariably result in the disconnexion of the visual regions not only from the speech area and from the somesthetic areas but also from the motor region.

This discussion illustrates again the principle that division of phenomena into aphasic, agnosic or apraxic often obscures underlying mechanisms. It is far better in general to specify the classes of response which are lost or are preserved in response to each class of stimuli.

XII. THE CONDITIONS FOR CONFABULATORY RESPONSE

I have advanced a theory of classical 'agnosic' disturbances which views them as naming disturbances which are frequently overlaid with confabulatory responses. Why do these confabulatory responses occur? Weinstein *et al.* (1963) have stressed certain features of this type of response. Let me paraphrase some of their major conclusions: (1) Confabulation does not occur in the absence of a defect; it is, indeed, clearly an attempt to fill in a gap of performance of some type. (2) Confabulation is less marked in the presence of aphasia; this seems reasonable since it clearly is a response of the speech areas. (3) Confabulation is more likely in the presence of some over-all impairment of awareness, such as generalized dementia or clouding of awareness.[10] To these three conditions

I would add a fourth (Geschwind, 1963): Confabulation is much more likely in the presence of disease of association cortex or association fibres (either commisural or intrahemispheric) than it is in cases of damage of the primary sensory pathways up to and including the primary sensory cortex. We can combine this with the third condition above to make the following statement: The more demented or clouded the patient, the more likely he is to show confabulatory response in the presence of any defect of whatever origin; the less dementia and clouding are prominent, the more confabulation depends on lesions of association cortex or fibres; the two mechanisms interact frequently. These, of course, are not the only conditions; the influence of personality prior to illness may also be highly important.

Many examples come to mind to support this contention. My personal experience of several cases of denial of blindness illustrates this rule. The first case was that of a man suffering from a septicaemia, markedly febrile, drowsy and confused, who was blind as the result of infection of the anterior chambers of both eyes. Another was a man with a large subfrontal meningioma with advanced optic atrophy and a classical severe frontal syndrome with dementia. In these cases denial of blindness was related to marked general impairment of function of the nervous system. I have seen a confused patient who gave bizarre confabulatory responses when asked to name objects held in the hand. He could not, however, select the objects correctly from a group afterwards. This therefore must have represented the confabulatory response to an actual sensory loss. The same patient gave confabulatory responses to visual field testing when there was no stimulus in the field but gave correct answers to stimuli within the bounds of a normal visual field.

In contrast with these cases, patients with bilateral occipital infarcts (which probably usually involve much visual association cortex) are likely to manifest denial of blindness without marked dementia. Blindness from peripheral lesions is far less likely to produce such confabulatory denial in the absence of dementia or clouding. Confabulatory denial of a lost limb is uncommon; although the patient frequently has a phantom limb, he nearly always treats it as a phantom. But denial of hemiplegia is common and probably more common in the presence of accompanying lesions of association areas; even if it were insisted that all of these patients show some dementia or alteration of awareness, it is obvious that it is far less than the degree of dementia needed for the denial of a peripheral defect. I have, of course, already cited many examples of confabulatory response in the earlier sections of this part of the paper.

There is some suggestive but not definitive experimental evidence for this hypothesis. Warrington (1962) carried out experiments on visual completion in patients with hemianopias. She found that those patients who were unaware of a hemianopia were likely to 'complete' forms across a field defect while those who were aware of their hemianopias did not. Thus, the first group would report a segment of a circle in the good field as 'a circle' while the second group would report it as an incomplete circle. She found that this tendency to complete forms across field defects was strongly associated with the presence of a parietal lobe lesion, i.e. in our sense a lesion of association cortex. The anatomical and other data are however not fully convincing. Thus, of the 11 patients who showed completion, 6 had tumors and 3 of the remaining 5 showed mental deterioration. Thus, 9 of 11 had evidence either of widespread deterioration or of a lesion likely to cause more than focal disturbance. (It must be admitted, however, that the six tumours were all parietal in location.) In the group of 9 patients with little completion only 2 had tumours (neither of them parietal) and only one other case of this group was said to have shown deterioration. Thus, only 3 of 9 patients in this group had evidence of dementia or of a lesion with possible widespread effects. It is therefore possible to use Warrington's data in support of a different thesis from the one I have advanced – that patients with dementia or tumours are more likely to develop unawareness of a hemianopia and to complete forms across a field defect. Further studies of this type with greater description of the degree of dementia should help to resolve the issue in favour of one or the other interpretation.

Another paper by Kinsbourne and Warrington (1962) is also suggestive evidence for the thesis I have presented. They found patients with right hemisphere lesions who tended to misread the left halves but not the right halves of words ('novel' for 'level', 'cucumber' for 'encumber'); 4 of these cases showed complete left hemianopias but 2 did not. The authors thought that all of these patients had right parietal disease. It would be most interesting if it could be shown that this type of confabulatory reading error resulted from right parietal association area lesions, but this would require a control group of patients matched for dementia and with proven nonparietal right-sided lesions who would not show the defect.

Why should such confabulatory responses be less common in lesions of primary sensory cortex and more common in lesions of association cortex? I cannot answer with assurance, but I can propose a highly hypothetical mechanism which is at least suggestive of how such a situation could arise. Since the speech area has in the normal no direct contact with the visual cortex proper, destruction of the calcarine cortex leaves the speech area still innervated by visual association areas. The association areas in this case receive no stimulus from the visual cortex and send the 'message' to the speech area that there is no visual 'message', i.e. that all is black. This presupposes that the association areas never fail to send a 'message' to the speech area and that they always send *positive* messages of light or darkness. The destruction of association cortex or fibres corresponds to a totally unphysiological state – one in which *no* message is received by the speech area. Since this is not interpreted (not being one of the normal states) the speech area begins to react to random messages received from subcortical pathways or indeed responds to its own spontaneous firing. It may respond to *incomplete* information coming over subcortical pathways which are inadequately extensive to carry all the information about the seen object; this may lead to errors which are less bizarre than those in which no information at all gets to the speech cortex. As I have stressed this is a highly speculative mechanism, but certainly not an impossible one.

XIII. VISUAL IMAGERY

Related to the problem of the agnosias is the problem of loss of visual imagery. Wilbrand's (1887) monograph on psychic blindness actually concerned itself primarily with patients in whom visual imagery was lost. I have already commented briefly in discussing Macrae and Trolle's interesting case on the patient who could not 'visualize' a route but could travel it correctly. The word 'visualize' has two connotations. We may, as it were, see a picture in 'our mind's eye', and can then say that we are visualizing. At a simpler behavioural level we may describe an absent scene. A classical notion was that visualization was an activity of the visual association areas. The evidence for this is suggestive but by no means complete. The peripherally blind can indeed describe absent scenes and say that they have internal visual imagery. Would a patient with a pure destruction of the visual cortex bilaterally still retain imagery? The answer to this question is not known.

Loss of the ability to describe an absent scene might reasonably result from a disconnexion of the visual regions from the speech area, as probably was the case in Macrae and Trolle's patient. That this failure to give a verbal account need not be associated with a loss of visual memories is also illustrated by the same patient who could in fact correctly drive to work although he could not verbally describe the route. Was this patient capable of developing inner pictures? There would be no way of finding this out by interrogation since if the region in which such pictures were taking place were disconnected from the speech area, we would be unable to get a verbal account from the patient. The question is not a purely academic one since the eventual development of more advanced physiological techniques may settle some of these questions without recourse to a verbal account from the patient. This would be analogous to recent work on dreams for which objective criteria are now being developed.

The question of dreams and their locus is related to that of waking visual imagery. Are dreams functions of the visual cortex or of the visual association areas? These problems of imagery and dreams deserve further study using both standard clinical testing procedures as well as some of the new techniques employed so effectively in the study of dreaming.

XIV. PHILOSOPHICAL IMPLICATIONS – THE WHOLE MAN

It is not my intention to embark on an extensive philosophical discussion in this section. It seems reasonable, however, to sketch rapidly the implications of many of the results discussed here for the philosophical foundations of study of the higher functions of the nervous system.

For the past forty years there have been schools of thought which have stressed the importance of thinking of the patient as a whole, of seeing his responses as those of an integrated unitary structure, even in the face of damage. The ramifications of this thinking in neurology, psychiatry, psychology and other fields must be well known to most readers. It should be clear from much of our discussion that this principle, while it may be useful in some cases as a stimulus, may be actively misleading when it is regarded as a philosophical law. When Edith Kaplan and I were studying our patient, we constantly found that many confusions about the patient in our own minds as well as those of others resulted from failure to do the exact opposite of what the rule to look at the patient as a whole demanded, i.e. from our failure to regard the patient as made of connected parts rather than as an indissoluble whole. We were constantly dealing with questions such as 'If he can speak normally and he knows what he's holding in his left hand why can't he tell you?' We had to point out that we could not say that 'the patient knew what was in his left hand' and that 'the patient could speak normally', since that part of the patient which could speak normally was not the same part of the patient which 'knew' (non-verbally) what was in the left hand. This is at first blush an odd way to speak – it is hard not to say 'the patient' and yet it is clear that this terminology is misleading.

We have little difficulty with the concept of disconnexion at lower levels. If the spinal cord is transected we are usually capable of treating it as separate from the higher centres. We may say of such a person, 'The patient urinated' since we know that in this case this means 'The patient urinated involuntarily'. Although 'the patient urinated' is at first glance ambiguous, in practice we understand and do not find it necessary to use clumsy locutions such as 'The patient's spinal cord urinated'. We get into difficulties, however, with disconnexions at higher levels since we do not expect highly organized activities dependent on learning to be carried on in disconnected parts of the brain. We must become accustomed to thinking in this way in order to understand some of the more complex disturbances consequent on lesions of the brain. I am not advancing 'the atomistic approach' as a basic philosophical postulate to replace 'the holistic approach', but I am rather suggesting that failure to consider the applicability of *either* type of analysis will in one situation or another lead to errors. It should be pointed out that the usefulness of sometimes considering animals or humans not as a unit but as a union of loosely joined wholes need not apply only to disease states although probably it will find its greatest use in that situation. Probably even in the normal person parts of the brain are so weakly connected as to make their interaction difficult. I have suggested that some connexions are normally present in adult man which are absent or of less extent in lower animals;

these connexions may take years to develop in some children. Perhaps an adult man is more unitary than a chimpanzee, but perhaps total unity is never obtainable because of the necessary separation of some structures.

XV. THE UNITY OF CONSCIOUSNESS

A corollary to the above discussion is that it forces us to be somewhat more precise than we have been about 'the unity of consciousness'. It would no doubt be startling to suggest that the patient of Geschwind and Kaplan had separate consciousness in each hemisphere; it would on the other hand be a little difficult to understand just what would be meant by saying that his consciousness was unitary. If the ability to give a verbal account is a prerequisite of consciousness then only the left hemisphere was conscious; if the ability to respond in a highly organized manner and to use the results of past experience constitutes consciousness then he had multiple consciousness. Perhaps there are better criteria of consciousness than these; this case and similar ones only re-emphasize the necessity of re-evaluating the idea of the unity of consciousness if it is to be at all useful.

I believe that Kurt Goldstein was perhaps the first to stress the non-unity of the personality in patients with callosal section and its possible psychiatric effects. Thus he wrote (Goldstein, 1927), "The separation of so large a part of the brain and the resulting impossibility of evaluating stimuli perceived with the right hemisphere... surely cannot be without effect on the total personality.... I have pointed out the presence in my patient of a feeling of strangeness in relation to movements of the left hand, which she described with such curious expressions [she would say that someone was moving her hand and that she was not doing it herself] that she was regarded at first as a paranoiac. It appears to me not to be excluded that on this basis and under certain conditions there may develop paranoid states, perhaps also the experience of doubled personality and above all the experience of being influenced from without...."

XVI. THE VALUE OF INTROSPECTION

Still another corollary to the observation that parts of the brain may be disconnected is the conclusion that introspection may be an extremely ineffective way of obtaining information about many of the patient's experiences. I have already made this point in the presentation of some of the syndromes of disconnexion, e.g. in respect of the colour-naming disturbance of pure alexia without agraphia. If a part of the brain is fully disconnected from the speech area it will not be possible for the speech area to give an account of what goes on in that part of the brain. The patient with a colour-naming disturbance can give only a poor account of his colour-experience or indeed none at all since his speech area has little or no access to information about the colour-experiences of the visual cortex. Other examples were cited illustrating the same principle. I have also discussed in some detail the tendency for the patient to show confabulatory response to demands for introspective observations. It is certainly well known that even in normals introspection may be misleading or incorrect.

Let me point out carefully that what is presented here is not the gross behaviouristic assertion that introspective information is in all situations useless nor that one should never listen to what the patient is saying. The conclusion being presented is the less extensive one that there are certain situations, particularly in the presence of lesions of the brain, in which the patient for simple anatomical reasons is incapable of recounting verbally the experiences of parts of his nervous system which are functioning at a complex discriminative level.

XVII. LANGUAGE AND THOUGHT

An old problem is the one of the extent to which language controls one's 'perception' of the world. Whorf (1956) suggested an extreme form of the view that language influences what is perceived. The view at the opposite extreme is the naive 'natural' one that language is an infinitely flexible tool for describing without prejudice the impact of the world on one's nervous system. The behaviour of our patient with colour-naming defect is of interest here. Despite his inability to name colours correctly, he had no difficulty in sorting, and did so by colour. This, of course, does not necessarily reflect the behaviour of a 'naive' nervous system since the visual region may have been trained by earlier verbal experience. It is clear at least that words need not be available for the sorting process to take place.[11] On the other hand, as I noted in my discussion of secondary sorting errors, the use of the incorrect word may lead to errors insorting under the conditions of certain types of instruction. This type of 'secondary' error might be more marked in tests where the subject is required to remember colours after an interval since it seems reasonable that such memory is frequently mediated verbally (Brown and Lenneberg, 1954). The study of patients with such specialized disturbances may aid in further exploring the relationship of language to cognition.

Harvard Medical School, and Neurological Unit,
Boston City Hospital, Boston

BIBLIOGRAPHY

Bailey, P. and Bonin, G. V.: 1951, *The Isocortex of Man*, Urbana.
Bonin, G. V. and Bailey, P.: 1961, in *Primatologia* (ed. by Hofer *et al.*), vol. II/2, section 10.
Brain, W. R.: 1941, *Brain* **64**, 43.
Brown, R. W. and Lenneberg, E. H.: 1954, *J. Abnorm. Soc. Psychol.* **49**, 454.
Butler, R. A.: 1953, *J. Comp. Physiol. Psychol.* **46**, 95.
Connolly, C.: 1950, *The External Morphology of the Primate Brain*, Charles C. Thomas, Springfield, Illinois.
Critchley, M.: 1953, *The Parietal Lobes*, Arnold, London.
Dejerine, J.: 1891, *Mem. Soc. Biol.* **3**, 191.
Dejerine, J.: 1892, *Mem. Soc. Biol.* **4**, 61.
Ettlinger, G.: 1960, *Behaviour* **16**, 56.
Ettlinger, G. and Wyke, M.: 1961, *J. Neurol. Neurosurg. Psychiat.* **24**, 254.
Flechsig, P.: 1901, *Lancet* **2**, 1027.
Geschwind, N.: 1962, in *Reading Disability* (ed. by J. Money), Baltimore.
Geschwind, N.: 1963, *Trans. Amer. Neurol. Ass.* **88**, 174.
Geschwind, N.: 1964, *Monogr. Ser. on Languages and Linguistics*, **17**, 155, Georgetown Univ. Press.
Geschwind, N.: 1965, *Brain*, **88**, 237 and 585.
Geschwind, N.: 1967, in *Research in Verbal Behavior* (ed. by Salzinger), Academic Press, New York, p. 423.
Geschwind, N. and Fusillo, M.: 1966, *Arch. Neurol.* **15**, 66.
Geschwind, N. and Kaplan, E.: 1962, *Neurology* **12**, 675.
Goldstein, K.: 1927, in *Handbuch der Normalen und Pathologischen Physiologie* (ed. by A. Bethe and G. von Bergmann), vol. X, Berlin.
Hécaen, H. and Angelergues, R.: 1963, *La Cécité Psychique*, Masson, Paris.
Kinsbourne, M. and Warrington, E. K.: 1962, *J. Neurol. Neurosurg. Psychiat.* **25**, 339.
Lange, J.: 1936, in *Handbuch der Neurologie* (ed. by O. Bumke and O. Foerster), Berlin, vol. VI, p. 807.
Lissauer, H.: 1889, *Arch. Psychiat. Nervenkr.* **21**, 222.
Macrae, D. and Trolle, E.: 1956, *Brain* **79**, 94.
Meyer, A.: 1895–1896, *Amer. J. Insan.* **52**, 118. Reprinted in *The Collected Papers of Adolf Meyer*, vol. I, 1950, Baltimore, p. 641.

Nielsen, J. M.: 1937, *Arch. Neurol. Psychiat. Chicago* **38**, 108.
Pick, A.: 1931, in *Handbuch der Normalen und Pathologischen Physiologie* (ed. by A. Bethe and G. von Bergmann), vol. XV (ii), Berlin, p. 1416.
Sugar, O., French, J. D., and Chusid, J. G.: 1948, *J. Neurophysiol.* **11**, 175.
Trescher, J. H. and Ford, F. R.: 1937, *Arch. Neurol. Psychiat. Chicago*, **37**, 959.
Warrington, E. K.: 1962, *J. Neurol. Neurosurg. Psychiat.* **25**, 208.
Weinstein, E. A., Cole, M., and Mitchell, M. S.: 1963, *Trans. Amer. Neurol. Ass.* **88**, 172.
Whorf, B. L.: 1956, *Language, Thought, and Reality*, New York.
Wilbrand, H.: 1887, *Die Seelenblindheit als Herderscheinung*, Wiesbaden.
Wittgenstein, L.: 1953, *Philosophical Investigations*, New York.

NOTES

* Some of the work reported here was done under grant MH 08472 from the National Institutes of Health to the Boston University School of Medicine. [Permission to reprint materials from the three earlier papers has been received from the copyright owners, to whom we wish to record our thanks. The full texts of these papers appear in Geschwind, 1964, 1965, and 1967. Ed.]

[1] For example, the theory deals with the problem of color-naming as a corollary. The theory predicts that color-naming, if learned in a certain way, should be affected by certain lesions of the nervous system. In a hypothetical language in which there were no color-names, but in which the color of an object was stated as the color of some well-known object, the theory would predict that this lesion should not have the same effect. A linguist is more likely to avoid the naive error that color-naming is done in the same way in all languages.

[2] This principle holds to its full extent in man, the monkey, and chimpanzee, and probably in the other primates, at least the higher ones. It does not hold completely in sub-primate forms; for example, it is clearly not true of the cat's brain. It is therefore important to distinguish the species involved and not to generalize too readily to the human brain from that of lower forms, particularly when these lower forms are sub-primates.

[3] An objection might be raised that Ettlinger's experiment actually relates to cross-modal *transfer*, not cross-modal *association*. This problem is too complex to discuss fully at this point. Transfer is probably the simpler of the two, and therefore the failure of transfer probably also implies difficulty in association. In any case other evidence, e.g., from conditioning experiments, also points to the difficulty of non-limbic inter-modal *associations* in subhuman forms.

[4] It is true that in adult man some visual-tactile associations may be done indirectly by the linking up of two separate cross-modal associations (visual-auditory and tactile-auditory). Each of these components is, however, itself an elementary non-limbic cross-modal association.

[5] There is a possible exception in the 'curiosity' drive which has been studied by such workers as Butler (1953) and which does not appear to be quite so 'primitive' as the other limbic reinforcers. Even this drive is very likely to depend on the limbic system.

[6] See for example the discussion in Critchley (1953).

[7] Dr. Edwin Weinstein and his co-workers have been instrumental in stressing in recent years in a series of careful papers the importance of confabulatory responses. The

stimulation of his work was a major factor in attracting my attention to this problem.
[8] Pick (1931) pointed out that sorting errors may result from difficulties in verbal mediation in patients with colour-naming troubles.
[9] A brief note is appropriate here on Nielsen's interesting and important review. In his introduction he points out that many authors have stressed the bilaterality of lesions in most cases of this disorder. He then goes on in his study to confine himself only to the unilateral cases on the grounds that "if in any case a unilateral lesion produced visual agnosia, there can be no point in citing cases of bilateral lesion of the occipital lobes". This argument is of course based on the reasoning that a bilateral case must only be a unilateral case with an extra unnecessary lesion. The fact that probably the *majority* of cases of visual 'agnosia' have had bilateral lesions militates against this thesis. Furthermore it should always be kept in mind that many different *combinations* of lesions may lead to the same effect when the physiological mechanisms are kept in mind. He also neglects the possibility of individual differences. The paper of Nielsen is therefore a contribution to the anatomy of the unilateral group of cases but not necessarily to that of the bilateral cases. Many of Nielsen's cases show the discrepancies between naming and other functions that I have cited above although his interpretation of much of the material is different from mine because he accepts the idea of defects of 'recognition'. He frequently uses as one of his criteria the failure of the patient to *behave correctly* in relation to the 'unrecognized' stimuli; at the end of this section on the agnosias I will discuss this problem briefly.
[10] Although I have leaned heavily on the work of Weinstein and his co-workers in these preceding statements, I would by no means wish to suggest that Dr. Weinstein would have used the phraseology I use here.
[11] Some informal experiments on children suggest that perceptual differentiations precede naming. My 4-year-old son misnamed colours but could sort correctly and could trace out correct numbers on the Ishihara test; although I could name the colours of dots, I did much poorly than he did on these tests. His difficulty was one of colour-naming, mine was partial colour-blindness.

ROBERT EFRON

WHAT IS PERCEPTION?

I. INTRODUCTION

Perception is man's primary form of cognitive contact with the world around him. As all conceptual knowledge is based upon or derived from this primary form of awareness, the study of perception has always had a unique significance for philosophy and science.

Unfortunately, the exact nature of perception has never been adequately defined or conceptualized, with the result that those who attempt to study disorders of cognitive function do not always know whether a disturbance of cognition is due to a defect in a sensory, perceptual or conceptual process. The inadequacy of the prevailing concepts of sensation, perception, and conception and the practical importance of resolving the abstract philosophical issues entailed in formulating precise definitions of these terms can be dramatically illustrated by a concrete neurological problem. Many important issues in the history of epistemology are raised by the following case history of a 24-year old soldier who developed a severe cognitive disturbance after an accidental over-exposure to carbon monoxide.

After the accident, Mr. S. was totally blind. After a period of some weeks his vision returned but with a striking peculiarity. He was not able to name any common object at which he looked despite the fact that his visual acuity was at least 20/100. His visual acuity was estimated by his ability to reach for objects, to point to small pieces of white cardboard on a black background (or the reverse), and to point to fine threads on a piece of paper.

If this were the only information available, one might think that Mr. S. had merely lost the ability to *name* objects he perceives and one might conclude that his brain injury had produced a generalized disturbance in conceptual function. Such an interpretation, however, is completely undercut by the discovery that Mr. S. *could* name the object if it were placed in his hand! When allowed to feel any object, he could identify it about as well as a normal blind-folded adult. Not only could he name the object, but he could also indicate by verbal or gestural means how the object is used. He had no loss of ability to identify objects by using his auditory system either: When shown some keys on a ring, he said that he could see "it", pointed in the direction of the key ring, but was only able to report that it was "shiny like silver". When the keys were jingled, however, he snapped out the reply – "Keys". He identified nurses, doctors and ward personnel by their voices. By vision alone he could not tell a circle from a triangle, pictures of his mother from those of his girl friend, or a man from a woman. He could not even recognize himself in a mirror!

On the basis of this evidence one might suspect that there was a major disturbance of visual perception. Yet, as already described, his visual acuity was not significantly impaired, and he was able to name colors correctly. He could tell whether two colored filters were of the same or different hue – even when the hue differences were quite subtle (7–10 mμ). He could also tell which of two objects was the larger and he could report if an object was moving, even if this movement was quite small. He could point to the brighter of two lights even when the difference in luminance was extremely small (0.1 log units). Thus, Mr. S. could isolate and name the attribute (color, brightness, or size) of an object which made it different from another object when all other attributes of the two objects remained identical.

In addition, Mr. S. was able to make accurate judgments of the attributes of objects, even when the two objects he was comparing differed in several respects simultaneously. For example, when he was shown two discs of light which differed in hue, luminance and size, he could readily point to the light which was the "brighter", the "larger", or the "deeper red". His ability to perform this task indicates that he had retained the capacity to isolate and compare various attributes of two discs of light even in the complex case in which the *other* attributes of the two objects to be compared are different.

In more physiological tests of visual function, such as spatial summation and flicker fusion thresholds (Figures 1 and 2), only minor deviations from normal performance occurred. These minor deviations are not, in terms of our present knowledge, sufficient in themselves to account for his failure to identify objects, drawings, pictures, geometrical shapes, or

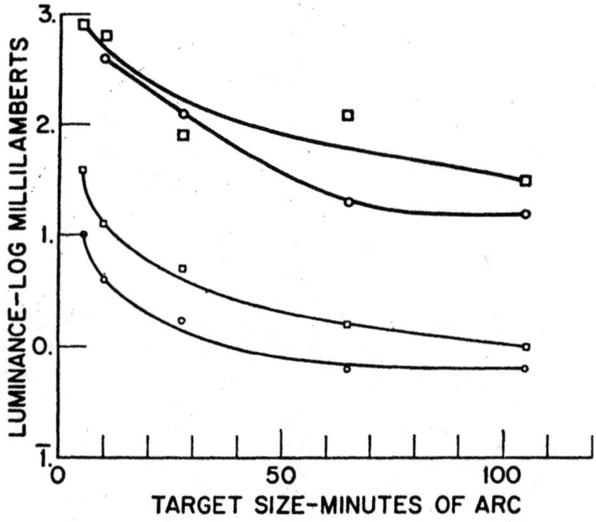

Fig. 1. Square symbols represent spatial summation curve from a point 10° from the fovea. Circular symbols represent curve from fovea. Light lines represent normal response. Heavy lines represent patient's response. Exposure time = 40 msec. Background illumination 1.0 millilamberts. Observe that the patient achieves full spatial summation (i.e., curve reaches plateau) at fovea at same target size as normal subjects. No significant difference in *summation* is found. The vertical separation of patient's curves from the normal curves is merely a measure of his lowered sensitivity to light.

letters and numbers by the use of vision. The constriction of his peripheral visual fields (Figure 3) and the small reduction in light sensitivity (Figure 4) are also insufficient to account for the defective recognition of these objects by foveal vision. Such defects are often observed in patients who have *no* defect in object identification.

Why then is Mr. S. unable to identify objects visually? If we ask *him* what the trouble is, he replies, "I can't tell what an object is because I don't see it clearly." Yet, using the full range of physiological tests of his vision currently employed by neurologists, we cannot find any defect which appears *sufficient* to account for his complaint of 'unclear vision'.

These unusual cases have been recognized by neurologists since 1876 when Hughlings Jackson described such a patient and coined the word 'imperception' to describe the cognitive defect. Other neurologists have used the descriptive term 'mind blindness'. The large majority of neurologists have accepted Sigmund Freud's reasoning and his descriptive

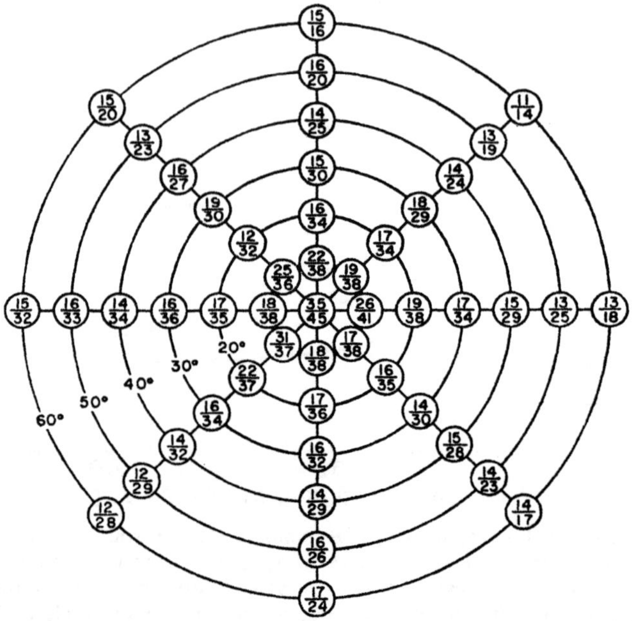

Fig. 2. Within each circle the upper number represents the number of flashes per second for flicker fusion required by the patient. The lower number represents the value obtained for normal controls of the same age group. The results plotted here are for the left eye. The right eye results were essentially similar.

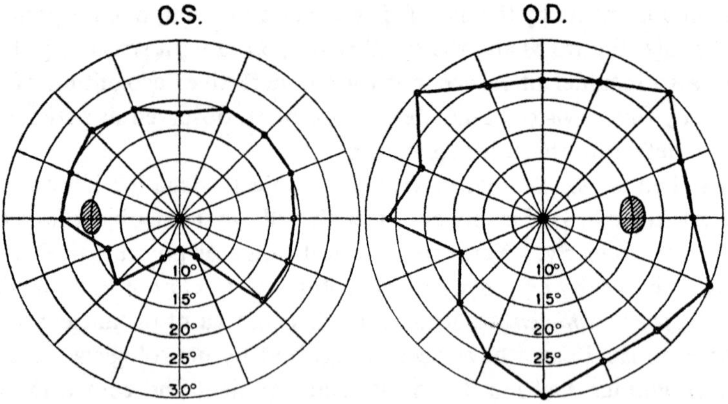

Fig. 3. Fields plotted on tangent screen with 36/1000 (approximately 2°) white target. This field size is normally obtained with a 2/1000 target. Appreciable fluctuation in responses on repeated trials make these fields unreliable. They can serve only as a rough guide to his degree of constriction.

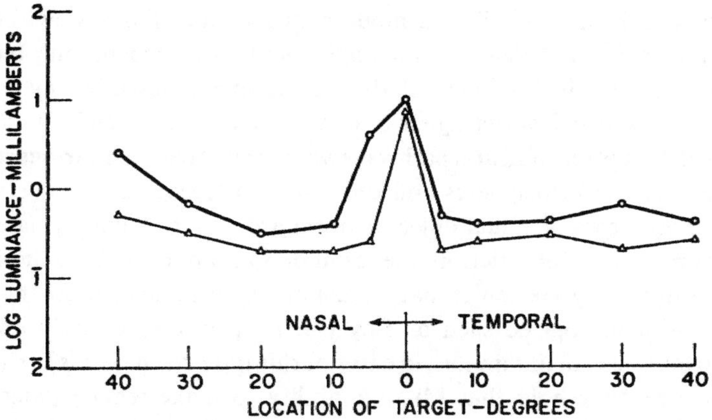

Fig. 4. The heavy line and the circular symbol represent the patient. The light line and the triangular symbol represent the normal values for the same age group. The target was 66 min in diameter. It was exposed for 40 millisec. The background luminance was $\bar{3}$ millilamberts. Observe that the shape of the patient's curve (including the rising portion in the nasal periphery) is similar to the normal. This gives evidence that ocular fixation was normal and that the patient's ability to understand and to follow the complex instructions of this task was within normal limits. The fact that reliable data could be obtained as far peripherally as 40° with a *smaller* target than was used to plot the fields in Figure 3, gives further evidence of the unreliability of the visual field measurements.

terminology – 'Visual object agnosia' – when they 'explain' these cases. They maintain that since patients like Mr. S. have either normal or at least adequate visual function, the cognitive disturbance cannot be attributed to a perceptual defect and thus must be attributed to a 'recognition defect', i.e., a specialized failure in one of the aspects of memory function. Some neurologists would maintain that Mr. S.'s defect is due to a loss of the 'stored visual memories' or 'engrams' of objects. Representative of those who hold the view that visual object agnosia is due to a 'recognition defect' are Dejerine (1914), von Monakow (1914), and Nielsen (1936). The last-mentioned author stridently asserted [1]: "I, therefore, insist that agnosia is a disturbance in recognition through one sense organ as von Monakow and Dejerine have stated".

Other neurologists maintain that in such cases the stored visual memories are intact, but that the inability to give the correct name to an object is due to a failure to 'associate' the perception of the object with the name. Geschwind (1965) is a modern proponent of this view. He has maintained [2] that there "is no single faculty of 'recognition', but that the term covers the totality of all the associations aroused by any object". He explains visual object agnosia as an 'association defect' due to anatomical disruption of neural pathways which travel between various regions of the cerebral hemispheres and the so-called 'speech areas'. He argues that 'most' cases of visual object agnosia are due to "highly isolated disturbances of *naming* which are the result of disconnexions from the speech area" (my italics). He denies that there is any perceptual disorder involved.

A few neurologists, such as Bay (1957), adopt an entirely different approach. They maintain that in cases of this type, the defect is *perceptual.* In essence, they hold that Mr. S.'s ability to make refined judgements of luminance, hue, size, and movement is evidence that he has 'visual sensations' but that his 'perceptions' are abnormal. In their view, a 'visual sensation' is the conscious experience of the attributes of light (hue, luminance, etc.) in contrast to 'perception' which, they say, is the awareness of objects.

A somewhat similar position is adopted by Gestalt psychologists who argue that patients of this type do not actually perceive objects – having lost the capacity to 'form Gestalts'. In their view, these patients have lost the ability to discriminate between the 'figure' (the object) and the 'ground' (the rest of the visual input).

Finally, there are other theorists who maintain that perception is normal in patients with visual agnosia, but that the defect is due to an inability to 'interpret' their visual perceptions. According to this view, patients with visual agnosia are similar, in essence, to patients who are born blind but who later recover vision. Such people cannot immediately recognize objects by the use of vision because they do not know how their perceptions are to be interpreted.

From this brief review of some of the current explanations put forward to account for the bizarre syndrome of visual object agnosia, it is clear that no single explanation has been accepted and that there is no uniformity in the way the terms perception and sensation are used. While some of these explanations may appear to be reasonable, there is no way to validate any one of them until there is a clear conception of the nature of perception, and of the distinctions between perception and other processes of consciousness.

The two questions to which this paper will be addressed are: (1) What is perception?, and (2) How is it to be studied? The answers to these questions will then be applied to the case of Mr. S. to see if they lead to a clearer understanding of the nature of his cognitive defect.

II. WHAT IS PERCEPTION?

The starting-point of the present discussion is the recognition of the fact that the term 'perception' denotes a form of awareness. The philosophical trend in contemporary biology is to dispense entirely with the concept of consciousness and its derivatives, and to attempt to study all the actions of living organisms, including man, from the standpoint of reductive materialism. As a consequence of this ideological orientation, many biologists today discard the idea that the term 'perception' denotes a form of awareness. They now use the term to refer to a *reaction* of an organism or of a part of an organism to a physical stimulus.

If perception does not denote a form of awareness and is defined only in terms referring to the physiological reactions to physical stimuli, how does the reductive materialist know *which* reactions to *which* stimuli are to be called 'perceptions'? Having discarded all reference to the concept of awareness, he has no principle or criterion by which to differentiate 'perception reactions' from any other physiological reaction. The ultimate logical consequence of this position is that he must accept *any* reaction to *any* stimulus in *any* organism or *any* part of an organism as a 'perception'.

Despite the absurdity of this definition of perception, reductive materialists continue to use the term perception as a synonym for 'physiological reaction'. For example, a plant physiologist recently published a paper with the title 'The Mechanism of the Perception of Gravity in Plants' [3]. He actually studied the relationship between the direction of plant growth and the direction of a gravitational field! Similarly, a molecular biologist recently claimed that "...the question of how cells *perceive* certain stimuli might be pursued on the molecular level by studying chemotaxis in bacteria" [4] (my italics).

The naivety and epistemological barbarism of defining and studying all actions of living organisms from the standpoint of reductive materialism has been pointed out by many philosophical critics, and I shall not debate the matter here. I have discussed this problem elsewhere [5] in a paper entitled 'Biology Without Consciousness – And Its Consequences'.

We begin, therefore, with the fact that the term 'perception' denotes all the various forms of direct, immediate, awareness of external reality which result from energy absorption by receptor organs. There is, of course, another category of direct, immediate, awareness resulting from energy absorption by receptor organs. This type gives us no direct information about external reality but informs us only of the physiological state or condition of our body. Examples of this second group are the experiences of nausea, hunger, pain, and cramp. The crucial *psychological* differentiation between these two categories of awareness is based on the fact that the latter are all directly and unambiguously experienced as 'inside' us while the former, those we call perceptions, give us direct awareness of what is 'out there'. The crucial *epistemological* distinction between these two categories of direct experience is based on the fact that the perceptions tell us about external *entities* while the other class tells us of our physiological condition.

Regardless of what decision we may take in respect to the proper name to apply to the second class (they are sometimes referred to as 'sensations', 'feelings', or 'enteroceptions'), the fundamental difference in the psychological and epistemological status of these two categories of receptor evoked conscious experience necessitates the use of separate concepts. For our purposes in this paper, however, it is not necessary to decide on the name of this second group because we will be concerned here only with perception.

While all forms of perceptual awareness are the immediate consequence of energy absorption, man does *not* perceive energy qua energy: He perceives *discriminated existents*. What do we mean by the phrase 'discriminated existent' and why do we use this particular phrase to denote the *content* of the various perceptual forms of awareness, i.e., that which is perceived? The word 'existent' merely means 'something which exists'. Within the context of a discussion of perception we use this term because it is the only word which is sufficiently abstract to encompass all the different kinds of things we perceive. We need a word to refer to the *objects* which we see or touch, to the *sounds* we hear such as the 'chirp' of a bird, to the *shadow* cast by an object, or to an *odor* or *taste* of an object. The term existent is applicable to *all* contents of the various perceptual forms of awareness. The terms 'object' or 'entity' are not. Indeed, they are somewhat misleading if applied to the 'chirp' of a bird or to a shadow. By convention, the terms 'object' or 'entity' are usually applied to spatially cohering collections of matter.

The word 'discriminated' is used in our definition of perception because it names and directs attention to the most fundamental characteristic of the perceptual process. This characteristic can be illustrated by the perception of the 'chirp' of a bird outside the window. The 'chirp' is an existent which is discriminated (isolated) from all the other existents ('sounds') reaching our ears at that time – the airplane 'buzzing' overhead, the street 'noises', and the 'rumble' of air conditioning. The 'chirp', the 'rumble', and the 'buzz' are consequences of a complex pattern of pressure waves distributed in space and time. The pressure waves initiated in the bird's throat are physically mixed with those waves emitted by the air conditioner and the airplane. Despite the fact that the pressure waves emanating from each of these objects is completely mixed physically, the remarkable feature of perception is that each set of waves is experienced as a separate, distinct existent. The most fundamental action of auditory perception is to isolate and treat as a 'unit' the set of waves emitted by each. It is only by virtue of this action that we can be aware of the 'chirp', the 'rumble' and the 'buzz' as distinct, separate *existents* to which we can direct our attention at will. An equivalent process of isolation is found in every other perceptual modality.

The action of perceptually isolating and treating as a unit some *part* of the total spatio-temporal stimulus configuration is not only the psychological pre-condition for man's perceptual awareness of reality: It is also the *defining* characteristic of perception.

We shall use the term 'discriminated existent' to apply to the segregated, isolated, cohering, 'thing' which is perceived. The objects we see or touch; the notes, tones, or voices we hear; the odors we smell, and the flavors we taste, are all 'discriminated existents'. The term 'discriminated existent' is used because it is sufficiently abstract to denote the object or content of *all* forms of perception. In short, the 'discriminated existent' is the 'it' that we perceive.

All discriminated existents are perceived by virtue of a gradient or pattern of energy impinging on a group of receptors. The terms 'gradient' or 'pattern', in this context, refer to any spatial and/or temporal variation in the quantity or quality of energy impinging on a specific group of receptors. If there were no gradients of energy reaching these receptors, we would not perceive discriminated existents. If we do not perceive discriminated existents, we are not perceiving. To illustrate this point, attempt to project what it would be like to wear translucent contact lenses and to live in an environment with no fluctuation in the quantity or quality of light passing through these lenses. The translucency would abolish all spatial gradients; the constancy of illumination would abolish all temporal gradients. Under such conditions you would 'see' nothing. That is to say, there would be no *content* to the perceptual form of awareness. Furthermore, it is doubtful if you would even perceive 'light' since the receptors in the eye quickly adapt to a constant level of energy absorption and return to a 'resting' rate of discharge. (It is not possible, for technical reasons, to actually achieve a complete abolition of visual gradients. But even with the unsatisfactory techniques now available, Dr. M. E. Wilson and I found that within a few minutes of exposure to a relatively gradient-less illumination, we became unsure if we were even aware of 'light'. A single eye blink (introducing a temporal fluctuation), however, quickly altered this state of uncertainty.) The phenomenon referred to as 'sensory adaptation' is too well known to require further discussion.

In sum, the existence of gradients or patterns of energy in space and time is the *physical* pre-condition of perception. If there were no gradients of energy impinging on our receptor organs, we would experience an undifferentiated, undemarcated, unlocalized, homogeneous *nothing*. That is to say, we would not be perceiving: There would be nothing to isolate or to treat as a 'unit'.

Both neurophysiological as well as psychological factors determine what parts of the entire energy input pattern will be isolated and will form the discriminated existent. Neurophysiological factors set the limits of what *can* be isolated. They determine the shallowness of gradients of energy which can be detected. Psychological factors determine what *will* be isolated within these limits. One of the most striking illustrations of this issue is seen in the 'hidden figures' of animals or objects. A picture of such hidden figures may be repeatedly exposed to the eyes. At each exposure the same gradient of energy impinges on the retina and yet we do not see the animal amidst the leaves of the tree. If we are told what to look for, or where it is, we will isolate that existent more readily. Once we 'see' it we may look at that region of the picture either as 'leaves' or as an 'animal'. The existence of the gradients, therefore, does not necessitate the perception. The *capacity* to detect the gradients is physiologically determined. Whether they will be used for the purposes of isolating a particular discriminated existent is determined by the context of the past experiences, current purposes and the attentiveness of the observer.

There are some philosophers and psychologists who seize on the fact that certain aspects of perception are psychologically determined and generalize this into the view that perception is entirely subjective. They often claim that man perceives what he wishes to perceive, what he is told to perceive, or what he is 'conditioned' to perceive. According to this view, perception is entirely disconnected from reality. The arguments given in this paper, and the example of hidden figures do not, as supposed by these philosophers, support such a position. The facts of reality, i.e., the gradients of absorbed energy, set the limits of what can be treated as a discriminated existent. We *can*, as a result of experience, or, for example, as a result of a conscious purpose disregard certain contours in time or space and use others. However, the particular contours we *do* use for the purposes of isolating any discriminated existent *exist*. They are not created by our desires, wishes or unconscious motives.

To answer the question 'What is perception?' requires, of course, that one give a definition by essentials. Summarizing the facts discussed in this section, we can define perception as *the direct, immediate awareness of discriminated*[1] *existents which results from patterns of energy absorption by groups of receptors*. This definition identifies and names the most essential characteristics of this form of awareness. It is the definition which will be used throughout this paper.

III. THE ATTRIBUTES OF DISCRIMINATED EXISTENTS

The science of perception is the rigorous and systematic analysis of the properties of the conscious experiences of external entities. It can be contrasted with the science of physics which is the rigorous and systematic analysis of the properties of the external entities themselves. As a science, the study of perception began in the late 19th century and from its inception was strongly influenced by the science of physics, which was quite advanced at that period. The influence of physics on the new science is revealed by its original name – psychophysics – as well as by its explicit purpose. The goal of the early psychophysicists was to establish a quantitative relationship between the forms in which we perceive certain attributes of existents and the magnitude of those attributes as measured in physical units.

The initial focus of the first psychophysicists was directed to those aspects of the perceptual experience of entities which are relatively easy to quantify. Some lights are 'brighter' than other lights: The relationship between the experience of 'brightness' and the magnitude of the energy emitted by the light source was readily established. Similar relationships were established between the loudness of sounds and the amplitude of air pressure waves; between the pitches of tones and the frequency of the pressure waves.

There are, however, other attributes of existents which are not so readily quantified and which the field of psychophysics has ignored, at least until recent times. Before we turn to an analysis of one of these, it is important to stress the fact that we are not aware of the attribute of an existent apart from the awareness of the existent itself. For example, we may perceive a discriminated existent in the auditory realm as 'a tone'. We can then direct our attention to that attribute of the tone which is related to the physical frequency of the air pressure wave (pitch). We may, if we wish, direct our attention to that attribute which is related to the magnitude of the pressure wave (loudness). It is the 'tone' which is the discriminated existent, not the pitch or loudness. If we did not perceive the 'tone' as the existent, we could never ask 'What is its pitch?'. A sound of any pitch which had no loudness would not be heard. Similarly, a light of any color which had no brightness cannot be seen; and a touch which had no locus (which is nowhere on the body surface) is not felt.

Although we do not perceive any attribute of a discriminated existent in isolation, we can attend to it selectively and mentally disregard the existent's other attributes. The psychophysicist exploits this capacity for selective attention to an attribute when he asks his subjects to perform such tasks as comparing the pitch of two tones of different loudness, or comparing the brightness or saturation of two lights of different hue. He must, however, always be alert to the possibility that his subject may be unable to isolate the particular attribute in question, or, if he *can* mentally isolate it, that he has failed to conceptualize his knowledge of that attribute. For example, most children and a surprising number of adults do not initially grasp the distinction between loudness and pitch, and are apt to call high-frequency as well as high-intensity sounds 'high' or 'loud'. It is clear that such subjects must be perceiving *both* attributes of the tone or they would not have heard it at all. Although they are perceiving the attributes, they have not isolated them, or have not conceptualized them adequately. At a more complex level, a trained musician can selectively focus upon a particular instrument in an orchestra which an inexperienced listener cannot do. Similar phenomena occur for each form of awareness: The subtle distinctions made by wine tasters, experienced chefs, and color experts is based on their ability to focus selectively on, to conceptualize, and to remember a larger number of attributes of the discriminated existent than the untrained person. The means by which we distinguish between a failure to perceive an attribute (as in the case of color-blindness), a failure to isolate it mentally (as in the case of children failing to distinguish between pitch and loudness), or a failure to conceptualize the attribute will be dealt with in more detail later. When we turn to the analysis of Mr. S.'s cognitive disturbance, it will be crucial to keep all three possibilities in mind.

IV. WHAT IS SENSATION?

If perception is "the awareness of discriminated existents", to what does the term 'sensation' refer? Unfortunately, there are so many different ways in which the term 'sensation' is used that it is impossible today to use this word without inviting confusion. The only reason for mentioning sensation at all in this paper is that the term is used in all discussions of perception, and is all too frequently given several different meanings by a writer within a single paragraph. I shall merely list the seven most common meanings.

(1) It is used to denote *forms of perception*. Many authors speak of the 'sensation of light', the 'sensation of sound', etc.

(2) It is used to denote the *attributes of discriminated existents*. Many authors speak of the 'sensation of brightness', the 'sensation of loudness', etc.

(3) It is used, as already mentioned, to denote *enteroceptions* as

contrasted with exteroceptions. Many authors speak of the 'sensations' of pain, hunger, cramp, or nausea to distinguish enteroceptions from exteroceptions.

(4) It is used to refer to those discriminated existents of which we are only *marginally aware* at any instant. Many authors speak of the 'sensations from the toe' of which we are marginally aware while we are perceiving something else. In this view, everything of which we are aware which is *not* the center of our attention are to be called 'sensations'. For example, in the process of focusing our attention on the attribute of pitch of a sound, its loudness would be a 'sensation'.

(5) It is used to denote the 'raw data' derived from receptor organs which has *not* been integrated into a perception of a discriminated existent. Some hold that men are *never* conscious of 'sensations' (as so defined). According to this view, the term 'sensation' does not refer to a phenomenon of *consciousness*, but to action potentials in nerve fibers (a phenomenon of physiology). Others hold that we only experience sensations (as so defined) in early infancy before a presumed chronological period at which man commences to perceive discriminated existents.

(6) It is used to denote forms of awareness of unintegrated facts of existence which are fleeting, unattended and which are not retained.

(7) It is used by some authors to refer to primitive, undifferentiated, unlocalized, forms of awareness. The experiences of those wearing translucent contact lenses might be called sensations by these writers.

It is not important for the purposes of this paper to decide on the most defensible meaning for the term sensation. Almost every use of this word refers to an actual phenomenon of consciousness. Therefore, we will retain the various referents for the term 'sensation', but will avoid the word itself. Instead of 'sensation', we will use the concepts of forms or modalities of awareness, enteroceptions, discriminated existents, attributes of discriminated existents, marginal awareness of discriminated existents, unretained perceptions, or undifferentiated perceptions (see below). We will ignore the fact that, historically, each of these concepts has been associated with the term 'sensation'.

V. HOW IS PERCEPTION TO BE STUDIED?

All methods of studying or describing perception quantitatively employ *measurements.* By the term 'measurement', we refer to the process of ascertaining the magnitude of some particular attribute, action or process of an entity by relating it to an appropriate standard or unit. For example, we measure the attribute of length of an entity in centimeters, its attribute of mass in grams, and its attribute of charge in electrostatic units. Similarly, we measure the attribute of heat in calories and work in gram-centimeters when we quantify an action or process of an entity. Observe that a different unit of measure is used for each type of attribute – a unit which is appropriate to the attribute being measured. *The attribute of the perceptual process which is quantified in all tests of perception is the capacity to discriminate.* (The units which are used in these measurements of discriminatory capacity will be discussed later.)

Entailed in the rigorous systematic study of perception is the quantitative analysis of the capacity to isolate existents and to discriminate differences between them. Regardless of the enormous variety of methods used to study perception, they all can be reduced, ultimately, to two fundamental and closely related questions which must be put to the organism in an appropriate form: 'Are you aware of an existent?' and 'Is existent A discriminably different from existent B?'

While these two questions pertain to fundamentally different aspects of the perceptual process, any act of perceiving always involves both. One cannot isolate without (implicitly) discriminating differences and one cannot (explicitly) discriminate differences between existents without having isolated the existents which are to be compared. The answer to the first question establishes the fact that the organism can isolate a discriminated existent, i.e., that it is perceiving something. The answer to the second question establishes its capacity to discriminate differences between the existents which have been isolated.

The various methods used to study perception all represent an adaptation, or translation of these two questions into a form suitable for the perceptual system and the organism being studied – a form in which a measurement can be made. The translation of these two questions into a convenient and quantifiable form frequently involves such a change in formulation that (a) it is not immediately apparent that the particular method used to study perception is actually reducible to one or the other question, and (b) it tests other cognitive functions at the same time.

A familiar example of such a case is the method of measuring visual acuity by use of a Snellen chart. The subject is asked to 'read the letters' on the chart starting with the largest. This task, at least on superficial examination, appears unrelated to either of the two fundamental questions. Yet, if we examine this task more closely, it can be seen that we have implicitly asked the subject to isolate an existent (a letter) and then to discriminate between or compare that discriminated existent with an internal retained standard for the appearance of each of the 26 letters. At letter sizes at which discriminatory capacity begins to fail, P's may be confused with F's and B's but not with O's, I's, or C's. With still smaller letter sizes no discriminations of shape can be made at all.

This example not only illustrates the fact that the Snellen chart is a measure of discriminatory capacity, but it also illustrates the point that its valid use depends upon or assumes the normality of other, non-perceptual, processes. In this instance it tests the visual acuity only in those subjects familiar with the Roman alphabet. For obvious reasons, such a chart cannot be used to test the visual acuity of animals, small children, Chinese – or patients with visual object agnosia.

While all methods of studying perception are ultimately found to contain the two primary questions, they rely on (and therefore test) other mental processes as well. It is the compound nature of all so-called 'tests of perception' which has masked the fact that the *perceptual* part of the test is always in the form of one of these questions and always represents an attempt to quantify an attribute of the perceptual process – the capacity to discriminate.

Having identified the particular attribute of the perceptual process which is to be quantified, the question then arises, 'What unit of measure is to be used?'

The action or process of isolation is most often measured by ascertaining how much energy of some particular form must be absorbed by the receptors to give rise to the awareness of a discriminated existent. These measurements are often referred to as absolute threshold determinations – and are always expressed in physical units (ergs). The absolute threshold is a measurement of the capacity to isolate a discriminated existent.

The action or process of discriminating differences between two or more discriminated existents is measured by presenting an observer with two identical existents and quantitatively varying a single attribute of one of these existents so that the two discriminated existents are just noticeably different. The amount of change in the attribute (as measured in the physical units) of the existent which is just large enough to make it noticeably different from the control existent is a quantitative measure of the capacity to discriminate *differences* of this attribute.

Observe one crucial aspect of this means of testing discriminatory capacity: It does *not* depend on the subject's capacity to mentally isolate the attribute which is varied, nor does it depend on his capacity to conceptually identify the difference between the two existents he is comparing. He is only required to indicate, by any method of signaling, if the two existents are the 'same' or 'different'. Because of the fact that the basic question put to the organism does not require any process of conceptual identification, such a test of perceptual capacity is suitable for use with animals, children, or with adults who have conceptual disorders (provided they can grasp and retain the nature of the task). This basic paradigm for the measurement of discriminatory capacity is referred to as a differential threshold.

The greater the perceptual capacity to detect differences between attributes, the greater are the number of discriminations or distinctions which can be made between discriminated existents. The greater the number of distinctions which can be made with a particular perceptual modality, the greater its cognitive and biological value. This can be illustrated by comparing the perceptual form which enables us to be aware of different volatile substances and the perceptual form which enables us to be aware of electromagnetic radiation. While we do not yet have a precise quantitative measure of the total number of discriminations possible to each perceptual modality, there can be little doubt that the number of distinctions (and the usefulness) of vision for man is vastly greater than the number of distinctions which can be made by olfaction. The phrase 'maximum perceptual capacity' will be used to refer to this, as yet undetermined, number of distinctions possible[2] to a perceptual modality.

While we cannot yet accurately measure the *total* number of discriminations which we are capable of making with any particular perceptual form, the use of differential threshold measurements provides an accurate and objective estimate of the capacity to discriminate differences in a *single* attribute of an existent *under a specific set of conditions*. The differential threshold is a valid and easily reproducible *contextual* measure of discriminatory capacity.

It is important to point out that the observer has no direct *perceptual* awareness that his discriminatory capacity (as measured by differential threshold) has altered under different observing conditions or on two separate observations under the same set of conditions. Since this issue is crucial to an understanding of Mr. S.'s complaint that he "cannot see clearly", it will be treated in some detail in the following example.[3]

Two different experimental conditions, designated A and B, are arranged in which a differential threshold for color discrimination is to be made. In condition A we find that the subject cannot discriminate a disc of 600 mμ (orange) light from discs of 598 and 602 mμ, but can discriminate a disc of 600 mμ from discs of 597 mμ and 603 mμ. That is to say, his differential threshold is 3 mμ. In condition B the area of the patch, or the duration of exposure to the patch, has been changed, and we find that the subject now cannot discriminate a patch of 600 mμ light from patches of 590 mμ and 610 mμ, but can discriminate a patch of 600 mμ from patches of 589 mμ and 611 mμ. His differential threshold under these conditions is 11 mμ.

The subject who is not informed of the results of this experiment would have no way of knowing that his differential threshold had altered under the two conditions. All *he* knows is that he can sometimes tell that the two colors are different and at other times they do not appear different. He has no *perceptual* cue as to the variation in his differential threshold. Yet, there *must* have been a difference in his form of awareness of the 600 mμ light disc under these two sets of conditions or there would be no difference between the two differential thresholds.

It is obvious that under both conditions the subject will perceive the 600 mμ stimulus as a disc whose color would be described by the word 'orange'. In case A, his awareness of that 600 mμ disc is such that he *can* differentiate it from an orange of 603 mμ but in case B, his awareness of the 600 mμ disc is significantly different: He *cannot* differentiate it from the orange of 603 mμ.

There is no term in the field of psychology or psychophysics which refers to this difference in the form of awareness which *underlies* the difference measured by the differential threshold. We will use the term 'specificity' to describe this difference and will say that the subject's form of awareness of the orange color of the 600 mμ light under condition A is more 'specific' than the orange color of which he is aware under condition B. We use the phrase 'more specific' because it conveys the essential idea that the form of awareness of orange under condition A is confused with fewer other shades of orange than the form of awareness of orange under condition B. The term 'more differentiated' could have been used equally well to describe this aspect of the form of awareness of an attribute of an existent.

The specificity of awareness of the attributes of existents varies along a continuum. In the visual realm, for example, the crudest, least differentiated experience is a vague awareness of 'light' with an imprecise locus, shape and having no color. Such a non-specific perception of light is experienced by normal observers at threshold conditions. It is also experienced by patients with impairment of the visual system at even higher levels of energy absorption [7]. A gradually progressive increase of specificity of the awareness of all attributes of the existent is readily observed in the normal subject as more energy is delivered to the receptors. In patients with visual impairment due to disease, a gradually increasing specificity of awareness occurs as recovery ensues [8].

While historically, many authors have referred to the crudest, least-differentiated perceptions as 'sensations', they actually lie on a continuum with the most highly differentiated perceptions which occur under optimal conditions. There does not appear to be any justification for the arbitrary segregation of these poorly differentiated perceptions into a separate category to be called 'sensations'.

In sum, the concept of specificity relates to the degree of differentiation of *any* attribute of an existent which the form of awareness permits. The specificity of the awareness of any attribute of an existent is inferred from the measurement of a differential threshold. The concept of specificity and the method of its measurement will now be applied to an attribute of existents which has not been studied by the neurologists and psychologists investigating patients with visual object agnosia. I am referring to the attribute of shape.

VI. THE ANALYSIS OF A CASE OF VISUAL AGNOSIA

In the introduction to this paper an unusual cognitive disturbance was briefly described. In summary, the patient is unable to identify objects, geometrical shapes, numbers, letters or people by the use of vision alone despite the fact that conventional clinical tests of his visual function do not reveal abnormalities of quality or degree which could account for this difficulty.

A satisfactory explanation for such a phenomenon as visual object agnosia must fulfill two basic requirements: it must account for the facts of the individual patient's disability, and it must not be contradicted by any of the activities which that patient can perform.

The analysis of Mr. S.'s cognitive disturbance will begin with a more detailed description of the actions he can and cannot perform:

(1) As previously mentioned, Mr. S. can point with his finger to an object which is held before him. He can do this efficiently only if the object is *moved* before his eyes. If it is stationary he does not appear to know *what* object he has been asked to look at; his eyes randomly scan the entire room, and he appears to be 'searching'. For example, if a toothbrush is placed on a white cardboard (so that he does not observe the procedure of placing it there) he may point to the white card, to the toothbrush, to the desk on which the card is placed, or he may simply look blank and bewildered. However, if the object which we want him to identify is moved, he can immediately point to it without going through the 'searching' pattern.[4] The movement enables him to *isolate* the object from the background. He is still unable to identify it.

(2) When he finally locates the object in question, he is unable to say what it is despite intense and prolonged scrutiny. When asked its color, he quickly replies by naming the *dominant* color (unless this color is black or white). If he is pressed by the further question "Does it have any other color?" he names the color with the second largest area. He usually reserves naming the 'colors' white and black until he has named all the other colors – even if the white and black areas are larger.

(3) When he is shown two squares on a piece of paper, and each of them is moved so that he 'sees' the objects in question, he can say which is the larger. His accuracy in this task was not tested quantitatively. By his performance, however, it was obvious that he had some difficulty in detecting small differences in size.

(4) After he has located an object, he can trace its contour with his finger. (The object was placed so that he could not touch it and thus

Fig. 5. The dotted line superimposed on this photograph indicates the contour traced by the patient when he was asked "Trace the object I am holding here."

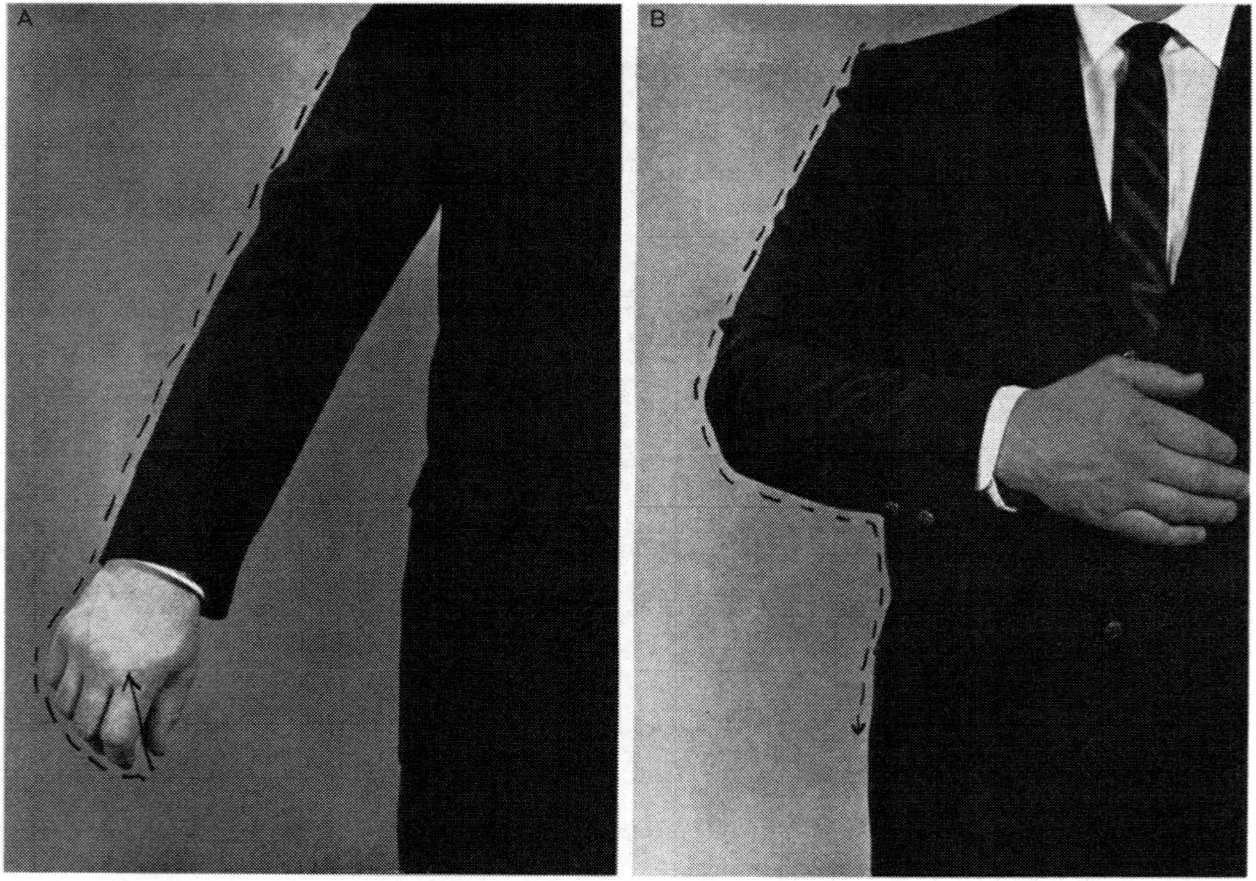

Fig. 6. The dotted line superimposed on Figure 6A indicates the path traced by the patient when he located the hand. The dotted line of Figure 6B indicates the path traced when the patient was unable to locate the hand.

obtain tactile cues.) There are several important qualifications which must be added to this description of his capacity to follow contours. In the first place, he has appreciable difficulty in performing this task because of a partial paralysis in hand movement which makes him clumsy, slow, and ataxic. Secondly, he will often go round a simple figure many times, not knowing that he has completed the task. Thirdly, in those cases in which he is asked to trace a complex object, he will almost always follow the contour of a single color area. This type of contour following is illustrated in Figure 5.

When asked to find the examiner's hand, or the hand in a color photograph of a person, Mr. S. traced the outer (silhouette) contour of the body until he came to the region of the hand (Figure 6a). On some occasions he stopped and indicated that he had achieved the goal; at other times he continued to follow the silhouette contour, apparently failing to recognize that he had passed the hand. On those occasions (see Figure 6b) in which the arm and hand were not outstretched, and thus did not appear in a silhouette projection, he never managed to locate them.

(5) He could not, as previously mentioned, give the correct name for any geometrical shape presented to him visually, but he could describe from memory the shape of coins, shoe boxes, a loaf of bread, and a playing card. From memory, he could also name the color of objects such as lettuce ("green and white"), tomatoes ("red"), banana ("yellow with some brown"), telephone ("usually black but may be other colors too"). He could also correctly report from memory which was the larger of two objects, such as a postage stamp or a pea, a book or a playing card, an apple or a grape.

(6) He was unable to count the number of coins placed on a piece of white paper. The number he reported was almost always larger than the actual number of coins because he unknowingly counted some coins twice. (More rarely he would miss a coin.) That this defect was not due to an inability to count was demonstrated by his ability to accurately count flashes of light, taps on the skin, and tone bursts.

(7) He was taught to apply a name to each member of a small (under 30) repertoire of objects which were presented to him one at a time on a piece of white paper. For example, after he was repeatedly shown the back of a red and white playing card and informed of its identity on each occasion, he was able on subsequent exposures to call it a "playing

card". After such a period of training he easily made this 'identification', even when he was exposed to the object for the duration of a few microseconds. (A gas discharge lamp was used as an illumination source in a dark room.)

His ability to learn the name of an object may suggest to some readers that Mr. S. perceives objects normally but has either lost the memory of its name, or lost the visual memory of its appearance. Such a conclusion is not supported by the following experiments which show that Mr. S. actually identifies these objects by a process of *deduction*. In the first place, after he had learned to apply the name "playing card" to the red and white playing card, he was shown a playing card identical in every way except that it was blue and white. He was unable to identify it. Secondly, when he was shown a red and white postage stamp, he called this object a "piece of a playing card". (The reader will recall that he could tell differences in size.) Thirdly, after he had been trained to call a brass key a "key", he subsequently called a brass gear a key as well. Finally, he was trained to call a pair of scissors "scissors" – but he called all other shiny, silvery objects such as knives, paper clips, and metal rulers, scissors as well. From these observations, it was quite clear that Mr. S. could learn to 'identify' correctly an object from a limited repertoire by its attributes of color, reflectance and/or size, and not by any other attribute. *Knowing the entire repertoire*, he could deduce which object was before him from the attributes he *could* identify.

(8) Even when trained to make the correct verbal reply, his inability perceptually to identify an object was strikingly demonstrated by his failure to name a now familiar object when it was placed before him on an unfamiliar background. Instead of showing him the object on a white cardboard, it was put on a color photograph. Under these conditions he failed to see the object at all, despite the fact that he scanned the color photograph (and superimposed object) minutely. However, when the photograph was shaken so that the object in question moved relative to the surface of the photograph, he immediately identified it by the name he had been taught. With a great deal of training on this task, he could apply the correct name to only a few of the limited repertoire of objects on a photograph when the object was *not* moved relative to the photograph.

(9) When shown Pseudoisochromatic test cards [10] for color vision (Figure 7), he always named the colors correctly but could never name the

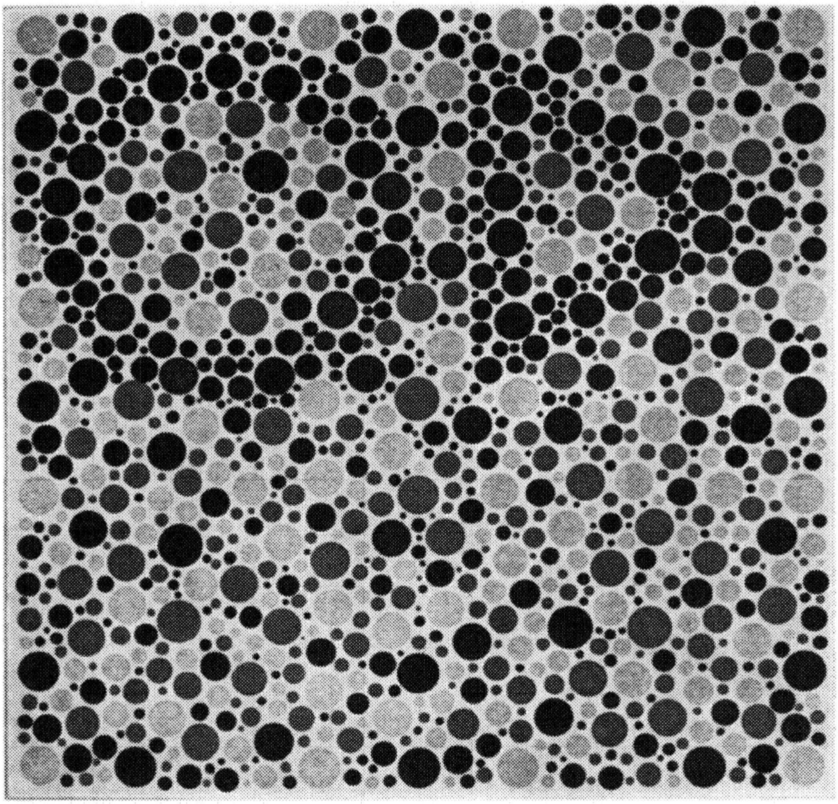

Fig. 7. A black and white reproduction of the Pseudoisochromatic test chart photographed in such a fashion that the shape of the triangle and circle are revealed by luminance differences. In the chart actually used with the patient no luminance differences, but only spectral differences were present.

shape – nor could he even trace the shape. He could touch, with a pencil point, the individual circles of color. On one occasion, when I outlined a circular figure repeatedly with a pencil he was apparently able to see the shape. For a brief instant his face lit up with pleasure and he claimed that he saw the circle. A few minutes later, under static conditions, he was unable to identify the same object.

When simple geometrical shapes were slowly drawn in front of him so that he could see the movement of the pen, he frequently named the shape of the figure. When he was shown the same figure a few moments later, he was unable to identify the shape.

(10) He responded, by blinking his eyes, when an object rapidly approached his face, and he could tell if an object was being moved toward him or away from him. However, he was unable to say which of three *stationary* objects placed on a white card in front of him was the farthest, nearest or middle one. When he was asked to touch the objects, he could then tell their distances, *but only after he moved his hand.* He clearly used the kinesthetic cues from arm movement to report correctly their respective distances.

(11) When asked if two large black strips of tape on a white paper were in the same or in different orientations, he carefully followed the contours of each by moving his head. Using this method, he frequently gave correct answers. However, when he was prevented from making head movements, he could no longer perform this task.

From this detailed description of Mr. S.'s performance, we can conclude that he *is* perceiving, i.e., that he is visually aware of discriminated existents. He can point to objects, trace their contours and isolate them from the rest of his perceptual field (albeit with difficulty). He is capable of selectively attending to their attributes, and he can discriminate between similar attributes of two existents. In terms of the analysis of perception given in the earlier sections of this paper, there is no fundamental action of perception which Mr. S. is unable to perform.

To say that Mr. S. can perform all the different *types* of perceptual action does not necessarily mean, however, that he performs them normally. To ascertain if his visual perception is normal, we must first know if he is aware of *all* the (visually perceivable) attributes of a discriminated existent. Then we must ascertain if the specificity of his awareness of each of these attributes allows him to make the number of discriminations which can be made by a normal observer under identical conditions. In short, we must ascertain if his perceptual capacity, as defined in the last section, is normal.

It is the failure to grasp that *all* aspects of the perceptual process must be studied and that perceptual capacity must be studied *quantitatively* which has led neurologists and psychologists to conclude that patients with visual object agnosia have 'normal' or 'nearly normal' perception. It should be emphasized that these neurologists are not entirely ignorant of this requirement: they adopt this methodology when they study certain *simple* perceptual disorders. For example, if a patient is able accurately to distinguish colors and brightnesses of objects, yet fails to recognize people on the street, few neurologists would jump to the conclusion that he is suffering from a 'recognition' or an 'association' defect. They would first test his eyes, discover that he suffered from severe myopia and conclude, if only implicitly, that his perceptual capacity was insufficient for the task of identifying faces at distances greater than 6 m.

Because they have not explicitly conceptualized their methodology for studying perception, neurologists and psychologists do not as a rule apply the *same* methodology to more complex and subtle disorders of cognitive function. As a consequence, they have concluded that cases of visual object agnosia are due to recognition, memory, naming or association defects without ascertaining if the failure to recognize objects could be accounted for by a diminished perceptual capacity.

The various tasks which Mr. S. cannot perform are the identification of objects, geometrical figures, letters, numbers and people. Is there one attribute of these various existents which must be perceived for their identification? The answer is inescapable: *They are all identified by their attribute of shape. The perceptual capacity to make discriminations between shapes is, therefore, a pre-condition for their recognition and identification.* Letters, geometrical figures, and numbers are identifiable *only* by shape. Most common objects, with very few exceptions, are identified by their shape; their color, size and brightness are non-essential attributes for the purposes of identification. People are identified by the shapes of their features, and only to a very minor degree by skin color and size.

While it is a fact that the five classes of objects which Mr. S. cannot recognize are all identified by virtue of their shape, it would be illogical to conclude from this that Mr. S. is unable to perceive shape. All that we can legitimately assert is that shape perception is a *necessary* precondition for these identifications. It is not necessarily the *sole* precondition. We must determine empirically if Mr. S.'s perceptual capacity to make shape discriminations is sufficient for him to identify objects. If it is not, it would be just as ludicrous to postulate the existence of a memory or naming defect as it would in the previous example of the myope who cannot recognize people on the street.

How do we measure this particular perceptual capacity? As pointed out in the previous section, we measure the capacity to discriminate an attribute of a discriminated existent by asking 'Can the subject discriminate between two existents which differ *only* in respect to this particular attribute?' We then determine the differential threshold for this attribute by repeatedly asking the subject the fundamental question 'Is existent A the same as B or different?'

The following experiment was performed on Mr. S. to test his perceptual capacity to discriminate differences between two existents when the only difference was in the attribute of shape.

A set of cards (Figure 8) was prepared, each card containing two objects; a black square and a black rectangle of *identical area* and *reflectance*. On each card, the shape of one of the objects was progressively elongated. Mr. S. was asked if the two objects on each card were the same or different. The cards were presented to him in random order and with the square randomly on the right or left side. Figure 9 is a graphical display of his performance on this task. When the rectangle was very long and thin (the ratio of lengths of the two sides being 8.65:1), Mr. S. reported only 90% of the time that the two objects were different. As the ratio of the lengths of the sides of the rectangle was made smaller, his difficulty in making the discrimination increased. The fact that he called physically identical existents "different" 60% of the time suggests that his 90% correct score may be contaminated by this bias. It is quite possible that the 90% score is spuriously high.

This experiment depends, as does every other test of discriminatory capacity, on a variety of non-perceptual processes. For example, its validity relies on the assumption that the subject grasp and retain the nature of the task he is supposed to perform. The fact that Mr. S. grasps and retains the instructions for all other tests of visual perception justifies this assumption. Even the results of the experiment justify the

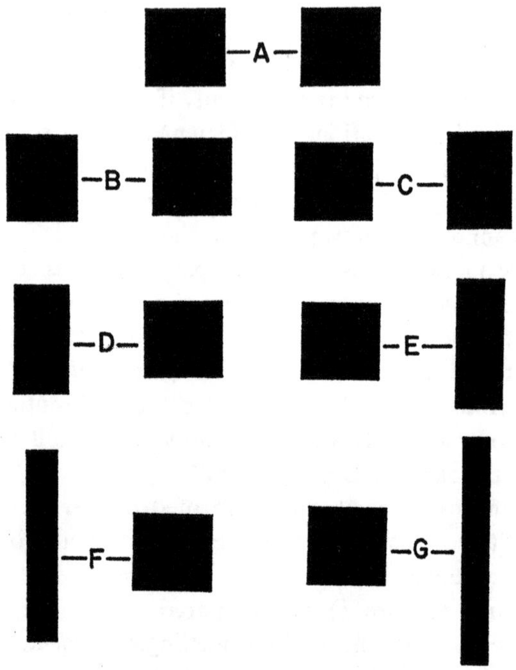

Fig. 8. The targets were observed binocularly with central vision at a distance of 18 inches. The square subtended a visual angle of approximately 6 degrees. In Figure 8A, the two figures are identical; the ratio of the lengths of the two sides was 1.0. In Figures 8B–G, the ratio of the lengths of the sides of one of the component figures is progressively increased.

assumption. As the difference between the two shapes increases, his percentage of correct replies increases. This experiment further relies on the assumption that Mr. S. is motivated to perform this discrimination and is attentive to the task. The fact that he was motivated to perform and to closely attend other visual discriminatory tests (such as color and brightness discriminations, sensitivity profile, flicker fusion and spatial summation tests) justifies this assumption as well as the earlier assumption.

On the other hand, this experiment does not depend on any long-term memory or association function since the two shapes to be discriminated are present simultaneously. He was not required to report in what respects the two shapes were different. That is to say, he was not required to mentally isolate and selectively focus upon the attribute of shape.

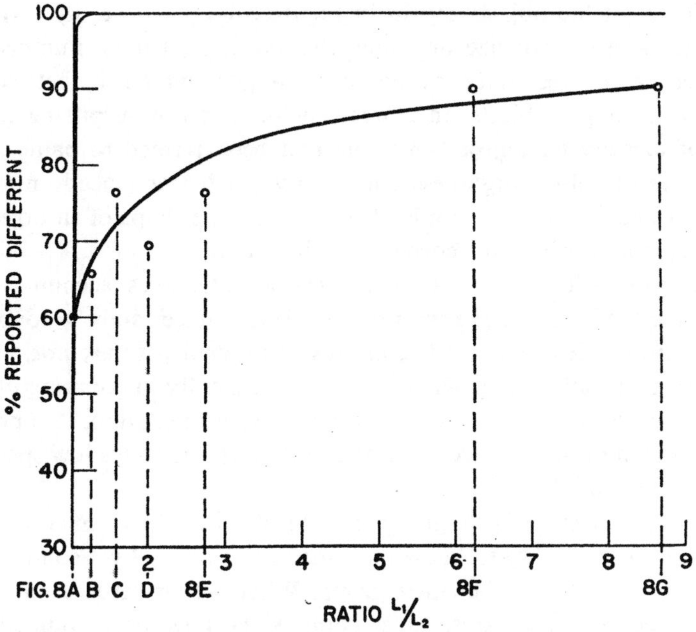

Fig. 9. The percentage of 40 trials on each card in which the patient reported that the figures on the test cards Figures 8A–G were "different". The fine curve which steeply rises to the 100% level represents the performance expected of a normal individual for whom this test is trivially simple with foveal vision.

Finally, this experiment does not depend on the ability to conceptually identify a shape. In this particular experiment the shapes which were used do have names ('square', 'rectangle'). However, in other experiments he had identical difficulty in detecting differences in shapes and patterns for which there are no standard names (curvilinear figures). In accordance with the principle stated in the last section, he was only asked if they were the same or different – a task which does not depend on the processes of classification, abstraction, or conceptualization. Since the brightness of the two objects (luminous flux per unit area) was identical, and since the total luminous flux was the same (as the areas of the two figures were always equal), the *only* way to distinguish between these two objects is by shape. It is clear that Mr. S.'s capacity for making this discrimination is severely restricted. If he cannot discriminate a difference between the shapes of the objects shown in Figure 8B–E, it is hardly surprising that he cannot recognize or name objects, faces, letters, numbers and geometrical figures. If he cannot discriminate differences between the shapes of simple objects on a white field, it is not surprising that he cannot identify an object (which he had been trained to name by the attributes of color, brightness and size) when it was placed on a color photograph. On such a complex background, the shape of an object is a more crucial attribute for correct identification.

Can this failure to discriminate shape differences account for *all* the defects of his visual performance? While it certainly accounts for his failure to identify these five classes of existents, it may not, at first, seem to be a sufficient explanation for his difficulty in counting objects. Yet, on closer reflection and with further experimentation, it is possible to show that even his inaccuracy of counting is due to his low specificity of awareness of shape.

To count accurately a group of objects, i.e., to enumerate them, depends on the ability to know that one has or has not yet counted each individual member of the entire group. When we are given a number of coins to count, we separate those coins which have been counted from those which have not. For example, we may move the coins from one pile to another. If the task must be performed without moving the coins, we are forced to group them mentally into some spatial configuration to guide us in the sequence of counting and to ensure that we have counted each object only once. If four coins are placed on a table and each is located at the corner of an imaginary square or rectangle, we 'see' the shape of the configuration and will have no difficulty in counting them. However, if 30 coins are placed in a circular array, we must count them in some rigorous order, making mental note where, on the circle, counting was started. It is very easy to overshoot the starting-point, especially if its locus is not clearly marked. Without *some* projected or superimposed schema of shape (spatial configuration), accurate counting of a number of spatially distributed coins would be impossible.

Since Mr. S. does not perceive the attribute of shape either of single, spatially cohering objects or arrays of objects (Figure 7), it is to be expected that his counting will be erratic, and that he could not know when or where to stop the count. However, if the objects to be counted are distributed in *time*, he should have no difficulty in obtaining an accurate count, because he need no longer rely on shape information to tell him when or where to stop counting. His ability to count accurately *temporally* distributed flashes of light and his inability to count *spatially* distributed objects would thus be entirely consistent with and was predicted by the hypothesis that his difficulty in counting objects is due to his defect of shape perception, and not due to a more generalized intellectual defect in counting.

The hypothesis that Mr. S.'s cognitive disturbance is due to his low specificity of awareness of the attribute of shape thus accounts for all the things he *cannot* do. Yet there is one thing he *can* do which superficially appears to contradict this hypothesis: He can follow a contour. Is there actually a contradiction involved here?

To answer this question we must first distinguish between the perception of a contour and the perception of a shape. A contour is a consequence of a spatial gradient in quantity or quality of the energy absorbed by the retina. A contour is the demarcation (transition) between two adjacent areas which differ in some physical attribute. The *shape* of an existent is an attribute which pertains to the relationship of its contours. It is perceived as a consequence of the integration of the entire contour which may be visible from a particular viewing angle. Mr. S. can detect a contour whether that contour is a consequence of a luminance or a wavelength difference between two adjacent areas. He isolates color and brightness areas in this way and thus perceives discriminated existents. (See item 2 in the detailed description.) However, as a consequence of his brain damage he apparently cannot integrate all the transitions into a perception of *shape*. For example, when shown the book illustrated in Figure 5, he traced the contour of the red area and the contour of the hand. (Precisely why he included the hand and excluded the cuff is not clear. Perhaps the difference between the color of the hand and the background (a light green wall) was of greater significance to him than the difference in color between the hand and the red area of the book.) While he can trace a contour, thus demonstrating that he obviously perceives transitions, he can no longer integrate or 'put together' the entire contour to form a discriminated existent with the attribute of shape. As illustrated in Figures 8 and 9, he cannot do this, even in those cases where simple shapes are used.

The detection of a transition and the tracing of a contour can be performed without any perceptual awareness of the shape. This possibility is illustrated by Figure 10, which suggests the means by which this may occur. The contours of the object in Figure 10A may be traced by a process of repeatedly scanning (with eye movements) *across* the demarcation line separating two adjacent areas of different luminance. Pursuing this strategy around the silhouette contour of an object will permit an accurate

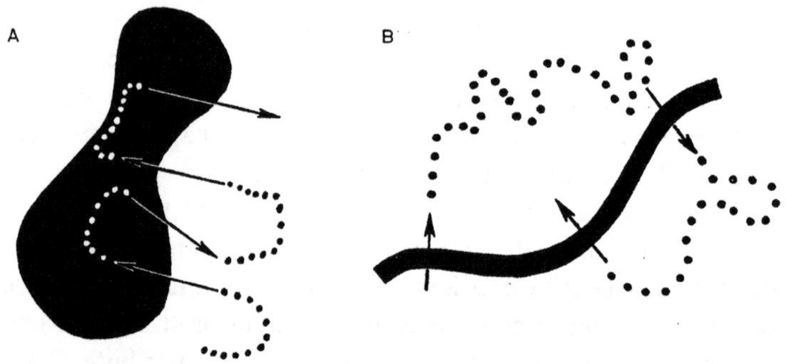

Fig. 10. See text for discussion.

contour tracing, but does not necessarily give any integrated perceptual awareness of the shape.[5] Some of Mr. S.'s clumsy or apractic hand movements during the contour tracing tasks may be due to the fact that he was pursuing this strategy. Even more striking evidence that this was the case comes from the observation that Mr. S. was *unable* to follow single curved black lines drawn on a piece of paper. Following a scan across such a thin line (see Figure 10B) he would apparently lose track of the location of the line, and would have to find it again, perhaps at a point quite remote from the point of last contact. The erratic nature of his tracing movements effectively precluded accurate tracing.

It should be stressed that in the example just given, the proposed scan was considered to be a sequence of eye movements. A similar process may exist in the form of a central neural mechanism which quickly performs the same kind of operation on the incoming pattern of energy. The fact that a normal person can identify shapes even with microsecond exposures which preclude the possibility of eye movement is consistent with such a central scanning procedure.

Once it is accepted that Mr. S. has a low specificity of visual shape perception, it is not difficult to understand why he cannot *identify* objects. It is less evident, however, why such a defect will also result in difficulty in *finding* an object located in a complex field, or why this difficulty is overcome when the object is in motion.

To clarify this problem it is first necessary to consider the actions of a normal observer attempting to find an object in the presence of many other objects. In such a case the observer is presented with a complex set of hue and luminance gradients. What (implicit) procedure does he follow to determine if a particular sub-set of gradients are to be isolated and seen as the 'object'? How does he know if there are two adjacent or overlapping objects, or only one object of a more complex shape and composed of different colored areas?

While we clearly do not know the detailed workings of the perceptual processes which permit this determination, we can be certain that the knowledge of the shape of objects is a crucial element in performing this task. This knowledge, derived from past experience, is compared with the various sub-sets of gradients until a 'match' is found. The impossibility of finding a 'hidden' object in a drawing if one has no previous knowledge of the shape of that object testifies to the importance of the role of previous experience in the act of finding an object.[6] We must know what we are looking for – and we must be able to compare our (implicit) expectations of shape with the shapes of objects we perceive.

Now let us consider Mr. S.: He has retained appreciable conceptual knowledge of the shapes of objects (see item 5 in the description). However, as he is now virtually unable to perceive the shape of an object, he cannot perform the operation of comparing his retained knowledge of shape with his perceptual awareness of shape. He thus cannot determine how the complex gradients are to be divided up. He has no basis on which to say, 'This is the object, and this is part of another object.'

However, if the object is moved in relation to the rest of the complex field, additional information is provided which gives him the basis for perceptually isolating the object. This phenomenon is seen even in normal people who are frequently assisted in perceptually isolating a particular object by virtue of its movement in relation to the background. For example, we may not perceive a mouse in the corner of a room – even if we are looking for it – until it moves. The movement immediately allows us to isolate the object. In contrast to the normal adult person,

Mr. S. is relying almost exclusively on the additional information provided by movement to isolate an object when it is located in a complex field of many objects.

In a somewhat metaphorical manner of speaking, Mr. S. lives in a nightmare world of hidden figures which remain invisible until they move. Even when they do move and become visible to him, Mr. S. does not know what they are because he cannot perceive their shape.

In sum, the hypothesis that Mr. S. suffers primarily from a defect in shape perception is sufficient to account for all the actions he cannot perform. It is also consistent with those actions which he can perform. In the face of the demonstrated inability to make distinctions between shapes – a necessary pre-condition for the identification of objects – there is no justification, in this particular case to speculate about a possible memory, naming or association defect.

Earlier in this paper, I pointed out that when Mr. S. is asked what causes his difficulty in recognizing objects, he adamantly maintains that he does not "see clearly". This complaint is made by many other patients with visual object agnosia reported in the medical literature – and with few exceptions it has been either ignored or dismissed because conventional tests of visual function are insufficient to account for the complaint. It has been interpreted as a rationalization used by the patient to relieve his fears of mental inadequacy. In view of the fact that Mr. S. has a demonstrated severe defect in visual perception, his complaint cannot be ignored or lightly dismissed as a psychological defense reaction. What does he mean by not seeing clearly? What might it be like to see unclearly in this way?

Now, as was previously pointed out in Section V, we have no direct *perceptual* awareness of the specificity of our awareness of an attribute of an existent. However, if as a result of past experience we would expect to be able to discriminate between two existents, and we discover that under the present circumstances we cannot make this distinction, we *conclude* that we are not now perceiving optimally. We are apt to describe this state as 'not seeing or hearing clearly'. It is important to grasp that we come to this conclusion by virtue of a retained (remembered) standard of the kinds of discriminations we have been able to make in the past. Entailed in the concept of 'seeing clearly' is to make as many discriminations as we are accustomed to under optimal conditions.

The only way in which a normal person could have any equivalent of Mr. S.'s perceptual condition would be to produce a state of low specificity of the attribute of shape, while retaining a normal specificity of awareness of the other attributes of visually discriminated existents. There is no known way to achieve all of these conditions experimentally. It is possible, however, to experience a state of low shape specificity by attempting to identify common objects by use of the peripheral retina. While color discriminations and visual acuity are also diminished in the peripheral retina, the low shape specificity which the normal person can experience in the periphery gives some insight into the reasons why Mr. S. may have used the word "unclear" to describe his visual difficulty. At 60–70 degrees peripheral to the fovea, the normal individual has about as much difficulty making discriminations of the shape of the objects in Figure 8 as Mr. S. has at his fovea [7], and he will, like Mr. S., complain that he cannot see clearly.

Just as the normal person will complain that objects in his peripheral vision appear 'unclear' (i.e., less specific) by virtue of an implicit comparison with the number of discriminations he is accustomed to make with foveal vision, Mr. S. complains that his foveal vision is "unclear" by virtue of an implicit comparison with the number of discriminations he was accustomed to make prior to his cerebral injury. If he had no retained standard of what he could expect of his visual perceptions, he would not be complaining of "unclear" vision.

VII. CONCLUSION

Mr. S. suffers from a severe restriction in his perceptual capacity to discriminate between shapes. His perceptual awareness of the attribute of shape is crude, undifferentiated or, as formulated in Section V, of low specificity. The case of Mr. S. illustrates, in a dramatic way, the fact that the specificity of perceptual awareness of one attribute of an existent may be low while the specificity of perceptual awareness of all other attributes may be normal or nearly normal. We must conclude from this that neurological disease may interfere with the perceptual processes in a highly selective fashion.

The fact that Mr. S. has a defect in shape perception does not permit the conclusion that *all* cases of visual object agnosia have the same psychological defect. There are a large number of *different* cognitive disturbances which are loosely referred to as visual object agnosia. For example, there is a very rare type of case in which common objects and people cannot be recognized but in which the recognition of letters, numbers and geometrical shapes is unimpaired. One author has even reported a case in which the patient was supposedly unable to recognize inanimate objects but could recognize animate ones. The similarities and differences between the case of Mr. S. and other cases of so-called visual agnosia will be discussed extensively in a forthcoming paper by D. F. Benson and J. Greenberg.

Klüver and Bucy [11] have described a syndrome of 'visual agnosia' in monkeys following bilateral removal of the temporal lobes. While the disturbances which they produced also include defects in tactile and auditory 'recognition', the visual behavior of these animals resembles certain aspects of human cases of 'visual agnosia'. Such monkeys do not appear to 'recognize' objects, faces (even simian ones), or food by vision. It is worth noting that there is some evidence that shape perception is disturbed: Klüver and Bucy found that with extensive training procedures involving hundreds of trials, these animals eventually could be taught to respond differentially to a circle and a square – a task which is accomplished by a normal monkey in a few trials.

The purpose of the present paper was not to maintain that all cases of visual object agnosia in man and in lower animals are due to a defect in shape perception. The case of Mr. S. was used merely to illustrate the fact (a) that by the prevailing methods of testing perception his perceptual defect would have remained undiscovered, (b) that, as a consequence of inadequate testing methods, we must seriously question previous claims that "perception was unimpaired" in all cases of specific cognitive disturbances due to brain disease not studied by adequate methods, and (c) that the inadequacy of the conventional testing methods derives, ultimately, from an inadequate conceptualization and definition of perception.

Observe that these conclusions are based upon a *psychological* analysis. That is to say, they pertain to the properties and characteristics of Mr. S.'s consciousness – his perceptual awareness of an aspect of existence. It is not generally appreciated (in fact it is often denied) that the psychological analysis of a cognitive disturbance due to brain disease must always *precede* the physiological analysis of that defect. Without such a psychological analysis, the physiologist attempting to explain the behavior of such a patient would not know which mechanism(s) to examine: He would not know where to begin because he would not know *what* he had to explain. If the psychological analysis is incorrect, the physiologist will study inappropriate neural mechanisms. For example, a psychological analysis which concludes that patients with visual agnosia have recognition defects will inevitably lead physiologists and anatomists to study those mechanisms and anatomical regions of the brain which are important for memory. On the other hand, a psychological analysis which shows that patients with visual agnosia have a defect in shape perception will lead physiologists to study those mechanisms which are responsible for the integration of contour information into the awareness of shape.

The failure to understand the nature of perception has led to a failure to develop a fully conceptualized methodology by which it is to be studied. The absence of an explicitly formulated, non-contradictory methodology for studying perception not only accounts for the wide variety of 'explanations' offered for such disturbances as visual object agnosia, but it also accounts for the fact that no 'explanation' once suggested has ever been abandoned.

The most fundamental purpose of this paper was to show that a precise answer to such an abstract and philosophical question as 'What is perception?' can lead to an explicit formulation of a valid methodology for its study. This methodology, when applied to an individual with visual object agnosia, has not only allowed us to test and to discard earlier explanations for this syndrome, but it will serve to guide the physiologist in his efforts to unravel the neural mechanisms which underlie normal perception. The phenomena of consciousness must be understood *before* one can hope to 'explain' them in terms of neural action. The attempt by many neurophysiologists to reverse this order – to study the neural mechanisms underlying perception *prior* to any adequate definition or conceptualization of perception – is doomed to failure.

ACKNOWLEDGEMENTS

The author is indebted to Dr. F. D. Benson of the Neurology Service,

Boston VA Hospital for referring this patient to me; to Dr. M. E. Wilson of the Neurophysiology-Biophysics Research Unit for performing the visual field studies (Figure 3); to Dr. E. Wolf, of the Retina Foundation, Boston, Mass., for performing the studies of flicker fusion (Figure 2) spatial summation (Figure 1) and sensitivity profile (Figure 4); to Miss B. Klein of Clark University for performing the experiments described in Figures 8 and 9; and to American Optical Co. for permitting the black and white reproduction of one of the H-R-R-Pseudoisochromatic test cards.

Neurophysiology-Biophysics Research Unit,
Veterans Administration Hospital,
Boston

BIBLIOGRAPHY

[1] J. M. Nielsen, *Agnosia, Apraxia, Aphasia,* Hafner Publishing Co., New York, 1962, p. 51.

[2] N. Geschwind, 'Disconnexion Syndromes in Animals and Man', *Brain* **88** (1965) 237–294, 585–644.

[3] L. J. Audus, 'The Mechanism of the Perception of Gravity by Plants', in *Biological Receptor Mechanisms,* Symposia of the Society for Experimental Biology, University Press, Cambridge, 1962.

[4] M. W. Nirenberg, in *Brain Function: RNA and Brain Function in Memory and Learning* (ed. by M. A. B. Brazier), University of California Press, Berkeley, 1964, p. 5.

[5] R. Efron, 'Biology without Consciousness – and its Consequences', *Persp. Biol. and Med.* **11** (1967) 9–36.

[6] R. Efron, 'The Duration of the Present', *Ann. N.Y. Acad. Sci.* **138** (1967) 713–729.

[7] M. E. Wilson, (a) *Spatial and Temporal Summation in Defective Fields of Vision,* Ph.D. Thesis, Univ. of London, (b) *Hue Similarity,* 1968. – (c) *Vision Research* (in press).

[8] H. Klüver, 'Visual Disturbances after Cerebral Lesions', *The Psychol. Bull.* **24** (1927) 316–358.

[9] N. K. Humphrey, and L. Weiskrantz, 'Vision in Monkeys after Removal of the Striate Cortex', *Nature* **215** (1967) 595–597.

[10] G. H. Le Hardy, G. Rand, and M. C. Rittler, Pseudoisochromatic Plates, Copyright by American Optical Co., 1957.

[11] H. Klüver, and P. C. Bucy, 'Preliminary Analysis of Functions of the Temporal Lobes in Monkeys', *Arch. Neurol. Psychiat.* **42** (1939) 979–998.

NOTES

1 The term 'discriminated' is recognized to be redundant in this definition. (How could

one be aware of an existent which was *not* discriminated?) It is used in this definition to focus attention on the phenomenon of isolation.

[2] We use the term 'possible' because the number of distinctions which may be actualized may be appreciably smaller than the number possible. The effect of experience and training (see Section III) will be to increase the number of distinctions actualized.

[3] This example is taken from an earlier paper by the author, [6].

[4] It is of interest to note that identical behavior has recently been described in a monkey whose striate cortex was reported to be completely excised [9].

[5] This strategy could give knowledge of the shape of an object only if there was also the capacity to retain and integrate the distances between and angular rotation of each scan across the transition zone.

[6] This statement is not to be construed as suggesting that there are no innate perceptual processes involved in the action of isolation. It is clear that some processes independent of experience must be operative or the new-born infant (or animal) would never have been able to isolate the *first* object and thus could never acquire any experience with objects.

[7] A central scotoma, reaching out to 60–70° renders a person functionally blind. Mr. S. does not resemble such patients because of his good foveal visual acuity and color vision, which he is able to exploit, at least to limited advantage. His capacity to discriminate other attributes (luminance, size, wavelength) is adequate for him to move his wheelchair around the hospital without running into people or objects.

HENRYK SKOLIMOWSKI

KNOWLEDGE, LANGUAGE, AND RATIONALITY*

Statement of the Problem

Epistemology is a perennial subject. Only when human knowledge ceases to grow can reflection on the nature of this knowledge come to an end. We must perpetually restate the problems of knowledge and constantly redefine the concept of knowledge in order to elucidate an everchanging knowledge-situation, and in order to draw conclusions from new departures in human knowledge. Changes in epistemology thus follow like a shadow (but not in any direct or functional way) changes in human knowledge. New and significant departures in human knowledge invariably cause a shift of the epistemological lenses. The development of semantics in the 20th century was an occasion for one of these shifts. The new epistemological vision resulting from this shift has been known as the 'semantic' concept of knowledge. This concept of knowledge attempts to establish the primacy of linguistics over epistemology, the primacy of the criteria of meaning over epistemological criteria; it claims that only by establishing clear criteria of meaning can we delineate the realm of significant knowledge.

Now, the first part of this essay, 'Knowledge and Language', is directed against the semantic concept of knowledge on the grounds that it is too narrow and in many respects mistaken. Indeed, in this essay I reverse the relationship between knowledge and language which is held by linguistic epistemologists. I argue for the primacy of epistemology over linguistics and attempt to demonstrate that the so-called pure linguistic criteria of meaning are loaded with epistemological assumptions, that the criteria of meaning follow from an implicitly assumed concept of knowledge, that the concept of language presupposes the concept of knowledge and is determined by it. It will thus become evident that there is a hidden circularity in the semantic concept of knowledge: it claims to elucidate the characteristics of knowledge in a novel way, whereas it itself is based on traditional and often obscure epistemological tenets. Perhaps the most representative example of the concept of knowledge as established by criteria of meaning is the concept developed by logical empiricism. For this reason a large part of my analysis is devoted to the theory of knowledge of the logical empiricists.

The second part of this essay, 'Knowledge and Rationality', is directed against concepts of knowledge which are too broad, and in particular against personal knowledge. Personal knowledge, and all other conceptions which tend towards the radical subjectivization of knowledge, are too broad because by unduly relaxing the criteria characterizing knowledge, they force us to accept as 'epistemological inquiries' all sorts of whimsical pursuits of unrestricted imagination.

In my examination of personal knowledge, rationality is at the heart of the discussion. The second part of this paper thus demonstrates that rationality is the touchstone of any coherent concept of knowledge. Furthermore, it demonstrates that rationality is presupposed by all explicit conceptions of *ir*rationality. Without rationality, the concept of irrationality cannot even be stated, let alone defended. The point is that once we decide to take part in a meaningful discourse, we decide thereby to uphold the principle of rationality; without it, even arguments against it are null and void.

Thus, the main two theses of my paper are:

(I) Epistemology is presupposed in all semantic systems, even in the semantic concept of knowledge.

(II) Rationality is presupposed in all cognitive systems, even in systems systematically defending irrationality.

I. KNOWLEDGE AND LANGUAGE

1. To begin with I shall briefly analyse prevailing attitudes to language. A clear grasp of these attitudes makes perspicuous the shortcomings of epistemological approaches based solely on the analysis of language.

I have argued elsewhere[1] that, in relation to language, philosophers can be divided into the four following groups:

(1) formists – those who radically mistrust ordinary language and consider it utterly unsuitable for the adequate formulation of ideas in science or philosophy, and because of this they *form* new artificial languages.

(2) reformists – those who slightly mistrust ordinary language and who insist that if ordinary language is reformed, it may serve as an adequate medium for the formulation and transmission of ideas.

(3) rectifiers – those who trust and obey ordinary language but not without qualifications; they think that although ordinary language works, with some tidying up it may still work better.

(4) codifiers – those who unconditionally trust and obey ordinary language and consider it the only adequate vehicle for the expression and transmission of ideas.

It would be misleading to consider the above categories as exhaustive or mutually exclusive. It may be that a more adequate division should include five types, with a group between reformists and rectifiers. On the other hand, it may very well be that one should distinguish nuances, that is subspecies, within particular groups. The main point of this classification is to escape from the drastically inadequate but so widely spread division into formalists and linguists. And also to emphasize that a philosopher may hold different attitudes to language at different times or with regard to different subject matters. It would be also positively misleading to contend that the attitudes analysed here are to be found only among philosophers interested in language *per se*, that is in analytical philosophers. Historians of philosophy, particularly the creative ones, while imaginatively reconstructing the doctrines of the past, are usually 'imaginative' in their use of philosophical language, that is they are by and large reformists, and sometimes rectifiers. It should be noted that creative scientists exhibit a similar attitude to language: by breaking away from existing usage or introducing new concepts they are constantly reshaping language, that is reforming language.[2]

My classification may be presented in a linear form.

formists	reformists	rectifiers	codifiers

Then the linear distance between particular groups indicates the degree of animosity or approval.

Now, the first two groups of my division are composed of those philosophers who consider ordinary language 'a swamp' and 'a mess', and because of this, they think it imperative to improve on it or to replace it by artificial languages. The latter two groups are composed of philosophers who think that ordinary language is basically all right, and that consequently it should be followed rather than violated.

The two extreme types, unconditional servants of ordinary language (codifiers) and unmitigated rebels against it (formists), began to be fully represented only in the twentieth century. Traditional philosophy is by and large composed of reformists and rectifiers.

We are told that we have the freedom to adopt any of these four attitudes. Indeed, every attitude is represented in contemporary philosophy. The choice of attitude seems to be a matter of preference or perhaps temperament. But is it really so? Are we in fact free to choose any of these attitudes? I submit that this is *not* so. It is not so if the object of our inquiry is the philosophy of science and particularly epistemological questions of the philosophy of science. I contend that it is highly undesirable to choose any of the two extreme attitudes if we wish to arrive at fruitful results in the philosophy of science or epistemology. I shall argue that the adherence to and consistent application of the codifier's attitude leads to a trivialization of the inquiry in the philosophy of science and epistemology; and I shall argue that the adherence to and consistent application of the formist attitude changes significantly the subject matter of our inquiry or else renders it vacuous.

It is perhaps trivial to observe that ordinary language contains in itself an epistemology. According to ordinary usage, the words 'chair', 'table', 'tree', and the like refer to or simply name concrete objects located in space-time, outside language and outside ourselves. Ordinary language thus presupposes the standpoint of uncritical realism in epistemology. Although this assertion may be trivial, it is nevertheless true. Now it is less trivial to observe that ordinary usage, if it is really *ordinary*, hardly allows for the formulation of such a highly abstruse idea as 'the *relation* between ordinary usage and epistemology'. If we question this uncritical realism or attempt to refine it, we must *ipso facto* depart from the canons of common usage. To talk about the *relation* between epistemology and common usage is to make the first breach.[3]

The shortcomings of ordinary language as the sole basis of inquiry are even more pronounced within the realm of the philosophy of science. If ordinary language is to be strictly observed, then the distinctions contained in it ought to be preserved and no new ones should be 'artificially' made. However, if we glance at the history of science, it is quite apparent that the progress of science almost invariably leads to the violation of or departures from ordinary language, either by introducing new categories or by changing the meaning of existing categories of ordinary language.

In the introduction to his *De Magnete*, William Gilbert wrote:

> We sometimes employ words new and unheard of, not ... in order to veil things with a pedantic terminology and to make them dark and obscure, but in order that hidden things which have no name and that have never come into notice, may be plainly and fully published.[4]

This is a classical statement and it has been amply illustrated by the succeeding centuries of the development of science. Newtonian physics would have been unthinkable if Newton did not endow with *new* meanings such concepts as: 'force', 'gravity', 'mass', etc. Lobatchevsky's geometry and all non-Euclidean geometries for that matter were brought forth partly through a radical violation of some concepts of ordinary language. It was clearly a linguistic outrage to *say* at that time that *infinitely* many lines can be drawn parallel to a given straight line through a given point. Without these violations and radical departures, the progress of science would not be possible. It thus follows that the acceptance and strict observance of the codifier's attitude leads to a barrenness or trivialization of epistemological inquiry. Now, if we agree that in science conceptual refinements, sharpening of meanings, stretching and adjusting of concepts *are* admissible, then we in fact agree that ordinary usage is not sacrosanct, that it can be improved upon. And then our attitude is not that of the codifiers – unconditional servants of ordinary language – but rather that of rectifiers or reformists.

The formist attitude – the other extreme position – seems to be equally unrewarding for epistemological inquiry. It is based on the assumption that since we successfully employ highly abstract and idealized concepts in science, we might be equally successful in all our cognitive endeavours if we introduce an idealized and abstract language, that is an artificial and formalized language. The concept of idealized languages is thus based on an analogy with idealized concepts of physics. There is now enough evidence to realize that this analogy unfortunately does not hold; at any rate, it does not hold in the way it was intended to. Idealized languages, conceived as super-laboratories for investigating reality, turned out to be so idealized that in order to make use of them, the subject of the inquiry, whether it be science or philosophy, had to be suitably idealized, that is appropriately tailored to fit them. (An example of this sort of procedure is the transformation of all philosophical questions into syntactical questions.) Consequently, what was being investigated through these languages was not living science as actually practised and as historically developed and not actual problems of philosophy, but some abstract constructs so arranged as to be easily manageable by means of suitable logical categories, but with little resemblance to the problems of science and philosophy.

However, not all idealized languages proved to be irrelevant to the actual inquiry. Some were found quite useful. But these were not strictly 'ideal', but fashioned on existing languages. They were found applicable and useful in so far as they were so to say 'contaminated' by the pattern of actual languages.

In a nutshell, if the formist persists in his approach, then he finds that his model-languages, although structurally and logically impeccable, are irrelevant and inapplicable to the actual problems; if these models are made useful, it is only because their ideal features are so modified that they resemble actual languages. In the latter case we cease to be pure formists, forming new artificial languages into which reality is to be fitted; instead we become *re*formists – reforming existing and actual languages, sometimes indeed significantly.

Our inability to devise an ideal language which would in advance prepare categories adequate for future inquiry reflects perhaps the very nature of human cognition. The conceptual refinement of language cannot be divorced from the actual process of cognition, cannot, in other words, be accomplished through the analysis of language alone.

To refine language in advance so as to make it a suitable and adequate medium for future inquiries would mean no less than to anticipate the process of discovery in science and the process of a new comprehension in philosophy.

Such anticipations would mean nothing less than making these new discoveries and gaining this new comprehension. Although logically this is not impossible – that is to say, to have conceived of new, powerful distinctions which explicate the future content of scientific discoveries and from which these discoveries clearly follow – empirically it has never, to my knowledge, been the case. That is, it has not happened yet that the purely linguistic activity of making subtle and numerous distinctions led to important discoveries in science or other realms of human thought. (If this were possible, poets, of course, would be best fitted for such an occupation and potentially would be the greatest scientists.) Newton's elucidations of the concepts 'force', 'gravity', and 'mass' are inseparable from the content of his physics. Elucidations of these concepts in this *particular* way are synonymous with bringing about the system of classical physics.

We have now very grievous misgivings about our understanding of space and time. If only we could elucidate these concepts 'in the right way', we might come to a new understanding of space and time, and indeed to new discoveries about space and time. Unfortunately, however, such discoveries are not made by wrestling with the meaning of concepts as an exercise in a linguistic game.

It should be obvious that the conceptual changes following important discoveries are not merely extrapolations or slight adjustments of the established usage; they are usually radical departures from this usage and sometimes they are indeed conceptual shocks. No one can say that their injection into the established usage with this usage blends harmoniously.

In conclusion, if it is true that necessary distinctions and elucidations of language are only possible at the time when new scientific discoveries are made, then no ideal language can be devised in advance for future inquiries.

To sum up, language is a flexible but at the same time a vulnerable medium and it requires a creative hand. No creator is worthy of the name if he leaves the material given to him exactly as he found it. And no creator destroys entirely the substance of his medium. The most fruitful approaches to language consist neither in leaving it exactly as it is nor in changing it beyond recognition, but in constantly reshaping it to serve our constantly changing cognitive and practical needs. Thus the most fruitful approaches are those of the reformists and rectifiers because through them we can capture and express the new insights about the world and ourselves. Intellectual activity of any kind justifies its existence by questioning the given and the established. Since language is a walking shadow of almost all of our intellectual pursuits, its given and established forms must be questioned and challenged along with our intellectual development.

2. I now come to the second issue which is the relation of the epistemological basis to its semantic superstructure. I have already mentioned that the criteria for so-called cognitive or meaningful statements are usually determined by the concept of knowledge implicitly assumed by a given semantic system. In this section I shall further analyse dependence of the theory of meaning on the theory of knowledge. By examining briefly three different concepts of knowledge in three contemporary philosophers: Carnap, Quine, and Popper, I will attempt to show in Section 3 that the attitude to meaning is a consequence of an epistemological commitment.

Any statement about knowledge is to some degree self-referential because it requires a concept of knowledge as its basis. Consequently an adequate analysis of knowledge is bound to include two processes: the external directed towards the world and resulting in what we call 'knowledge'; and the internal directed towards the process of analysis itself during which *a* concept of knowledge is implicitly assumed. We could (I am not quite certain whether we should) carry this process a bit further by attempting to analyse the concept of knowledge which is assumed in our internal analysis, that is when we analyse the process of analysis itself.

What do we mean by the process of analysis, that is the internal process? The fact that epistemological problems are formulated in a language, and more specifically related to a given semantic framework. If language used for the formulation and expression of our problems carries with itself an epistemological standpoint, if, in other words, our semantic basis is 'contaminated' by an epistemology, then the mere formulation of problems carries with itself some implicit solutions. I shall illustrate this point more amply while analysing the criteria of meaning.

There is no doubt that many contemporary epistemological disputes take the form of semantic disputes. Often they are torn out from the epistemological context to which they belong. I shall try to reconstruct this context and relate these disputes to traditional epistemological problems. This will enable us to trace the roots of contemporary controversies and to pin down some of the sources of confusion. I take it that the four basic questions of traditional epistemology are:

(1) How do we acquire knowledge?

(2) What kind of knowledge is most reliable or important?

(3) Which are the most effective ways to search for knowledge? And finally,

(4) What is knowledge?

Although the question 'What is knowledge?' is independent of the other three and perhaps logically prior to them, answers to this question have often been provided by the simultaneous intrusion of questions (2), or (1), or both. And this is true not only in relation to traditional philosophy. Even in contemporary philosophy the answer to the question 'What is knowledge?' is often derived, although not always consciously, from the analysis of other questions, such as 'What is the origin of knowledge?', or 'What kind of knowledge is most trustworthy or most reliable?'.

To return to the semantic form of contemporary epistemological discussions. Most frequently the foci of concern are the criteria of meaning. I shall begin with one of the classical criteria of meaning, the verifiability criterion of meaning which in its classical formulation states: only experientially verifiable statements are meaningful; all others are meaningless, save tautologies. This heroic formulation of the criterion of meaning did not last long. Why? Because it was found inadequate. Inadequate with regard to what? With regard to that knowledge which, although not verifiable empirically, was expressible by means of statements which could hardly be thought meaningless. Now, if criteria of meaning determine the cognitive character or meaningfulness of statements, then, if some statements do not meet these criteria, they should be rejected as meaningful or cognitive. If, however, certain statements which violate a criterion are nevertheless accepted and modify this criterion in turn, this means that there is something prior to this criterion, something on which it is based. I maintain that it is our concept of knowledge, or more precisely, our contention as to what knowledge is genuine and reliable that determines these criteria. In other words, these criteria are extracted from a concept of knowledge implicitly assumed and not the other way around.

There is a definite conception of knowledge at the bottom of the verifiability criterion of meaning. This conception is based on the assumption that all that occurs objectively can be described in science, that is in the physical sciences[5]; hence the contention that all genuine knowledge ought to be reduced to the physicalistic basis; and hence the idea of the unified science.

Another component of this conception of knowledge is the assumption that knowledge, which is genuine, must be certain. And consequently that all knowledge to be recognized as such must be reduced to a basis of certainty. Since that which is given to the senses appears to be the most certain knowledge, a language of sense data ought to serve as the basis for the certainty of knowledge.[6] To sum up, the concept of the *empirical statement* as expressed by the verifiability criterion of meaning is based on two premises of traditional epistemology. One is concerned with the origin of knowledge, that is with the question 'How do we acquire knowledge?' (The answer to this question is a classical one – the only certain knowledge is that directly acquired by the senses, the standard position of empiricism.) And the other is concerned with the validity and genuineness of knowledge, that is with the question 'What knowledge is reliable and trustworthy?' (The answer to this question is again traditional – empirical knowledge, that is knowledge of the natural sciences, is reliable and only that knowledge.) In the verifiability criterion of meaning, the answers to both these questions are lumped together.

An even more interesting fact is that they serve as a basis for determining another question, 'What is knowledge?' As we can see, although the debate was seemingly about semantics, that is about criteria for recognizing certain types of statements, in actual fact it was an epistemological debate *par excellence*.

In addition, this debate was to determine the scope and nature of science. So, in actual fact the verifiability criterion of meaning was to take care of yet another basic problem, 'How should we conduct our search for scientific knowledge?'. This clearly methodological question does not logically follow either from the empiricist programme of the reconstruction of all knowledge or from the assumed certainty of our sense experience. It turned out to be a misfortune for logical empiricism to attempt to derive a working methodology of science from the empiricist programme.

When too glaring discrepancies began to show between the actual search for scientific knowledge, that is between methodology as practised by scientists, and the methodology derived from the empiricist programme, this gave rise to an endless process of reformulations of the principle of meaning, of the empiricist programme, and of the concept of knowledge.

Some chose the thorny path and successively reformulated the principle of meaning. In the process the principle changed first its form, and then its content. Although on the surface it has been a process of evolution and conceptual refinement, the end-product is so significantly different from the original that one wonders whether it makes sense at all to talk about 'the evolution of the verifiability criterion of meaning'.

Some chose to give up the verification criterion of meaning altogether as unsound and instead introduced a probabilistic theory of knowledge paralleled by the probabilistic theory of meaning. According to this theory, a statement is accepted as meaningful if it is possible to determine its weight on the basis of given observations.

Some others, desiring obstinately to preserve the original concept of the verifiability criterion, went so far as to suggest that physical laws should not be regarded as general statements, but rather as rules for derivation of singular statements which was quite ingenious but rather startling, particularly for scientists.

The metamorphoses of the verifiability criterion and its successive embodiments illustrate admirably the dependence of the criterion of meaning on the concept of knowledge, and also illustrate the propensity for confusing different epistemological problems even by most rigorous thinkers. For Locke and other British empiricists, the problem of empiricism consisted in showing the *empirical origin* of all our knowledge. For twentieth century empiricists, the problem consists in the *logical reducibility* of all knowledge to empirical elements. For the British empiricists the immediacy of the given was the guarantee of the certainty of knowledge. For 20th-century empiricists the problem of certainty has become a *problem*.

It is rather extraordinary that 20th-century empiricists with their logical acumen and highly developed critical faculties did seriously consider sense data and the sense data language as the basis of certainty for science and all cognitive knowledge. The peculiarity of sense data seems to lie in the fact that they are *un*suitable as a basis for inter-subjective knowledge, scientific knowledge in particular. The apparent certainty of sense experience stems from its immediacy. To be immediately experienced, a given experience can be experienced only *once*. A statement which describes an immediate experience represents a subjective and unrepeatable event. The statement 'A rock has smashed the wind-shield of a car' is directly experiential for me if and when I am witnessing the event; it can be an experiential statement for a few other possible witnesses. It will not be directly experiential for me a day after the event, or for other people who were not witnesses. Consequently, knowledge based on direct experience is unique in the sense that a statement expressing it (a directly experiential statement) cannot be tested by other people; and furthermore cannot be tested by the repetition of the event which gave rise to the experience. Thus, knowledge obtained from immediate experience and expressed by experiential statements is neither repeatable nor intersubjective.

Now, if we agree that science and cognitive knowledge are to be characterized by their intersubjectivity and their testability, if in other words scientific statements are testable by other individuals, and if we further agree that this testing can be repeated many times over, then it is obvious that directly experiential statements do not satisfy the criteria characteristic for scientific statements, and consequently that the sense data language cannot serve as a basis for cognitive knowledge. Put differently, directly experiential statements as such are not scientific statements and consequently as such they do not belong to science.[7] However, experiential statements can *become* scientific statements. If they do, it is because they are validated intersubjectively; they cease then to be directly experiential, that is subjective and individual. Their scientific status is thus dependent on intersubjective validation and not on the immediacy of the given. Briefly, whenever statements attain scientific status, they are no longer directly experiential; whenever they are directly experiential, they are not yet scientific.

In view of the above, it is indeed strange that logical empiricists chose the sense data language as the basis for the certainty of knowledge. It remains strange that after they gave up the sense data language, they continually referred to the certainty of empirical knowledge as rooted in immediate experience.

The programme of the empiricist reconstruction of knc wledge is worth a candle only if there is a rock bottom of knowledge which is indubitable. Other kinds of knowledge acquire then their certainty and indeed their recognition in so far as they are firmly supported by this basis. If the basis of certainty is removed, if, in other words, the rock bottom is shattered or shown to be non-existent, then the indubitableness disappears; and then the empiricist reductions to the basis of certainty are misconceived; and then the criterion of meaning for cognitive knowledge as based on indubitable empirical elements is an arbitrary convention. And since the rock bottom of knowledge is admitted now to be non-existent even by empiricists themselves[8], empiricist reductions and the validity of the empiricist criteria of meaning loose their *raison d'être*.

3. Another serious dilemma of contemporary empiricism resulted from the confusion of the empiricist programme with a working methodology of science, that is lumping together the question concerning the reliability of empirical knowledge with the question 'How should we search for knowledge?'. If we are engaged in doing science, it is of great importance to know whether the theories we devise are but poetic fantasies or whether they can contribute to the growth of knowledge. To illustrate this I have devised the following theory of human behaviour. I claim that my behaviour and the behaviour of all other people is controlled by two kinds of invisible fluids. The first is the active fluid; the second is the passive fluid. The active fluid is in operation and in control of my behaviour whenever I stand, walk, or run. The passive fluid is in charge of my behaviour whenever I sit or lie. Now, I maintain that there is a constant struggle between these two fluids: whenever the first fluid, the active one, takes over, I must stand, walk or run; whenever the passive fluid defeats the active one, I must sit or lie. At this very moment while I am writing these words, I am sitting because my passive fluid is in control of my behaviour. If in a moment I suddenly get up, it will be because my active fluid has taken over. Every instance of movement or non-movement only confirms and supports my theory. Needless to say, I can claim that the theory has been confirmed on countless occasions by the past behaviour of mankind.

As a theory aspiring to be scientific, it is no doubt a ridiculous theory. From the standpoint of the empiricist criteria of meaning, however, my theory ought not to be rejected. It is not meaningless. In addition, as I have mentioned, I can claim it to have been confirmed innumerable times.

It is obvious that if a given theory of meaning obliges us to accept clearly ridiculous theories, it may be useful in many ways, but it is not useful as a part of the working methodology of science. A more general conclusion is this. The inquiry into epistemology and the philosophy of science cannot be reduced to the inquiry into semantics and the theory of meaning. There are many more problems in epistemology and the philosophy of science than semanticists can dream of. No system of semantics, regardless of how superior its structure, can guarantee the fruitfulness of our search for new knowledge.

To show the dependence of the theory of meaning on the concept of knowledge even more vividly, I shall very briefly consider a concept of knowledge (as represented by Quine) which assumes that there is no hard core of certain and irrevocable knowledge, but rather that human knowledge is centrifugal, that in other words it forms one tightly connected system of which some parts are more central whereas others are peripheral, but that nevertheless any part may be subject to revision. The peripheral parts are, in actual fact, the parts which are most accessible to experience. The system thus impinges on experience only at its edges.[9] If statements from the experiential periphery clash with reality, it does not necessarily mean that they have been invalidated; by making some thorough conceptual adjustments in more central parts of the system, we may reconcile those statements with reality. We thus possess a wide latitude of choice as to which statements of the system to preserve and which to revise.[10] In other words, statements describing our knowledge about the external world face the tribunal of sense experience not individually, but collectively, as a corporate body.[11] If this is the case, then criteria for distinguishing individual empirical statements are but arbitrary conventions. And consequently, the sharp dichotomy between the analytic and empirical statements (so much cherished by those who believe in the hard core of certain and unshakable knowledge) must be seen as artificial; such a dichotomy must be dissolved if we are to describe the system of knowledge adequately. To repeat, if statements are tied to the testimony of sense experience in a holistic way, which defies any statement-by-statement testing[12], then it is pointless to distinguish a special class of statements which are solely empirical and separate these statements from other statements. In short, blurring the boundaries between the analytic and the empirical is a necessary consequence of the centrifugal concept of knowledge.

It may be observed parenthetically that the tenacious defence of the analytic-empirical distinction by prominent representatives of logical empiricism is a reflection of their deeply rooted belief that there is a body of knowledge which is unshakable (and certain) and that this is empirical knowledge. I maintain that, in spite of their overt admission nowadays that there is no rock bottom of knowledge (originally assumed in the programme of the Vienna Circle)[13], logical empiricists have never recovered from their former spell. It is this belief, it seems to me, that motivates the staunch and unending defence of the analytic-empirical distinction.

Also parenthetically I should like to observe that the dispute over the analytic, when divorced entirely from its epistemological background, becomes curiously void and purposeless. Indeed, many shots in the controversy concerning the analytic were misfired because they were unrelated to any theory of knowledge.

Preoccupation with the theory of meaning and the criteria of meaning seem to be characteristic of those epistemologists and philosophers of science who cherish the idea of the empiricist edifice of knowledge and who assume that there must be a rock bottom of knowledge. Both these contentions are a legacy of epistemological inquiries of some 200 years before. It was no doubt another step towards the liberation of philosophy from its theological tethers to have observed that a great deal of our knowledge has its origin in our senses. And, in this context, an overall empiricist programme was not a whim of some ill-informed individuals, but an inevitable development. As to the other contention, it was also quite natural if not inevitable to have assumed after Newton that there is knowledge which is unshakable and certain. However, to subscribe to these tenets in the 20th century is less natural because in the meantime the sense data theory and the sense data language have not only been extensively employed but also extensively and effectively criticized; and because in the meantime there was Einstein.

Consequently, we may and perhaps we should start with different basic contentions about the nature of knowledge. If we begin with the premise (as Popper does) that there is no *one* source of knowledge, but that there are many diverse sources and that we can never trace *all* the sources, and that consequently the traditional problem of the origin of human knowledge is misconceived[14]; if we come to realize that science does not rest on any rock bottom but rather that it is like a building erected on piles which are submerged in a swamp[15]; and if we are concerned not so much with idealizing the features of human knowledge, but rather with the acquisition of new knowledge, then the *methods* of acquiring scientific knowledge become of prime importance, and then the problem of the 'foundations of knowledge' is the problem of driving our piles into deeper layers of the swamp so that they can carry the edifice of knowledge with our most recent additions and adjuncts. In this task the theory of meaning which is concerned with the logical reconstruction of knowledge will be of little use and therefore of little significance.

Now, among the three concepts of knowledge briefly examined here, the first is the concept determined by the pre-established criteria of meaning. Its advocates want to show that knowledge can be presented as a neat and logically tidy edifice with clear and sharply defined departments. But human knowledge, as contained in many sciences, has proved, and is constantly proving to be excessively resistant to all attempts at forcing it into neat logical boxes pre-arranged by excessively tidy-minded philosophers; and no amount of adjusting these boxes makes the results of this work satisfactory. The main issue in this approach to knowledge is to postulate what knowledge *ought to be* in order to satisfy our criteria. This is a *normative* approach to knowledge.

The second concept of knowledge does not start from any assumption concerning what knowledge *ought to be*, but rather attempts to grasp what it *is*. The centrifugal concept of human knowledge viewed from the standpoint of the theory of meaning leads to a dissolution of the sharp distinction between analytic and empirical statements. This approach to knowledge can be called *descriptive* as it attempts to grasp what knowledge is and then to acknowledge or not to acknowledge appropriate kinds of statements.

The third concept of knowledge is not concerned with knowledge as it *is* but rather with its *growth*. Knowledge is taken here to be a so to say living organism. A deeper understanding of what it is can only be accomplished by the study of its growth and development. The main issue in this approach to knowledge is to understand the nature of its growth and above all to devise methods for the further and more efficient acquisition of knowledge. From this vantage point, neither an epistemology which *describes* a static state of knowledge (as if it were a frozen piece), nor in particular a semantic theory which *postulates* criteria which knowledge ought to obey, are of much relevance.

It is thus clear that the importance or unimportance of the theory of meaning, and of the empirical criteria of meaning is a function of the concept of knowledge. We should, of course, distinguish here the genetic problem from the epistemological one. Genetically it is no doubt true that we must have had language 'of a sort' before we began to develop epistemologies. But this 'sort of' language was quite unlike the language we use in contemporary epistemological discourses. The epistemological side of the problem is this. The relation of language to epistemology is epistemological, not linguistic in character. We have to have a notion of epistemology in order to comprehend it. Epistemology is thus prior to linguistics because in considering their mutual relationship, language becomes an epistemic category.

When we consider the *growth* of knowledge, we may grant that human cognition is inseparably linked to language, that cognitive knowledge must be represented by cognitive statements. The criteria for these types of statements, however, cannot be established once and for all. In particular, they cannot be established by an *ad hoc* consideration of the structure of language. But, they are to be refined and re-defined according to the increasing depth of our comprehension and according to new insights into the nature of knowledge.

To summarize: reflection upon epistemology is epistemology, whereas reflection upon linguistics is not linguistics but again epistemology. Epistemology is thus prior to the theory of language because any conceptual analysis of the two must be performed within an epistemological framework. We say, in logic, that the propositional calculus is prior to the calculus of names because in order to develop the latter, some assumptions (even if they are tacit assumptions) of the former must be made. Such is the case in the relation of the theory of language to the theory of knowledge.

II. KNOWLEDGE AND RATIONALITY

1. In the second part of this essay I wish to examine some of the claims of the proponents of personal knowledge. Personal knowledge attempts to offer a more adequate description of the process of the acquisition of scientific knowledge. In particular, it accentuates the subjective and personal elements in the process of *knowing*. Since science is grounded in the subjective and private elements of human experience – the argument goes – no true description of science can safely ignore these personal and unique components of human knowledge. Thus, an adequate understanding of science, and all cognitive endeavours for that matter, must be ultimately grounded in the subjective sphere of human experience. This conclusion and the whole concept of personal knowledge is, in my opinion, ill-founded.

In one and a rather trivial sense of the word 'personal' all knowledge is necessarily personal knowledge. Our knowledge is human knowledge. This means that it is knowledge for you and for me. This means, in other words, that it would not have been knowledge if it were not knowledge for some persons. Or still differently, that knowledge not possessed by human beings is not knowledge. In short, if, by personal knowledge we mean knowledge which is apprehended by human beings, then personal knowledge is synonymous with human knowledge. In this sense, *im*-personal knowledge would be knowledge beyond the reach of human beings which clearly would not be human knowledge.

If, however, by personal knowledge we mean individual knowledge uniquely possessed by a particular individual, then there is no such knowledge. Uniquely individual knowledge would have to be the sole property of the individual and as such could not be shared. If it can be shared, then it ceases to be individual and unique. To share any piece of personal knowledge means to recognize criteria of 'impersonal sharing', which is another way of recognizing intersubjective criteria of cognitive knowledge. By spelling out the concept of personal knowledge, we annihilate its privacy; we make it public. The fact that we can argue about this species of knowledge simply means that it is not personal knowledge in the sense of being individual and unique.

The concept of personal knowledge seems to result from the confusion of knowledge with the act of knowing, that is the content of comprehension with the process of comprehending. Although the latter is an individual psychological act, it does not mean that the former (the content of our knowledge) is but a collection of subjective states of the mind. It is precisely because of this confusion between the act of our thinking and the result of our thoughts that nineteenth century psychologism could claim that all science in the final analysis is reducible to psychology. And it is precisely owing to Husserl's sharp distinction between the act of thinking (being the psychological and individual process) and the result of thinking (a set of statements in the logical sense) that psychologism was defeated as it seemed forever. Consequently, an attempt to formulate the concept of human knowledge as subjective and personal, because there are subjective elements in the process of *knowing*, is but an attempt towards a restitution of nineteenth century psychologism. It is but a denial of the indisputable fact that the results of our thinking can be expressed in statements possessing logical content and intersubjective meaning.

It is rather obvious that a rescue operation by the introduction of private languages, conceived as the media for expressing personal knowledge, will be of little help. How can these private languages be characterized and distinguished from public ones? Presumably by emphasizing their intentional departure from ordinary language. Would this not imply the use of the common intersubjective language as a point of departure and as a vehicle in which we formulate this departure? Hence it follows that characteristics and peculiarities of private languages are accounted for on a meta-language level in no other way but by assuming an intersubjective common language. In short, the concept of private languages is exposed to the same perils as the concept of uniquely individual knowledge.

It should be made emphatically clear that when we talk about the intersubjective character of knowledge, its testability and communicability, we do not intend to attribute to these terms superhuman meanings. The meanings of these terms were established by human beings and for human usage. The 'testability', 'communicability', and 'objectivity' of human knowledge does not mean more than it does. No doubt such terms as 'objective', 'communicable' and 'testable' have a margin of imprecision. If we talk about 'objectivity', we do not mean absolute objectivity because there is no such thing; if we talk about 'communicability' we again do not mean absolute communicability because such a thing does not exist. By attributing to terms like 'testability', 'objectivity' and 'communicability' an absolute, that is a super-human meaning, the advocates of personal knowledge postulate in fact some super-human concept of knowledge. But our concern is with *human* knowledge with all its limitations and imperfections, and not with absolute knowledge which, although perfect in every respect, has only one imperfection, namely that it does not exist. Consequently, attributes of human knowledge must be taken in their human, not in their absolute sense. Otherwise, we shall have ideal, but non-existent knowledge. Many criticisms directed against traditional concepts of knowledge lose their *raison d'être* when these simple truths are plainly realized. If, on the other hand, advocates of personal knowledge wish to emphasize that our knowledge is simply human knowledge and thus burdened with imperfections, then they are perfectly right, and there is no argument here. But then there is nothing novel about their thesis.

2. Further and more specific arguments against personal knowledge lead us to consider the issue: rationality *versus* irrationality. The concept of rationality has been and still is crucial for an objective, that is a non-personal concept of knowledge. It is little wonder, therefore, that it has been the centre of attacks by those who aim to establish a subjectivistic and relativistic concept of knowledge. Demands for recognition of the subjective, private and irrevocably unique elements in our concept of knowledge form together a demand for recognition of an irrationalistic basis for human knowledge. If any variety of irrationality is to be recognized as a basis for knowledge, rationality then must be sent to oblivion. Thus, rationality is, in a sense, the key element in the dispute concerning the objective *versus* the personal character of knowledge.

Aristotle once said: "Whatever is not possible, does not exist; whatever exists, is possible." So it is with rationality. If we do not confuse the justification of rationality with its possibility, then the possibility of rationality cannot be seriously doubted; this is a fact grounded in the existence of science and in the existence of all cognitive philosophies. In practising science, we constantly apply and thereby implicitly assert the principle of rationality. Science is a child prodigy, an epitome and at the same time 'a final justification' of rationality. Without the rationalist attitude, there would be no science. Although this attitude is hard to define precisely, some of its components can be easily discerned and these are: to divide, to analyse, to abstract. To divide reality into various compartments in order to give to phenomena natural as opposed to supernatural explanations. To analyse these subdivided compartments simply for the sake of knowing more about them; and to abstract from the whole in analysing a part of the whole. Although some further components could be distinguished within the rationalist attitude, I take it to be a *sui generis* attitude, not reducible to other phenomena[16] and one of the defining components of Western man.

Can rationality justify itself rationally? This question is supposed to be pernicious to the rationalist attitude and supposedly it 'justifies' irrationality as the end of our final commitment. However, irrationality, in order to pretend to be the final arbiter of our commitment, must be able to assert, without enjoying any support from rationality, what it is and on what grounds it is based. This modest task is quite beyond its ability. Why? Because irrationality, whatever its denomination, is an impoverished and distorted version of rationality; because it derives its sense and existence from rationality itself.

A fundamental mistake that advocates of rationality make is to take a defensive position and to assume that the assaults of irrationalists have genuine grounds, as if all questions were answerable questions, and as if all objections were valid objections. Now, can we not ask for the identity card of irrationality? More specifically, let us ask: 'What does it mean to be an irrationalist if we do not have the concept of rationality at the back of our minds? Is it not clear that without the aid of rationality, irrationality could not even be stated?' The point is that it is on the basis of and in reference to rationality that we conceive of and formulate the idea of irrationality. *Therefore, irrationality is not a denial but a perversion of rationality.* But perverse rationality is still rationality. A sexual pervert (*sit venia verbo*) does not become non-sexual through his perversion. Sexual perversion can only be explained when we possess the basic notion of sexuality. The variety of irrationalistic perversions can only be comprehended when we assume the concept of rationality.

To use another metaphor, let us take two concepts: beauty and non-beauty. Non-beauty does not automatically become ugliness *unless* we independently introduce the concept of ugliness and equate it with that of non-beauty. Let us imagine a beautiful woman, and let us agree which features of her body make her beautiful. Now, what can we say about a non-beautiful woman? That she lacks some or all the features of the beautiful one. We can perhaps imagine that her bust is not sufficiently developed or her hips are overdeveloped, that her forehead is too small or that her nose is too large. But this can only be done if we know what the right proportions of her hips to her bust are; if we know what the size of her forehead and her nose should be; in short, if we know exactly what she is like when she is beautiful. A 'non-beautiful woman' becomes a significant phrase only when we imagine a beautiful one and when the features which make her beautiful are nullified one by one. Thus, we can easily see that the concept of non-beauty is derivative and solely dependent on the concept of beauty.

To recapitulate, we cannot talk about the concept of non-beauty as meaningful unless we have planted firmly in our minds the concept of beauty or the concept of ugliness. The same is true of irrationality. We cannot talk about it meaningfully unless we have the concept of rationality in the background. It can be easily seen that we do not possess any concept which would correspond to irrationality in the sense in which ugliness corresponds to non-beauty. And it can be easily seen that rationality is not a concept which describes a shape of the body, but a concept which describes a state of mind. Consequently, since rationality cannot be visualized and presented pictorially, irrationality in turn cannot be conceived in terms of an image, that is cannot be presented as an image of rationality with some or most of the features of rationality eradicated.

Thus, the understanding and significant denial of rationality *ipso facto* leads to the recognition and acceptance of rationality. Beyond the scope of rationality, irrationality cannot be claimed to have any sense. Therefore, it is limited and twisted rationality, or it is nothing at all. Irrationality is but a rationalization of our intellectual pitfalls.

Because they do not perceive that irrationality has no identity in itself, rationalists have often experienced what is called the crisis of integrity. According to the opponents of rationality, rationalists do not live up to their own principles, particularly because the principle of rationality cannot be rationally justified. Consequently, the integrity of the rationalist is at stake. There is indeed a more or less permanent crisis of integrity within the rationalist tradition acknowledged even by rationalists themselves. Various remedies have been offered to cure this crisis. Some attempts towards resolving the crisis of integrity were made by giving to the rationalist a new, more coherent identity.[17] In my opinion the crisis of integrity in the rationalist tradition follows mainly from a failure to perceive that irrationality lacks identity. There is no crisis of integrity, if there is no identity in the first place. This is why so-called irrationalists have no problem with integrity.

It should be clear now that the main argument of irrationalists, who insist that in the final analysis the irrationalist can claim to have as much integrity as the rationalist, misses the most important point, namely that in order to talk about integrity, one must first possess identity. The loopholes in someone else's integrity do not constitute our own integrity; even less can they guarantee our own identity.

3. Believers in irrationality often find a refuge in an apparently innocuous concept of 'commitment'. They argue that even if it is granted that rationality is an attitude, a *sui generis* attitude, it is nevertheless a commitment; it is a commitment to an attitude whatever its characteristics. Then they go on to say that by opting for rationality, we are committing ourselves to a *value*. So, on a higher level, since it is a commitment to a value, it is an irrational commitment. This seemingly natural conclusion, as I will attempt to show, is based either on a perversion of language or on a devaluation of the idea of value.

First of all, what could it possibly mean to say that on 'a higher level' a commitment to rationality is an irrational commitment? Could it be that the rational attitude, when considered on a 'higher' level, changes itself into its opposite? If rationality is to mean anything, it means one and the same thing. One therefore wonders what the term 'on a higher level' means, and furthermore, how meanings can be compared on various levels? Clearly, if talk about various levels and particularly 'higher levels' (higher in relation to what?) serves to confuse the meaning of terms so as to turn them into their opposites, then such talk is 'nothing but sophistry and illusion' as David Hume would say; or, to put it in a modern idiom, it is but a perversion of language.

A commitment is not necessarily a blind leap of wild emotions which can be neither controlled nor guided. A decision to embark on a course of action, a line of pursuit or, indeed, a philosophy of life may be the result of a process of scrupulous reasoning and may be based on logical arguments. This simply means that some of our commitments are arrived at in a rational way and are therefore rational commitments. A commitment to rationality may be thus a rational commitment. Since irrationality does not exist in its own right, the only reasonable commitment we can make is inevitably to rationality.

In summary, the dilemmas of rationality are the dilemmas of our language, are the dilemmas of our discursive way of probing the world. Once we decide to take part in a human dialogue, we cannot help but lean heavily on the principle of rationality because it is built into our discursive language. Even if we wish to denounce rationality, we are bound to accept it; because without rationality the argument against it would be void and incoherent.

To conclude, in order to practise the philosophy of science in an illuminating manner, and in order to arrive at significant results in epistemology, we must be able to free ourselves from the restraining bounds of a too narrow concept of knowledge, such as logical empiricism without, at the same time, falling into the amorphous embrace of a too broad concept of knowledge such as personal knowledge. A most promising concept of knowledge, is the concept which regards knowledge in its growth and not the one which considers knowledge as something static and frozen. Because knowledge constantly grows, it is impossible to force its changing content into neat logical boxes as logicians have often attempted to do. On the other hand, this process of growth should not be considered as something unintelligible, as something entirely subjective and idiosyncratic. The systematic study of this process is a most important part of the philosophy of science.

REFERENCES

* I wish to express my gratitude to Professors Mario Bunge, Morris Engel, Imre Lakatos, Stephen Toulmin and J. O. Wisdom for reading the manuscript of this paper and for making many valuable suggestions.

[1] See Henryk Skolimowski, *Polish Analytical Philosophy*, Routledge & Kegan Paul, London, 1967, Chapter VI.

[2] No one perhaps showed this more clearly and forcefully than Imre Lakatos in his 'Proofs and Refutations', especially Part IV, *The British Journal for the Philosophy of Science* **14**, Nos. 53, 54, 55, 56 (1963–64); see also my *Polish Analytical Philosophy*, Chapter I.

[3] One may of course deny, as some Oxford philosophers would, that the adherence to the codifier's idiom commits us to naive realism. One may indeed deny this, but the fact *is* that the codifier's position *implies* naive realism none the less.

[4] William Gilbert, *De Magnete*, London 1600; English trans., New York 1958), p. L.

[5] As late as the 1960s, Carnap insisted that "all that occurs objectively can be described in science; on the one hand the temporal sequence of events is described in physics; and, on the other hand, the peculiarities of man's experiences with respect to time, including his different attitude towards past, present and future, can be described and (in principle) explained in psychology". Clinching the matter in the style so characteristic of his early Vienna days, he said, "Since science in principle can say all that can be said, there is no unanswerable question left" (*The Philosophy of Rudolf Carnap*, (The Library of Living Philosophers), ed. by P. Schilpp, Open Court Publishing Co., Chicago, Ill., 1965, pp. 37, 38).

[6] In his intellectual autobiography in the above-mentioned volume, Carnap wrote: "I believed that the task of philosophy consists in reducing all knowledge to a basis of certainty. Since the most certain knowledge is that of the immediately given, whereas knowledge of material things is derivative and less certain, it seemed that the philosopher must employ a language which uses sense-data as a basis." Then without any justification, Carnap immediately went on to say: "In the Vienna discussions my attitude changed gradually toward a preference for the physicalistic language" (*The Philosophy of Rudolf Carnap*, p. 50).

[7] My position here coincides with the position of K. Ajdukiewicz in his essay 'Subiektywność i niepowtarzalność metody bezpośredniego doświadczenia' [The Sub-

jectivity and Unrepeatability of the Method of Immediate Experience], in *Język i poznanie* (PWN, Warszawa 1966), Vol. II.

[8] Referring to the heyday of the Vienna Circle, Carnap said: "According to the original conception, the system of knowledge, although growing constantly more comprehensive, was regarded as a closed system in the following sense. We assumed that there was a certain rock bottom of knowledge, the knowledge of the immediately given, which was indubitable. Every other kind of knowledge was supposed to be firmly supported by this basis and therefore likewise decidable with certainty" (*The Philosophy of Rudolf Carnap*, p. 57). Then referring to the early 1950s and his discussions with Einstein who challenged the notion of an "absolutely certain basis of knowledge", Carnap in the same volume wrote: "I explained to Einstein that we had abandoned these earlier positivistic views, that we did no longer believe in a 'rockbottom basis of knowledge'; and I mentioned Neurath's simile that our task is to reconstruct the ship while it is floating on the ocean" (*op. cit.*, p. 38). See also note 13 below.

[9] W. V. Quine, *Methods of Logic*, Holt, Rinehart and Winston, New York, 1961, p. XII.

[10] *Ibid.*, p. XII.

[11] *Ibid.*, p. XII.

[12] W. V. Quine, 'Mr. Strawson on Logical Theory', *Mind*, October 1955, p. 434.

[13] There is little doubt that the idea of the 'rockbottom of knowledge' was abandoned by logical positivists at least partly as the result of Popper's forceful criticism launched in the early 1930's, especially in *Logik der Forschung*.

[14] Karl R. Popper, *Conjectures and Refutations*, Basic Books, London, 1962, especially Chapter I.

[15] Karl R. Popper, *The Logic of Scientific Discovery*, Hutchinson, London, 1959, especially Chapter V.

[16] And particularly not reducible to faith and passions as advocates of personal knowledge wish us to believe. One might ask in this context: if rationality is to be 'justified' and accounted for in terms of human passions and faith, how are these elements, in terms of which rationality is explained, to be justified themselves? One would probably be told that they are unanalyzable, that they indicate certain basic attitudes which explain other things but which require no further explication themselves. If so, the same is true of rationality; it is a distinctive way of viewing things, of grasping things, of discerning and separating them from other things.

[17] W. W. Bartley III, for example, in his essay 'Rationality Versus the Theory of Rationality' in *The Critical Approach to Science and Philosophy* (ed. by Mario Bunge) argues that the crisis of integrity in the rationalist tradition is mainly due to a neglected crisis of identity. By improving the rationalist identity, Bartley wishes to improve the rationalist's integrity. This new identity, according to Bartley, is "comprehensively critical rationalism" which stipulates that all beliefs of the rationalist, including his standards and his basic philosophical position, are left open to criticism. The focal point is on criticizibility of whatever view is entertained by the rationalist. Any position deemed rational must be open to criticism and consequently replaced by another position as the result of the criticism. Theoretically speaking, the rationalist may argue himself out of his own position.

STEPHEN TOULMIN

COMMENTS: LANGUAGE AND KNOWLEDGE

The invitation to comment on Henryk Skolimowski's paper this evening was particularly welcome. As early as 1959, Skolimowski came on a visit from Poland to England, and presented to us at Leeds some of the material recently published in his book on *Polish Analytical Philosophy*; and ever since then I have been following with admiration his pioneer work in such areas as the philosophy of technology – a subject that, among English-speaking philosophers, he still has just about to himself. It is an excellent thing for him to keep reminding us that the field of 'pragmatics' can be made the subject of serious and profound enquiries (what his teacher Kotarbinski calls 'praxiology'), and that we need not, like the logical empiricists, dismiss it as an intellectual slagheap or dump – without form and philosophically void.

Nor was it any surprise to find that, philosophically speaking, Skolimowski's heart is in the right place: that he insists on the *developing* character of science, and rejects any analysis which treats science (either in content or in method) "as if it were something static". For the great divide in 20th-century philosophy of science is that between the Platonisers who seek to find in science (or to impose on it) some *permanent* 'structure' or 'system', and those of us who see the crucial problem, instead, as being to discover the rationale of scientific *change*. On this issue, Skolimowski (as I understand him) stands squarely with the Aristoteleans – those philosophers whose concern is to understand (so to say) the morphogenesis of live sciences, rather than the skeletal anatomy of dead ones.

Over fundamentals, then, I believe Skolimowski and I are natural allies. But that is just about as far as our agreement will go this evening. In philosophy as in politics, it is always painful to watch your allies undermining their positions – and your own – by presenting arguments weaker than the occasion demands. That (in my view) is what Skolimowski has done in his present paper; and I shall criticise his arguments, less from any desire to demolish his basic positions, than in the hope of showing how they could be put on firmer foundations.

Boston Studies in the Philosophy of Science, IV.

Let me enter a preliminary warning. My criticisms may in fact be irrelevant to the thesis Skolimowski is really attempting to establish, for – after reading his paper carefully several times – I still cannot be sure what that thesis is. He describes it in several ways: to show "that the criteria of meaning follow from an implicitly assumed concept of knowledge", to establish "the primacy of epistemology over semantics" or "the logical priority of the concept of knowledge over the concept of language" – those are only a few of the ways he puts it. Yet, in this context, what do such terms as 'primacy' and 'logical priority' signify? We are not doing pure mathematics, where one might perhaps debate whether the concept 'rational number' was prior to the concept 'real number' or *vice versa*; nor are we doing physics, where it could no doubt be said that the concept 'light-ray' was logically prior to the concept 'refractive index'.

Possibly I am being very stupid, and Skolimowski has made these things perfectly clear to the rest of you; but I must just do my best with what I *do* understand. So I propose to comment – fragmentarily, and in succession – on three separate points in his paper, all of which he puts some weight on. These are (1) his analysis of the possible philosophical approaches to 'ordinary language'; (2) his bewilderment about the logical empiricists' reliance on 'sense-data' as an epistemological foundation for science; and (3) the relationship, within the growth of science, between the extension of knowledge and the development of language. The remarks I shall have to make about the last of these points, at any rate, will hint at the reasons why – if someone *compelled* me to answer the question, "Which is logically prior, the concept of knowledge or the concept of language?" – I should be just as unhappy about answering that question in Skolimowski's way as I would be about answering it in the way to which he objects.

First, then, the question of 'ordinary language': my comments about this may appear – at first hearing – to be picayune, but the additional distinctions I shall insist on bear directly on the central subjects of Skolimowski's paper. In this connection (you will recall) he classifies philosophers into 'formists', 'reformists', 'rectifiers' and 'codifiers', and concludes the first section of his paper with the assertion that "the most fruitful [philosophical] approaches [to ordinary language] are those of the reformists and rectifiers". Now: given once again a *forced choice* between the alternative approaches as Skolimowski defines them, his preference for the two less radical approaches is understandable. But we can be compelled to make this particular choice *only* if his categories are in fact – as he apparently intends them to be – collectively exhaustive. This I dispute. On the contrary (I shall argue) *none* of the attitudes, as defined in his paper, provides a proper account of the effect which novel scientific theories and discoveries have on our established linguistic usages.

Skolimowski cites Newton and Lobatchevsky to demonstrate that scientific advances frequently involve "radical violation of some concepts of ordinary language", and so to refute the 'codifiers' who (on his definition) "consider ordinary language the only adequate vehicle for the expression and transmission of ideas". On the contrary (he claims) these illustrations show that ordinary language must either be 'reformed', so as to "serve as an adequate medium for the formulation and transmission of ideas" or, at the very least, that it must be 'rectified' or 'tidied up' so that "it may work better". Yet his examples drive us to no such conclusion, since they are entirely compatible with a fifth philosophical approach to language – one for which I do not intend to invent a name, but which is clearly different from all of the four he discusses.

The crux of the matter lies in this question: "When we introduce a new linguistic usage for scientific purposes, in what respects – if any – does this innovation react back onto our established usages?" (Notice that I deliberately speak about 'established usages', rather than about 'ordinary language'. By this time, I suspect, the idea of 'ordinary language' has become something of a philosophical Aunt Sally – committing one to all sorts of horrors, such as "the standpoint of uncritical realism in epistemology" – and so represents a position which anyone can safely take pot shots at, because nobody ever adopted it, least of all the people who *speak* 'ordinarily'.)

As Skolimowski states his position in this paper, any extension of language in scientific contexts necessarily involves its reform or rectification in extra-scientific contexts. Yet this is surely incorrect. To use an example of Wittgenstein's: after studying the physics of electromagnetic radiation, and recognising the analogies between radiation within the visible range of wavelengths and radiation of other wavelengths, we may start referring to ultraviolet and infrared radiation using the phrases 'invisible light' and 'invisible colors'. These neologisms are natural, intelligible and legitimate. Yet what bearing need they have on our earlier, established *usages* – as distinct from our earlier beliefs? None at all: afterwards, day will still be light and night will still be dark, however much infrared radiation there may prove to be around at midnight. And, if you subsequently ask your tailor for some cloth of an 'invisible color', he will still – however much physics he knows – suspect you of talking nonsense. For the new usage involves not the *reform* or *rectification* of our previous usages, but merely an *extension* of them.

Linguistically, this extension is on a par with those involved in the Newtonian redefinition of the terms 'force' and 'mass', or the Lobatchevskian redefinition of geometrical terms – but the example has the merit of being less portentous, and so less mystifying. And one can well afford to underline the lesson it teaches us, even at the expense of appearing pedantic. Just because the advance of science creates occasions for *extending* language, it does not necessarily follow that philosophical defenders of previously established usages are mistaken in seeking to codify and preserve them. For the purposes underlying our established usages may well be different from those which give a meaning to the neologisms of science; and it may, on appropriate occasions, be highly important to defend the *minutiae* of established usage against well-meaning attempts to reform or rectify them in the name of scientific discovery.

There is no time now to document this point fully. But let me just refer in passing to the recurrent debate about the concepts of 'responsibility' and 'freewill'. We have all encountered those arguments in which new advances in physiology or psychology are cited as blanket reasons for thinking that the concepts of 'responsibility' and 'freewill' are obsolete. Faced with such arguments, we can do much worse than copy the late J. L. Austin, by drawing attention to the complex battery of distinctions crystallised in our existing usage (between actions performed willingly / wittingly / voluntarily / deliberately / on purpose / knowingly / designedly / premeditatedly / with forethought, and so on, as contrasted with those done by mistake / accidentally / unwillingly / unwittingly / carelessly / recklessly / unthinkingly / on impulse / under a misapprehension, and the rest) and by questioning whether anything in contemporary neuroscience or psychology reflects genuine doubt on the legitimacy and significance of these particular – and crucial – distinctions.

In short, we need not choose between being formists or reformists or rectifiers or codifiers *outright*. Rather, we are at liberty to be eclectic – to codify established usage, where our purposes justify codification; or to rectify it, in cases where some specific and demonstrable confusion of thought calls for some specific and demonstrable piece of 'tidying up' – and, very occasionally, we may even have to act as 'formists' or 'reformists' – though, as a matter of praxiology, something must have gone very wrong indeed before scientists can be justified in denouncing an established usage completely.

So let me counter Skolimowski's arguments about 'ordinary language' by claiming that he mis-states the problem: it is never a matter of rejecting or reforming or rectifying or codifying 'ordinary language', for 'ordinary language' never has been and never will be a homogeneous or clearly definable entity. And let me emend the first conclusion of his paper, which reads:

> The most fruitful approaches to language consist neither in leaving it exactly as it is nor in changing it beyond recognition, but in constantly reshaping it to serve our changing cognitive and practical needs.

Instead, I have argued, he should have said:

> The most fruitful approach to language consists in constantly *extending* it to serve new cognitive and practical needs, but without supposing that these extensions compel us to reshape pre-existing usages, except where such reshaping is specifically forced on us.

This emendation (as I said) may sound picayune, but its significance for Skolimowski's general position will become clear later.

Next, let me say something about Skolimowski's criticisms of the 'sense-datum' theorists: notably, the early logical empiricists who accepted the philosophical program of Ernst Mach and based their account of scientific evidence on the idea of *Empfindungen* or 'sensations'. Once again, let me declare that I reject their program as completely as Skolimowski does, and agree further that the plausibility of the 'Verification Principle' was closely tied up with the acceptance of Mach's epistemological ambitions. Yet, once again, Skolimowski's arguments against the sensationalist position are weaker than they need have been, and this weakness reacts back onto his more general position.

His key argument against the logical empiricists is that 'sensations' cannot serve as evidence on behalf of 'cognitive scientific claims'; and the reason he gives for saying this is the fact that direct experiential reports can be asserted only as an individual basis, while cognitive scientific claims are necessarily intersubjective:

> whenever statements attain scientific status, they are no longer directly experiential; whenever they are directly experiential, they are not yet scientific.

Skolimowski comments: "It is rather strange that logical empiricists were not aware of this fact." One can only reply: "It would indeed have been rather strange, if they had not been aware of it!" For of course they were; and of course any attack on their position based on the *evidential inadequacy* of sense-data can be countered. What Skolimowski calls "intersubjective validation" they will call "pooling the evidence of our individual sense-experiences"; and the intersubjectivity of scientific statements they will re-interpret by analysing any cognitive claim that *p* as a general propositional function of the form, "For any observer in a relevant situation, at any place and time, the evidence of his senses was, is or will be that *p*." And, if this leaves us dissatisfied, they will reply that, since individual sensation-reports provide the only data about which we can be *absolutely certain*, the complexity of the logical operations we must go through to get from these to intersubjective scientific claims about the physical world is neither here nor there. As in Pascal's argument about the doctrine of Salvation, the infinitude of the prize cancels out any finite doubts about the intervening argument.

No: if we are to justify dismissing the 'sensationalist' epistemology (as I, too, would certainly want to do) we must criticise it more radically. What we must demonstrate is, not the evidential weakness of sense-data, but rather the internal inconsistencies *within* the sensationalist program: notably, the contradictions involved in attempting to make the sensory observations on which we rely for evidential purposes serve, at the same time, as uninterpreted descriptions of the contents of our sensory fields. To the best of my knowledge and belief, there is only one really cogent set of arguments to this effect: namely, those employed by Kant against Hume, by J. L. Austin against A. J. Ayer, and by the later Wittgenstein against Moore, Russell, and the Vienna Circle.

In each case, the conclusion of these arguments is, not that sensations are *inadequate* witnesses, but rather that – of themselves – sensory inputs cannot meaningfully be said to 'testify' at all. Such idiomatic phrases as "the evidence of our senses" seem to suggest that our sense-fields can volunteer indubitable self-descriptive truths, in a self-explanatory language of their own. But if we consider the procedures by which – if at all – statements describing the contents of our sense-fields would have to be given a meaning, we shall see that this cannot be so. "Percepts without concepts are blind", as Kant put it; or, in Austin's words, "Sense-data are dumb." The 'evidence' of our senses comprises merely the answers we can get by looking, listening, sniffing, tasting, or feeling, to questions about the *world* that *we* have formulated in *our* language.

If this is right – if the sensationalist program can be undercut only by considering what kinds of *meaning* (language being what it is) can be given to statements about sense-data – this has important consequences for Skolimowski's central thesis. For, in that case, he cannot hope to subordinate the theory of meaning to the theory of knowledge as completely as he set out to do at the beginning of this paper. No doubt it would be very clean and tidy if we could just say, "Knowledge first, language afterwards"; but this would leave us precisely where Moore and Russell and the logical empiricists would have us stay – back in the philosophical world of David Hume – and it would mean ignoring Kant and all that came after him, including the most significant parts of the later Wittgenstein.

At the beginning of these comments, I questioned what it meant to talk of the concept of knowledge being "logically prior to" the concept of language, or *vice versa*. In doing so, perhaps I exaggerated my mystification; and, on second thoughts, I do have some idea what Henryk Skolimowski was getting at – at any rate, enough of an idea to *contradict* him. And curiously, I cannot help suspecting that, in contradicting what he says, I am very possibly asserting what he means. For the whole burden of my remarks is this: that the concept of knowledge is logically neither prior to, nor subordinate to, the concept of language. And this can be brought home by touching very briefly on the last of the points I raised – namely, the relationship within the growth of science between the extension of knowledge and the development of language.

The relevant trains of argument can be put, epigrammatically, as follows: (1) Discovery arrives at knowledge by way of insights. (2) Insights yield knowledge (in the required sense) only when what has been discovered can be stated. (3) We can state only what we have *words* to state. (4) So insights that require the reshaping of our concepts can yield knowledge *only* when we have developed the new linguistic means for their expression. (5) So, at the level of fundamental discoveries, the growth of knowledge and the development of language necessarily proceed hand-in-hand. (6) So the concept of knowledge and the concept of language are *correlative* concepts – neither being logically prior to the other.

Let me end with one example from the history of science, which illustrates the part played by language in the creation of new scientific knowledge. Lavoisier's work on the theory of the chemical elements is sometimes expounded by emphasising, above all, the chemical experiments he did – e.g. the well-known experiment on the calcification and reduction of mercury. There are two objections to this procedure. Firstly, in point of logic, there is no way of building a coherent argument by which the unvarnished evidence of Lavoisier's experimental observations can be made to validate the fundamental positions of his theory; and secondly, in point of history, Lavoisier's own argument proceeded quite otherwise. Having embarked with Berthollet, Fourcroy and Guyton de Morveau on the task of producing a new system of chemical *nomenclature*, he found that he could deepen and refine the terminology of chemistry only as an integral part of the wider task, of deepening and refining also his understanding of chemical phenomena. For Lavoisier began as a follower of that 'linguistic philosopher' the Abbé de Condillac; and he continued to stand by Condillac's maxims – such as, that "Languages are true analytical methods", and that "The art of reasoning is nothing more than a language well arranged".

The position which I have advanced in opposition to Skolimowski's is, in fact, put very clearly in the preface to Lavoisier's *Traité Elémentaire de Chimie*:

As ideas are preserved and communicated by means of words, it necessarily follows that we cannot improve the language of any science without at the same time improving the science itself; neither can we, on the other hand, improve a science, without improving the language or nomenclature which belongs to it. However certain the facts of any science may be, and, however just the ideas we may have formed of these facts, we can only communicate false impressions to others, while we want words by which these may be properly expressed.

Knowledge (to paraphrase) comprises insights expressed in meaningful language. So the theory of knowledge, the theory of meaning and the theory of language are not three theories but one theory. And, though Condillac treats the relation of science to its language as one of *homoousia* where I should prefer a doctrine of *homoeousia*, a Trinitarian position of one kind or another seems to me inescapable.

BOGUSLAW WOLNIEWICZ

A PARALLELISM BETWEEN WITTGENSTEINIAN AND ARISTOTELIAN ONTOLOGIES

I

Different ontologies adopt different notions of existence as basic. Aristotle's paradigm of existence is given by the equivalence:

(A) to be = to be a substance.

On the other hand, the paradigm of existence adopted in Wittgenstein's *Tractatus* is given by the parallel equivalence:

(W) to be = to be a fact.

Now, an Aristotelian substance is the denotation of an individual name, whereas a Wittgensteinian fact is the denotation of a true proposition. It seems therefore that the notions of existence derived from these two paradigms should be quite different, and one might readily expect that the metaphysical systems erected upon them will display wide structural discrepancies.

It turns out, however, that in spite of this basic difference there runs between these two systems a deep and striking parallelism. This parallelism is so close indeed that it makes possible the construction of a vocabulary which would transform characteristic propositions of Wittgenstein's ontology into Aristotelian ones, and conversely. To show in some detail the workings of that transformation will be the subject of this paper.

The vocabulary mentioned is based on the following four fundamental correlations:

	Aristotle	*Wittgenstein*
(1)	primary substances (*substantiae primae*)	atomic facts
(2)	prime matter (*materia prima*)	objects
(3)	form (*forma*)	configuration
(4)	self-subsistence of primary substances (*esse per se*)	independence of atomic facts

Boston Studies in the Philosophy of Science, IV.

Aristotle's ontology is an ontology of substances, Wittgenstein's ontology is an ontology of facts. But concerning the respective items of each of the pairs (1)–(4) both ontologies lay down conditions which in view of our vocabulary appear to be identical. To show this let us confront, to begin with, the items of pair (1): substances and facts.[1]

(The interpretation of Aristotle adopted in this paper is the standard one, to be found in any competent textbook of the history of philosophy. Therefore, with but one exception, no references to Aristotle's works will be given here.)

II

Relatively to the system involved substances and facts are of the same ontological status. Aristotle's world is the totality of substances (*summa rerum*), Wittgenstein's world is the totality of facts (*die Gesamtheit der Tatsachen*). For Aristotle whatever exists in the basic sense of the word is a primary substance, for Wittgenstein – an atomic fact. Moreover, both ontologies are MODAL ones, allowing for different modes of being (*modi essendi*); and both take as basic the notion of 'contingent being' (*esse contingens*), opposed to necessary being on the one hand, and to the possibility of being on the other. Both substances and facts are entities which actually exist, but might have not existed. The equality of ontological status between substances and facts is corroborated by the circumstance that both are PARTICULARS, there being – as the saying goes – no multiplicity of entities which FALL UNDER them.

Substances and facts stand also in the same relation to the ontological categories of pairs (2) and (3). Both are always COMPOUND entities, a substance consisting of matter and form, and a fact consisting of objects and the way of their configuration. But in neither of the two systems is this compoundness to be understood literally as composition of physically separable parts or pieces. The compoundness (*compositio*) of a substance consists in its being formed stuff (*materia informata*), and the compoundness of a fact in its being a configuration of objects.

In view of correlation (4) we have also an equality of relation which a substance bears to other substances, and a fact to other facts. Self-subsistence is the characteristic attribute of primary substances: *substantia prima* = *ens per se*. If we take this to mean that each substance exists independently of the existence or non-existence of any other substance we get immediately the exact counterpart of Wittgenstein's principle of logical atomism stating the mutual independence of atomic facts. It should be noted that thus understood the attribute of self-subsistence or independence is a relative one, belonging to a substance – or to a fact – only in virtue of its relation to other substances – or facts.

From a Wittgensteinian point of view Aristotle's substances are not things, but hypostases of facts, and thus their names are not logically proper names, but name-like equivalents of propositions. (By that term we mean roughly either a noun clause of the form 'that *p*', or any symbol which might be regarded as a definitional abbreviation of such clause.) Surely, from the Aristotelian point of view it might be easily retorted here that just the opposite is the case: substances are not 'reified' facts, but on the contrary – facts are 'dereified' substances. Without passing judgement on these mutual objections let us note in passing that their symmetric character seems to be itself an additional manifestation of the parallelism discussed.

III

The Wittgensteinian counterpart of Aristotelian names denoting primary substances are true elementary propositions stating atomic facts. Let '$F(M)$' be a symbol capable of both interpretations. On the Aristotelian account it will be the name of some definite primary substance S, the name-forming functor '$F(\ldots)$' representing here the form of S, and the constant 'M' representing its matter – that parcel of matter (*materia quantitate signata*) which shaped into form F yields the substance S. Thus we have the definition:

(A) $S =_{\mathrm{df}} F(M)$.

On the Wittgensteinian view, however, the symbol '$F(M)$' is not a name, but an elementary proposition stating the fact that some parcel of matter is shaped in some definite way, or – to put the same alternatively – that some objects stand in some definite configuration. The constant 'M' represents now the set of all the objects involved in the fact S, and the sentence-forming functor '$F(\ldots)$' represents the way they are configurated. Thus we have the definition:

(W) $S =_{\mathrm{df}}$ that $F(M)$,

which shows the symbol 'S' to be a name-like equivalent of a proposition abbreviated into a single noun.

For the sake of illustration let us suppose that the matter of S consists of just three material points a, b, and c, and that the form of S is the shape

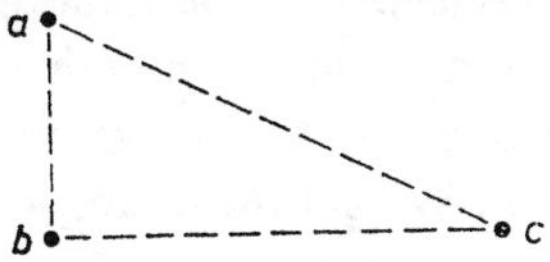

Fig. 1.

of a right-angled triangle like Figure 1. In this particular example we have:

$$M = \{a, b, c\}.$$

'$F(\ldots)$' = 'Objects ... form a right-angled triangle'. The Aristotelian reading of the symbol '$F(M)$' is the phrase 'this right-angled triangle', which might be regarded as an individual name of the substance S. But on the Wittgensteinian view that phrase is really a name-like equivalent of the proposition '$F(a, b, c)$'. For we must not write:

$$S =_{\mathrm{df}} F(a, b, c),$$

which is obviously a syntactically ill-formed formula with the identity sign having a name to its left and a sentence to its right. Instead we have to write:

$$S =_{\mathrm{df}} \text{that } F(a, b, c),$$

which shows the phrase 'this right-angled triangle' to be a disguised STATEMENT to the effect that the objects a, b, and c – referred to by gesture or by their proper names – form in fact the configuration of a right-angled triangle. Certainly, on both accounts they might have formed a quite different one: Aristotle's matter is capable of taking various forms, and Wittgenstein's objects are capable of entering various configurations.

IV

Let us consider next an item of Aristotle's ontology whose Wittgensteinian counterpart is not so easily seen. In Aristotle's system there are in addition to the primary substances also the so-called 'secondary substances' (*substantiae secundae*). Their relation to the primary ones has been the subject of much dispute, but for our limited purpose it seems sufficiently clear. Secondary substances are universals, and they are somehow derivative with respect to the primary ones. We are told, moreover, that the latter FALL UNDER the former. Thus a secondary substance is the correlate of a GENERAL concept, or the denotation of a general name. It corresponds to a Platonic 'idea', though differing from it by not existing on its own account, but only as inherent in some primary substance (*universale in re*). We get the concept of a secondary substance by ABSTRACTION, that is by detaching in thought some components of the form of a primary substance.

To correlate this distinction to a distinction within Wittgenstein's system let us start with their conceptual or linguistic counterparts. The linguistic counterpart of a primary substance is an individual name (like 'Socrates' or 'this man'), that of a secondary substance a general one (like 'man' or 'a man'). Wittgenstein's counterpart of an Aristotelian individual name is a true elementary proposition. Therefore his counterpart of an Aristotelian general name should be simply a propositional FUNCTION derived from some elementary proposition '*s*' by changing some component of it into a VARIABLE. If the phrase '*this* right-angle triangle' is the individual name of a primary substance, then the phrase '*a* right-angled triangle' will be the general name of a secondary substance. And likewise, if '$F(M)$' is an elementary proposition, then by substituting a variable 'X' for the constant 'M' we get the propositional function '$F(X)$' whose denotation is the exact double of an Aristotelian secondary substance.

Going back to our example, we transform the proposition '$F(a, b, c)$' into the propositional function '$F(x, y, z)$' representing what is common to all right-angled triangles determined by any three material points x, y, z. To Aristotle's operation of abstraction there corresponds in Wittgenstein's system the operation of substituting a variable for a constant. And since the function '$F(x, y, z)$' is satisfied by any three objects standing in the relevant configuration we may say as well that there are many atomic facts FALLING UNDER this function.

V

It is not so easy to find an Aristotelian counterpart of the distinction drawn in Wittgenstein's system between ATOMIC and MOLECULAR facts.

A primary substance being the counterpart of an atomic fact, what then is the Aristotelian counterpart of a molecular fact? This is a difficult question, and we must confine ourselves to a hint only. To a Wittgensteinian molecular fact there might correspond in Aristotle's system a NON-SUBSTANTIAL WHOLE – that is, a whole which is not a substance (*unum per se*), but an aggregation of substances (*unum per accidens*). Thus a sheep is a substance, but a flock of sheep is not.

The essential point of this comparison is that two substances taken together are not another substance, and likewise two atomic facts taken together are not another atomic fact. But it is obvious that this idea would work at best for conjunctive composition of facts only.

VI

Let us have now another look at pair (3). To Aristotle's form there corresponds in Wittgenstein's system the way objects are actually configurated in some fact. This is called by Wittgenstein the 'structure' of the fact (2.032). Now, different substances may have the same form, and likewise different facts may have the same structure. Hence both Aristotle's form and Wittgenstein's structure are UNIVERSAL, in contradistinction to the particularity of substances and facts.

Moreover, in Aristotle's world all change consists in the change of form only. But the same holds of Wittgenstein's world where all change is change of configuration.

Even more striking is the parallelism between Aristotle's form and Wittgenstein's structure if we look at it from the epistemological point of view. Substances are perceptible entities, things given in experience. But a substance is knowable only in virtue of its form, and the acquisition of empirical knowledge consists in the senses taking over the form of substances without taking over their matter. This is, to be sure, the point of Aristotle's parable of the seal and the wax, the seal being the simile of a perceived substance, and the wax that of the perceiving mind. The wax takes over only the shape of the seal, not its metallic stuff, and thus the same form becomes actual in two quite different materials.

According to Wittgenstein what is given in experience are facts (5.552).[2] His counterpart of the Aristotelian mind is SYMBOLISM, and that of empirical knowledge – the description of facts by meaningful propositions. Description of a fact consists in another fact – the propositional sign – reflecting by its own structure the structure of the fact described. And both structures have to be identical.

VII

Aristotle's prime matter is the persistent substratum of phenomena. Taking various shapes it gives rise to the transitory substances as their 'material cause', but itself it stays forever unchanged. Wittgenstein's objects constitute what he calls 'the substance of the world' (2.021). This substance is also unaffected by the changing configurations of the objects composing it (2.024).

Aristotle's prime matter is absolutely shapeless and indeterminate. If it were not formed into substances it would have no definite properties at all. The same is true of Wittgenstein's objects (2.0231, 2.0232). They acquire what he calls 'material properties' only by being constituents of some facts, that is by actually entering some configuration with other objects.

Wittgenstein's objects cannot be described, only named, for all description must be already description of some fact, not of a single object. This indescribability of Wittgenstein's objects squares well with the unintelligibility of Aristotle's prime matter. And both are equally enigmatic from the commonsense point of view.

Most important here is the circumstance that Aristotle's prime matter is the POSSIBILITY of form. Wittgenstein's objects are likewise the possibility of configuration (2.012, 2.0123, 2.0124). Moreover, Aristotle seems finally to arrive at the conclusion that matter and form are just two aspects of one and the same reality, matter being the possibility of form, and form being the actuality of matter (*Metaphysica*, 1045^{b}). And this might be true of Wittgenstein too in view of his startling statement that the substance of the world is 'form and content' (2.025). It goes without saying that the 'form' Wittgenstein is speaking about is not to be confused with that of Aristotle.

VIII

So far we have passed over in silence the fundamental distinction drawn by Wittgenstein between a 'fact' (*Tatsache*) and a 'state of affairs'

(*Sachverhalt*). This was so because we wanted to keep as close as possible to the original variant of Aristotelian ontology which has no clear-cut counterpart of that distinction. Turning now to the variety of Aristotelianism represented by Thomism we find at once what is needed. In Thomistic philosophy there is the well-known distinction between the 'essence' of a thing (*essentia*) and its 'existence' or 'being' (*esse*). To correlate this distinction to that of Wittgenstein we have only to split up both items of pair (1) into three components:

	Aquinas	*Wittgenstein*
(1.1)	a being (*ens*)	a fact (*Tatsache*)
(1.2)	essence (*essentia*)	state of affairs (*Sachverhalt*)
(1.3)	to be (*esse*)	to be (*bestehen*).

A Thomistic substance is a DOUBLY compound entity.[8] As consisting only of matter and form it is merely a POSSIBLE entity, an 'essence' capable of existence, but possibly not existing at all. If it actually exists there is added to it a new factor in virtue of which it is a 'being' (*ens*), an existing thing. This new factor is called in Thomistic terminology 'the act of being' (*esse*), and the relation involved here is sometimes expressed by the quasi-equation:

(T) *ens* = *essentia plus esse.*

If one assumes, as we do, that a Wittgensteinian *Sachverhalt* is, like a Thomistic essence, only a possibility of existence, not necessarily something 'actual' or 'factual', then the parallelism of both distinctions becomes evident immediately. A Thomistic essence is a possible composition of form and matter, and a Wittgensteinian *Sachverhalt* is a possible configuration of objects. And the relation between a fact and a *Sachverhalt* might be too expressed by a strictly analogous quasi-equation:

(W) a fact = the existence of a *Sachverhalt.*

IX

In view of the thoroughgoing parallelism between the two ontologies discussed the question arises of what might be the Wittgensteinian counterpart of Aristotle's distinction between ESSENTIAL and ACCIDENTAL predicates of substances. To answer this question we take as our point of departure *Tractatus* 2.0201: "Every statement about complexes can be analysed into a statement about their constituent parts, and into those propositions which completely describe the complexes."

Let us go back again to our triangle example and consider the situation given in Figure 2. Let the triangular configuration of the three material

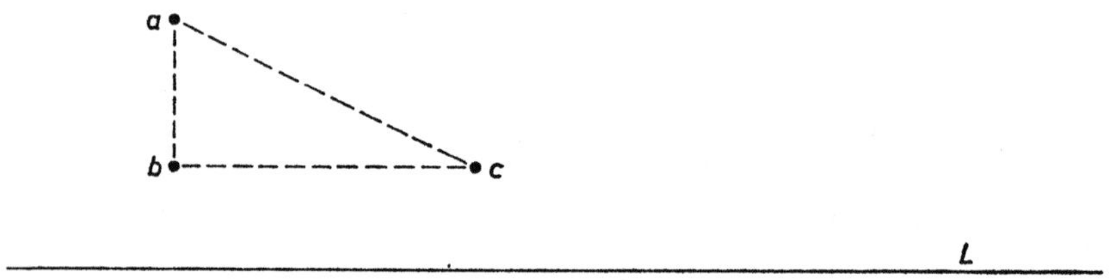

Fig. 2.

points *a*, *b*, and *c* be a COMPLEX in Wittgenstein's sense. Then '*S*' is the individual name of that complex, and since every complex is a fact this name can be defined as before:

$$S =_{df} \text{the fact that } F(a, b, c).$$

Let the statement about this complex be the predicate '... is above line *L*', or '$L(...)$' for short. According to 2.0201 the proposition '$L(S)$', being a statement about a complex, can be analysed in the following way[4]:

$$L(S) \equiv L(a) \wedge L(b) \wedge L(c) \wedge F(a, b, c).$$

By assumption the complex *S* is an Aristotelian primary substance. (This assumption is not so unrealistic as it looks, for on the one hand Socrates is certainly an Aristotelian primary substance, and on the other it might be plausibly argued that Socrates is just a relatively stable configuration of atoms.) Therefore '$L(...)$' is an accidental predicate of *S*, for the conjunction:

$$L(a) \wedge L(b) \wedge L(c)$$

might be false – *S* could be below *L*, or bisected by *L*, etc. – without *S* ceasing to be what it is, namely the triangular configuration of the objects *a*, *b*, and *c*. But the predicate '*F*(...)' is an essential one, for the proposition:

$$F(a, b, c)$$

cannot be false without *S* loosing its self-identity. And this is just what Aristotle wants an essential predicate to be.

University of Warsaw

REFERENCES

[1] Correlation (2) has been first pointed out by Professor Irving M. Copi in his memorable paper 'Objects, Properties, and Relations in the *Tractatus*', *Mind* **67** (1958) 145–165. In fact Copi's remark that "Wittgenstein's objects are substantial in the later sense of substrata, and correspond more closely to Aristotle's prime matter than to his primary substances" (p. 164) was the starting-point in developing the parallelism presented here.

This writer is happy to acknowledge also his indebtedness to the writings of Professor Gustav Bergmann and his school, especially to the following papers: G. Bergmann, 'Strawson's Ontology', *Journal of Philosophy* **57** (1960) 601–622 (wherefrom the terms 'ontology of facts' and 'ontology of substances' have been taken over); E. B. Allaire, 'Existence, Independence, and Universals', and H. Hochberg, 'Elementarism, Independence, and Ontology' (both published in E. B. Allaire *et al.*, *Essays in Ontology*, Iowa-City – The Hague 1963).

[2] Cf. G. E. M. Anscombe, 'An Introduction to Wittgenstein's *Tractatus*', London 1959, p. 165.

[3] Cf. Aquinas *Summa contra Gentiles*, II, 54: "In substantiis autem compositis ex materia et forma est duplex compositio actus et potentiae: prima quidem ipsius substantiae quae componitur ex materia et forma; secunda vero ex ipsa substantia iam composita et esse."

[4] It needs stressing that there are great difficulties concerning the proper interpretation of 2.0201, especially with regard to the question why is 'statement' (*Aussage*) being used there in the singular, and 'propositions' (*Sätze*) in the plural. Moreover, in the present context it should be stipulated that *S* is to be an ELEMENTARY complex, that is, that '*F*(*a*, *b*, *c*)' is to be an elementary proposition.

HENRY RUF

WOLNIEWICZ ON WITTGENSTEIN AND ARISTOTLE

In *Process and Reality*, Alfred North Whitehead characterized the European philosophical tradition as consisting in a series of footnotes to Plato. In the *Philosophical Investigations*, Wittgenstein declares that the concept of an object, which he used in the *Tractatus*, is modeled on the concept of a primary element, which Plato used in the *Theaetetus*, where Plato argued that there must be primary elements which are simple and therefore nondescribable but only nameable because the essence of speech is the composition of names; names become descriptive language by being compounded together. Professor Wolniewicz has argued today that Wittgenstein's footnote to Plato is really an Aristotelian footnote.

In response to Professor Wolniewicz's claim, I will argue that the parallelism which he finds existing between Aristotelian metaphysics and the ontology of the *Tractatus* holds only given a certain interpretation of the *Tractatus*, an interpretation which stands in conflict with major themes of the *Tractatus*, an interpretation which attributes to Wittgenstein a position which is basically incoherent and which, therefore, cannot even be offered as a statement of what Wittgenstein should have said. In presenting this argument, I am not going to claim that the *Tractatus* does present a totally consistent or coherent theory, but that the difficulties, which do exist in Wittgenstein's conclusions, exist there because of the arguments he used and which he believed compelled him to hold the position he did. Comparing one philosopher's position to that of another can be helpful in illuminating the position of the less well understood philosopher (although I am not at all sure that the concepts of prime matter or abstraction are well understood at all), but such a comparison is of little value, if one overlooks the difficulties that exist within one of the positions, and, doubly so, if the conclusions which one does select out for comparison themselves form an incoherent theory. Professor Wolniewicz is concerned with some of Wittgenstein's conclusions. When Wittgenstein's arguments are introduced again for consideration, I suggest that we will discover in the *Tractatus* many Platonic features which are inconsistent with the Aristotelian interpretation which Professor Wolniewicz has given.

Professor Wolniewicz quite correctly characterizes the ontology of the *Tractatus* as being an ontology of contingent facts, each as conceptually atomic as are Aristotle's primary substances. Since, however, atomic facts are configurations of objects, the crucial issues will turn on what we take objects to be and what we take configuring to be. In drawing out the supposed parallelism between Wittgenstein's and Aristotle's metaphysics, he compares objects to prime matter and configurations of objects to form and thus atomic facts to primary substances. Simply in setting up the comparison this way, Professor Wolniewicz shows a major difference between the two ontologies, one which makes the supposed Wittgensteinian position incoherent. On this interpretation, the *Tractatus* is committed to the position that all properties are relational properties, that all monadic predicates must be analyzed out into at least dyadic predicates, that there ultimately are no qualities but only relations between bare quality-less objects. Objects, says Professor Wolniewicz, are shapeless and indeterminate. Like prime matter they are pure possibility. "If (prime matter) were not formed into substance, it would have no definite properties at all. The same is true of Wittgenstein's objects." An object qua being an object is only the possibility of being configured.

Does this, however, make sense? If an object is a pure possibility and not even a certain sort of possibility, if an object is prime matter and not a form or a primary substance, how then can one talk of a plurality of such objects. A piece of mere prime matter is a logical impossibility. If an object is like prime matter, what does it then mean to talk of configurations of such absolutely indeterminate objects, to talk of possible and actual configurations? How can a pure possibility of being configured either be possibly or actually configured with another such mere possibility of being configured? X might possibly be to the right of Y or possibly to its left but how can the possibility of something being to the *right* of something else be either possibly or actually to the right of or to the left of the possibility of something being to the left of something?

In Professor Wolniewicz's own explication of the concept of configuration he drops the whole idea of objects being absolutely indeterminate and talks instead of parcels of matter being geometrically configured into a rectangular triangle. He speaks of objects as being referrable to by gestures or proper names. A piece of matter, however, is not prime matter for it has a definite form and a whole host of empirical properties: mass, impenetrability, etc. A piece of matter, which we could gesture at or give a proper name to, is a primary substance and so for Wittgenstein would be a fact and not an object. The granular atomism of the *Tractatus* is an atomism of facts, not of particulars. The granular contingent world must be seen against a background of an organic world of necessarily related objects. Objects cannot be particles of matter. As Russell points out in his *Philosophy of Logical Atomism*, "... each particular that there is in the world does not in any way logically depend upon any other particular. Each one might happen to be the whole universe."[1] That is why Aristotle called particulars primary substances.

As a matter of fact, Wittgenstein does not claim either that objects are pure possibilities without any definite properties at all or that all properties are relational properties holding between bare particulars. What is missing in Professor Wolniewicz's interpretation of objects and configuration, either as pure possibilities or as material particles, is Wittgenstein's distinction between the essential or internal properties of an object and the accidental or external or material properties of an object. Objects have material properties, but only by means of a proposition are they represented, only by a configuration of objects are they produced. Whether this means that all material properties are relational or not depends upon whether objects are taken to be particulars or not. Internal properties, however, on any interpretation, are not relational. They are the properties I must know to know an object. They constitute the essence of the object, the form of the object. We specify this form by specifying the states of affairs in which it is possible for the object to occur. The material, external properties of an object are determined by its actual configurations and the internal properties by its possibilities for configuration. Different objects have different internal properties. It makes no sense to say that love does or doesn't weigh ten pounds. This then means that objects must be seen as at least being secondary substances or Platonic forms and not as being prime matter. Of course, what this does to Wittgenstein's theory that all logically necessary truths are truth functional tautologies is obvious.

Professor Wolniewicz wants to parallel Aristotle's secondary substances or Plato's forms to Wittgenstein's propositional functions and thus also wants to parallel Aristotle's operation of abstraction to Wittgenstein's operation of substituting variables for names. To get a general concept we take a proposition $F(a, b, c)$ and substitute variables for constants so as to get $F(x, y, z)$. Of course, we get a general concept this way, but what is it that we then have? We have a possible configuration of objects, and that is, by Wolniewicz's own definition, an object. Professor Wolniewicz believes that an object's essence is determined by its possible configurations which we discover by turning actual propositions into propositional functions, but in fact it is the essential nature of the object, the meaning of its name, which determines the object's possible configurations, and thus what is meant when it is said that it is configured with some other object. We do not have to say that *a*, *b*, *c* are configured in a certain way to determine that some *x*, some *y*, some *z*, can be so configured. We only have to know the meaning of *a*, or *b*, or *c*. The essential nature of an object, the meaning of a name, is not discovered by discovering all the propositions it can enter into. Unless we know the meaning of the names combined in a proposition we cannot know the meaning of the proposition. Wittgenstein claims that only if the meaning of a proposition is determined by the meaning of the names composing it can we be able to understand propositions which we have not encountered before. It is to account for this phenomena, as much as for any other reason, that Wittgenstein wrote the *Tractatus*, and drew upon Plato's theories of names naming forms and forms blending with forms.

As David Keyt put in it his article, 'Wittgenstein's Notion of an Object'[2],

> Although Wittgenstein's objects are not self-subsistent with respect to one another, they are self-subsistent with respect to atomic facts. There can be no atomic facts unless there are objects, but there can be objects without there being any atomic facts. 'Substance is what there is independently of what is the case' (2.024). ... The self-subsistence of Wittgenstein's objects is similar ... to the separate existence of Plato's Forms from his particulars. The Forms are internally related one to another, and thus one Form cannot exist separately from other Forms; but the world of Forms as a whole exists whether there are any particulars or not. On the other hand, there can be no particulars without there being forms.

Keyt goes on to point out, of course, that this is not the whole of Wittgenstein's account of objects. "Plato's Ideas provide the form but not the content of the world. Wittgenstein's objects, on the other hand, are both form and content. In providing the content of the world the objects are similar to the atoms of Leucippus and Democritus."[3] It is this second dimension of Wittgenstein's thought that Professor Wolniewicz is accenting today, but without also asserting the Platonic dimension of his thought one cannot begin to understand the questions which Wittgenstein is trying to answer.

Wittgenstein took from Plato a referential theory of meaning and attempted to take the objects which are the meanings of names, and which, therefore, give to propositions their meaning, and turn them also into the objects which we refer to when we assert that something is the case and which determines whether or not what we say is true or false. Objects and configurations of objects are to meet, at one and the same time, the demands of a Platonic referential theory of meaning with its organic picture of logical space and the demands of a radically empirical and positivistic referential or correspondence theory of truth, where the only truths that entail other truths are empty truth functional tautologies, all saying the same thing, nothing. The difficulties in Wittgenstein's theory of objects is due to the difficulties present in trying to meet these two sets of demands this way.

By presenting his position the way he does Wittgenstein gets himself into a great deal of trouble, but it is his trouble due to his way of trying to solve certain problems and it is not Aristotle's troubles with prime matter and abstraction. In Wittgenstein's system, the internal properties of an object only define a particular sort of thing, only define a form, and not a particular. Wittgenstein admits that two objects may have the same form and then, apart from their material properties, the only difference between them would be that they are different. If this be the case, then actual configurations of objects cannot be separated from possible configurations and then all facts become necessary facts, even as the blending of forms did for Plato in the *Sophist*. One cannot rescue Wittgenstein from his difficulties by making Professor Wolniewicz's Aristotelian claim that just as prime matter is only an abstraction and never exists unformed, so objects are always as a matter of fact configured with other objects. This move is comparable to Pitcher's claim in his book, *The Philosophy of Wittgenstein*, that all objects must have some material properties because the supposition that they do not is unintelligible, it being impossible to specify which particular one is referring to otherwise.[4]

These rescue attempts will not succeed for a number of reasons. First of all, Pitcher's arguments assume that objects must be particulars. Secondly, we have Wittgenstein's explicit claim (2.013) that although knowing an object requires knowing all of its possible occurrences in states of affairs, the space of possible states of affairs in which it is, although I cannot imagine the thing without the space, I can imagine the space empty, i.e. none of the possibilities being actual. Reversing Wittgenstein at this point would undercut the whole ontology of the *Tractatus*, for into a world of contingent and logically independent atomic facts would be thrust this non-tautological necessary fact that every meaningful term of every proposition must refer to some particular with some contingent material property or other. Third, making such changes simply would not eliminate the problems of identifying what is configured, or explaining what it means to be configured, or explaining what determines the possibilities of an object's configurations. To say that *a* is blue if *a* is configured with *b* is meaningful only if we know how to identify *a* and *b* independently of knowing whether the relation holds between them, because the relation, if it holds at all, only holds contingently, for surely it is not necessary that an object have an accidental, material property. The problem is, as Aristotle saw, that rules of individuation seem to be able to function only when there are also rules of classification. One cannot identify simply a being, but only a being of a certain sort. Neither can a specification of possible relations give the classification criteria needed. As a matter of fact, unless we know what sort of thing the individual is we cannot specify the possible relations into which it may enter. The possibility of relational properties between individuals seems to require that the individuals have some qualities. Likewise, different objects must have different sorts of qualities if we are going to be able to divide up objects into subsets so that membership in one subset means that each is yellow and membership in another means that each is red.

It should not have surprised us when Professor Wolniewicz chose a geometric relation as his example of a way of configuring. A proposition is composed only of names and not of names plus a sign denoting the manner of configuration which one is saying holds between the objects named. If the manner of configuration is not already determined by the meaning of the names, which is not possible it seems if the names name material particles or points, and if we don't have to learn for each different type of elementary proposition what configuration is being pictured, which seems necessary if the picture theory is not to be vacuous, then the manner of configuration must be shown by the way the names are combined. In a written language, geometric variations are the only variations really possible. One name comes before the other or on top of the other, etc. Then, however, the only relations that can hold between objects are spatial relations, for that is all that can be shown by such a picture. All properties then are spatial properties. It looks as though names must not be modeled after proper names after all. Knowing the meaning of the names must already settle the question of the manner of their configuration when they are configured. I would suggest that when we consider the question of meaning rather than the question of truth, names should be modeled after terms in a proposition. We know what fact is being asserted when we say that the cat is black, if we know that the expression 'the cat' is functioning as a referring term and that the expression 'is black' is functioning as an ascriptive term, which we do know if we know the meaning of the language words used and the grammatical rule which governs their use in this serial order.

For all of these reasons, it seems to me that Max Black is quite correct when he says in his book, *A Companion To Wittgenstein's Tractatus*,

> For the proper appreciation of Wittgenstein's solutions it is essential to hold steadily in mind that his pattern of analysis seemed to him necessitated *a priori*; he did not reach it by generalization from cases... (The existence of an 'elementary proposition') is guaranteed only by metaphysical inference. We can produce no elementary propositions and would not recognize them if we had them. Consequently, we can form no clear idea of what genuine 'names', the constituents of elementary propositions, are like, nor any clear conception of the nature and logical form of objects for which those names stand. The arguments in which some writers have engaged about whether objects are to be counted as particulars or universals, whether elementary propositions are relational or not, and so on, are bound to be futile.[5]

Wittgenstein, I would suggest, attributed to objects all the properties which must be attributed to them in order that we, simply by referring to such objects, can explain how language can be used to describe the world, how we can understand the meaning of propositions which we have not encounted before, how there can be true and false meaningful propositions. Objects must have internal properties of being possibly configured because unless simply understanding the concepts used in asserting a proposition enables us to understand the meaning of the proposition we cannot understand the meaning of propositions which we have not previously understood. Logical space must be organic with each object related to each other in possible states of affairs because terms in propositions are incomplete symbols whose meaning is determined by their functional relation to the meaning of the proposition as a whole. It must be possible that all objects have only internal properties and thus that no objects are actually configured because the meaningfulness of a contingent proposition can not guarantee truth or there could be no false statements. If there are any contingently true propositions, then some objects must be actually configured, because the only way that language can reach out to reality is through the naming and picturing relations.

Where this argument goes wrong is another story to be told at another time. What needs to be stressed now is that appreciation of the *Tractatus* cannot be gained by an attempt to directly characterize its ontology, but can only be gained by starting with Wittgenstein's considerations of language and thought and with the arguments which he believed compelled him to affirm the ontology he did.

Boston University

REFERENCES

[1] B. Russell, 'The Philosophy of Logical Atomism', in *Logic and Knowledge*, London 1956, p. 203.
[2] David Keyt, 'Wittgenstein's Notion of an Object', in Copi and Beard, *Essays on Wittgenstein's Tractatus*, New York 1966, p. 292.
[3] *Ibid.*, p. 292.
[4] George Pitcher, *The Philosophy of Wittgenstein*, Englewood Cliffs, N.J., 1964, p. 122.
[5] Max Black, *A Companion to Wittgenstein's Tractatus*, Ithaca, N.Y., 1964, p. 11.

F. J. CROSSON

THE COMPUTER AS GADFLY

Computers seem as importuning, to some of us at least, as Socrates ever was. They can produce in us the same amazement and the same shamefaced puzzlement about the most elementary of human things; they evoke reactions ranging from jokes with an undertone of uneasiness to denunciations of 'computerized societies' (though unlike Socrates they are too useful to be threatened with extinction); and despite the fundamental questions about ourselves they pose – or reflect – to us, they force us to work out the answers for ourselves: or at least, they respond only to precisely formulated queries. Moreover, one often has the feeling that the dialogue is aporetic rather than dogmatic, that it moves toward a clarification of our ignorance rather than toward an epistemic answer. If this is the case, the question is whether it is a learned ignorance or sheer nescience. I want to try to state a case for the former, on the basis of a recurring difficulty in several areas of artificial intelligence.

If we have no blueprint for the 'labyrinth inside our skulls', we have at least a model, that of the neural networks proposed by McCulloch and Pitts. Guided by the formal similarity between logical functions and neuronal interconnections, they developed a model of neural networks which showed that in principle a network could be constructed – by appropriate interconnection and threshold determinations of elements – to compute any specific input-output function. The only limitation on the scope of this conclusion is that the function be specifiable, i.e. logically describable. To quote John Von Neumann:

> any functioning in this sense which can be defined at all logically, strictly and unambiguously in a finite number of words can also be realized by such a formal neural network.[1]

Not only did the theorem foreclose earlier efforts to specify neurological functions which were 'too complex' for machines to perform, it also undercut all future objections of the form: 'One thing a machine can never do is φ', provided the objector can tell us precisely what behavior he has in mind.

Boston Studies in the Philosophy of Science, IV.

But what is the import of this last phrase ("provided the objector can tell us *precisely*")? It refers of course to the limitation mentioned, that the function be definable unambiguously in a finite number of terms. That this limitation may constitute a problem is generally recognized, e.g. by Donald MacKay:

> What is doubtful, of course, is whether we shall ever be clever enough to specify, even in principle, an exhaustive test for human-like behavior. It is this aspect of the problem ... that I believe is of more than trivial philosophical interest.[2]

But the question is about the meaning of our inability: for such an inability, if based on sheer ignorance, has no demonstrative import. My colleague, Ken Sayre, has taken it in this sense. Any act of which all men are capable, he argues, either is or is not clearly understood. "If it is not clearly understood, then no clear meaning can be attached to the claim that all men but no machines can accomplish it", while if it is clearly understood, i.e. logically specifiable, then it can be computed by a suitable finite automaton.[3]

An ordinary language philosopher might dispute the Procrustean use of the term 'clearly understood' here, by pointing to cases where its use does not entail formal-logical specifiability. However, I want to examine the dilemma in the context of work in cybernetics, specifically in the field of simulation.

Part of the reason for this choice is that some contemporary philosophers have referred to work in this field to corroborate particular mind-body theories, work which it seems to me does not provide very strong ground. (E.g., Wilfrid Sellars speaks of the "decisive importance" of the "analogy between speech and what sophisticated computers can do and... between computer circuits and conceivable patterns of neurophysiological organization".[4]

I want to look briefly at two areas of simulation work, handwriting recognition and language translation.

Human beings, one might say, spend their cognitive lives in discriminating patterns: visual, musical, linguistic, etc. The way to begin simulating these achievements, it would seem, is by determining the features or criteria on the basis of which we recognize, say, a written number or letter. Alpha-numeric characters may be poorly inscribed, but presumably there are some invariant features by which we recognize the inscription. It quickly becomes apparent that such criteria, however easily we may use them in recognizing, are not easy for an individual to specify, unless the range of possible variations is limited. Think of the varieties of handwriting that teachers must decipher!

Nevertheless, if we cannot determine how we do do it – i.e., if we cannot say how we do it – we might try to discover invariant features by means of which we could classify the various patterns. For example, we could try such features as closure, concavity or convexity, extension of lines above or below a 'base line envelope' and so forth. The attempt could then be made to characterize letters on the basis of the presence of a differential set of such definable characteristics. Actual writing often deviates considerably from any such pattern, but if we take this into account by weighting the probability of finding certain features for a given letter and making the computer's decision depend on such a probabilistic assessment, we can ameliorate the effect of such departures. Programs have been written which exhibit 80 to 90% success in recognition in this way, especially if some limitations are placed on the inscriptions, e.g. that the lines be confined within a well-defined space.[5]

But there is a paradox which awaits the production of any successful program, even – indeed especially – one that achieves 100% success. For in fact a letter which anyone would judge to be a *u*, say, if it were isolated, might very well be seen as a *v* if it were in the context of a word whose meaning and hence whose spelling required a *v*. We constantly unconsciously 'adjust' such discrepancies in recognizing all sorts of patterns about whose overall sense we have no doubts. On reflection, it seems clear that we regularly recognize letters not only on the basis of the presence of invariant features, but by assessing features in the context of the word and even the sentence in which they are imbedded. Moreover, the context referred to is not only statistical but is in part determined essentially by the meaning of the word.

If these last observations are correct, then the problem of programming a computer to recognize a letter-pattern as a sum or set of discrete features, however successfully it may be solved, will constitute a simulation of human letter-recognition at best only for isolated letters. In actual reading of script, the recognition of a word will not be through a one-by-one identification of the letters plus summation, but rather deviant letter-shapes will be assimilated in reading to what the context requires. Since the context is not only alphabetic, but also syntactic and semantic, we we are led on to the area of language translation.

In the 1940's, inspired by the seeming analogy between decoding and translating, high hopes were entertained for machine translation of Russian texts into English. Warren Weaver wrote that a Russian text could be regarded as "really written in English" but "coded in some strange symbols". This view implicitly assumed that some kind of one-to one correspondence existed between syntactic and semantic units in the two languages, and that these correspondences could be made explicit, i.e. discovered and specified. It assumed, that is, that the computer could parse the sentence, identify the individual word-roots (e.g. separating off inflected endings) of the input language, search a mechanical dictionary in its memory and finally combine the thus translated units into an output language sentence via the syntactic rules of the latter language.

Instead of this, the great achievement of the work in automatic translation has been the disclosure of a remarkable imbrication of syntax and semantics. Pairing a meaning with an identified word requires the determination of its syntactic status, and this has proven impracticable because of a high degree of syntactic ambiguity in natural languages. Moreover, frequently the only way to resolve this ambiguity is by recourse to the meaning of the sentence as a whole, in context, which means that a circularity is introduced into the program as described.[6]

For example, consider the sentence, 'Time flies like an arrow'.[7] The word 'time' here may be either a noun, an adjective or a verb, yielding three different syntactical interpretations. (To see how it could be one of the latter two, which are perhaps less perspicuous, compare: 'Fruit flies like a banana' and 'Time runners with a stop-watch'.) The only way to choose among these possible interpretations is by recourse to the semantic context, either that of the surrounding sentences, or the much vaguer and vaster one of general information (e.g. there are not any things called 'time flies'). This means, of course, that we cannot treat the semantic level as simply dependent on the syntactic.

As A. G. Oettinger comments,

> No techniques now known can deal effectively with semantic problems of this kind. Research in the field is continuing in the hope that some form of man-machine interaction can yield both practical results and further insight into the deepening mystery of natural language. We do not yet know how people

understand language, and our machine procedures barely do child's work in an extraordinarily cumbersome way.[8]

Even if the task of determing a unique syntactic and semantic structure were solved, further problems lie on the third level of the program described above, that of transposing the input sentence into a syntactically correct output language sentence. A word by word translation of one natural language into another frequently if not generally results in either obscurity or loss of meaning. 'Nous la leurs donnons' cannot be rendered as 'We it them give', the latter must be transformed into English form. The exploration of transformation rules by Noam Chomsky and his associates in recent years has resulted in a clear distinction between 'surface grammatical structure' and 'deep structure', which raises serious questions about the possibility of mechanical translation.

Surface structure analysis is the sort which most of us learned as paradigmatic in grammar school, and which analyzes a sentence by successive divisions or bracketings nested within one another: (1) subject phrase, (2) predicate phrase; (1a) noun phrase, (1b) prepositional phrase; (2a) verb, (2b) prepositional phrase, and so on down to nouns, adjectives, articles, etc.

Chomsky has remarked,

> With such a conception of language structure, it made good sense to look forward hopefully ... to such projects as machine translation. But these hopes have by now been largely abandoned with the realization that this conception of grammatical structure is inadequate at every level, semantic, phonetic and syntactic.[9]

When sentences are transformed grammatically, e.g. from active to passive, the neat bracketings of surface structure are broken up. Successive transformations reveal structures which are hidden in the surface analysis, which are 'deeper'. Moreover there is experimental evidence which suggests that sentences are stored in memory in terms of these deeper structures.[10] Chomsky has been led by his work to a return to the 'classical rationalist' account of knowledge and language, which attributed to the mind a system of innate rules for the generation of linguistic and perceptual organizing schemata. Our knowledge of language, on this view, is not arrived at simply by exposure to repeated use of words: "Serious efforts have been made in recent years to develop principles of induction, generalization and data analysis that would account for knowledge. These efforts have been a total failure."[11]

It has been impossible, that is, to construct an associative-genetic characterization which is both formal and sufficiently flexible or powerful to cover the range of linguistic patterns which we recognize or acknowledge as analogous. The theory of language which Chomsky and his associates are developing represents a hypothesis about the generative rules underlying such analogous extensions, rules the knowledge of which is the endowment of every linguistically competent person. But these rules are known by such persons only 'tacitly', they cannot 'spell them out' on demand. More strongly put, the innate rules which the theory of language attempts to articulate need not ever be accessible to consciousness in order to be exercised.[12]

A similar difficulty in the field of pattern recognition has been formulated by John Von Neumann. Nobody, he says, would attempt to define, in any practicable compass, the "general concept of analogy which dominates our interpretation of vision", e.g. as we move from badly drawn triangles to convex and concave triangles, to triangles composed of dots, of different kinds of physical objects, etc. Consequently, he argues, it may be that

> it is futile to look for a precise logical concept, that is, for a precise verbal description, of "visual analogy". It is possible that the connection pattern of the visual brain itself is the simplest logical expression of this principle.
>
> Obviously [he adds], there is on this level no more profit in the McCulloch-Pitts result.[13]

The Protean variety of what we call 'visual analogy' resists being derived from or reduced to formally simple and specifiable criteria. For Von Neumann, this is because the indefinitely large number of variant features reflects the difference in order of magnitude of the neural networks and poses a qualitative and practical obstacle. For Katz and Chomsky, 'linguistic analogy' has been impossible to spell out because its recognition is based on a tacit or 'unconscious' exercise of innate rules.

The latter difficulty is of course not insuperable in principle: a theory of language can be formulated by hypothesis and verification on the basis of comparative linguistic descriptions. With respect to machine simulation, what is moot is (1) whether such a theory can be made sufficiently complete or comprehensive without running into the Von Neumann limit, and (2) whether even if complete it could be used in a non-tacit way. For a simple comparison, let me borrow one of Polanyi's examples. Learning to ride a bicycle involves learning to turn the front wheel into a curve whose radius is proportional to the square of the velocity divided by the angle of imbalance, in order to keep going. Of course, few bicycle riders know this rule, but one might want to say they know it 'tacitly'. Certainly their behavior in riding exemplifies it. Moreover, we could certainly build an automaton with a gyroscope, programmed to ride precisely according to the explicit calculation or correlation of the quantities involved. So, in this case, simulation by explicit steps of a behavior governed by a tacitly known rule is possible: the variables are few and their relation is expressed by a relatively simple function.

On the other hand, as the number of relevant items of information increases (e.g. in translation), or where there is no closure for the set in sight (e.g. visual analogy), step-by-step simulation becomes, as Von Neumann suggests, impractical and perhaps impossible in principle, although solutions for restricted ranges of behavior continue to be produced (e.g. Evan's program for computer solutions of analogy tests[14]).

But is not the human brain a kind of 'existence proof' for the possibility of neural networks capable of all such behavior? Certainly it is for the biologists who are working on the synthesis or organisms on the basis of the DNA code: if we can produce men from 'raw materials' or from germ cells artificially programmed, we would expect human behavior as an indirect result of replication of structure. But for simulation of such behavior on a computer or computer-based system, similar confidence does not seem warranted.

The reason for this latter claim lies in the fact that the human 'use' of the brain does not appear to employ the same kind of step-by-step operations which the computer – or any machine, if to be a machine is to instantiate a formal system – does employ. I refer here, in the first instance, to the thesis of Katz *et al.* that knowledge of linguistic rules is tacit, i.e. non-explicit.[15]

To be made more plausible, this reference may be generalized to include other evidence – and here I follow at first Polanyi, who has drawn attention to some current work in psychology.[16] The import of this work is that subjects can be trained to integrate into a behavioral pattern interoceptive stimuli of which they are not and cannot become focally aware. Hefferline, who worked with a minuscule muscular twitch at the base of the thumb, is willing to say that the subject can become 'aware' of the twitch "in the ordinary sense of the word [aware]", even though "he can say nothing else about it, such as describing his criteria for naming it" when it occurs.[17]

But in at least one ordinary sense of the term 'aware', it is misleading to apply it under these circumstances. In contrast to the phenomenological dictum that "all consciousness is consciousness of... (some object)", we lack here an awareness of the stimulus as a specific object. The subjects said that they could not identify what they were reporting nor how they were doing it.

It should be noted that in all of these experiments – and Razran has reviewed many different kinds – "contrary to assertions, autonomic reactions can be modified by subsequent reinforcement in operant conditioning; [but] that such reinforcement is effective only when cognition is present".[18] This is why I have referred to integration into a behavioral pattern: it is only by relating the interoceptive stimulus to a datum of some sort (perceptual, imaginative, affective) that it can begin to be discriminated and come to play a role in the subject's conscious activity.

Thus while there is no awareness "in the ordinary sense of the word", it is impossible to term the stimulus 'unconscious' or 'non-conscious', because by attentive striving its occurrence can be reported and it can be voluntarily correlated with objects of which one is focally aware. Razran prefers to speak of the 'inferable conscious', which expresses the paradoxical character of the phenomenon. In the language of intentionality, there is no focal awareness *of*, but there is still some kind of awareness. Moreover, it is not like that of 'raw feels', whose non-intentional character Sellars and others have stressed, for these are data and can be characterized.

Recent work by a West Coast researcher, Dr. Joseph Kamiya, has extended these experiments to the brain, and succeeded in training subjects to become tacitly aware of and to control their own alpha-wave frequency. If the awareness here, as before, is not focal, it nevertheless is an awareness which, precisely because it shades off toward the sheerly absent, blurs a certain kind of distinction between the physical and the mental. It represents, that is, an infra-level of interaction between the two which is neither a "clear and distinct idea" nor a simply non-conscious neural process, empirically correlatable with but logically distinct from clear and distinct ideas.[19]

My argument, then, rests on a paradox which is puzzling: namely, that while the brain appears to be (or at least to be simulable by) a neural network, and hence to instantiate a formal system, the mind seems to work informally with tacit data, and thereby to bypass the difficulty posed for simulation by Von Neumann. It does this, apparently, by means of a capacity to take account of information without having to or indeed in some cases without being able to take explicit account of it – or as Plato put it long ago in the *Meno*, by being able to know and not know at the same time.

The force of this argument depends, in part at least, on whether all of the *Gestalten* or perceptual wholes which we can discriminate are equally accessible to sequential synthesis as well as to tacit integration. I take it to be the case that many visual and acoustic forms are grasped without any explicit notice of their parts, and moreover through ignoring of irrelevant elements which could weaken the closure of the forms. In contrast to this, I take it to be the case that a computer must enumerate the parts to form any whole.

Claude Shannon acutely recognized this problem several years ago:

> Efficient machines for such problems as pattern recognition, language translation, and so on, may require a different type of computer than any we have today. It is my feeling that this computer will be so organized that single components do not carry out simple, easily described functions. ... Can we design ... a computer whose natural operation is in terms of patterns, concepts, and vague similarities, rather than sequential operations on ten-digit numbers?[20]

Let me try to explicate this distinction in terms of the contrast between tactile and visual preceptions.

Von Senden's work on patients born blind who were able to see after operations leaves no doubt that forms, *Gestalten*, are immediately discriminated by the patients in their visual fields. Although recognition, even of simple geometrical shapes, was slow and required learning, visual wholes were spontaneously apprehended. It appears true, in general, that in vision the total form represents the starting-point for any analytic articulation of the parts. The total form guides the visual analysis into primary and subordinate parts and makes possible the discrimination between parts and 'noise' (in the information-theoretic sense).

In tactile perception, on the other hand (no pun intended), quite different conditions hold, as Révész has emphasized in his notable study *Psychology and Art of the Blind.* In Haptics, "the structure, the spatial arrangement and mutual relationship of the parts have to be brought into consciousness intentionally by strenuous work".[21] Subordinate parts with special importance may become the center of attention because of their size or material peculiarities, since the parts are given sequentially and are more or less structurally independent. 'Noise' cannot be excluded as irrelevant until an overall organization is achieved by synthesis. Indeed for this reason Révész' blind subjects proved completely incapable of deciphering relief sculptures.

On the basis of his experiments, Révész was led to distinguish between two kinds of perceptual wholes: synthetic or summated wholes and formal wholes. Since the summated whole "sums up the spatial arrangement of the examined parts as a result of intuition and intellectual work, that summing up presupposes an act of conscious arrangement, as becomes manifest in the eventual product". The formal whole however is "given spontaneously as an individual unity and does not require any conscious arrangement of structure", but is rather "conditioned by unconscious formative factors".[22] Succession and synthesis govern tactile perception, simultaneity and spontaneous unanalyzed wholes dominate visual perception.

(From a physiological point of view, one might want to argue that, presumably, successivity is involved in vision also, perhaps something like scanning of the visual cortex. But this would not advance the task of simulation (as distinct from replication of the biological structure), since it leaves one with the problem of dealing with summated wholes and explicit criteria of recognition in order to discriminate wholes which we perceive physiognomically. As before, the question is whether we can do this without encountering the limitation envisaged by Von Neumann.)

Some of the most interesting programs which have been written – e.g. Evan's program for the recognition of analogical figures of the type used in intelligence testing – clearly exemplify what has been described above as tactile perception.[23] The topological relations of the figures are spelled out by explicit statements and similarities (partial identities) are matched. Marvin Minsky, in describing the program, remarks that "rules or procedures of the same general character are involved in any kind of analogical reasoning".[24] I agree with this, but I think it is just the reasoning (i.e. ratiocinative) character of the process which embodies its weakness as well as its power. Let me try to state what I mean here as clearly as possible.

I am not arguing that human beings reason in a better or 'higher' way than a machine can. If anything I am rather inclined to the contrary view that insofar as machines approach human performance in recognition it will be through faster and more precise logical processes than men are capable of. But precisely this seems to me necessary to supplement the multitudinous character of sequential or 'tactile' input. And just this constitutes the intellectual challenge and excitement of simulation work, trying to find a structural and formal equivalent for tacit processes.

In discussing a subject suffering from lesions in the occipital region and deficiencies of vision, M. Merleau-Ponty remarks on his inability to understand simple verbal analogies unless he can find and state a middle term which establishes the identity of the two relationships. But, he adds, "if we describe analogy as the apperception of two given terms under a coordinating concept, we should be giving as normal a procedure which is exclusively pathological". ('Exclusively' overstates the case, since even in normal subjects explicit analysis may be necessary to confirm or discern analogies in complicated cases.) Merleau-Ponty's reason is that "it is easier for the normal patient to understand the analogy than to analyze it, whereas the patient manages to understand only when he has made it explicit by recourse to conceptual analysis".[25]

The fundamental problem, involved in this contrast between tactile and visual perception, and underlying all simulation work in the basal area of recognition, is that of assigning instances to a category. Until we can conceive of a machine which can deal with 'vague similarities' (as digital computers cannot) there will be an elementary difficulty facing the machine simulation of such assignments.

This difficulty can be put as follows. Any attempt to specify explicit criteria for membership in a class C on the basis of a comparison of instances of C must rely on the pre-criterial collection of such instances. Consequently, no set of criteria can ever be declared to be complete, since by virtue of the way in which it is constituted, it leaves open the possibility (short of stipulative definition) that other instances may be adduced on the same implicit grounds as those of the original collection. Recognition has an 'open texture'[26] which resists the closure implied by criterial classification.

W. V. Quine, in a recent discussion of this "inscrutability of categoricals" seems to follow Kant in attributing to undefined native powers the ability to identify specimens of a class.[27] More generally, Wittgenstein, arguing at length for the inexactness of our language, claimed that we cannot justify our uses of terms by grounding those uses on necessary and sufficient criteria. It is possible, he thought, to be justified in saying, e.g., on the basis of certain behavior that a man has a toothache, even though no strict entailment holds between the behavior and having a toothache.[28] Such justification appears to be based, in his view, on a fact of natural history: "Here we strike rock bottom, that is we have come down to conventions."[29]

R. W. Newell has explored the logical ground of this type of justification and argued persuasively for its validity, despite its lack of either inductive or deductive form.[30] But at best, it seems to me, he only establishes that we *do* accept as reasonable in such cases the offering of considerations which while not a proof in a strictly logical sense are still (to quote J. S. Mill) "capable of determining the intellect either to give or to withhold its assent"[31], and that we cannot not do so without abandoning our "form of life".

It is the merit of Polanyi to have faced the fact of our capacity to recognize specimens of an empirical class as something requiring an explanation and to have proposed one. Rejecting (as does Merleau-Ponty) the suggestion that identification which goes beyond the given guarantees (i.e. which is not strictly entailed by the explicit observational data) is grounded only on conventions or is only a "verbal recommendation", he contends that a multitude of unspecifiable clues may enter into an act of identification. Such an act represents a particular instance of 'tacit knowing', which always involves the contribution of subsidiary elements to the meaning of the object to which we are attending. In many cases, the subsidiary elements cannot be isolated and rendered explicit – hence the term 'unspecifiable'.

I cannot review here the wide-ranging examples and arguments by which Polanyi seeks to validate the comprehensiveness of his analysis.[32] Nor is it necessary to do so for the sake of my argument here. The difficulty alluded to above concerning the manner in which a set of criteria is established stands on its own, I believe, whatever its explanation, and the evidence for the role that implicit information can play in behavioral discrimination is secure, whatever its scope may turn out to be.

In sum, then, both on the side of the discriminated perceptual whole and on the side of the subject, tacit elements appear to play an essential role in the human patterning of information. It is not inconceivable that as self-adapting automata increase in complexity, they will reach a point where they begin to display the capacity for this kind of operation. Indeed it would be baffling if they did not, if the brain is only a neural network. But at present, it seems to me, we do not begin to know how to respond to Shannon's question, i.e. how to design an automaton which can deal with similarities without reducing them to identities. It appears far more likely that artificial persons will come first from the biology laboratories, and that a more modest expectation should be directed toward cybernetics: perhaps something like Aristotle's natural slaves, creatures which display some rationality but not enough flexibility to be considered complete persons.

Having returned to the Greeks for this comparison, let me conclude by recalling that Socrates' wisdom – or learned ignorance – was displayed in his ability to reveal to his interlocutors how little they really knew discursively about what it is to be a human being. Computers have certainly exercised that chastening function for many.

University of Notre Dame

REFERENCES

[1] J. H. Von Neumann, 'The General and Logical Theory of Automata', in *The World of Mathematics* (ed. by J. R. Newman), New York 1956, vol. IV, p. 2090.

[2] D. M. MacKay, 'On Comparing the Brain with Machines', *The American Scientist* **42** (1954) 263.

[3] K. M. Sayre, 'Philosophy and Cybernetics', in *Philosophy and Cybernetics*, Notre Dame 1967, p. 25.

[4] W. Sellars, *Science, Perception and Reality*, New York 1963, p. 30.

[5] Actually the problem as described has been shifted from that of recognition to that of classification. For a critical discussion and a survey of the field, cf. K. Sayre, *Recognition: A Study in the Philosophy of Artificial Intelligence*, Notre Dame 1965.

[6] As Mortimer Taubes comments, "In ordinary grammatical studies a sentence is

analyzed to show *how* the meaning has been expressed, not to determine *what* the meaning is" (M. Taubes, *Computers and Common Sense*, New York 1961, p. 36).

[7] For the computer print-out of the analysis of this example and discussion, cf. A. G. Oettinger, 'The Uses of Computers in Science', *Scientific American* **215** (Sept. 1966) 168.

[8] *Ibid.*, p. 169. A recent survey has even called into question the practical value of man-machine interaction in translation: D. B. Orr and V. H. Small, 'Comprehensibility of Machine-Aided Translation of Russian Scientific Documents', *Machine Translation* **10** (March & June, 1967) 1–10.

[9] N. Chomsky, 'Language and the Mind', *Psychology Today* **1**, No. 9 (Feb. 1968) 66.

[10] H. Savin and E. Perchonock, 'Grammatical Structure and the Immediate Recall of English Sentences', *Journal of Verbal Learning and Verbal Behavior* **4** (1965) 348–53.

[11] Chomsky, *art. cit.*, p. 68. His argument is developed with reference to earlier theories of language in *Cartesian Linguistics*, New York 1966.

[12] J. Katz, *The Philosophy of Language*, New York 1966, p. 183 *et passim.*

[13] *Art. cit.*, p. 2091.

[14] Cf. M. Minsky, 'Artificial Intelligence', *Scientific American*, Sept. 1966.

[15] My reference here will be to the awareness or introspection of reporting subjects. There is also neurological evidence for the claim that the 'logical depth' of the brain's operation is far less than that of computers.

[16] M. Polanyi, *The Tacit Component*, London 1967.

[17] R. Hefferline, 'Learning Theory and Clinical Psychology – An Eventual Symbiosis?', in *Experimental Foundations of Clinical Psychology* (ed. by A. J. Bachrach), New York 1962, p. 130.

[18] G. Razran, 'The Observable Unconscious and the Inferable Conscious', *Psychological Review* **68** (1961) 122.

[19] J. Kamiya, 'Trained Self-Control of the EEG Alpha Rhythm', in *Altered States of Consciousness* (ed. by C. Tart), Wiley, New York (in press); 'Conscious Control of Brain Waves', *Psychology Today* **1**, No. 11 (April 1968) 56–60.

[20] Cf. *Computers and the World of the Future* (ed. by M. Greenberger), Cambridge, Mass., 1962, p. 310.

[21] G. Révész, *Psychology and Art of the Blind*, London 1950, p. 84. Even where the object is small enough to fit in the hand, successivity of perceptions dominates: p. 61.

[22] *Ibid.*, pp. 86, 87.

[23] Cf. note 14, above.

[24] *Ibid.*, p. 250.

[25] M. Merleau-Ponty, *Phenomenology of Perception*, New York 1962, p. 128. Note the protocols of the subject's identification of an object: he "runs his fingers over an angle several times: 'My fingers', he says, 'move straight along, then stop, and then move off again in another direction; it is an angle, it must be a right angle.' – 'Two, three, four angles, the sides are each two centimetres long, so they are all equal, all the angles are right angles ... It's a dice'" (For a similar report, cf. *ibid.*, p. 131), p. 107n. I have frequently observed such a spontaneous verbalizing of the analysis in experiments of my own with blindfolded subjects.

[26] Cf. F. Waismann, 'Verifiability', *Proceeding of the Aristotelian Society Supplement*, 1945.

[27] In S. Koch (Chm.), 'Logic and Psychology'. Symposium presented at the meeting of the American Psychological Association, Wash. D.C., Sept. 1967. Cf. Kant, *Critique of Pure Reason* A141.

[28] Cf. R. Albritton, 'On Wittgenstein's Use of the Term "Criterion"' in *Wittgenstein: The Philosophical Investigations* (ed. by G. Pitcher), New York 1966, p. 246.
[29] L. Wittgenstein, *The Blue and Brown Books*, Oxford 1958, p. 24.
[30] R. W. Newell, *The Concept of Philosophy*, London 1967.
[31] *Ibid.*, p. 132.
[32] For a recent statement, cf. M. Polanyi, 'Logic and Psychology', *American Psychologist* **23**, No. 1 (Jan. 1968) 27–43.

LUCIEN GOLDMANN

THE SUBJECT OF CULTURAL CREATION

If I have chosen this theme, it is because it seems most apt to bring to light both the profound convergences and the fundamental divergences that exist between sociological and dialectical study of cultural creation on the one hand and psychoanalytical study on the other.

Indeed, it is just as important not to ignore what psychoanalysis, even within the perspective of a sociologist, can bring to the understanding of man and of cultural creation as not to blur the oppositions in order to arrive at a kind irenicism, as eclectic as vague, which could only hinder positive research.

To begin with, what do the common elements consist in? Dialectical sociology and psychoanalysis both start from a common affirmation, namely, that on the human level, nothing is ever without meaning. This does not mean, as it has often been claimed of Hegelianism, that dialectic is a panlogism, especially since, given the contemporary development of formal logic, we would run the risk of giving this term too narrow a meaning. Perhaps it would be better to create a term starting, not from 'logic', but from 'signification', and to speak rather of pansignificance.

On the condition, of course, (and this is an idea I have already had occasion to develop with Paul Ricoeur) that we remain aware of the fact that signification does not begin with man and less still with thought and language, and especially that it is not always conscious.

If there were in this room a more or less hungry cat, and if a mouse were creeping along the wall on the left, the fact that the cat would turn in that direction and catch the mouse would be a perfectly significant action adapted both to the cat's problem, that of satisfying his hunger and finding food, and to the context in which the problem arises, that of a mouse creeping along the wall on the left toward which the cat must turn to catch it.

Now, although we do not know much about the psychology of animals, it is not certain, it is even very unlikely, that the cat is conscious of the problem and of the process she is using to solve it. Yet her behavior is

Boston Studies in the Philosophy of Science, IV.

nonetheless significant in the sense that *signification* means the biological, corporal, and implicit solution to a problem arising in a given situation.

Whatever the case may be, however, with the consciousness of the cat, what is important to note is the fact that since man's apparition, signification for him always passes through (true or false) consciousness, through communication, speech and language, so that we find this signifying consciousness wherever we have to do either with a contemporary human reality, or with a past human reality which has left enough traces and witnesses to enable us to study it.

But what thinkers like Hegel, Marx, Lukacs, Freud, and I think I could add Piaget, have in common is the affirmation that each time we find ourselves before a human behavior, a linguistic expression, a written phrase, some indication of communication or other[1], whatever may be the immediate impression otherwise produced by this fragment or this behavior, and even if we do not see its rationality at first sight, the manner in which it could contribute to the solution of a problem, it is still in fact a fragment of meaning which, if we succeed in integrating it to the whole of which it is a part, *will reveal itself as significant.*

Also, the analyses of Marx as well as those of Freud, whether it be a matter of economics, studies of ideologies, political history, literary history, philosophy, religion, and scientific thought, or whether it be analyses of dreams, neuroses, or *lapsus,* all result in bringing to light the significant character – and this means both structural and functional – of any sample of human behavior which seemed at the beginning to be more or less, and perhaps even entirely, lacking in signification.

That is the first common element.

The second is found in the way that Hegel, Marx, Lukacs, and Freud manage to restore the meaning starting from a fragment that is not of itself significant or that seems, at first sight, to have a signification other than the one which dialectical or psychoanalytical research will end up bringing to light.

The way of arriving at this meaning for all these thinkers consists in integrating the object under study into a more vast relative totality called structure, social life, network of images, or unconscious psyche. Permit me to add that, in this perspective, the notion of *polysemia* which has been talked of a good deal becomes quite acceptable and means simply the possibility of validly integrating the object under study into many different structures, on the level of consciousness as well as on the level of historical life and perhaps also (I do not have the competence to say with certainty) on the level of biology.

A third common point: the idea that the structures are not invariable and permanent but constitute the term of a genesis; also we cannot understand the significant character of a structure except by starting from an ensemble of actual situations inside which it was born as an attempt of the subject, *already structured itself by its previous becoming*, to modify the ancient structures in answer to the problems arising from these situations, attempts at answering which in the future, in conjunction with the external influences or the behavior of the subject and his action on the surrounding world, will transform the situations and raise new problems and progressively modify the actual structuration of the subject.

Briefly, the thought of Freud, as well as that of Marx (and to be sure, these two names have more general value here and signify any dialectical and positive sociology, and any psychoanalysis of Freudian inspiration) are genetic structuralisms.

This being said, we still have also to insist on the differences that separate Marxism and psychoanalysis.

These seem to me to be situated first of all at the point I intend to deal with today: that of the subject of human behavior and, starting from there, of signification and of significant language, and, within this behavior and this language, the purpose and importance of the subject of cultural creation.

The essential difference between any dialectical sociology and Freudian thought seems to me to reside in the manner of conceiving this subject. This for two reasons that have become mutually reinforcing. Freud, it seems to me, thought that the subject was always and everywhere an individual. He did so, first of all, to the extent that he was still situated in the continuation of the thought of the Enlightenment[2] and that this philosophy, which constituted for many centuries the predominant form of thought in the Western world always started, in one way or the other, from the individual. Cartesian or Husserlian *cogito*, sensations or protocol propositions for the empiricists, always it was the individual that this outlook saw as the only possible subject of action, of thought and of behavior.

All this, besides, sometimes led to forms both as paradoxical and as revealing as the one I have referred to in one of my books: a grammar of Bruneau for third or fourth grade in which we could read, as if it were a matter of evident truth, that 'I' is a pronoun which has no plural; *we* signifies 'I' and '*you*'. In short, in the mind of the Enlightenment and for Freud after it, there are only individuals; the subject is always an 'I'.

Now this individualist position was reinforced in Freud by the fact that, even where his great discoveries are concerned, especially that of the unconscious, and where thanks to this he was going in an essential way beyond the thought of the Enlightenment and turning toward a dialectical conception of personality, he still found himself in the very first instant confronted by the biological aspect, or that immediately derivative from the biological in personality, by sexuality, by what has been called desire here, but which it would perhaps be better to designate, to avoid any confusion with Hegel, as libido.

But, even if the structure of libido or desire is not especially genital – and Freud was the first one to discover and to teach that it certainly was not – I think we can still define it in a rigorous enough way by the fact that it exclusively embraces drives the subject of which is biological and individual, and for it, other individuals could only be objects and, more exactly, objects of satisfaction or obstacles to it – such as, for example, the Mother and the Father in the Oedipus complex.

This being said, it still remains evident that, once integrated into a personality capable of symbolic thought and language, these tendencies themselves become more complex and acquire certain new characteristics, to wit, the possibility of integrating themselves into a reflexive consciousness, of thinking the I, and from there of making it the object of a drive. Thus we come to narcissism, which is a human particularity and has, as a characteristic, that it presents at the level of the individual subject – which according to us is only a secondary factor in cultural creation – one of the most important particularities among the summits of collective consciousness: the identity of subject and object.

Yet, whatever the reasons may be, it is a fact that the Freudian analyses of cultural creation are a kind of rigorous transposition, with hardly a modification except on some secondary points, of the analyses of individual behavior and of individual libido.

But if structuralist psychology is incontrovertibly founded, up to a certain point – we emphasized this ourselves at the beginning of this conference – nevertheless the transposition of the individual subject from the biological and libidinal domain to social life and cultural creation seems very highly questionable to us and we are afraid that it will bring into question any positive and scientific value of these analyses.

Before entering the essential aspect of the problem, however, I would like to recall that a certain number of particularities in Freudian thought both important and contestable, seem to flow from this position. I will restrict myself to mentioning only one: the lack of the category of the future.

The future of the individual is limited in fact: it stops at death; hence it would be difficult to make it a fundamental category in individualist thought.

Furthermore, the absence of the category of totality in individualism also brings about, if the thinking is consistent, the loss of the idea of time and its replacement by the two other equivalent and atemporal categories of *instant* and *eternity*. It is not by change that this atemporality characterizes the two great rationalist systems: that of Descartes and that of Spinoza. Similarly it could be argued that, even though the political involvement of the thinkers of the Enlightenment in the 18th century led them to strive for a better future, it was still no less difficult to ground this idea of future (which very often they thought of under an atemporal form) in their systems.

Thus also there is nothing surprising in the fact that Freud's thought, in spite of its genetic character, ignores the future and seems to evolve in a two-dimensional temporality: the present and the past, with a marked predominance of the latter. Unless I am mistaken, the word future is found only once in the title of one of his works: *The Future of an Illusion*, where precisely it is argued that this illusion has no future.

But let us return to the problem that concerns us: Freud himself raised it in *Civilization and its Discontents*. There he observes, as a matter of fact, that the free satisfaction of the libidinal tendencies that have as objects the first beings encountered by the child (more specifically, as has been so often noted, the Mother and the Father), would seem to result in the creation of very small autonomous groups and prevent any constitution of a broader society.

Yet, men have created such broader societies and to do this, among other things, they have forbidden the satisfaction of the most intense libidinal drives, those which correspond precisely to the Oedipus complex. The forbidding of incest is one of the most widespread and most general social institutions we know of. Freud asks himself what could have brought men to freely accept such a serious and such a painful frustration in order to create social life and civilization, and he claims that this is one of the most important problems for the human sciences; and without any previous serious hypothesis. Now his answer, which it is true brings into question the whole of Freudian individualism, had long been formulated in Marxist thought.

With the development of the symbolic function, language, and communication, completely new and revolutionary means of satisfying man's other fundamental need, other than libido, namely the protection of life – protection against hunger, cold, etc. – came to the fore. We summarize the totality of these behaviors corresponding to this second need under the term 'mastery of nature'.

But, in spite of all the development and modifications brought about by the apparition of consciousness, of symbolic function and language, libido always remained individual, and the behavior corresponding to the need of mastering nature in order to ameliorate the conditions of life was changing through and through; for, with communication and language, there was developing the possibility of a division of labor which in turn was reacting on the symbolic function, and so on – this is what Piaget has called the rebound shock (*choc en retour*) – engendering something completely new and unknown up until now: *a subject constituted by many individuals.*

If I pick up a very heavy table with my friend John, it is not I who am lifting the table, nor is it John either. The subject of this action, in the strictest sense of the word, is constituted by John and me. (And it goes without saying that for other actions we would have to add other individuals in much greater numbers.) This is why the relations between John and me are not subject-object relations, as in the domain of libido or the Oedipus complex, for example, nor are they intersubjective relations, as the philosophers who take individuals as absolute subjects think, but they are what I would propose to designate by a neologism as *intra-subjective* relations, that is, relations between individuals each of whom is one of the partial elements of the true subject of the action.

But for us to be able to pick up the table together, we have to be able to designate it, and to designate a whole lot of other things; there must, therefore, be some theoretical thought. Thus, everything that will be said on the theoretical plane will constitute (to the extent that it remains tied to the behavior that has, as its object, either other human groups or the surrounding natural world) a domain where the subject will be transindividual, and where all communication between John and me about the table that we are carrying will remain a communication inside the subject, an intrasubjective communication, as we have just said.

I think this is where the fundamental break between dialectical sociology and psychoanalysis is situated. For Freud, who discovered the domain both of the unconscious drives and also of the behaviors meant to satisfy them, also naturally saw the domain of the drives created, or at least assimilated, by society; and he saw that their satisfaction is essentially tied to consciousness, that is, fundamentally, the domain of behavior directly or indirectly oriented toward the mastery of nature and cultural creation.

Unfortunately he did not register the change in the nature of the subject that takes place in the passage from one to the other, and that is why he always attributed drives to an individual subject. It is highly typical that he designated all drives under the global term of *Ichtriebe*, i.e. 'drives of the Ego', whereas what characterizes the appearance of man, the birth of civilization and, connected with it, the appearance of consciousness and of the division of labor, is precisely to have made possible the development of a sector of life and behavior with a subject that is transindividual and infinitely extensible, a subject acting, as we must recall, not only on the natural world, but also on other men or on other groups of men, who then constitute the object of its thought and its action.

The true opposition is not, as Freud thought, between the drives of the Id, an individual subject predominantly unconscious and biological, and the drives of the Ego, also an individual subject but predominantly conscious and socialized. It is situated between the drives of the Id and those that structure the consciousness of a being who, while remaining biologically an individual, now represents, as a conscious and socialized being, *only one partial element in a subject that transcends it.*

Let us add that, in this perspective, there is no difficulty in admitting that the energy spent in social behavior finds its origin in the transformation of libidinal drives. The truth or error of this claim is a problem of psychology.

It remains for us to emphasize that the conception of the subject as individual or transindividual is not merely a problem of terminology – in which case we would not give it any importance – but a crucial problem for any research in human sciences. It is indeed a question, no more nor less, of knowing with reference to which subject is situated the fundamental intelligibility of every behavior with even a partially conscious nature.

For psychoanalysis this intelligibility remains always individual, any further social intelligibility having only a secondary, derivative character, ultimately imposed from the outside, though subsequently interiorized.

For dialectical thought, on the other hand, it is intelligibility with reference to the collective subject that is primordial, while further intelligibility with reference to an individual subject has, as long as we do not have to do with irrational phenomena, such as mental illness, dreams and even the *lapsus*, a subordinate and secondary character.

To be sure, there is no question in all this of a collective consciousness that would be situated apart from individual consciousnesses, and there is no consciousness other than that of the individuals. Only certain consciousnesses of individuals are found in, not intersubjective, but *intrasubjective* relations one with another and thus constitute the subject of every thought and every action of a social and cultural character.

And so, the meaning discovered by psychoanalysis in the human manifestations that appear absurd at first glance (*lapsus*, dreams, neuroses) and the objective significations discovered by sociological analysis behind the apparent significations or the apparent absence of signification in sociological, historical, and cultural facts are situated by reference to different subjects. An individual subject coinciding with the biological subject in one case, a transindividual, or if one prefers, a plural subject in the other.

To avoid any misunderstanding, let us add that, under certain conditions, the plural subject elaborating theoretical thought and world visions can also elaborate an individualist vision; but then such a vision is no less collective than any other form of thought. For being isolated on his island, Robinson Crusoe is no less a collective creation than the visions and the forms of thought that deny any reality to the individual.

Of course the form of intrasubjective community that I have described in order to simplify the relations between two persons that aim at picking up a table is idyllic and quite remote from the operative social reality.

Still it was enough to illustrate the problem for I do not have time to insist upon the many forms of social pathology analyzed first of all by Marx and Marxist thinkers, but also by many other sociologists and historians, notably by Theodor Adorno and the Frankfurt school.

It will be necessary in concrete studies to analyze the different forms of social pathology and especially for contemporary Western societies, to analyze reification, the replacing of the qualitative and the human with the quantitative, and the pathologies of bureaucratic and technocratic organization. But whatever may be the forms of social pathology, they are fundamentally different from the pathological forms of libido; they are, in fact, pathologies of the transindividual subject, of cooperation and of the division of labor, while the others are pathologies of the individual.

I come to the essential question I would like to ask Paul Ricoeur. I am quite in agreement with him when he points out that every genetic structuralism, whether it be psychoanalytical or dialectico-sociological, is threatened by the same danger: that of reductionism, the *nothing more than*. This painting is *nothing more than* the expression of a libidinal desire; the work of Valéry *nothing more than* the ideological expression of 'petty bourgeois' thought.

I am also in agreement with him in thinking that every genetic structuralism must explain how a point of departure has been surpassed to arrive at a certain higher complex creation, and not how such a creation is reducible to its point of departure. But I would ask him whether, to the extent that it brings everything back to the libido and the individual subject, the *least reductionist* psychoanalytical thought *in its proper domain* does not necessarily become reductionist when it approaches cultural creation, whether it will not always be lacking the possibility of relating it to a collective subject, and whether society does not remain for it simply the medium through which the individual subject expresses himself. Now I think that this is one of the most problematic and least empirically satisfying points in the psychoanalytic perspective.

What are the relations between interpretation and explanation? I think I could give an answer. This important question, it seems to me, has been treated more often than not in a highly contestable manner. Indeed, *explanation* has been artificially opposed, as belonging to the realm of the physico-chemical causal sciences, to *interpretation*, something proper to the human sciences which would then belong to the realm of participation, dialogue, and sometimes affectivity.

For dialectical thought, the problem arises differently. To understand is an intellectual process.[3] The description of a significant structure according to what is essential and specific to it. To bring to light the significant character of a work of art, of a philosophical work, or of a social process, the immanent sense of their structuration, is to *understand* them, by showing that they are structures having their own proper coherence. To *explain* is to situate these structures, as elements, in more vast structures that contain them. *Explanation always refers to a structure that contains and surpasses the structure being studied.*

If I analyze the internal coherence of Pascal's *Pensées*, I understand them by means of a strictly intellectual activity. But if I situate these same *Pensées* inside extremist Jansenism or Jansenism in general, then I *understand* the latter and I *explain* the genesis of the *Pensées*. Likewise, if I insert the structure of Jansenism into the totality of the class relations in seventeenth century France, or in the *noblesse de robe* of that period, I *understand* the evolution of the *noblesse de robe*, I *explain* the birth of Jansenism, etc....

But then an essential question for our problematic arises:

To interpret a dream, to find its signification (I use this example, but I could just as well speak of a neurosis), the psychoanalyst cannot in any case stop at an immanent interpretation, at simply bringing to light its structure; he must appeal to the unconscious drives, insert the dream into something more vast, something beyond its simple, manifest content, and account for the work of transformation. The question therefore arises: why cannot we understand and interpret dreams as we might, ultimately, understand and interpret the *Phèdre* of Racine or the *Orestes* of Aeschylus? I have said that it seems senseless to me to admit the existence of an unconscious for Orestes; Orestes is no more than a literary figure attested by the text and has no existence whatsoever apart from this. But as we have just said, psychoanalysis cannot interpret a dream without going beyond its manifest content, that is, without falling back on explanation. In a sociological analysis, I admittedly almost always go beyond the text, but that is in order to *explain* it and not to understand it, whereas, to find the sense of a dream, the psychoanalyst is *obliged* to explain. There is no such thing as an *immanent analysis* of dreams.

The reason is that a dream does not constitute of itself a significant structure and is, as a conscious manifestation, only one *element* of such a structure (whether biological or individual), whereas social logic creates significant structures with a relative autonomy and a particular meaning. To be sure, there is an *explanation* for dreams as for conscious structures, but what distinguishes one from the other is their respective situation along a continuum going from the purely biological (without signification apart from explanation) to the great cultural work (capable, in principle, of a comprehension different from and independent of its explanation). In concrete studies, it is true, the explanation always helps comprehension and inversely.

This explains why, whenever we want to situate a dream with reference to conscious logic, that is *interpret it* or give it a meaning, we must have recourse to the unconscious as an *explicative* factor for the deformations, and we find its latent signification through the *explanation* of the distortions of meaning with reference to social logic. In short, what distinguishes a cultural creation from a dream is that it is situated at the level of signification with reference to the collective subject; not that psychoanalysis finds no libidinal significations, for there is no collective consciousness apart from individual consciousnesses, but every individual consciousness is composed both of libidinal elements whose subject is individual and conscious elements which pertain to the level of cultural creation and for which the subject is transindividual.

Of course, there are not two separate sectors in consciousness, but in this interpretation it is possible that the collective element might preserve its autonomy and its proper laws and thus create something quite significant with regard to the transindividual subject acting, laboring, and elaborating culture. In this case the elements of individual satisfaction can come into account only to the extent that they are adapted to this logic without changing it. They will be able to account for the fact that it is precisely this individual, Racine, who wrote the play that we know, and not some other. But the meaning of the plays, the absolute exigency, the silent characters and spectators who, for the hero, are represented by the existence of an insoluble contradiction, the distance that separates the tragic hero from the exacting beings, all this is a transposition of the categories elaborated by the Jansenist group, and the only question to which psychoanalysis can bring some information that is both valid and valuable is why these categories manifested themselves in the individual Racine with a special force, how it is that they coincided with his individual problems to such a point that this individual succeeded in giving such a coherent form to tendencies present with more or less intensity and coherence in all the other members of the group. If on the contrary, this coincidence between the individual and the intrasubjective does not come about, if the individual drives succeed in obscuring the logic and the structure of the collective signification, we then progressively go from the work of Racine to the average consciousness and, at the other extreme, to dreams and to pathology.

But to the extent that psychoanalysis tries to bring the totality of consciousness back to the Ego, the individual, the libidinal, it is destined inevitably to wipe out the differences between these various types of expression by abandoning all criteria that would enable it to distinguish the sick person from the genius.

When the psychoanalyst finds himself before a work or a painting, by his very method he situates them on the same plane as any other expression of the same nature by a sick or alienated person, and he is right from his own viewpoint, since it is highly probable that the work has an analogous function for its author, for example, analogous to that of a painting for the lunatic. Hence the psychoanalyst probably is not without justification in saying that in writing his play, Racine was expressing this or that unconscious and libidinal aspiration.

The question remains, however, of knowing what is the relation between this aspiration and the signification of the work as a whole, which could be referred only to a transindividual subject.

And it is only when the former succeeds in expressing itself without clouding in any way the latter, that it can reinforce its cultural value – literary, in this case.

We must always keep in mind the fundamental affirmation of structuralist genetic sociology, that, the significant (and this means collective) coherence of works of art, far from being more individual than the coherence of the thought and writings of average individuals, on the contrary reaches a much higher degree of socialization.

For our part, whenever we have to do with an important cultural text or an historical event, we find ourselves before an object of study wherein the transindividual, or if you prefer, the collective subject, has found expression at a much higher level of coherence than that reached by the consciousness of average individuals, yours or mine for example, and that means at a level where positive study can set aside the individual subject. Not that libidinal satisfaction did not exist or constitute an important link in the genesis of the work, but it is especially difficult to lay hold of and contributes very little to the understanding of the object under study.[4]

On this point, I recently had a discussion in Montreal which I would like to evoke here, since it revealed to me a misunderstanding to be avoided. After a presentation on sociological aesthetics in which I had mentioned the example of 17th-century French tragedy, certain members of the audience, who were incidentally professional historians of literature, raised an unexpected objection: 'Everything you say is very good, and we are quite prepared to admit it, but can your sociological categories apprehend the aesthetic fact? To do this would not you have to bring in specifically literary categories?'

But it has never been my intention to use sociological categories *for understanding a work*. The aesthetics of a work depends first of all on its richness, on its significant coherence and the coherence between its universe and its form in the strict sense of the term. Only, to bring this internal signification and coherence to light, I have to make use of *explicative* procedures which imply an insertion into a more vast structure, that is into a social structure. But in doing this I am by no means, and not in any case, trying to find sociological elements inside the work. The work is nothing else than a text with, or without, a coherent structure: when it is a question of interpreting the *Orestes*, we can, no doubt, appeal to explicative procedures by inserting it into a structure that contains it, into the psychology of Aeschylus, for example, or into Athenian society, but we have no right to add a line or a word to what is written. The character of Orestes therefore can possibly be explained by *Aeschylus' unconscious* or by the social structures of Athens, but we could never attribute a particular unconscious to him unless we find in the text a phrase explicitly affirming the existence of such an unconscious, nor could we still less introduce sociological categories into the text.

In short, in cases where there is a predominance of the libidinal signification, the researcher is obliged to have recourse to explanation in order to interpret. In the case of cultural creation, interpretation and explanation are complementary procedures which mutually facilitate one another as research advances, but nonetheless they remain different procedures. The individual cannot penetrate into a work of art without weakening or destroying it; he can penetrate only to the extent that it is integrated into the collective signification. The result of this is that, when the psychoanalyst comes to study such a work, looking for individual significations in it, he will undoubtedly find some, and sometimes many of them, but almost always he will fragment it and leave to one side its total structure and its essential problematic.

Whether it be the Moses of Michelangelo, or the smiles of Saint Anne and of Mary in the Holy Family, the essential thing is not to know what, in the life of Leonardo, in his relations with the Pope or with his father, could have led him to paint these two as he did, for analogous libidinal relations could have existed at another instant and in another society, but what brought it about that this expression of individual desires could be inserted into a structure and a work of art which had, at the level of painting, a very high degree of significant coherence. Relations between brother and sister analogous to those between Blaise and Jacqueline Pascal have existed perhaps by the thousands. It was at a certain moment, and in a certain context, that such a relation proved particularly favorable for the expression, at an extreme level of coherence, the level of a philosophical system, of a whole vision that had been elaborated in Port-Royal, in Saint-Cyran, and beyond this, inside a particular social group, that of the summit of the *noblesse de robe* in France.

This brings us to a particularly important problem: that of the nature of aesthetic satisfaction, since, as seems quite evident, this satisfaction entails an element of pleasure. Someone may ask me, 'What do you make of the pleasure felt in the presence of a work of art? After all it is of the same order as that of which Freud speaks in the realm of libido.'

Yes and no, as is always the case in the relation between psychoanalysis and dialectics.

Yes, to the extent that there exists a close kinship between the social function of the work of art and the individual function of the imaginary, of dreams and of madness, such as it was described by Freud. Both the one and the other, indeed, arise from the inadequate aspirations of the subject with regard to reality. To bear up under the frustrations that reality imposes, one is obliged to compensate by the imaginary creation which, incidentally, as long as it is a question of a normal and not a pathological psychology, favors its insertion into the surrounding world.

No, to the extent that, at the individual level, these frustrations almost always concern an object (and, most often, a human being functioning as object) that the individual subject was unable to possess. Inversely, at the level of the transindividual subject, the aspiration does not concern, or at least does not primarily concern, an object but a significant coherence, the frustration being constituted by the fact that reality imposes upon everyone of us a certain degree of incoherence and compromising.

The latter result, not only from the relation between the collective subject and the surrounding world, but also from the very structure of this subject composed of individuals who belong to a large number of diverse social groups and in whose consciousnesses libidinal elements still intervene (Freud has shown this to our satisfaction). These individuals thus constitute mixtures, and the transindividual subject a group which tends toward a coherent signification without ever succeeding in attaining it in reality.

Thus the most important function of the literary and artistic creation seems to us to be to bring, on the imaginary plane, that coherence from which men are frustrated in real life, precisely in the way that, on the individual plane, dreams, delusions, and the imaginary provide the object or the substitute for the object which the individual could not possess in reality.

There is nevertheless a great difference between the coherence of a conscious structure referred to a collective subject, a coherence which by the way is not always reducible to the explicit meaning, and the latent coherence of a libidinal structure referred to the individual subject.

I have already said that, in one case as in the other, the imaginary creation serves the function of compensating for a frustration, only in the case of the individual subject and the libidinal frustrations studied by Freud, it is always a matter of getting around the censor of consciousness, of surreptitiously introducing elements in the latter that it has been refusing to admit and had repressed.

In the case of cultural creation, on the contrary, the aspiration to coherence constitutes an explicit or an implicit tendency of consciousness, a tendency that is in no way repressed. Here creation reinforces consciousness in its immanent tendencies whereas libido more often than not tries to get around it and to introduce into it elements that are foreign and contrary to its nature.

This difference, furthermore, should not surprise us, given the fact that consciousness is intimately connected with the collective subject or, if you wish, the transindividual subject, while it appears, on the other hand, in the libidinal domain only to the extent that, when libido manifests itself *in man*, it is obliged to incorporate this general element of the human structure which is the existence of a conscious life, while trying still to maintain its own proper structure by assimilating it or at least by assimilating certain of its elements to its own proper needs.[5]

To be sure, unconscious elements are almost always integrated into the coherence of a global structure, either by deforming it into a *lapsus*, a dream, or madness, or by conserving its explicit and clear structure but adding to it a further libidinal type of determination (*surdétermination*).

In this last case it is possible that to satisfactions of the transindividual type, provided by the cultural work, are added individual satisfactions and pleasures common to all men (I would be willing to call them general satisfactions), and when this takes place, the work is more likely to be well received. But for all this improving the chances of being well received, such coincidence is not necessary and must be examined in different particular cases.

What is important to be emphasized, however, is the fact that, when there is question of interpreting a cultural work in what is specifically cultural about it, the system of thought that refers to the individual and libidinal subject can only play a secondary role and could even, especially in the case of an important work, be completely left out. We must also add that, given the extreme difficulty of knowing an individual consciousness, especially when it is that of an author whom we cannot analyze for months or who has been dead for many centuries, it is above all the work and its transindividual coherence, in the measure that the sociologist succeeds in bringing it to light, that bring forth the decisive elements for understanding the author. This was, for example, the case in large part during my studies of Pascal and Racine.

Let us add that the study of the conflicts and interpenetrations between intrasubjective coherence and libidinal coherence raises another problem that has been emphasized particularly by the Frankfurt school (but it should, it seems to me, be studied in a more concrete way, better situated historically, and especially in the context of contemporary society at the same time), the problem of discovering to what extent transindividual coherence, with everything that it entails in the practical order, the economic, the social, the political, and also the cultural, engenders important frustrations in the individual and libidinal life of the subject. It is well known that, according to Freud who said so categorically enough, all social life implies frustrations of libido, so that there will always be some discontents in culture. Yet this remains extremely general for, as Marcuse, Adorno and their colleagues saw very well, these frustrations can have a more or less intense character, and especially be spread out in uneven fashion among the individuals and among social groups. And it is a fact that they are not distributed equitably, and homogeneously, among the different social classes.

Now the question arises today of knowing to what extent the high technical level reached in the advanced industrial societies would allow these frustrations for each individual to be reduced as low as possible, if the social order were organized in a more efficient way, and to what extent the concrete forms that contemporary society has taken, especially the technocratic society of organizational capitalism, might be at the origin of the particularly intense degree of frustrations that most men of today must accept.

Whatever the case may be, I do not believe it enough to say that labor can become a pleasure and that today, given the high level of standards of living, we could effectively confer upon it the characteristic of a game. The social organization actually existing, unquestionably has the tendency to blot out the individual within the transindividual subject, as it organizes a kind of spontaneous and implicit brainwashing.

Thus the most important practical problem of our epoch is precisely to know in what direction to take action, what attitude to assume so as to contribute something to social evolution, an orientation different from the one it seems to be taking spontaneously, a direction that would permit us to modify an evolution on the verge of suppressing the qualitative element and the human personality, while considerably raising the standard of living and the possibilities of consumption for individuals. Thus we might hope to create a situation on the level of culture whose paradoxical element I once characterized by writing that we are in danger of coming to an extensive production of illiterate people with university degrees and doctorates, to replace it with an orientation toward a social structure capable of effectively assuring the harmonious development of both the libidinal subject (who today has the right and the possibility of obtaining much greater and more intense satisfactions than those that were possible in previous societies that were living under the pressure of penury and scarcity) and the intrasubjective and socialized personality; the harmonious development of the individual and the personality.

But this brings us, in psychoanalysis as well as in sociology, to the problem of what attitude to take before contemporary society and what alternatives now present themselves, a problem whose importance I am far from underestimating but which does not constitute my immediate theme.

I am not sure that everything has been said when one emphasizes (as some have done) that to replace the quantitative by the qualitative leads to a return to archaic values; I am not so sure that it is *only* a matter of archaism here, nor also, on the contrary, of the future and novelty. Since human needs are by nature bound to the qualitative aspect of objects, it is possible that the re-appearance *in the consciousness* of men who enjoy qualitative relations with things and with other men, might be both, at least from a formal viewpoint, a return to archaic values and also a real and essential orientation toward possibilities of human development in the future.

Finally, I would like to use a particularly suggestive example, which I have read in Freud's work on Leonardo, that the construction of the flying machine is intimately connected with the libidinal symbolism that the psychoanalyst often finds in the dreams of his patient during his therapeutic work. Not being a psychoanalyst, I could not enter upon a technical discussion of this affirmation, and I am willing to accept it. What interests me is the hypothesis according to which, each time someone constructs a flying machine in reality or in imagination, and notably Leonardo who long before such a thing was technically possible imagined such a machine, it is a libidinal element that comes into play with preponderance; ultimately, it would seem to have led to the development of our contemporary technology of aviation.

It seems to me that, even if we admit the three points of departure in this analysis, namely: the fact that men often dream and see themselves flying; the fact that such dreams are connected with certain libidinal drives and constitute sublimated satisfaction; the fact that Leonardo did imagine a flying machine and that since then, the technology of flying has asumed very great importance in human society and the life of men; we could not admit the connection Freud tries to establish and that psychoanalysis tends to reinforce.

It is possible – if Freud is right – that from all times and at every epoch, people have dreamed of flying, but at the moment when Leonardo constructs his models of flying machines he finds himself at a certain, precise level in the development of science and technology, and the flying machines in his case were only one attempt among many others. It would be difficult to sharply separate the model of the flying machine, which supposedly has a libidinal signification, from all his other attempts which, for the historian of technology and science, have an absolutely homologous character.

This is to say that, for all attempts to situate this work in the history of scientific and technological thought, the fact brought to light by Freud and the connection he tries to establish, without in any way necessarily denying the existence of this connection, have only a secondary and even negligible character, and that, if we wish to understand the real and objective nature of the phenomenon, we must, above all, come to the order of the historical transindividual subject.

This example seems particularly pertinent to illustrate the problematic I have tried to treat here.

In conclusion, I think we must both accept and reject psychoanalysis, accept it in any case on the level of psychological individual research and of clinical therapeutics and also allow it some not-negligible place in the analysis of the psychological processes in cultural creation, but also avoid any attempt to bring the objective signification of such creation (and when I speak of objective signification it would perhaps be better to say *specific* signification, for it is a matter of the literary signification of literary works, pictorial signification of pictorial works, philosophical signification of philosophical systems, theological signification of theological writings), back to an individual subject, a procedure which would necessarily and for methodological reasons lead to a dangerous reduction and even to the complete eclipse of this signification.

Paris

REFERENCES

[1] This affirmation also extends to biological behavior, for, at the human level, the biological itself becomes signifying at the symbolic level; it becomes at least, as Sartre said, "consciousness of self"; and subsequently it can become reflexive consciousness.

[2] This explains among other things Freud's deep-seated hostility to religion of which Ricoeur has spoken (that is a very frequent trait among the philosophers of the Enlightenment), and the short-circuit in reasoning which caused him to reduce religion to illusion and ideology.

[3] This does not mean a purely theoretical attitude in the sense that any theoretical attitude is at the same time theoretico-practical.

[4] This was the case, for example, when we tried to understand Racine's tragedies, Pascal's *Pensées*, and the Jansenist movement to which both of these were connected.

[5] When we speak of transindividual or collective subject, we must always keep in mind that there is no question of what Durkheim's school used to designate by these terms, namely a collective consciousness situated apart, above, or beside individual consciousness, but on the contrary it is a question of a collective subject in the sense that Bastide gave this term, namely the relations between the Ego and the others in a situation where the other is not for the Ego an object of thought, of desire, or of action, but a subject in the process of elaborating an awareness and of performing an action in common with the Ego.

GAJO PETROVIĆ

DIALECTICAL MATERIALISM AND THE PHILOSOPHY OF PRAXIS

I

Discussion on Marx as philosopher is not reduced to the question: 'What are the main themes and theses of Marx's philosophy?', 'What is the essential meaning of his philosophical thought?', 'What is the historical value and importance of his philosophical work?'. The question 'Is Marx a philosopher at all?' still arouses controversy. There is nothing wrong with the question. But this does not mean that every possible answer to it is equally good.

The thesis that Marx is a philosopher has been disputed not only by many 'experts' and 'critics' of Marxism, but also by many prominent Marxists. During the Second International it was disputed by orthodox revolutionary Marxists (e.g. F. Mehring), by centrist opportunists (e.g. K. Kautsky) and by open revisionists (e.g. E. Bernstein). It is disputed by many contemporary Marxists as well. But all who agree that Marx is not a philosopher, disagree on what he is. The thesis that Marx is a philosopher is most often countered with one of the following: 'Marx is a non-philosopher', 'Marx is an anti-philosopher', 'Marx is a trans-philosopher'. Each of these theses has its particular attractions and its particular 'arguments'. Only these are false attractions and shaky arguments.

One view, once widely accepted and still encountered nowadays, is that Marx was not an opponent of philosophy, but that he himself was never seriously concerned with it, or at least did not make any major contribution to it. Those who share this view differ only on the question why he did not achieve much in this 'field'. His critics are rather inclined to deny him any feeling and ability for philosophy; his followers and supporters frequently insist that he was continuously overloaded with other, more urgent work and so could never find enough time for philosophy. Whatever one may think of these 'explanations', their common presupposition seems acceptable. With the help of Marx's own statements it is

Boston Studies in the Philosophy of Science, IV.

possible, for instance, to show that he never realized some of his philosophical 'wishes' and 'plans'. But which author has ever realized all his wishes and intentions? Whatever the amount of his unrealized philosophical plans, one thing is indisputable: Karl Marx, Doctor of Philosophy, left, beside his doctor's thesis, several expressly philosophical works (*A Contribution to the Critique of Hegel's Philosophy of Right*, *Economic and Philosophical Manuscripts*, *The Holy Family*, *German Ideology*, etc.).

In his youth, some people say, Marx was a philosopher, but already in his *German Ideology* he broke with philosophical speculation to become a critic and opponent of philosophy as such from the positions, and in the name, of empirical, positive science. This is borne out by his explicit statements ("Phrases about the world cease and real knowledge has to take their place. When reality is depicted, philosophy as an independent branch of activity loses the medium of existence"); this is also testified by the fact that after the *German Ideology* he stopped working on philosophy and devoted himself to economic, historical and political studies and practical revolutionary activity. Facts are facts, of course, but it is also a fact that Marx's 'mature' 'non-philosophical' works do not contain 'real knowledge' but also philosophical 'phrases about the world' (*Das Kapital* is written throughout in 'philosophical phraseology'). It is also a fact that Marx's 'non-philosophical' ('economic', 'political', 'historical') works are in their deepest sense philosophical, for they are directed at the essence of the modern world and man, they present a radical criticism of an alienated society and a 'non-scientific' vision of a really humane community.

If Marx is not an a-philosopher or anti-philosopher, perhaps he is a trans-philosopher? Perhaps the most essential thing is Marx's idea that philosophy should not be simply rejected or ignored, but that, through being realized, it should be superseded? Is it not Marx's 'ideal' that reality should become philosophical, and that philosophy as such should disappear? The idea is very attractive. But for this very reason it must be carefully examined. Convincing arguments do not preclude still more convincing counter-arguments.

One argument says Marx taught that the proletariat cannot be abolished without the realization of philosophy. But how can philosophy realize itself, if it does not abolish itself? – To this one could reply with counter-arguments: Philosophy can be abolished only by being completely realized.

And only a completed, closed philosophy, by which the completed man thinks of himself in his completed world can be completely realized. An incomplete, open philosophy can be realized but not definitively. The complete realization of philosophy presupposes the end of history. History can be broken for ever by a cosmic catastrophe or a thermo-nuclear war. But history cannot say to itself: Whatever has been has been, and now I am going to sleep. Completed history is a *contradictio in adjecto*. There is no evidence that Marx was at any time ready to accept such a contradiction.

The second argument runs: Marx's 'philosophy' is not set out according to traditional philosophical disciplines. There are no such things as Marx's ontology, epistemology, logic, ethics, aesthetics, etc. But this is not all. Marx's 'philosophical' thought is so deeply merged with his 'sociological', 'economic', 'political' and other thought that one cannot tell where one ends and another begins. Where philosophy is not separated from non-philosophy and dismembered into disciplines, there is no philosophy. – The reply offers itself here: The essence of philosophy does not consist in its formal separation from science, art, religion, nor in a schoolish grouping in disciplines. Plato and Nietzsche did not break up their philosophies into disciplines, nor did they separate them so strictly from, say, the arts. Aristotle and Hegel may be said to have expounded their philosophies according to 'disciplines'. But nevertheless it would be going too far to maintain that the distribution into disciplines is the essence and value of their philosophies.

The third argument may seem to be the most convincing: Marx reproached the philosophers with having only explained the world, when the point was to change it. Does not this reproach reveal the meaning and essence of Marx's 'philosophical' thought? Is it not Marx's basic intention to give up trying to explain the world philosophically and instead to embark upon its practical revolutionary change? To this one could reply with the counter-question: Is the essence of philosophy as such a mere 'explaining' which refrains from revolutionary change? Did not, say, the French thinkers of the Enlightenment consciously aim at a revolutionary change of the world on the basis of their philosophy? And surely even those philosophers who wanted to keep exclusively to explaining, in fact by their very 'explanations' encouraged and took part in a revolutionary change of the world?

Finally, somebody may remark: Granted that Marx was a philosopher after all, does this not diminish the essential novelty of his message? Does this not make him just one in an endless series of philosophers? However, if Marx is one among many, it does not follow that he is equal with everybody else. And the novelty of his thought, and his thought in its integrity is certainly more endangered by the demand that philosophy should become definitely 'realized' and 'abolished'. A definite reconciliation of thought with reality is possible only as a definitive capitulation of revolutionary thought before reactionary reality. A definitive abolition of philosophy can be imagined only as a definitive victory of blind economic forces or political violence. Thus it is unimaginable.

II

The thesis that Marx really is a philosopher does not solve the question about Marx as a philosopher. Inevitably the question imposes itself: Wherein consists the essence and importance of Marx's philosophy? The meaning of this question is not: What are the general or common characteristics of all the philosophical conceptions which have appeared under the name of 'Marxist' or 'Marx's' philosophy? This name has been used of conceptions which have almost nothing in common; their systematic exposition may be highly interesting, but does not solve our question. Nor is the meaning of the question: What are the philosophical conceptions that Marx always stood for? What philosophical conceptions did he express most frequently? What philosophical conceptions did he consider to be the most important? These and similar questions may be interesting; to some it is probably possible to give a precise answer, and to some not.

The question of the essence of Marx's philosophy cannot be reduced to the question: What has Marx's or Marxist philosophy been so far?; nor can a reply to it be supplied by an empirical description or summing up of what Marx or some Marxist wrote about philosophy. The question of the essence of Marx's philosophy aims at what makes Marxist philosophy what it really is: the fundamental possibility for thought in our times, the critical humanistic thought of modern man about himself and his world. Therefore the answer to this question cannot be a report on what has been or is still going on, but participation in the creation of something that can and should exist. A correct answer cannot be obtained by a detailed comparison of quotations from Marx, but only by creative thinking in the spirit of Marx, by co-thinking with Marx and by trying to think through Marx's guiding ideas.

Could it not be said that the essence of Marx's philosophy is that it is a 'philosophy of action', 'philosophy of deed' or 'philosophy of praxis', that it is such a 'theory' as does not remain just theory but also demands the act of changing the world and at the same time participates in this work? This could certainly be said, and this definition would not be difficult to substantiate and explain. There are so many critical statements made by Marx against 'theory' separated from 'praxis', while Marx's whole life offers evidence that cabinet thinking for thinking's sake was not his ideal.

But it is not enough to say that Marx's philosophy is a philosophy of action or a philosophy of deed or praxis. This thesis can be understood in various ways and so it must be explained in greater detail to preclude wrong interpretations.

When we say that Marx's philosophy is a philosophy of action (deed, praxis) this cannot mean that action is added to this philosophy from outside by a coincidence, or on the basis of a special decision of the philosopher. If we characterize a philosophy as essentially a philosophy of action (deed, praxis) this must mean that action (deed, praxis) follows from the essence of its theoretical content, that the transition from theory to praxis is its essential 'thesis'. Thus if we characterize Marx's philosophy as a philosophy of action, we must explain what are its essential theses that prevent it from remaining pure theory and cause it to become the action of revolutionary transformation of the world.

The reply to this question is seemingly simple: Marx's philosophy is the philosophy of revolutionary action because its nucleus is the conception of 'naturalism-humanism' ('naturalistic humanism', or 'humanistic naturalism'), the conception of man as a being of praxis who by his free creative deed moulds and changes his world and himself. There is no doubt that in his early works Marx developed such a conception. And there is no ground for disputing the view that Marx did not give up this conception in his 'mature' works, but continued developing and concretizing it. It would be more justified to dispute the expressions quoted than the conception itself, but one should keep in mind that these expressions can be arrived at on the basis of some of Marx's texts. When for instance Marx says that "communism as a fully developed naturalism is humanism and as a fully developed humanism is naturalism"[1] he obviously means to say that naturalism and humanism, which in their incomplete "not fully developed" form may be different and even contrasting, become identical in the completed, 'fully developed' form. The terms 'naturalism-humanism', 'naturalistic humanism', 'humanistic naturalism', are naturally suggested as suitable for such a conception. Since the two composite parts of these terms were understood in various ways both up to Marx's time and afterwards, the combined term may also be interpreted in different ways. It should therefore be kept in mind that in the text below the term 'naturalism-humanism' is used to denote that philosophical conception by which Marx tried to overcome the opposition between naturalism and humanism.

However, under the name of 'Marx's philosophy' a conception is often encountered which is more appropriately called 'dialectical materialism', a conception which was developed by Engels, Plekhanov, and Lenin, further elaborated by Soviet philosophers between the two World Wars, and canonized by Stalin in a simplified form. In the form given it by Stalin this conception is definitely discredited but many still think that it is 'good' in the form given it by Engels and Lenin. Even many who are not quite satisfied with the form which dialectical materialism was actually given by Engels and Lenin consider that in the works of the classics of Marxism there is room for a "more creative, better diamat than we have today".

Under the name of 'Marx's' or 'Marxist' philosophy we then encounter at least two philosophical conceptions; one of them we have conditionally called 'naturalism-humanism', while the other is usually described as 'dialectical materialism'. We must then consider the question: What is the relationship between the conception of 'naturalism-humanism' and the conception of 'dialectical materialism'? Are these two names for one and the same thing, two aspects of the same conception, two conceptions one of which is part of the other, two different conceptions which complement each other, two different conceptions irrelevant to each other, two conceptions which exclude each other? For each of these theses one could probably put forward some argument or other, but a detailed consideration of all theoretically possible hypotheses would take us too far. We shall content ourselves with saying a little more only of the thesis which seems to be most correct, that 'naturalistic humanism' and 'dialectical materialism' are two different conceptions which logically do not complement each other, nor are simply indifferent to each other, but at least in certain essential points are mutually exclusive.

III

In favour of the thesis that Marx's 'naturalism-humanism' is incompatible with 'dialectical materialism' one could quote several statements by Marx, as for instance that "consistent naturalism or humanism is distinguished from both idealism and materialism, and at the same time constitutes their unifying truth".[2] But such statements are not decisive; it is more important to consider the main theses of dialectical materialism and see whether they can be reconciled with the basic ideas of Marx's humanism.

The basic ontological thesis of traditional dialectical materialism is the thesis of the primacy of nature in relation to the spirit, of matter in relation to mind, of the physical in relation to the psychical. This is not exclusively Stalin's thesis. It is well known that Engels regarded as the great basic question of all philosophy the question of the relationship of thinking to being, of the spirit to nature, and, according to the way in which they replied to this "paramount question of the whole of philosophy", to the question "which is primary, spirit or nature?", he divides all philosophers into "two great camps": the idealists who insist that the spirit is primary, and the materialists who regard nature as primary.[3]

Engels's thesis on the relationship of thinking and being as the basic question of philosophy was also endorsed by Plekhanov, only he often formulated it in terms which were much more customary in the philosophy of his time, for instance as "the question of the relationship of the subject to object"[4] or the "question of the relation of 'I' towards 'Non-I'".[5] Although he did not classify all the replies to this question simply as materialist or idealist, taking into consideration non-monistic solutions as well, as for instance dualism, Plekhanov too believed that the fundamental philosophical trends were materialism and idealism, and the only correct solution – materialism.

Engels's formulation of the basic question of philosophy and his thesis of the primacy of matter, or nature in relation to consciousness or spirit, was endorsed by Lenin as well, at least during the phase in which he wrote his *Materialism and Empiriocriticism.* As can best be seen from the titles of two sections of this work, Lenin considered as essential components of the basic question of philosophy two 'more concrete' questions: "Did nature exist before man?" and "Does man think by means of brains?" Both these questions Lenin answered in the affirmative, and we can agree with his replies; what alone is disputable is whether Lenin's sub-questions exhaust the meaning of Engels's 'basic question' and whether this question of Engels is really the 'basic question of philosophy'.

I do not propose to maintain that the basic philosophical question, as understood by Engels and Plekhanov and concretized by Lenin, is meaningless. But everything that is meaningful is not 'basic'. Besides, every question is based on certain assumptions, and it will not be unwise to ask whether Engels's 'fundamental question' does not already contain certain assumptions. Does it not assume that in a certain way the world is divided or split into two main 'parts', 'sides', 'aspects' or 'forces': matter and spirit? If the object of philosophy is the world in which we live, and if matter and spirit are the two fundamental realities of this world, it is indeed the fundamental task of critical thought of the world to determine the relationship between these two fundamental realities.

But is the world really divided in essence into 'nature' and 'spirit', 'matter' and 'consciousness', 'being' and 'thought?' No reply is really possible if we do not know what 'nature', 'spirit', 'matter', etc. are. Let us assume for the present that 'matter', 'nature', 'being' are the same and that we know what they are; but what about the other term of the relation, spirit?

Spirit can be conceived in various ways, among other things as non-human, independent of man, objective and absolute, populary called 'god'. If we assume the existence of such a spirit, the question of the relationship of nature or matter towards absolute spirit will naturally emerge, and the various aspects of this question call for an answer. However, there are few people nowadays who take this assumption seriously, while those who have advocated dialectical materialism, have as a rule rejected it outright. But if there is no such thing as absolute spirit, then there is little sense in asking about its relationship to matter and/or nature.

'The basic question of philosophy' can also be interpreted as not concerning the relationship of nature to absolute spirit, but the relationship of nature to man's spirit. But man is not merely spirit; 'spirit' is but one 'part', 'side' or 'aspect' of man. It would be strange if the question of the relationship of one part of man to matter or nature were to form the basic question of philosophy. Is not the relationship of integral man towards the world in which he lives a more important question?

Faced with this objection we might try to save the 'basic question of philosophy' approximately as follows: The spiritual side of man's being may be one of the two terms of the basic relation in which philosophy is interested, for the other term is not matter or nature in general, but man's matter or nature. The object of philosophy is man, and in man we can distinguish the 'material' or 'natural' and the 'ideal' or 'spiritual' side. The division into 'matter' and 'spirit' is the fundamental internal division of man, and thus the question of the relationship of matter and spirit in man is the basic question of philosophy.

However, such rescue efforts are of no great use. Man is a united being whose integrity does not exclude internal differentiation. His practical activity is differentiated in many forms, but it is very difficult to regard these forms as divided into two main kinds, material and spiritual. Hardly any activity of man is exclusively material or exclusively spiritual. Let us take, for instance, political activity: is man, when acting politically, active as matter or as spirit? Or let us take man's artistic activity: is the activity of a painter or sculptor spiritual or material?

Consequently, division into matter and spirit is not the basic division of the world we live in, nor is this basic division within man. How then can the basic question of philosophy be the question of the relationship between matter and spirit? Is not the question possible only given certain dualistic assumptions which Marx's 'naturalism-humanism' excludes?

It is very difficult to say if there is a basic question in philosophy and how it could be formulated most adequately. But is it not rather plausible that the question of man's relationship to the world is wider and more fundamental than the question of the relationship of spirit to matter (or subject to object) and that the latter is but a distorted form of the former?

IV

Dialectical materialism cannot simply be reduced to the thesis of the primacy of matter over spirit. A vital component of this conception is the theory of reflection which has been discussed in the greatest detail in Todor Pavlov's well-known book *Theory of Reflection.* Stalin too was strongly in favour of this theory, but it does not belong exclusively to Stalin and Pavlov. In his book, Pavlov merely 'elaborated' the thoughts contained in Lenin's *Materialism and Empiriocriticism*, while Lenin only followed up Engels who in his reflections on the basic question of philosophy maintained that this question had yet another side: "Is our thinking capable of the cognition of the real world? Are we able in our ideas and notions of the real world to produce a correct reflection of reality?" This formulation of the question clearly indicates Engels's reply.

The theory of reflection does not belong to dialectical materialism only because it was endorsed by Engels and Lenin among others, but also because it seems to be the most adequate complement to the materialist thesis of the primacy of matter over consciousness. But this is not to say that this theory deserves a place in Marx's philosophy as well.

The theory of reflection appears in two basic variants. According to one, which we find in Lenin and Todor Pavlov, our entire spiritual life is in essence a reflection, and all forms of our consciousness are just different forms of the subjective reflection of objective reality. But it is not consciousness alone that is a reflection; matter too possesses a property akin to sensation, the property of reflecting. Reflection is a general property of the material world which in its higher form makes up the essence of our entire spiritual life.

In this version then, this is a theory of the consciousness or thinking which does not aspire at solving by itself the problem of knowledge and truth, and so introduces a supplementary theory for the solution of this problem. For Engels, Plekhanov and Lenin this is what is called the theory of adequation or correspondence, a theory according to which our reflections are true when they correspond to reality. Thus in Lenin's interpretation of the theory of reflection every consciousness is reflection, and reflection can, although it need not, correspond to what it reflects. If there is correspondence, agreement between reflection and reality, the reflection is true.

According to the second version of the reflection theory, it is reflection that is the very essence of truth and knowledge. The reflection theory need not be 'supplemented' by any theory of adequation or correspondence. To say that a proposition is true, simply means that it reflects reali y. This version is in fact encountered more frequently among those who oppose the reflection theory than among its advocates.

The two versions must certainly be distinguished. This does not mean, however, that one is better than the other.

In its both main versions the reflection theory is incompatible with Marx's conception of man as a creative being of praxis. But what is more important, the reflection theory cannot easily be reconciled with the phenomena of consciousness, knowledge and truth.

When I say that it is irreconcilable with the phenomenon of consciousness I have in mind the fact that it cannot explain to us why, in what sense, and in what way all our conscious acts reflect reality. For what do the will and emotions reflect? Are love, hatred, envy and malice only different forms of the reflection of the external objects towards which they are directed?

The reflection theory appears more acceptable if regarded only as a theory of knowledge and truth. But if we study it more carefully we shall see that it is not satisfactory even as a theory of the true proposition. We all state true propositions daily, whose truthfulness we do not doubt in the least, although it is impossible to say what they may reflect. A negative existential proposition, for instance, is true if what it denies does not exist. How can such a proposition be interpreted as the reflection of objective reality? The whole system of mathematical propositions is a system of true propositions for which it is difficult to say what they reflect. And what is reflected by propositions about the past, the future, about possibilities or impossibilities? The reflection theory seems to be unable to survive, even as a theory of the true proposition.

I have pointed out that the reflection theory is irreconcilable with Marx's conception of man as a creative, practical being. In saying so I do not mean to say that this theory cannot be found at all in the works of Engels and Lenin or even of Marx himself. The reflection theory exists partly even in Marx's 11th thesis on Feuerbach. May I here recall this thesis which is so well known to everybody: "The philosophers have only interpreted the world in various ways; the point, however, is to change it." Let us ponder what this thesis means.

The thesis has two parts. The first says what the philosophers have done so far. Consequently, this is a special historical thesis on philosophers. The second part says: "The point, however, is to change it." This then is a kind of programme, something that should be done, in contradistinction to what the philosophers have done. I do not want to enter into a discussion of this thesis as a whole – it contains very many problems, but I should just like to draw attention to its first part. What does it mean "The philosophers have only interpreted the world in various ways?"

First of all, what is the meaning of "ONLY interpreted"? This 'only', as indicated by the second part, means: they have not changed it; i.e., the philosophers have only interpreted the world, they have not changed it.

There has been a great deal of discussion on whether this is a right historical evaluation and whether Marx here is just towards other philosophers. I would not discuss whether Marx is 'just' or 'unjust', but I ask: what is the theoretical basis of such a historical evaluation, if we understand it literally?

The assumption is that it is possible to interpret the world without changing it.

In my opinion this assumption is in contradiction to the essence of Marx's philosophy. An interpretation of the world which is not a changing of the world is logically impossible. First of all, when man interprets the world, he changes, by this very fact, at least his conception of the world, and in changing his conception of the world he cannot help changing his relationship to the world as well. In changing his conception and his behaviour, he influences the conception and actions of other people with whom he is in different relationships.

The practical question is how much and to what extent individual theories change the world. In principle, however, it is impossible for a philosophical theory not to change the world at all. This is also impossible because every philosophical theory and every interpretation of the world also means a specific creating of the world. Philosophers create different theories, different interpretations of the world. What about these interpretations? Are they non-existent or do they exist somewhere outside the world? Or do they perhaps themselves form a new, newly created part of the world? We sometimes regard philosophical interpretations as something non-existent. But have not people for centuries lived and died with such interpretations, for them and in them? Did not people live with Aristotle or Thomas of Aquinas, do they not live today with Marx? Is Marx's interpretation of the world somewhere outside the world? Where could that be? Is man divided into two parts, one of which is in the world and the other somewhere outside? Is he in the world only when he eats, sleeps and carries out his animal functions, and outside the world when he thinks and interprets the world?

Elements of the reflection theory are sometimes encountered even where we would not expect them at all, for instance in the first part of Marx's thesis on Feuerbach – when viewed in isolation. But this theory is in contrast to Marx's entire conception of the world and man. According to Marx man is a being that has a certain, quite specific mode of being, and this mode is praxis. However, according to Marx, human praxis is a free creative activity. And if we accept the conception that man is a creative, practical being, it is natural to ask how thinking – as one of the forms of man's creative, practical activity – can be a mere reflection of reality.

In order to save the reflection theory, certain Marxists have maintained that reflection is a creative act. However, the term 'reflection' derives from the sphere of mechanics. Reflection is a necessary and exactly foreseeable consequence of collision. There is nothing creative about it.

If, in contrast to the original meaning of the word, we interpret reflection as creation, we get what I should call an 'improved' variant of the reflection theory. However, it is a 'reflection theory' in name only. Such a seeming reflection theory is probably of no use to anybody.

In his *Philosophical Notebooks* Lenin writes: "Man's consciousness not only reflects the objective world but also creates it." To a careless reader it may seem that he thus advocates the theory of creative reflection. In fact when he states that man's consciousness NOT ONLY reflects the world BUT ALSO creates it, Lenin obviously does not identify reflection and creation, but distinguishes between them and even contrasts them.

But if the reflection theory is untenable – in what direction should the Marxist theory of thinking and cognition be developed? I think one should start from the conception of man as a being of praxis, viewing thinking as a form of man's practical activity.

Thinking is not something beingless, but one of the forms of man's being, one of the ways of changing and creating the world. Let us add only that this is not a 'lowest' form of praxis, that in his spiritual creativity man is perhaps more creative than in anything else. The products of spiritual creation are often more lasting than those of purely material activity. Aeschylus' and Sophocles' plays, Aristotle's and Plato's works, continue to live, while many products of ancient material culture have disappeared without trace.

V

While of the materialist 'aspect' of dialectical materialism one can say with considerable certainty that it contradicts Marx's conception of man as a being of praxis, things are neither so clear nor so simple when the dialectical aspect is considered. Very different things have been understood by 'dialectics' in the history of philosophy; even among those who consider that dialectics is the essence or the essential aspect of Marx's philosophy, there are very different views in this respect.

Regardless of all these different viewpoints adopted nowadays, one thing is fairly certain: dialectics as understood by the founders of dialectical materialism – Engels, Plekhanov, Lenin – is neither only a method nor only logic or theory of knowledge, but also ontology. Its essential aspect or element is the conception that there are certain very general 'dialectical' laws according to which everything that exists, changes and develops. No matter how precisely we may formulate and systematize these laws, it seems legitimate to pose the question to what extent the idea of the inevitable, exceptionless general laws of every being can be co-ordinated with Marx's idea about man as a free creative being of praxis. If all that exists is subjected to dialectical 'laws', how can man be exempted? And if man is not excepted, how can we speak of his freedom and creation?

These thoughts and remarks by no means solve the question about the relationship of Marx's humanism and dialectical materialism, but they warn us that at least certain essential theses of dialectical materialism are full of difficulties and are irreconcilable with Marx's humanistic conception of man.

The criticism may be put forward that the difficulties have emerged because we have considered dialectical materialism 'non-dialectically', first only in its materialistic aspect, and then only in the dialectical aspect. The materialistic thesis on the primacy of matter should be examined in dialectical form, i.e. not simply as a thesis on the primacy of matter but as a thesis on the unity and inter-action of what is material and what is spiritual, with matter having the primacy, but with the spirit possessing a relative independence and importance.

To this possible criticism we can reply that the dialectical-materialistic solution of the question on the relationship of 'matter' and 'spirit' is certainly 'better', 'more flexible', etc., than the vulgarly materialistic one; but this does not change the fact that both these variants of materialism start from the same question based on dualistic presuppositions. In the same way, regardless of whether we interpret the most general laws of dialectics materialistically or idealistically, the fundamental difficulty will remain: how to reconcile the existence of exceptionless laws of being with the demand for free creative being.

In pointing to these difficulties of the inherited conception of dialectical materialism and contrasting it with the conception of Marx's humanism, I do not mean that Marxist philosophy should be limited to humanistic anthropology. Marx's philosophy is not only pure anthropology, and pure anthropology is not possible. When we say that the essence of man is praxis, we are making use of the 'metaphysical' concept of 'essence', and when we say that man is a being which realizes its human (or non-human) potentialities, we use modal categories ('possibility', 'reality', 'necessity'). When speaking about the present, the past and the future we refer to time. Marx's entire work as a struggle for man who will be realized as a being of praxis is at the same time a call for putting into effect both in theory and in practice the distinction between authentic and non-authentic being. And the meaning of being, the question of what 'to be' means is involved here.

Thus what is called Marx's conception of man is not narrowly anthropological, but a conception which rises from the question about man to fundamental questions about the meaning of being. Is it not a fundamental feature of Marx's philosophy that it has such continuity and constant 'interplay', such a constant relationship between essential questions of ontology and anthropology (to use the traditional language)? And is it not for this very reason that Marx's philosophical conception is what it is: a radical humanistic criticism of a non-human world? Is not the root of its criticism of non-humaneness in its humanistic conception of man?, and is not the root of its radicality in that it does not limit itself to man only or try to isolate him, but rises to the question of the meaning of being in general.

In conclusion I would like to say that if I have spoken critically here of the conception of dialectical materialism, it was not in order to diminish the importance of the personalities who developed this conception, personalities such as Engels, Plekhanov, Lenin, who made great contributions to the revolutionary movement and to Marxist theory. Nor is my criticism directed against those who in our times have tried, or are still trying, to connect Marx's humanism with dialectical materialism. Nor yet has it been my intention to belittle dialectical materialism: on the contrary, just because I regard it as a serious philosophical conception I think it should be seriously considered and criticized and not declaratively accepted or tacitly discarded.

University of Zagreb

REFERENCES

[1] Erich Fromm, *Marx's Concept of Man*, with a translation from Marx's *Economic and Philosophical Manuscripts* by T. B. Bottomore, New York 1961, p. 127. – Cf. *op. cit.*, p. 129, the reference to the "realized naturalism of man and the realized humanism of nature", and p. 181 referring to "consistent naturalism or humanism".
[2] *Op. cit.*, p. 181.
[3] K. Marx and F. Engels, *Basic Writings on Politics and Philosophy* (ed. by L. S. Feuer), Doubleday & Co., Garden City, N.Y., 1959, pp. 206–207.
[4] G. V. Plekhanov, *Sočinenija* [Works], vol. XVII, p. 18.
[5] *Ibid.*, vol. XVII, p. 296.

LEON J. GOLDSTEIN

THEORY IN HISTORY*

INTRODUCTION

It is obvious enough to those who read what philosophers writing in English have to say about problems of history that the main focus of interest and point of contention is the nature of historical explanation. And even within that general theme, interest is confined more narrowly to the question of whether or not explanations in history rest upon general laws or theories, "covering laws" in the increasingly used term of William Dray. On the one hand, we have the view, often associated with the names of Hempel and Popper, yet including among its supporters such writers as Brodbeck and Gardiner, and actually rooted in writings earlier than those of any of these, according to which explanation must always presuppose general laws for only these can warrant an inference from one particular state of affairs to another particular state of affairs. And on the other, we have writers such as Donagan, Dray, Gallie, and Scriven offering a variety of arguments to the opposite effect, and in particular, making much of the claim that historians do indeed explain or offer explanations which are complete and reasonable, yet make no use of general laws. Some may think that the debate has grown stale and tedious, and that essentially the same arguments – though occasionally with greater subtlety – are repeated time after time. And some may deem it a matter of regret that neither side seems able to reach the other, or if you like, that neither side seems amenable to persuasion. But surely, most who follow the debate have no doubt that both sides are attempting to cast light upon the same problem, that of how historical events are to be explained.

But this view of the matter is mistaken, and far from being engaged in a discussion of a common problem, each of the sides to the 'debate' is concerned with something quite different from that which engages the attention of the other. One side, those who argue in defense of the claim that explanation in history must rest upon general laws, are concerned with methodology of inquiry and the logical problems of validating claims to knowledge. For the most part, these interests do not arise out of preoccupation with problems of *historical* knowledge, and I think it can be shown that any number of the contributors to the development of the general-laws point of view are not particularly sensitive to the epistemological idiosyncracies of history. Rather, proponents of the general-laws point of view tend to ascribe to history a conception of explanation in general, and while all of them would acknowledge that the particular features they look for in a logically sound or complete explanation are not always, or even usually, to be found in the explanations that historians actually offer, they nevertheless argue that in principle these might be supplied. If not, the defenders of the general-laws position would not allow that a proffered explanation could pass muster as an explanation. In order to show that their position is neither arbitrary nor unreasonable, some of these writers construct examples to illustrate what they take laws in history to be, but these examples are almost always immediate generalizations of the given particular explanation. Thus, if some historian – or philosopher in quest of an example – suggests that some particular *a* is explained by some particular *b*, the general law allegedly called upon is '*a*-like events are caused by *b*-like events' or '*a*-like events are usually caused by *b*-like events' or some such thing. And this, unfortunately, leads historians to wonder about the relevance of such discussions to their own work. In addition, these examples give a somewhat distorted idea of the purpose of historical inquiry, the relation of historical research to other kinds of social inquiry, and the function of theory in the realization of the goals of all such research. I shall talk about all this in what follows, but before I do so, I want to make some passing comments on the other side of the 'debate' over the nature of historical explanation.

Strange as it may seem, the writers on the other side are not interested in history either. They are, rather, very much taken with the elucidation of modes of discourse and, for whatever reasons, have been attracted to the discourse of historians. The constant pre-occupation of the writers who oppose the conception that explanation in history presumes general laws is with what they find in historical narratives. It seems assumed that any historical narrative is complete in itself in every possible respect, as if the given narrative has and could have no relationship to other matters. And, more particularly, in dealing with the logic of explanation in history, these writers are of the view that in any given case we are to be limited to the idiosyncratic intentions of the given historian – presumably because he is the author of the narrative to be elucidated. Since the well-known book of William Dray[1] is likely the best-known contribution to this point of view, I shall draw several examples from it, but there is evidence of this attitude in the writings of others[2] as well.

On p. 28, in rejecting the possible claim that "a person cannot *mean* anything by the explanation if the denies the law", Dray says that "The historian has denied that he arrived at his explanation by means of a law", as if to suggest that if the given historian did not himself derive his explanation by deduction using general laws such laws are irrelevant to the explanation. On p. 36, he talks about an argument which "may lead to misunderstandings of the structure of typical explanations in history"; on the following page, he contends "that the historian, when he sets out to explain the French Revolution is just *not interested* in explaining it as a revolution...; he is almost invariably concerned with it as *different* from other members of its class", and on p. 57, in criticizing "The Humean assumption that nothing but 'regularity' can justify a 'because'" and the consequence this has for the arguments of Gardiner, he writes: "If the historian does not use a precise 'rule', then a vague one *must* be found; if no universal law is available, then a qualified one *must* have been assumed. The alternative which is too much to accept is that, in any ordinary sense of the word, the historian may use *no law at all.*" It seems clear that in each of these passages, the general-law point of view is rejected as not being an account of what is actually written in works of narrative history, and that 'the historian' of whom Dray writes is one who makes use of a certain mode of discourse – not one whose work raises certain epistemological problems or whose explanations lead to further questions of fact and theory. And thus, when, on p. 58, he says of possible claims that the covering law is a sufficient condition of an explanation that it, too, "is artificial and misleading", it seems reasonable to take it to mean that it fails as an elucidation of the historian's discourse. On p. 76, Dray says "Hempel's formulation begins by laying down the logical structure of explanation as he believes he finds it in physics", which would suggest that he takes Hempel's task to be the elucidation of the language of physicists, rather than attending to the philosophical questions that physics raises. It is hard to believe that Hempel would himself subscribe to this conception of what he does, but one need not be surprised that Dray should think it.[3]

In sum it should now be clear why I say that writers like Dray are not actually interested in history – the discipline – and its problems, but only in the language in which historians formulate their finished product.[4]

SOME GENERAL LAWS

If the elucidators of modes of discourse seem to suggest that there are no problems of history or historical knowledge and by implication treat history as a literary genre only, it does not follow that the view of that discipline implicit in some of the writings of their opponents is more useful. I do not think it profitable to seek to ascribe to history any particular set of goals, and it is clearly the case that historians are engaged in a variety of activities. Yet *implicit* in what is found in the statements of any number of those who subscribe to the view that historical explanation must always presuppose general laws is a conception of the historian's task which is both restrictive and untrue to the actual practice of that discipline. I do not suppose the writers I have in mind would actually wish to maintain such a view, but I think that it does emerge from what they write, particularly from the kinds of example they make up when they seek to defend their point of view. I think it best to present some examples of what philosophers have supposed might serve as general laws applicable to explanation in history and then draw from them what I think they imply. I shall first present examples used by Hempel, Gardiner and Nagel, all of whom subscribe to the general-laws thesis; then one from Popper, who shares their view on the matter, but, for reasons which are not relevant here, thinks that the laws in question must and ought to be trivial and so might not be bothered by some of the consequences I shall be drawing; and, finally, one from Michael Scrivin, who does not subscribe to the general-laws point of view at all but who nevertheless seems to agree with those who do about the character of laws in history if they were to be had.

In the course of what is most likely the best-known individual expression of the general-laws thesis, Hempel says the following:

Consider, for example, the statement that the Dust Bowl farmers migrate to California 'because' continual drought and sandstorms render their existence increasingly precarious, and because California seems to them to offer so much better living conditions. This explanation rests on some such universal hypothesis as that populations will tend to migrate to regions which offer better living conditions.[5]

Hempel goes on to write that it would "be difficult accurately to state this hypothesis in the form of a general law which is reasonably well confirmed by all the relevant evidence available", and it is clear enough from his discussion that he puts the hypothesis forward as an example of the *sort* of thing he has in mind and is not to be held responsible for *it*. Yet one may wonder just what evidence would be "all the relevant evidence available" to test a 'general law' which seems to be little more than a direct generalization of the historical event it is being called upon – or might be called upon – to explain. As we proceed with our catalogue of examples, it will become clear that all of the proffered laws have this same feature: they simply generalize what they are supposed to be explaining. What this implies about the discipline of history and the goals of historical explanation is precisely what we have to deal with in this present section of the paper.

At the beginning of a list of explanations culled from historical writings, Gardiner quotes the following from a work of Charles Seignobos: "Louis XIV died unpopular ... having caused France to lose ... the incomparable position she had gained by the policy of the cardinals."[6] Later, in another context, he asks: "But are we able to interpret ... the historian's explanation of the unpopularity of Louis XIV in terms of the generalization that rulers of nations become unpopular when they pursue a policy resulting in the decline of the country over which they rule?"[7] This makes it seem that in Gardiner's view one who held that Seignobos' explanation rests upon a general law would be committed to a law which was hardly more than a generalization of what was to be explained. In this case, that a king became unpopular because he pursued policies detrimental to his country is explained by appeal to a generalization to the effect that rulers who practice policies detrimental to their countries become unpopular. That this, indeed, is Gardiner's view becomes unmistakable a bit further on when he says: "Let us return again to the explanation given as to why Louis XIV died unpopular. The positivist interpretation would be that Louis XIV represented a case of the law 'Rulers are unpopular whenever their policies prove detrimental to the fortunes of their countries' and that the explanation in question was deduced from this law taken together with circumstances of the case."[8]

Our next example is the sort of general law Ernest Nagel thinks one might call upon in order to justify F. W. Maitland's explanation of the first Queen Elizabeth's use of 'et caetera' in her title, rather than stating her title in full. According to Maitland, the reason for this seemingly odd way of styling herself was that she hoped thereby to avoid taking a public stand on the issue of the separation of the Church of England from Rome by not proclaiming herself head of the church.[9] About this, Nagel says the following:

Now an examination of Maitland's discussion shows beyond doubt that the event he sought to explain does indeed logically follow from the explanatory premises, provided these include an assumption essentially as follows: Whenever anyone acquires a position of great political power, is faced with an issue fraught with peril, but is required to announce immediately a policy, then such a person will make a statement that is momentarily noncommittal.[10]

The assumption seems little more than Maitland's original explanation with the proper name and specific circumstances removed and replaced by 'variables'.

The example from Popper takes but one long sentence in his *Open Society*: "If we explain, for example, the first division of Poland in 1772 by pointing out that it could not possibly resist the combined power of Russia, Prussia, and Austria, then we are tacitly using some trivial general law such as: 'If of two armies which are about equally well armed and led, one has a tremendous superiority in men, then the other never wins.'"[11] I must admit, that in a sense it is not really fair to include Popper in this list. Popper holds that in the socio-historical field we cannot have comprehensive laws and must, in consequence, make do with trivial ones. The writers mentioned above might hope we could have serious laws, but Popper apparently has no such expectation. Thus, unlike their cases, that his is but another example of a trivial law is not just a bit of bad luck. But if what I shall soon be saying about the implication of these immediate generalizations for our conception of the discipline of history is sound, then we shall have to take issue with this example as well, and possibly, for the sake of fair play, we might agree to include Popper's generalization in our list of bad examples but omit Popper himself from any list of authors of bad examples.

Our last example comes from Michael Scriven and is taken from his account of how one committed to the general-laws thesis might set about justifying an explanation of Cortez' third expedition to Lower California by reference to his cupidity. He offers three generalizations from which one might choose: "All confident wealth-seeking people undertake any venture which offers very great wealth", or "All confident people seeking very great wealth undertake any venture which offers very great wealth", or "All confident people with Cortez' background of experience, seeking very great wealth, undertake any venture involving the hazards of this one, which offers very great wealth."[12] Scriven, to be sure, is opposed to the general-laws thesis, but that is not immediately to the point. What is of interest is that he, too, seems to think that the only kind of generalization we could have in history, if we could have them at all, would turn out to be directly generalized from the explanation it was called upon to justify. Clearly, if the alternatives offered by Scriven to those who would seek nomological justification for the explanation of Cortez' expedition were, in fact, all we could choose from, the formation of theories relevant to explanation in history would be a trivial enterprise indeed.

In the course of a paper largely favorable to the claim that theory is relevant to explanations in history, Samuel H. Beer takes a number of opportunities to oppose what he calls the "dogma of universality".[13] The name of dogma he seems to derive from what he has read in the writings of some of the authors I have just been mentioning. Thus, he cites Popper to the following effect: "To give a causal explanation of a certain *specific event* means deducing a statement describing this event from two kinds of premises: from some *universal laws*, and from some singular or specific statements which we may call the *specific initial conditions*." From Hempel, he quotes a characterization of a general law as being "a statement of universal conditional form which is capable of being confirmed or disconfirmed by suitable empirical findings".[14] The word 'universal' figures prominently in both these statements, and one may suspect that it is this which led Beer to the name he gave the noxious doctrine he wrote to attack. But it is not these cited passages alone which figure in Beer's conception of the dogma. It is the examples that are used to illustrate universal laws, and only a few lines after he quotes the passages given above, he produces the examples of Hempel and Popper which are included in our sample. The dogma of universality against which Beer quite properly protests is the view that explanations in history are to be justified by appeal to general laws which are arrived at by the simple expedient of generalizing – or universalizing – the particular explanation to be justified. It is, of course, not unreasonable to suggest that Beer has not fully caught the drift of the writers he deals with and does not realize that the theories of physics which they have in mind when they offer their accounts of what general laws are, are not at all the simple-minded generalizations of the particular causal explanations to be justified. Rather, what they have in mind are highly complex theories which may be applied to a wide varity of initial and boundary conditions in order to explain a wide variety of kinds of occurrence. Thus, it would seem that the very writers cited as the source of the 'dogma of universality' may not themselves subscribe to it. But Beer thinks they do because he makes the mistake of taking seriously the examples of historical generalization they make up.

But before we go on to the next section and its presentation of examples which point to the possibility of theory in history without the dogma of universality, it may be worth our while to take our bad examples seriously enough to wonder what they would imply about history if they were not as bad as Beer and I take them to be. I think that one consequence of taking them seriously would be the isolation of the discipline of history. That is, history would be seen to have no particular connection or contact with the work of other disciplines that seek to make sense of human social existence – the social or behavioral sciences – and, indeed, it would even seem that the particular problem of any given historian at any given time of his career would have little if anything at all to do with other particular problems of historical research. I suppose that this seems a most extraordinary thing to say, and for two reasons. First, it surely seems odd to say that proposed examples of general laws have any sort of isolating effect when it is clearly the case that the subsumption of an historical explanation under a general law would seem to have just the opposite effect, namely, to show that it is one of a class of similar explanations and is in consequence not unique or conceptually isolated. And, second, it is one of the main theses of a well-known paper of Morton White that there are no essentially historical laws or theories, and that explanation in history must make use of theories which have been formulated by the special social sciences.[15] This, too, suggests just the opposite of the isolation I see lurking in our bad examples.

Yet, it seems to me that the 'dogma of universality' that Beer has extracted from these examples has precisely the effect of isolating historical work, because of the way in which it seems to make each historical explanation something complete in itself. We are concerned, say, to make sense of certain internal American migrations in the 1930's which saw farmers from the Dust Bowl settle in California. We discover that the migrants were people whose existence had been rendered precarious by certain natural occurrences and who were in a position to hear about the more promising possibilities in California. And so, it appears as if we have an explanation for the migration: finding life in the Dust Bowl no longer tenable, the migrants moved west in hopes of improving their fortunes or because they believed that life would be easier for themselves there or some such thing. But for the usual Humean reasons, which we need not review here, it is concluded that that which is explained and that which explains it are not essentially bound to each other, since each may be conceived by itself, and so we are required to justify the explanation by appeal to a general law. According to the 'dogma of universality' which is suggested by the examples reviewed above, this is done simply by generalizing the given case. This satisfies the logical need revealed by our Humean analysis. But it also isolates the case in that only cases pretty much like itself – few if any more of which we are likely to turn up – has any bearing upon it or its justification. That seems to be one of the lessons drawn by critics of the general-laws point of view. They not infrequently point out that any threat to the general law offered in justification of some explanation would be met by spelling out in increasing detail just what sort of event the general law is applicable to until it becomes entirely clear that only the situation confronting us to begin with can be explained by its means. The point of the critics is that such 'general laws' are not worth bothering about. But what makes them not worth bothering about is precisely the isolating character that they have. They isolate the explanation and the kind of event it is about from every other possible interest of students of history and society.[16]

But it is certainly not the case that the only option open to those who would seek nomological justification for historical explanations is to apply the dogma of universality. In the next section of the paper, I shall present a number of examples of the sort of thing I have in mind taken from the historical literature. Here, I shall briefly sketch what this involves. Clearly, if the dogma of universality is to be avoided, what we are required to do is *not* quickly to conjure up a general statement which will enable us to deduce the explanation we wish to justify.[17] Rather, we focus upon the particular explanation, which is, in a way, the starting-point of our theoretical enterprise – that this particular situation, state of affairs or events came to be because of these other particular situations, states of affairs or events – and seek to discover in it what we might think of as its general features. There are no rules for this, though I believe that what this involves will become clearer in the course of the discussion of the next section. But this kind of procedure seems not peculiar to the historical or social studies. Consider the following statement of N. R. Hanson:

> In the thinking which leads to general hypotheses, there are characteristics constant through the history of physics from Democritus and Heraclitus to Dirac and Heisenberg. Kepler did not *begin* with the hypothesis that Mars' orbit was elliptical and then deduce statements confirmed by Brahe's observations. These latter observations were given, and they set the problem – they were Johannes Kepler's starting point. He struggled back from these, first to one hypothesis, then to another, and ultimately to the hypothesis of the elliptical orbit.[18]

Hanson then takes us through the stages of Kepler's investigation during which he proposed and tested a number of hypotheses. He concludes this part of his discussion as follows:

> Kepler typifies all reasoning in physical science. Would it have required so much time, and genius, to 'observe' the elliptical orbit in Tycho's data? *De Motibus Stellae Martis* is more than a compendius expression of Brahe's observations. Nor is it concerned with deducing geometrical consequences from the elliptical orbit hypothesis, 'thought of' Kekulé-fashion. Kepler's task was: given Tycho's data, what is the simplest curve which includes them all? When he at last found the ellipse his work as a creative thinker was virtually finished. Any mathematician could then deduce further consequences not included in Tycho's lists. It required no genius to take Kepler's idea and try it for other planets.[19]

What Hanson seems to say is that what Beer calls the 'dogma of universality' plays no role in the formulation of theories in physics, but he does not wish to suggest that it is not a goal of that science to formulate theories. I think that this way of looking at the problem has merit in history, too, not only as a possibility, but as something that is actually to be found.

We are now about to see that this is the case, but I think that one point is worth emphasis before we do. Some of those who take offense at the general-laws point of view think it outrageous that writers who are rarely historians should presume to suggest that this or that well-regarded explanation is, in fact, defective since its proponent has failed to provide general laws for it. I would like to suggest that this is not actually a consequence of subscribing to the general-laws point of view. Hanson makes it clear that Kepler's problem was to explain Tycho's observations which latter were not to be rejected or in any way disesteemed. I think that the same consideration obtains here. A sound historical explanation, say, one which excites the historical community and wins its acclaim, is not to be deemed a bad one or one not worthy of serious respect simply because it is not attached to some general law or other. Rather, it may very well serve as the starting-point for a subsequent theoretical inquiry the initial point of which is to explain *it*, or to justify it. From the standpoint of the present paper, then, one function of the search for theory is precisely to preserve all good *typical* historical explanations.[20]

SOME PROMISING EXAMPLES

According to Hanson, one difference between what he was trying to do in the book to which we have referred and what some other philosophers of science have written about scientific explanation is that "they have regarded as paradigms of physical inquiry not unsettled, dynamic research sciences like microphysics, but finished systems, planetary mechanics, optics, electromagnetism and classical thermodynamics".[21] Thus, they tend to see these sciences from the viewpoint of the present, when they are complete, and not from that of Newton, Kepler and Galileo, for whom they were sciences in the making. If advocacy of the general-laws conception of explanation in history is in the hands of those others, who write about it with these paradigms in mind, it is quite easy to see that it might very well seem to take the form of defending the dogma of universality.

There is nothing in history which even remotely resembles what mechanics has become, and it is this, I suppose, as much as anything which makes the 'dogma' seem dogmatic. Yet, there is a great deal of historical work which points to possibilities for development of theory if only scholars wish to pursue it. For some of it, the pursuit has, in fact, been attempted; for others possibly not. In what follows, I shall discuss a number of examples which illustrate what I am trying to say. In no case can we say that the required theories have been reasonably established. Yet, in each of them we could hope to have them – not by the immediate generalization of what we must justify, but through the trial-and-error testing of plausible hypotheses. We will begin with Max Weber's famous study of the spirit of capitalism.

The Capitalist Ethos

Though Max Weber is widely regarded as one of the major contributors to the development of sociology in the early decades of the present century, and one whose influence is still said to be felt, it is nevertheless the case that his famous essay, *The Protestant Ethic and the Spirit of Capitalism*[22] is an historical and not a sociological work. It is concerned with capitalist enterprise in Europe and America, and is, more particularly, concerned with the answer to one question about it: How did Western capitalism come to have the character it had? Thus, it is an instance of one fairly typical sort of historical writing, that concerned with making sense of some particular something or other. I do not say that this is the only sort of thing historical books do, but it is one fairly frequent purpose of authors doing historical research. Some, of course, want to describe or recreate a given course of events, but others are interested in more limited happenings. In one famous example, Henri Pirenne sought to delineate certain factors which affected the development of cities in the Middle Ages.[23] In another, about which I shall have some things to say later on, Mrs. Woodham-Smith tries to understand the factors which led to the destruction of the Light Brigade in the Crimean War.[24] In just the same way, Weber seeks to answer a particular question about Western capitalism, and, as is known, he traces what he takes to be distinctive of it to certain Calvinist doctrines which he believes tended to encourage certain modes of behavior favorable to the development of 'capitalist'-personality types.

There are any number of passages in Weber's essay which support the view that the problem it deals with is historical and not sociological (cf. pp. 27, 36, 46 and 55), and even though *The Protestant Ethic and the Spirit of Capitalism* is the opening essay of a series called *Gesammelte*

Aufsätze zur Religionssoziologie, Weber makes it clear that his comparative sociological studies of non-Western economic and religious ideas and institutions were undertaken because of his desire to clarify the relationship between such ideas and institutions in the West: "for only in this way", he writes, "is it possible to attempt a *causal evaluation* of those elements of the economic ethics of the Western religions which differentiate them from others, with a hope of attaining even a tolerable degree of approximation." And he adds that the comparative studies are "definitely oriented to the problems which seem important for the understanding of Western culture".[25]

If one looks at the character of the voluminous literature which has come out of the scholarly reconsideration of the Weber hypothesis one sees that it is both historical and sociological. The formulation of general laws in terms of which to justify Weber's original *historical explanation* does not in itself settle the historical question. And, certainly, a generalization formulated in the spirit of the dogma of universality would only be distrusted by historians and confirm them further in the view that general laws have nothing to do with explanation in history. Scholars concerned with the historical questions discussed in Weber's essay, continue to seek answers to the question "How did Western capitalism acquire the character it had?" And scholars whose interests are sociological or nomothetic continue to seek general propositions which would, among other things, make intelligible the answers of the historians. The theoretical problems emerge from scrutiny of the historical explanation and by focusing upon the particular relationships which the historical explanation emphasizes. If there is a sense in which the formulation of theories leads to a more complete statement of the historical explanation, it in no way involves that the particular historical explanation is in any way something inferior. For just as the explanation is justified by the theory, the theory is supported by the explanation. Reflection on the explanation raises questions for theory, but no theory can escape the need to accord with the claims of particular explanations which are accepted as sound by the community of historians.

Slavery in North America

One of the more interesting contributions to recent historical writing is

Stanley Elkins' collection of studies of American slavery.[26] The first of the four studies deals with slavery as a problem of historiography, and the last with the ways in which intellectuals have responded to slavery. From the point of view of our interest here, it is the middle two studies which attract our attention. The first of these two studies takes it for granted that historians have some clear conceptions of the institutional character of slavery in the southern states and is concerned to make that character intelligible. The other study is possibly based upon a less firmly established historical conclusion. It seeks to account for the personality of the slave, but Elkins' conception of that personality emerges rather indirectly, and it is not unlikely that other historians do not share it. I do not say this with the intention of being adversely critical, for what we know about such matters must be constructed out of sources written when sensitivity to and techniques for determining personality types were far less developed than psychologists have made them in recent decades. Yet for our purpose here, the study in question illustrates the same sort of thing as the others mentioned: given an historian's conception of the character of the personality of the slave, how are we to make sense of it?

For reasons which had best be assessed by other historical and psychological scholars and cannot detain us here, Elkins concludes that the so-called 'Sambo' stereotype was characteristic of American Negro slaves. But he gives no comfort thereby to those who would see this as evidence of racial peculiarities or inferiorities of Negroes, for, as he says, "One searches in vain through the literature of the Latin-American slave systems for the 'Sambo' of our tradition – the perpetual child incapable of maturity."[27] And this sets the problem which is to be explored in his third study, how is this difference between the personality types of North- and South-American Negro slaves to be explained. Some scholars have sought to account for the characteristics in question in terms of African culture, but against this it is argued that the Sambo type was not to be found in Africa and, too, that more elements of their African cultures were preserved by slaves in Latin America than by those in the United States.[28]

The explanation Elkins finally offers has to do with the psychodynamics of the reaction to capture, enslavement and the passage from Africa to North America.[29] A series of shocks is said to have had the effect of detaching them from their previous cultural moorings and to have made them amenable to the thought and personality re-formations out of which the Sambo type emerged. To put the matter simply, Elkins claims that there is a particular historical phenomenon which requires to be explained, namely, the existence of a certain personality type among some people which was not to be found among other people related to them. And he explains this by appeal to certain antecedent happenings. In this, his explanation resembles a great many others recognized as typically historical. But in *typical* instances, the explaining happenings are generally felt to be of the sort that *could* explain the explained phenomenon, and I should think that even historians who might disagree with a given explanation *could usually understand how someone might believe* that the proposed cause was the cause of the phenomenon to be explained. Why that should be, is not germane to our present interest, and I shall not bother to deal with it. But I do think it clear that the explanation offered by Elkins is not of the typical political or economic sort. And it is this which requires that he interpose between his phenomenon-to-be-explained and his explaining causes a psychodynamic mechanism which is thought to connect them. But except for the atypical addition of the particular character of the connection which he takes to bind what he explains with what he thinks explains it, Elkins' explanation seems to be rather like a good many others that historians offer.

But to say that the connection is psychodynamic is certainly not to end the matter. The relevant psychological theories have to be spelled out and their relevance to history clarified. None of this is done in Elkins' book, nor was it his intention to do so. But we do get some indication from Elkins of what might be involved in this, and here, too, the procedure seems to be to seek out general features of the subject of the inquiry which might be discovered in other contexts. And this leads him, possibly to the surprise of many of his readers, to a consideration of the experiences of those people who were imprisoned in Nazi concentration camps and the effects that experience had upon them. He is particularly interested in pointing to the possible similarities between the shock-and-detachment phenomenon he hypothesizes for the captured, enslaved and transported Africans and the adaptations of the camp prisoners to their new environments. The one procedural difference that may distinguish Elkins from Weber is that it seems that Elkins simply spotted the similarities in the adaptation of the two kinds of people he discusses – perhaps he was led to the shock-and-detachment hypothesis through reading memoiristic and psychological accounts of the camps – whereas Weber set about systematically to discover whether or not certain relationships he discovered in the course of his investigations of the capitalist ethos may be found in other times and places. But it is the similarity between the two that concerns us here. Both seek to justify their particular historical explanations by discovering in the situations with which they deal general features which may be discovered at other times and in other contexts. Both point to the need to justify their explanations theoretically, yet neither is led to the dogma of universality.

Counter-revolution in the Vendée

Following his critical remarks on the dogma of universality, Beer proceeds to offer two examples of historical work which he takes to be entirely acceptable yet they do not satisfy the requirements of the dogma. Actually, what he does is to discuss the two papers immediately following his own in the journal in which it appears. One of these deals with the counter-revolution in the Southern Anjou section of France during the 1790's and was written by Charles Tilly.[30] This is the only one of interest to us here, and our interest will be less in Tilly's paper than in Beer's account of it.[31]

Like the others we have dealt with, Tilly's, too, is a straightforward historical problem of the usual kind, though his procedure reflects more explicitly the influence of sociology. His problem is to explain how it was that the counter-revolution broke out where it did, rather than in other parts of France, and one would expect that a historian interested in it would seek out conditions peculiar to the counter-revolutionary areas and less in evidence in other parts of the country. According to Beer, it leads Tilly to

> systematic comparison of localities in the Vendée where counter-revolution took place with those where it did not. He asks what other differences characterized the two classes of localities and initially finds that revolution and counter-revolution correlate with two types of social organization, the major contrast being between localized, subsistence agriculture in the areas of counter-revolution and market-oriented rationalized agriculture in the areas where revolution was accepted. Thus, ... Tilly ... arrives at the generalization ... that in the Vendée at that time in history, if (and only if) traditional agriculture prevailed in an area, counter-revolution occurred.[32]

This is only one stage of Tilly's account, and in the course of his investigation he seeks to narrow the focus of his attention to factors relevant to the emergence of counter-revolution still further through additional comparison. To quote Beer again, "he observes that the correlation is not simply with the traditional farming areas, but with localities where there were important nuclei of trading and manufacturing set in the midst of traditional farming".[33] Tilly himself says that "such a social situation is more favorable to violent local conflict between 'old' and 'new', 'backward' and 'progressive' than is a uniformly advanced, or a uniformly backward social setting. In fact, it is not far off the mark to say that throughout the West, the peaks of counter-revolutionary activity were not in the backward sections so much as the junctions of rural and urban ways of life."[34]

In his account of Tilly's work, Beer calls the sort of thing we have been talking about "the method of comparison and generalization", clearly pointing to the fact that Tilly does both. He compares the different sections of France with respect to certain characteristics he suspects may be relevant to his inquiry and arrives at conclusions not about specific localities in the Vendée or elsewhere, but more generally about all the localities in revolutionary France having the characteristics specified. Beer distinguishes this method of Tilly's from another one he finds in the same writer's essay, "the method of hypothesis and verification". The point of this method is to explain the conclusions arrived at by means of the first, by the formulation of suitable theories or hypothesis, though Beer seems to restrict it to the choosing of available hypotheses from the writings of others.[35] Beer's own example will make this clear. We have already noted that Tilly, through his method of comparison, had concluded that revolutionary and counter-revolutionary inclinations were associated with the extent to which urbanization had infected a given area. This leads him to suspect that theoretical writings of sociologists concerned with urbanization would be relevant to understanding the question of his particular historical interest.[36] In effect, what Beer claims is that influenced by sociologists, Tilly came to see the material he was working on as rather like some others and attempted to see if the theoretical solutions proposed to questions raised about those others would work in his case as well. In this respect, his procedure reminds us of Elkins' attempt to find an analogy between the experience of concentration-camp prisoners and of captured Africans in the course of their journey to slavery in America.

Actually, there is nothing about the logic of this procedure which requires that its historian practitioners limit themselves to selecting among theories already formulated by other social scientists. Any historian who had some suspicions as to how he might explain the finding with which he had emerged might offer his own formulation which he and others might investigate. It is hard to believe that Beer would not agree to this, and one may suspect that his account of the "method of hypothesis and verification" was intended to describe its use in Tilly's study – which is most likely the way it will usually be used.

There are a number of steps in the course of development of history and theory to be found in Tilly's study – and its elucidation by Beer – and it may be well to close my discussion of it by summarizing what they are. To begin with, Tilly is clearly interested in the answer to a typical historical question, for he is concerned to account for something that happened in a particular place or country and at some given time. The philosophical writers whose interest is in the elucidation of modes of discourse and who have taken the historical narrative alone to be the typical product of typical historical research might not be pleased with this, but the concern with the particular has long been taken to be characteristic – some would make it the defining characteristic – of historical writing, and I cannot see how one could deny that Tilly's essay is concerned with a particular notwithstanding the fact that it is not a narrative, tells no story. In any event, Tilly answers the question not by presenting an account of the plans and actions of leading figures, but, rather, by way of a systematic delineation of the ways in which counter-revolutionary sections of France differed from other sections of the country. The procedure, adapted from the work of sociologists, leads to the formulation of general propositions albeit very much restricted in their spatial and temporal range of application, as Beer points out.

Actually, whether or not statements of the sort "In France during the 1790's, sections having characteristics x, y and z will also have characteristic r", are general would depend upon what they are about. My own inclination is to say that they are about France during the 1790's, hence rather more particular than general. Some might wish to argue that they are about sections having or not having certain characteristics hence having or not having certain others, and, thus, being about a good many sections, are rather more general than particular. They might correctly point out that such statements are exactly like many formulated by social scientists on the basis of the systematic comparison of social units or particular sections of a given country having and not having characteristics being studied. Since such statements are not about particular social units or particular sections of the country, it is usually assumed that they are general, and the sociologists who formulate them almost always, perhaps even always, take them to be general. Yet it seems to me that a proper understanding of the *ceteris-paribus* clause, which must always, if only implicitly, be attached to such statements, would show that *as they are formulated* they are applicable to the *particular* society of which these units or sections are part, and in consequence, must be taken as more particular than general.

But whether particular or general, this method which Tilly has taken over from the sociologists enables him to discover that in the France of the period of his investigation certain empirically determinate characteristics of sections of the country were or were not associated with a tendency to counter-revolution. The attempt to explain this, to answer why it should be so, leads him directly to the theoretical part of his study, the part Beer calls "the method of hypothesis and verification", in which he seeks to adapt the theoretical conclusions of sociological investigators to his task. At this stage of his work it becomes very clear indeed that answers to particular historical questions implicate general social theory. One seeks to explain the specific causal relationships uncovered by Tilly's research by appeal to general laws of one sort or another, and, at the same time, the kinds of data which emerge from the earlier stage of the investigation are clearly evidence of the sort with which the laws in question are to be tested.

The Charge of the Light Brigade

Cecil Woodham-Smith's *The Reason Why*[37] is precisely the sort of historical work that the elucidators of modes of discourse and others who tend to take a dim view of any effort to "make history a science" take to be typical of the field. And it certainly is typical of a very common sort of historical writing, though it is by no means a common one of that common sort. On the contrary, it is written in a most striking English prose – narrative history at its best. Nor is it the coldly 'objective' sort of work which advocates of scientific history may be thought to demand. It is, rather, a passionate book, moved by injustice, angry at social evil, and, if I may judge from myself, capable of arousing the same feelings in its readers. Surely, it may be thought, no one who subscribes to the general-laws point of view could think much of the book *as history*, however much he may appreciate it *as art*.

The purpose of Mrs. Woodham-Smith's book is to *explain* the suicidal Charge of the Light Brigade in the Crimean War. In the course of her account, she tells us a good deal about how the British army was managed during the period preceding the war – the war itself and its patent mismanagement were major factors in the subsequent reform of the army – and the way this was related to the power, attitudes and conceits of the ruling classes. In the tradition of narrative history, the book focuses particular attention on the two men who figure most prominently in the events with which it is concerned, the earls of Cardigan and Lucan. Through her treatment of these two, the author is able to make clear to us certain typical features of the military careers of noble lords who could rise to the highest ranks of the army, where they would make decisions affecting the lives and welfare of fellow human beings without ever being more than rank amateurs in matters military or having field experience in lesser commands. But the two earls are not only typical representatives of recurrent types. They are individual human beings, and Mrs. Woodham-Smith's narrative does not fail to account for the effects of their 'idiosyncracies' upon what happens. They were brothers-in-law, and they hated each other, in itself an evil omen for an enterprise in which the paranoic Lord Cardigan was subordinate to Lord Lucan. But without adding details, let me simply say that the book is a well-woven account in which the typicalities of ordinary expectation are meshed with the particularities of individual behavior in order to give a convincing explanation of the tragic debacle of the Light Brigade.[38]

If Mrs. Woodham-Smith is right about what factors led to the climactic events at Balaclava, it is clear enough that there must be social and psychological regularities in terms of which it is these factors rather than others that are relevant. These are matters to be investigated further, not in a spirit of claiming that the historical account presented in inadequate, but rather with a view to making theoretically intelligible what seems to be a very sound historical explanation.[39] This, to be sure, must proceed by way of detailed investigation of determinate relationships, not in the manner of the dogma of universality. And if Mrs. Woodham-Smith herself is not to be expected to pursue any of the special social and psychological disciplines which are implicated by her book, there is a sense in which her book may be taken as contributing to all of them. To the extent that her account is correct, it contributes relevant empirical evidence with which they must come to grips.

POSTSCRIPT

Comments I have had from readers of the foregoing lead me to think that I ought to add some concluding words of clarification about some of the matters I have dealt with.

(1) I hoped it would be clear that my discussion depends upon distinguishing between the explanations historians offer of whatever it is that they wish to explain and the theories we would require to have if those explanations are to be justified. Thus, in the example taken from Nagel, we can clearly distinguish three elements: (1) the event to be explained, namely, Elizabeth's signature; (2) Maitland's explanation of it; and (3) Nagels' generalization. Some would say that the explanation offered by Maitland is incomplete and Nagel's generalization is needed to complete it. But this obscures the fact that Maitland's explanation may be presumed to be compatible with an indefinitely large number of other generalizations as well. It seems more correct to think of him as having offered one specific explanation, rather than something which turns out to be the common component of an indefinitely large disjunction of explanations.

But while my way of putting it focuses on the justification of an historian's explanation, most of those who defend the general-laws point of view talk about the requirements for the explanation of an event. Let T refer to the statement of theories or general laws relevant to some given case, and e to the event to be explained. And let c refer to what most covering-law theorists think of as the initial or boundary conditions, but what most historians might think of as the set of factors which explain the event. My way of thinking about the problem might then be symbolized as $T \supset (c \supset e)$, while the usual view would be symbolized as $(T \cdot c) \supset e$. On the usual interpretation of the '$\supset$' symbol, these two are logically equivalent, but I think that the former is better nevertheless. For it makes clear that the starting-point for theoretical inquiry is the historian's explanation, which it takes seriously and actually represents in the symbolism within the parentheses. But the latter seems to suggest that history without theory is absolutely nothing, that the usual explanations of historians are not explanations at all. It is hard to doubt that the attitude that historians read into the sort of discussions the latter symbolism represents contributes much to the unwillingness of so many of them to take it all seriously. The latter symbolism seems to tell the historian where he has gone wrong; the former how his results may lead to further developments. While he may well resent the first of these, he may equally welcome the second.

This leads me to just what it is that I classified above as good and bad examples. Someone wondered what I objected to in the Maitland explanation. And someone else did not see how I could deny that various examples in Dray's writings were perfectly good and typical – I shall return to 'typical' in a moment – or how I could claim that they were inferior to the examples I seemed to favor. But in my list of bad examples, it is not the historical event that I object to, nor the historian's explanation of the event. My objection is only to the generalization offered to account for the explanation. Each of them is simply an immediate generalization of the explanation it is called upon to justify. While *logically* it serves to provide what a Humean analysis finds missing in the usual accounts of historians, it is too simplistic and trivial to be interesting or even true. What distinguishes, then, the two samples is not the character of the historical events in them or the ways in which historians might explain those events, but rather the ways in which the justifications of the explanations are offered. In the first set, each justification is an instance of the dogma of universality. In the second, even when no generalization is actually offered we can see how painstaking attention to what seem to be the crucial features of the event and its explanation may result in subsequent development of theory.

(2) The word 'typical' and its kin are found with some frequency in this paper, and one reader found my conception of a typical historical problem unclear and did not see how I could use it to distinguish between work I find promising and work I am dubious about. But I have no special conception of typical historians' activity, and I did not intend to use such a conception to make the distinction in question. What his-

torians are wont to do is typical of their craft and certain kinds of questions are typical historians' questions. Historians typically ask why something or other happened – why Elizabeth used 'et caetera' or why the Light Brigade charged or why Dust Bowl farmers moved to California. Or they typically ask why something took the form it took or had the character it had – Western capitalism or the personality of the North-American Negro slave. These are typical questions for historians as other kinds of activity are typical activity for tailors or shoemakers or barbers. My point is that the question, 'Why did the counter-revolution begin in the Vendée and not in some other part of France?' is a typical historian's question, notwithstanding the fact that Tilly uses procedures not typically used by historians to answer it. And the question, 'How did the Sambo personality type come to be as it was?' is a typical historian's question even though it led Elkins to an atypical concern with psychodynamic factors and away from the historian's typical preoccupation with social, political and economic matters. It is not a typical historian's concern to formulate general laws or to seek out the theoretical justifications for the explanation he offers. But that in no way entails that what historians typically do does not implicate general theory.

(3) This leads to the final point of this postscript. It is clear enough from my remarks early in the paper that I am out of sympathy with the elucidators of modes of discourse. Yet, someone challenges, the real issue confronting the parties to the present-day discussion of explanation in history is precisely whether there is a distinct historical mode of discourse. Perhaps 'mode of discourse' is not a sufficiently clear term, and my challenger thinks the question 'Is there a logic of explanation peculiar to history?' is essentially the same as 'Is there a mode of discourse peculiar to history?' I think that these may well be separate questions, though it is clear enough that some writers, W. B. Gallie for one[40], do not. F. J. Teggart once argued that the historians' tradition of casting their conclusions in the form of narratives is a stumbling block to the achievement of their goals[41], but if the narrative mode of discourse is precisely what distinguishes history from other intellectual disciplines, poor Teggart was only babbling. (He was not.) Similarly, we would have to admit that Tilly's problem was not historical after all, notwithstanding its concern to make sense of a particular happening. In my view, *history is not a mode of discourse but a way of knowing*, a way of warranting asser-

tions about past events. It is not the case that such assertions can only be made in narrative form if they are to be treated as history, just as it is not the case that every assertion about past events is historical.[42] It would be out of place at the end of this paper to pursue what follows from these views. I have only stated them here in order to make clear that my earlier remarks about the elucidation of modes of discourse were made not because I have any objection to that sort of activity as such, but because it does not seem to me that it casts any light at all on the problems of *historical knowing*.

State University of New York at Binghamton

REFERENCES

[*] This paper was also published in *Philosophy of Science* **34** (1967) 23–40.
[1] William Dray, *Laws and Explanation in History*, Oxford 1957.
[2] See the papers of Alan Donagan, W. B. Gallie and Michael Scriven in Patrick Gardiner (ed.), *Theories of History*, Glencoe, Ill., 1959.
[3] All the italics in this paragraph are Dray's.
[4] Cf. May Brodbeck's distinction between 'language as communication' and 'language as description' in *Minnesota Studies in the Philosophy of Science*, vol. III (1962), pp. 236ff. For a rather extreme consequence of the pre-occupation with modes of discourse, see Kai Nielsen's assertion that the issue is about, not how historians explain, but how they use the word 'explanation' (in Sidney Hock (ed.), *Philosophy and History*, New York 1963, p. 319).
[5] C. G. Hempel, 'The Function of General Laws in History', in Gardiner (ed.), *op. cit.*, p. 349f.
[6] Gardiner, *The Nature of Historical Explanation*, London 1952, p. 65.
[7] *Ibid.*, p. 87.
[8] *Ibid.*, p. 89; Gardiner's italics.
[9] *Frederic William Maitland, Historian*, selections from his writings (edited, with an introduction, by Robert Livingston Schuyler), Berkeley and Los Angeles, 1960, ch. 12. The editor, in his introduction to the essay, informs us that Maitland was mistaken in thinking Elizabeth was the first to use the 'et caetera' in this way, but this has no bearing on our point.
[10] Ernest Nagel, 'Determinism in History', *Philosophy and Phenomenological Research* **20** (1960), 291–317; 303.
[11] Karl R. Popper, *The Open Society and Its Enemies*, Princeton 1950, p. 448.
[12] Michael Scriven, 'Truisms as the Grounds for Historical Explanations', in Gardiner (ed.), *op. cit.*, p. 454.
[13] Samuel H. Beer, 'Causal Explanation and Imaginative Re-enactment', *History and Theory* **3** (1963) 6–29.
[14] *Ibid.*, 7. The quotation of Popper is from *The Poverty of Historicism*, Boston 1957, p. 122; that of Hempel from *op. cit.*, p. 345.

[15] Morton White, 'Historical Explanation', in Gardiner (ed.), *op. cit.*, pp. 356–73.
[16] It is interesting to note that those who would elucidate the historian's mode of discourse would also have an isolating effect upon historical work. Each narrative is treated as a totality in itself, the elucidator being concerned with what the narrator says, and why, in the particular story. This is quite explicitly affirmed by W. B. Gallie, who says: "The fact is that the kind of explanation that I claim to be characteristic of histories cannot be confirmed, or even be preferred against other possible explanations, except *via* the acceptability of the narrative which it enables the historian to reconstruct or resume. If the narrative has now been made consistent, plausible, and in accordance with all the evidence, if it is the best narrative that we can get, then the explanation that helped us to get it is the best explanation as yet available" (*Philosophy and the Historical Understanding*, London 1964, p. 124).
[17] For the distinction between explaining events and justifying explanations, see the first part of the 'Postscript' to this paper.
[18] N. R. Hanson, *Patterns of Discovery*, Cambridge 1958, p. 72; his italics.
[19] *Ibid.*, p. 84.
[20] Though I do not wish to deal with it here, another important function of theory is to aid us in choosing between seemingly plausible, but not always mutually consistent, explanations of the same historical happening. Harry Eckstein – in 'On the Etiology of Internal Wars', *History and Theory* **4** (1965) 133–63 – gives a striking list of explanations of the outbreak of the French Revolution (p. 137f.), and argues that only theory can enable to decide among them (pp. 139 and 163). cf. my 'Evidence and Events in History', *Philosophy of Science* **29** (1962) 175–94; 190f.
[21] Hanson, *op. cit.*, p. 1.
[22] Max Weber, *The Protestant Ethic and the Spirit of Capitalism* (transl. Talcott Parsons), New York and London 1930; references will be to the third British impression, 1950.
[23] Henri Pirenne, *Mohammed and Charlemagne*, London 1939 and 1954.
[24] Cecil Woodham-Smith, *The Reason Why*, New York 1953. (Time Reading Program edition, 1962.)
[25] Weber, *op. cit.*, p. 27; italics added.
[26] Stanley M. Elkins, *Slavery: A Problem in American Institutional and Intellectual Life*, Chicago 1959.
[27] *Ibid.*, p. 84.
[28] *Ibid.*, p. 89ff.
[29] *Ibid.*, p. 98ff.
[30] Charles Tilly, 'The Analysis of a Counter-Revolution', *History and Theory* **3** (1963) 30–58.
[31] Beer, *op. cit.*, p. 10ff.
[32] *Ibid.*, p. 11.
[33] *Ibid.*
[34] Tilly, *op. cit.*, p. 55.
[35] Beer, *op. cit.*, p. 12.
[36] *Ibid.*, p. 12f.
[37] See note 24.
[38] It is, of course, possible that the author was mainly interested in describing some aspect of 19th-century English social history, a subject she has been concerned with in other writings, and that the tragedy of the Light Brigade was only a device toward that end. But even if Mrs. Woodham-Smith was not primarily motivated to explain

the tragedy, her account does in fact do so, and that is all that concerns us here.
[39] This, of course, is only an assumption of my using it as an example.
[40] See note 16.
[41] Frederick J. Teggart, *Theory and Processes of History*, Berkeley and Los Angeles 1949, paperback ed. 1961; pp. 18–29, 51–66.
[42] In the sense of 'historical' as that which pertains to the discipline of history, not the course of human events. Only the former sense of 'history' is of concern in this paper. Even if there is every reason to believe that a certain proposition about a past event is true because its character as divine revelation is beyond doubt, that proposition would not be historical in the former sense. Similarly, statements I make about my own past because I remember this or that – say, what I had for breakfast this morning – are not historical in the given sense either. Neither of these kinds of statement is established in the historical way.

MICHAEL MARTIN

UNDERSTANDING AND PARTICIPANT OBSERVATION IN CULTURAL AND SOCIAL ANTHROPOLOGY

There was a time when cultural and social anthropologists did not do participant observation. Sir James Frazer, famous anthropologist of yesteryear, was once asked if he ever lived amongst savages. It is reported that he held up his hands "as though to ward off even the thought" and answered "God forbid!"[1]

Today things are quite different; field work and in particular participant observation seems, as one critic has put it, an initiation rite into the cult of professional anthropology.[2] How accurate this anthropological characterization of anthropology is I do not know. But it does seem to be true that great stress is laid on the method of participant observation in the profession today.

We will not consider here the anthropological significance of this phenomenon; after all this is a job for anthropologists or sociologists and not for philosophers. The philosophical questions connected with participant observation are rather different. First we want to know what 'participant observation' means. As we shall see the term is not clear and several things have been meant by it. Secondly, we want to know whether there is any methodological justification for participant observation in any of its various senses in cultural and social anthropology. After all there may be good methodological reasons why anthropologists have stressed participant observation and good reasons for its having the significance it does seem to have within the profession. In particular, we shall try to determine the relevance of participant observation in the various senses analyzed for understanding a community. This is necessary because the method of participant observation seems to be closely connected in anthropological thought with the goal of understanding a community. Indeed, so important is the notion of understanding for the evaluation of participant observation that it will be necessary to analyze it before we begin on participant observation.

The paper will thus be divided into three parts. In the first part we will consider the meaning of 'scientific understanding' in anthropology taking some time to separate out other concepts that may be confused with such understanding. In the second part of the paper we will consider the meaning of 'participant observation' in anthropology, clarifying and separating different senses to be found in anthropological literature. In the third part of the paper we will consider the methodological relevance of participant observation for a scientific understanding of a community.

I. UNDERSTANDING A COMMUNITY

A. *Scientific Understanding*

Traditionally one of the major tasks of cultural and social anthropology has been to understand particular communities. In this respect, it has been suggested, anthropology is like history.[3] Cultural and social anthropology and history, it is said, are ideographic, not nomothetic sciences; they do not aim at discovering laws or theories but in understanding particular phenomena.

Now whether such a description of cultural anthropology is accurate we need not decide here. Nor need we enter into two other closely related controversies: the question of whether anthropologists, although they may not aim to discover laws, nevertheless use laws in understanding particular communities; the question of whether anthropologists should aim to discover laws even if they do not do so.

Whatever the answers to these two questions and whatever the accuracy of the description of anthropology as an ideographic science, it does seem to be true that anthropologists have often aimed at understanding some particular community.

What does *understanding* a community mean? Although the term 'understanding' may be used in many senses in science and ordinary life, there is one sense that stands out. This might be called scientific or factual understanding. In this sense to understand something – a person, a subject, a community, and so on – is to know certain facts about the person, theory or community.[4] Thus in the case of persons expressions like, 'Jones understands Smith' are often reducible to expressions of the form 'Jones knows that P_1 and P_2 and ... P_n', where $P_1, P_2, ..., P_n$ refer to certain facts about Smith. What sort of facts these are will usually be clear from the context. Sometimes the facts concern Smith's motivation, his purposes or his character; at other times the facts will be facts of his social or cultural background and their relation to his present behavior. I interpret the term 'fact' quite broadly here. One fact about Smith that Jones might have to know in some context to understand Smith is that Smith's behavior is subsumable under certain laws. The present analysis remains neutral with respect to the question of whether knowledge of such facts would be necessary.[5] In any case, conceived in this way, understanding a person seems to be reducible to knowing certain facts about him. We call this type of knowledge propositional knowledge.[6] Propositional knowledge is simply knowledge that something is the case. In the case of Jones and Smith, Jones' understanding Smith consists in Jones' having certain propositional knowledge of Smith, i.e., knowledge that certain things are true about Smith.

Where the object of understanding is not a person a similar kind of analysis seems possible. To say that Jones understands psychoanalysis or Newton's Theory is usually to say that Jones has some particular kind of propositional knowledge. Again the context usually makes clear what this propositional knowledge will be. The suggestion is often that Jones' propositional knowledge consists in knowing that psychoanalysis or Newton's Theory has certain important structural properties or that the parts of these theories fit together in a certain way; at other times the suggestion is that the propositional knowledge consists in knowing certain facts about the history of psychoanalysis or physics.

The same sort of thing could be said about understanding a community. Understanding a community seems to consist in having certain propositional knowledge about this community. Again what this knowledge will be will depend on the context of the inquiry. Sometimes understanding a community may consist in knowing that certain aspects of the community are related to one another in certain ways. Thus to say that John Beattie understands Bunyoro[7] may mean only that Beattie knows that certain social institutions in Bunyoro are related to each other in certain ways, forming what has been called a functional system. In other contexts understanding a community may consist in knowing that the community has such and such a historical development.

It should be clear that the present analysis is opposed to the view that there is only one set of facts which, if known, gives understanding of a community. In anthropology, understanding a community will be relative to the perspective taken by the anthropologist, the theory utilized in the investigation, and the purposes of the investigation. A community can be understood functionally, historically, and in other ways[8], although given certain purposes some ways may be better than others. However all these different ways of understanding on the present analysis come down to having propositional knowledge about the community.

B. *Being Understanding*

Another notion of understanding which may not be clearly separated is anthropological thought from factual or scientific understanding is that of being understanding toward a community.[9] Now being understanding toward something or someone involves manifesting a certain attitude; one is understanding toward a person or group, for instance, if one is sympathetic, patient and tolerant towards the person or group.

It may be true that nowadays part of the ethical creed of the anthropological profession is humanitarianism and that part of this humanitarian attitude may be a sympathetic and tolerant attitude toward the community studied; that anthropologists are expected to be understanding toward the people they study. It must be stressed that however strong this commitment may be among field anthropologists and however justified it is on ethical grounds, there is no logical connection between this ethical posture and factual understanding of a community. As far as the logic of the two concepts is concerned, an anthropologist could fail to be understanding toward a community he studies and yet from a scientific standpoint understand the community very well; moreover, he might be understanding toward the community he studies yet have little scientific understanding of the community. To be sure, although there is no logical connection between these two concepts there might be some close practical connection. This might be true for a number of reasons: a failure to be understanding toward a community might be manifested in the anthropologist's behavior; this behavior might cause offense; as a result, the anthropologist might be denied access to vital information; this denial of vital information would in turn prevent scientific understanding of the community.

I say "might" because the degree of intolerance a community can tolerate from a visiting anthropologist without closing down necessary sources of information is not completely known. It may well vary from community to community and from anthropologist to anthropologist. We may suspect at least that in some communities a high degree of intolerance and lack of sympathy can be expressed by the anthropologist and yet a high degree of scientific understanding of that community still be achieved. Malinowski's recent diary[10], for example, indicates that Malinowski often felt and expressed great hatred toward the Trobriand Islanders; he was often not understanding toward them. Yet his study of these people[11] is usually considered to be a landmark in ethnographic understanding, i.e., he is considered to have scientific understanding of the community.

C. *Empathy and Understanding*

A discussion of understanding in anthropology is surely not complete without some mention of the notion of empathy. It is important to discuss the notion since empathy is often confused with the two notions of understanding I have already analyzed.

1. *The Adoption Sense*

I take it that on one common view, to say that someone empathizes with a person or group is to say that he 'puts himself' in the other's place, i.e., he adopts the attitudes, views, emotions, thought patterns, and so on of a person or group; he sees the world and emotionally reacts to the world in the same way as the person or group sees it and reacts to it. Thus an empathizing visiting anthropologist would see and think what the natives saw and thought, he would feel what the natives felt, and so on. Let us call this sense of 'empathy' the *adoption* sense of empathy. It might be said that only by this sort of empathy can the anthropologist understand the community at issue.

To evaluate this position one must sharply distinguish it from another position that may easily be confused with it. It has been maintained by some students of anthropological methodology that it is only by knowing that the natives have the beliefs, goals and attitudes they do have that one understands them. But it is one thing to say this and an entirely different thing to insist that it is only by seeing, feeling, and believing what the natives see, feel and believe that understanding can be achieved.

In short, there is a vast difference between *having* knowledge that people have such and such beliefs, goals and attitudes and *adopting* such beliefs, goals and attitudes. To make this point clear consider the following two expressions:

(1) Beattie knows that the Nyoro have a suspicious attitude towards Europeans.

(2) Beattie adopted the same suspicious attitude toward Europeans that the Nyoro have.

Notice that only in the second expression is there any question of empathy in the adoption sense. Notice also that these two expressions are logically independent. Beattie might know that Nyoro have a certain attitude without adopting that attitude. Beattie might, for instance, be psychologically incapable of being suspicious of Europeans although he knew perfectly well that the Nyoro are suspicious. Moreover, Beattie might adopt this attitude without knowing that the Nyoro have this attitude. For instance he might lack sufficient evidence that the Nyoro are suspicious. This would preclude his knowing that they do have such an attitude on most analyses of propositional knowledge. But it would not prevent him from adopting this attitude.

This example illustrates that propositional knowledge of people's beliefs, goals and attitudes is possible without empathy with the people in the adoption sense. And it is clear that this is true in general: scientific understanding as we have analyzed it in terms of propositional knowledge is possible without empathy in the adoption sense.

Indeed, empathy in this sense is at times incompatible with scientific understanding. For to empathize with a community in this sense may involve adopting their beliefs about their own culture and themselves. However, these beliefs may well be false and distorted and thus preclude scientific understanding. For example, suppose an anthropologist empathized with the natives of some community. This empathizing, let us suppose, involves adopting one of their false beliefs. For example suppose John Beattie in empathizing with the Nyoro believed that not *p* where not *p* is some false proposition about Bunyoro. On one standard analysis of propositional knowledge Beattie could not know that *p* if he believed that not *p*. If his knowing that *p* was essential to his understanding Bunyoro, his empathizing would have precluded his understanding Bunyoro.

Furthermore, empathizing with a community may at times be incompatible with being understanding toward the community. Recall that to be understanding toward a community is to be sympathetic and tolerant toward the community. But it is possible that an anthropologist could not in certain communities at one and the same time be both sympathetic and empathetic in the adoption sense. For to empathize with the natives in the present sense is at least to adopt the attitudes of the natives. But one attitude found in certain communities may be an attitude of intolerance and lack of sympathy toward everyone. Surely one could not at the same time be both sympathetic toward the natives and unsympathetic toward everyone.

I conclude that empathy in the adoption sense is usually irrelevant to the scientific understanding of a community and that under certain conditions it is incompatible with it. I conclude also that empathy in the adoption sense is at times incompatible with being understanding toward a community.

2. *The Assimilation Sense*

There is another sense of 'empathy' that must be distinguished from the adoption sense. This sense is suggested by a statement by Nadel. He remarked: "Much in this process of 'empathy' can be reduced to simple empirical facts. Complete knowledge and ready use of the vernacular; long sojourn in the community; and above all, the cumulative effect of constant enquiries into everything that is observed – all these built up a familiarity with the native society which in more poetic language, one might call 'entering into the soul of the savage'. When you find yourself making jokes at which the people laugh (no mean achievement, this); when you have mastered their ways of linking concepts and adducing proof, or when you can predict with fair accuracy what X will do and feel in such and such a situation, you have accomplished an adequate 'assimilation'."[12]

I take it that in this passage Nadel is using 'empathy' in a sense other than the adoption sense. In this new sense to say that someone has empathy with a group is to say that the person can get around in the group easily, that he speaks their language, is on joking terms with them, and so on. In this sense of 'empathy' empathy seems to consist in the utilization of a complex skill. To use Ryle's notion, empathy in this sense is the utilization of a type of knowledge how, or, as it has recently been called, procedural knowledge.[13] Borrowing from Nadel, one might call this sense of empathy the *assimilation* sense.

These two senses of 'empathy' distinguished here are logically independent. One might move about smoothly in a group, be on joking terms with them, and so on without adopting the beliefs, goals, and attitudes of the members of the group although it might be necessary at times to act as if one had adopted some of these beliefs and goals and attitudes. One might also adopt a people's beliefs, goals, and attitudes and yet not move smoothly in the group. For instance the group might not believe that one had adopted their beliefs, goals, and attitudes. This belief might prevent assimilation.

As we shall see in a moment the assimilation sense of 'empathy' is closely connected with one sense of 'participant observation'. Because of this close connection, the relevance of the assimilation sense of 'empathy' to scientific understanding will be discussed later when we discuss the relevance of participant observation for scientific understanding.

There is no logical connection between the assimilation sense of 'empathy' and being understanding toward a community. One could have empathy with a community in the assimilation sense and yet not be understanding toward the community – although one might have to act *as if* one were understanding toward it in order to be assimilated. Indeed, the process of becoming assimilated in a community may in certain cases breed contempt rather than sympathy. (Malinowski's experience is perhaps an example of this.)

Conversely one can be understanding toward a community without being assimilated in the community. Indeed, being understanding toward a community is possible over great periods of time and spans of space. Thus some contemporary classicists are understanding toward classical Greek civilization although these men could hardly be said to be assimilated in ancient Greek society.

To be sure, under certain conditions assimilation in a community may be conducive to being understanding toward a community; and conversely, under other circumstances not being assimulated in a community may prevent understanding toward the community from developing. However, under what conditions this may happen is unclear and is a matter for empirical investigation.

II. THE MEANING OF PARTICIPANT OBSERVATION

Although the term 'participant observation' is widely used in social science literature, little attempt has been made to separate the different strands involved and to clarify issues. There have been, however, some attempts to specify different senses of the term. In a well known article Becker and Geer differentiate different senses of 'participant observation' and we will consider some of these here.[14]

A. *Participant Observation as Being a Member of a Group*

1. *The Native Sense*

In one of Becker and Geer's senses to say that someone is a participant observer of a community is to say that one is a member of the community. This definition is hardly helpful until we are clearer on what being a member of a community involves. For example, is John Beattie a member of the African community of Bunyoro if he lives in this community or are only the Nyoro members of Bunyoro. To clarify this situation let us differentiate between two senses of being a member of a community: the sense in which the Nyoro are members of Bunyoro, let us call this the 'native sense' of being a member of a community; the sense in which Beattie was a member, let us call it the 'living-in' sense of being a member of a community.

It should be noted that although most of the Nyoro are members of the Bunyoro community in the living-in sense, not all are; presumably some of the Nyoro are away and are not actually living in the community at some particular time. Thus in the living-in sense Beattie was a participant observer of Bunyoro in 1951 although some of the Nyoro were not participant observers in 1951 in this sense. Moreover, all the Nyoro are participant observers in the native sense of participant observer of Bunyoro by virtue of *their* cultural background, although Beattie could never be a participant observer in this sense by virtue of *his* cultural background. It seems apparent that anthropologists are not interested in the native sense of participant observation; it is the living-in sense that has relevance for them.

2. *The Living-In Sense*

But what does living in a community involve in anthropological field work? Does it involve living in a community as the natives live? The answer is that it does but only to a certain extent. The extent will depend on the anthropologist and the culture. For Malinowski living in a community involved isolating oneself from white men, taking residence in the village, eating native food, seeking out native entertainment and conversation. Beattie did all of this and more: he gave banana beer parties, and even did some native work. On the other hand, for many anthropologists living in a community does not include dressing in native garb, engaging in native sexual practices, giving up certain Western customs. Thus it is not strictly speaking true, as Frazer suggests, that Malinowski "lived as a native" in his study of Trobriand Islanders.[15]

Why anthropologists should adopt some of the practices of a people and not others is an interesting question and we will consider it in a moment.

3. *The Assimilation Sense*

However, one needs to make a finer distinction in 'the living-in' sense of participant observation in order to do justice to anthropological method. For most anthropologists are not merely interested in living in a community in the sense just specified. One might live in a community in this sense and yet be unable to operate in the community with ease and comfort and familiarity: one might not know one's way about with respect to the customs of the people. It was only after some time that Malinowski said that he learned how to behave and to a certain extent acquired 'the feeling' for the natives' good and bad manners.[16] To say that Malinowski was a participant observer in a community is to say that he lived in a community and knew his way about the community socially. In this sense to be a participant observer is to utilize a certain social know-how while living among the natives.

It should be noted that this sense of participant observation is identical to the assimilation sense of empathy specified earlier. Indeed, let us call this sense of being a member of a community the *assimilation* sense of being a member of a community. Notice that in this sense, the extent to which an anthropologist is a member of a community and thus the extent to which he is a participant observer in this community will be a matter of degree.

It should also be noted that the extent to which an anthropologist follows native customs and practices may be closely connected with the degree of social know-how he develops and utilizes in the culture. A following of native customs and practices to a certain degree seems essential for developing social know-how. On the other hand the following of particular native customs and practices may bring ridicule and hostility and actually prevent social know-how from developing. The extent to which the customs and practices must be followed in order to achieve the maximal social know-how will undoubtedly vary from culture to culture and anthropologist to anthropologist. In any case, the possibility of adverse effects of following some customs of the natives does provide some methodological grounds for not going completely native.

Notice, however, that although it might be necessary for an anthropologist to follow some of the customs of the natives in order to gain social know-how and to avoid adopting others in order to maintain this know-how, it would not be necessary that the anthropologist adopt any of the natives' *beliefs*, *goals*, or *attitudes*. To be sure, it might be necessary for the anthropologist to act *as if* he had adopted some of them. But this is a different matter. Thus having empathy with the natives in the adoption sense of 'empathy' specified earlier, is not a necessary condition for participant observation in the assimilation or living-in senses just specified.

Of course, some methodologists have defined 'participant observation' in such a way that it includes being empathetic in the adoption sense.[17] On the other hand, others have suggested that the expression 'participant' excludes empathy in the adoption sense.[18] It seems to me that this latter view comes closer to the truth. As we shall see in a moment some of the standard senses of 'participant observation' in social science literature would tend in some cases to exclude empathy in the adoption sense.

Taking our point of departure from Becker and Geer we have thus far characterized two senses of 'participant observation' that seem quite relevant for anthropology and yet do not involve empathy in the adoption sense. First the living-in sense: in this sense a participant observer is a person who lives in a culture and adopts some of the customs and practices of the people. Second, the assimilation sense: in this sense a participant observer is a person who lives in a community, develops some social know-how, and is assimilated to a certain extent in the community. We have emphasized that these two senses are closely related to one another. To gain social know-how in a culture involves living in the community and following certain customs and practices of the natives. However gaining social know-how may also involve not following other customs and practices.

So far we have simply tried to clarify and expand one of Becker and Geer's condensed characterizations of participant observation. However, even with this clarification and expansion there is a fundamental omission in Becker and Geer's account. It leaves out all mention of the fact that the person living-in the culture and having certain social know-how has as one of his purposes observing the culture. A person who lived in the culture and had certain social know-how of the culture would normally not be considered to be a participant observer unless he had as one of his purposes observing the culture, i.e., studying it. Missionaries and traders would not normally be considered participant observers of a culture even if they lived in the culture and gained a certain social know-how unless they were attempting to study the culture as well as save souls or sell merchandise.

Thus the living-in sense of participant observation must be amended as follows: A participant observer is a person who lives in a culture, adopting some of the customs and practices of the natives, and has as one of his purposes while living in the culture observing the culture. The definition of 'participant observation' in the assimilation sense would have to be modified along similar lines.

Once it is admitted that this qualification is needed, empathy in the adoption sense would in certain cases be ruled out. To empathize with a person or group in the adoption sense is to adopt all the goals, attitudes and beliefs of the person or group. Moreover, since the basic idea is to put oneself in the other person's place, it would presumably involve repressing or eliminating some of one's own goals, attitudes and beliefs; namely those goals, attitudes and beliefs that are different from those of the person or group being empathized with. But then if a participant observer is a person who by definition has as his purpose the observation of the members of the group and the members of the group do not have this as one of their goals, being a participant observer of a group and being a complete empathizer with the group in the adoption sense at one and the same time is ruled out. This, of course, does not rule out being a participant observer in these senses at certain times and empathizing with a group at different times. Nor does it rule out adopting some of a group's goals, attitudes, and beliefs, i.e., *partially* empathizing, and being a participant observer in these senses at the same time. But it does seem to rule out *complete* empathy and participant observation in these senses at the same time, in certain cases.

B. *Participant Observation as Posing as a Member of a Community*

Becker and Geer differentiate another sense of participant observation. In this sense to say that someone is a participant observer is to say that the person is posing as a member of a group or community. To achieve participant observation in this sense would be difficult, if not impossible, in many typical field situations. Consider the native sense of being a member of a community. The anthropologist as a participant observer would have to pose as a member of the community under investigation. Beattie in this sense of participant observation would have to pose as a Nyoro, a most unlikely prospect! As Herskovits has noted, the white anthropologist in a black man's culture has a "high degree of social visibility"[19]; it is rather difficult to think of Beattie or Malinowski attempting to pass themselves off as one of their black subjects by disguises, make up, and so on.

Consider the living-in sense of being a member of a community. It is difficult to see how an anthropologist could pose as someone living in a community to the community, when the community is small and isolated as is typical in anthropological studies. How for example could Beattie have posed as a person who was living in the small Bunyoro village he wished to study to the members of the village? It might have been possible for Beattie to pose as someone living in a small Bunyoro community to his friends at Oxford and even to members of another distant village but hardly to the members of the small village he did study.

I think it can safely be said that participant observation in this sense has such severe limitations in most anthropological field work that it can be ignored, and we will so ignore it. In any case, this sense also would have to be modified in terms of the purposes of the person posing as a member of the group: the person posing as a member of a group would have to have as one of his purposes the observation of the group to be a participant observer.

C. *Participant Observation as the Role of an Observer in a Community*

Becker and Geer distinguish another type of participant observer. In this sense to be a participant observer one joins a group as "one who is there to observe". Although this kind of participant observation is not completely clarified, what they seem to have in mind is roughly this: on certain occasions a social scientist joins a group and makes it known that he is there to study the group. He takes the role of an observer or student of the group. Let us call this sense of 'participant observation' the *observer-role* sense of participant observation. One gathers that it was this type of participant observation that Becker and Geer themselves used in their study.

But what does joining the group consist of in the observer-role sense? Does it consist in living in the group, being assimilated to a certain extent, gaining a certain social know-how? Does this sense include the living-in sense and assimilation sense of participant observation specified above? Sometimes it does. Beattie, for instance, came to his Bunyoro village as an observer. He announced to the assembled villagers at the beginning of his stay:

"I have come to your country to learn your language, and about your history, your traditions and your customs and the way you live. I have come from a big school in Europe where grown-ups are taught, including some who come to Africa..."[20]

This speech and his subsequent behavior placed Beattie in the role of observer of the village while in the village. Nevertheless this did not prevent him from living in the village, gaining a certain amount of social know-how and to a certain extent being assimilated.

Although sometimes – as in Beattie's case – being a participant observer in an observer-role sense includes being a participant observer in the living-in and assimilation senses, there is no necessity in this. These senses of 'participant observation' are logically distinct. One could join a group in the role of an observer and yet never gain any social know-how, never be assimilated to any extent into the group, and adopt very few of the customs of the group.

Anthropologists who are participant observers in the living-in and assimilation senses of participant observation do not always take the role of an observer. Some argue that the fact that they are studying the community they are living in should be hidden from the natives. Florence

Kluckhohn[21], for example, stresses the necessity for the participant observer not to disclose that he is studying the group while living in the group. She argues that such a disclosure might distort the information received while living in the community and indeed prevent the anthropologist from obtaining information that he might otherwise have received. Kluckhohn, it should be noted, was in a fortunate position: for reasons we need not go into here she was able to live in a New Mexican village and study the people without any need for disclosing that she was studying them.

Whether Kluckhohn's approach would be possible or even desirable in all cases is uncertain. Could Beattie, for example, have been a participant observer in the assimilation sense in his Bunyoro village without having taken the role of an observer of the group? What pretext could he have used to start to live in the village? What would have happened if it had been discovered that he was lying? Moreover, is it really true that the observer-role he did take distorted his information as Kluckhohn seems to maintain? We do not know the answers to these questions. Much more information about human psychology and culture will have to be secured before they can be answered with any confidence.

One suspects, however, that Kluckhohn's view is much too simple. Whether or not observer-role participant observation will distort will depend on many factors including the type of information one is after and the culture under study. For instance, there might be certain types of information which could be gathered by observer-role participant observation with much more success than by other types of observation. Moreover, the very role of an observer may loosen tongues and provoke confessions in certain people under certain circumstances. A similar phenomenon has been noticed by some anthropologists: it has been noticed that in some cases anthropologists have access to certain information because they are non-natives. Thus far from their outsider role being a handicap, it may under certain conditions be a help.

III. THE JUSTIFICATION OF PARTICIPANT OBSERVATION

Now that we have discussed several different senses of 'participant observation' we will consider some possible methodological justifications for participant observation.

A. *Participant Observation as a Sufficient Condition for Scientific Understanding of a Community*

One possible methodological reason for the use of participant observation in anthropology could be that participant observation in a community is a sufficient condition for achieving scientific understanding of the community. It seems to me, however, that this thesis is mistaken no matter what sense of 'participant observation' one considers.

Consider the living-in sense of participant observation. Surely living in a community and following some of the customs and practices of the natives with the purpose of observing them does not guarantee scientific understanding of the community in the sense defined above. This is true for any number of reasons. For instance, while living in a community one may have to adopt the customs or practices of one particular social class and this may alienate one from another social class in the community, in turn preventing one from getting needed information. Under precisely what conditions such alienation and information blockage would occur is not completely clear at the present time. We do know that it sometimes happens. Beattie, for example, had the following choice in Bunyoro: if he lived with the King and his court, he would alienate the people; if he lived with the people, he would alienate the King and his court. Beattie chose the latter course of action and as a result much of the inner workings of the King and his court remained unknown to him. There seems to be no doubt that living with the people and adopting their customs actually prevented Beattie from gaining complete scientific understanding of a certain major social relation in Bunyoro.[22] It should be noted that this was indeed a loss for Beattie since because of his theoretical orientation he was committed to studying all the major social relations of the Bunyoro. The curious fact is that Beattie was also committed to doing participant observation in the living-in sense. The commitment to this method conflicted with full realization of his theoretical goal.

But even if there were no problem of alienation other problems remain. I have already mentioned that merely living in a community and adopting some of the customs of the natives will not necessarily bring about assimilation. Whether it will or not depends on many factors. And without this assimilation it may be impossible to get the sort of information one needs to understand the community.

Now consider the assimilation sense of 'participant observation'. At least one of the problems of the living-in sense remains. In becoming assimilated to one class of a community one may alienate another class; this in turn may block some important sources of information, in turn preventing scientific understanding.

There is another problem. Being assimilated in a community involves moving easily in the community, having a feeling for the manners and mores of the natives, and so on, but it does not in any way guarantee that scientific understanding of the community has been achieved. Scientific understanding of a community as we have specified it above consists in having certain propositional knowledge about the community. Being assimilated need not and often does not involve such propositional knowledge. One might be assimilated in a community and yet be ignorant of facts about the structure or history of the community which it would be necessary to know in order to understand the community given one's purposes and perspective.

Much the same sort of thing can be said for the observer-role sense of 'participant observation'. Announcing to the natives that one is there to observe them may be the kiss of death to one's research project. Even if one is allowed to stay after the disclosure, necessary information may be blocked by it. However, aside from these problems the mere fact of being in a certain role in a community, i.e. the role of an observer of that community, does not guarantee that any propositional knowledge has been obtained. And since scientific understanding involves obtaining such knowledge, being in the role of an observer does not necessarily mean that scientific understanding has been achieved.

B. *Participant Observation as a Necessary Condition for Scientific Understanding of a Community*

It may be granted that participant observation in a community in the senses specified is not a sufficient condition for scientific understanding of the community. However, it may be maintained that participant observation is a necessary condition for scientific understanding. This certainly seems to be Beattie's view. He argues that "only by at least some participation in the community life"[23] can the anthropologist understand the community and that the "observer must live in and with the community he is studying; it is not enough to stay in a comfortable hotl or rest-home and to visit them for a few hours daily or less often ".[24]

Before we can evaluate this contention it is necessary to clear up a certain ambiguity in the thesis. Is the claim that no person could scientifically understand a community unless the person himself was a participant observer in the community? Or is the claim rather that no person could scientifically understand a community unless some person or other was a participant observer in the community? Presumably it is the latter claim that is at issue since the former claim seems to have implications that are obviously false: no mere reader of anthropological literature could ever have scientific understanding of a community; an anthropologist who lived in a community could never communicate his scientific understanding of the community in his books and monographs. But surely this is mistaken. It may be true that anthropologists sometimes fail to give their readers any scientific understanding of a community. But they are hardly doomed to this failure; there seems to be no *a priori* reason to suppose they could not succeed.

The latter claim must be taken more seriously, however, but still we have a problem of clarification. What sense of 'participant observation' it at issue: the living-in sense, the assimilation sense, the observer-role sense? Let us interpret the thesis to be this: for all cultures someone's participant observation in one of the relevant senses is a necessary condition for understanding the culture. The thesis stated in this way has some difficulties.

It is worth pointing out initially that Beattie and most British social anthropologists investigate cultures that are still existing; and it is well known that British social anthropologists often have a definite ahistorical or even anti-historical orientation.[25] American cultural anthropologists, because of the types of cultures they have investigated, have been forced sometimes to use different methods than their British colleagues. Thus the alternative to participant observation is not always, as Beattie seems to suggest, traveling from a comfortable hotel or rest-home to "visit the natives a few hours daily or less". American anthropologists have sometimes wanted to investigate Indian civilizations that are virtually extinct. They have used old native informants and not participant observation[26] in their work; sometimes they have not even left their hotel. The informant has come to them.

Cornelius Osgood[27], for example, reports his investigation of the old culture of the Northern Athapascans primarily by the use of the remarkable Indian informant Billy Williams. Williams, a man of remarkable intelligence and photographic recall, was interviewed by Osgood 8 hours a day for weeks on end. Williams on his own initiative would travel to remote settlements to quiz old Indian women and men on details he did not know and reported the results to Osgood. Williams would make models of the old buildings and of manufactured objects for Osgood in his spare time and would spend hours relating the intricate and complex old ceremonies and dances to Osgood. It is hard to accept the view that Osgood, because he was not a participant observer in the old culture, did not come to have a scientific understanding of old Northern Athapascan culture through the informant Williams.

Indeed, I am far from convinced that even in Beattie's own work participant observation in any of the senses at issue was necessary. Beattie's procedure was as follows: After studying all the written documents on the Bunyoro culture, Beattie went to Bunyoro and became a participant observer in the living-in, assimilation and observer-role senses in a small Bunyoro village; he learned the language, adopted some customs, observed the social relations, took notes and after 6 months returned to Oxford. In Oxford he studied his notes, continued to study the language, consulted with his teachers and after 4 months returned to Bunyoro. One gathers that on this visit he used very little participant observation. Primarily he used other methods: he used informants, studied court records and official documents, took surveys, gave questionnaires, had natives write essays, and so on.

One naturally wonders whether his first 6 months' visit as a participant observer in a small Bunyoro village was really necessary. Would it really have been impossible to proceed without this initial contact? Is there any particular piece of information that Beattie would not have obtained without the first 6 months of participant observation?

These questions must not be confused with another type of question. Was Beattie's stay in a Bunyoro village an enriching and aesthetically satisfying experience? Could this experience have been achieved in any other way? Could it have been achieved by non-participant observer methods? It is no doubt true that participant observation is often an enriching, moving and aesthetically significant experience for anthropologists. And it may also be true that such an experience is difficult or even impossible to achieve in other ways. But this, it seems to me, shows nothing about the necessity of participant observation for achieving scientific understanding. The question we are concerned with here is only: is there any good reason to suppose that every fact Beattie learned about Bunyoro's social structure – this was what Beattie was primarily interested in – could not have been learned from good informants and other non-participant observer methods? We are not interested in whether Beattie will always cherish his 6 months in a small Bunyoro village or whether this moving experience could not have been achieved in other ways.

It must be emphasized that Beattie advances very little in the way of explicit argument to substantiate his view that his 6 months as a participant observer was necessary for a scientific understanding of the Bunyoro. As far as I can discern, his belief that it was necessary rests upon two major considerations:

(1) It is necessary to learn the native language of a culture to understand the culture. Unless he knew the language he would not have been able to check his informants and interpreters, compose questionnaires, gain insight into the natives' thinking. Six months alone as a participant observer where he was forced to use the language was an ideal way to achieve this linguistic knowledge.

(2) In his 6 months as a participant observer he got a certain feel for the culture. Without this initial exposure he would not have been able to ask his informants the right questions or compose an insightful questionnaire. Thus the first 6 months of participant observation could not have been eliminated.

These considerations, however, are hardly persuasive. Consider the first point.

Whether learning the native language is necessary for achieving understanding of a community we need not decide here, but it should be noted that not all anthropologists have believed that learning the language is essential in field work.[28] Let us grant, however, that learning the native language is essential. The most Beattie's argument shows is that the *best* way to learn the language would be through participant observation. It does not prove, however, that participant observation was a necessary condition for learning the native language and hence a necessary condition for achieving understanding. And it is surely mistaken to suppose that participant observation was a necessary condition for learning the native language in Beattie's case. Thus although Beattie did master the essentials of the native language while a participant observer presumably other methods were available to him, e.g., hiring a native tutor. Moreover, it is not completely obvious that participant observation is even the best way to learn a native language. So little is known about language learning that one might well take a skeptical attitude to this contention. For all we know at the present time a native tutor might well be a much better method.

In order to criticize Beattie's second point one need not belittle Beattie's idea of getting an initial feel of the culture. This initial and germinal exploratory period may indeed be necessary for all empirical research. The question is whether participant observation is necessary for gaining this feel. It should be recalled that anthropologists like Osgood who used a single informant like Billy Williams managed to get a feel for the old culture without participant observation. By asking the informant questions – perhaps stupid questions or perhaps by just letting the informant talk at random – one may get a feel for the culture. After a while more penetrating and intelligent questions can be asked.

It might be objected, however, that it was still necessary to have a participant observer in order to obtain any scientific understanding of, for example, the old North Athapascans' culture; that Billy Williams himself was the participant observer. But Williams was not presumably a participant observer in any of the senses at issue. Recall that to be a participant observer of a community in the senses under discussion one must have as one of one's purposes to observe or study the community at issue. Now although Billy Williams was a man of remarkable observing powers there is no good reason to suppose that he engaged in a purposeful study of the culture in which he lived while he lived there. It is true that Williams was a participant observer in the native-sense of participant observer. But as we have seen, this sense has little relevance to anthropological discussion of participant observation. Williams was an informant and the use of informants is *contrasted* with participant observation in the relevant senses in anthropological discussions.

Moreover, if we consider cultures that have long since perished, the dubiousness of the necessity of reports of participant observers even in the native-sense of participant observation, e.g. informants, becomes apparent. It does not seem to be true that one can have no scientific understanding of a culture without reports of natives. In reconstructions of prehistorical cultures archaeologists do not base their reconstructions on the reports of participant observers in any of the senses specified – even the native sense. Material artifacts such as coins, pottery, art work and so on of the old cultures may be the only source of evidence available. Thus Larco Hoyle[29] reconstructed the ancient Mochica culture of Peru primarily on the basis of scenes portrayed on grave pottery. Other investigators have reconstructed ancient cultures on similar evidence.[30]

It is important not to neglect the limitations of these kinds of reconstructions; on the other hand it is equally important not to belittle what they can achieve. And what they can achieve will depend on many factors. The type of material artifacts that are used, the kinds of cultural facts that are inferred, the strength of analogical reasoning that can be brought to bear, will all affect the understanding that these reconstructions can give.

It would be incorrect to say that these reconstructions always give us a great understanding of ancient cultures. But it is equally incorrect to say that they give us no understanding of these cultures. It seems correct to say that they do sometimes give us a moderate degree of understanding, although it may be easily admitted that different sorts of evidence could increase our understanding. The evidence obtained from informants and participant observation might well be evidence of this sort.

Since we have just made this small concession to the method of participant observation it is time to make another. There is a much weaker version of the thesis that participant observation is a necessary condition for understanding a culture. Instead of claiming that for *all* cultures someone's participant observation in one of the relevant senses is a necessary condition for anyone understanding the culture, the claim might rather be that for *some* cultures someone's participant observation in one of the relevant senses is a necessary condition for anyone understanding the culture. This claim is not so obviously mistaken and indeed may even be true. It may be impossible at certain times in certain cultures to get reliable and trustworthy informants, direct non-participant observation may be impossible, no written history or literature may be available, archaeological reconstructions may be out of the question, and so on. Under these conditions participant observation may well have to be used if any understanding is to be achieved.

How often such conditions are realized is difficult to say. In any case, it remains to be shown that these conditions occur often enough to justify the great stress placed on participant observation by the anthropological profession.

C. *Participant Observation as the Most Efficient Way of Obtaining Information of Certain Kinds*

If we reject the view that participant observation is always a necessary condition for obtaining scientific understanding of a community, we may attempt to justify the widespread use of participant observation on other grounds. One might argue that participant observation is the most efficient way of obtaining certain kinds of information in a community and that these kinds of information are necessary for a scientific understanding. This tack is taken by Morris Zelditch.[31]

Zelditch argues that there are three types of information that social scientists in community studies are interested in obtaining: information about frequency distributions, e.g. the number of married women in the community; information about certain incidents in the community, e.g. a recent wedding in the community; information about rules or norms in the community, e.g. marriage rules. Moreover, there are three methods that social scientists use to obtain these three types of information: surveys, participant observation, informants.

Zelditch maintains that these three methods should be judged by two criteria: (a) informational adequacy, and (b) efficiency. Judged in these terms he finds the following: Survey method has the greatest informational adequacy and most efficiency with respect to frequency distributions; participant observation has the least. On the other hand, participant observation has the greatest information adequacy and most efficiency with respect to incidents; survey methods has the least. Use of informants has the most informational adequacy and most efficiency with respect to rules and norms; participant observation, although it has good informational adequacy, is inefficient in obtaining this kind of information.

Zelditch's contention, if correct, would indeed provide some justification for the widespread use of participant observation in anthropology. There are, however, certain questions one can raise about Zelditch's views.

First of all, it is unclear what Zelditch means by 'participant observation'. What sense of 'participant observation' is being used in his claim that participant observation is the best method to obtain information about certain incidents in a community? Certainly this will make a difference since as we have seen certain types of participant observation may have particular problems in certain communities. For example, the role-observer type of participant observation may be difficult to carry out in certain types of communities and may be performed only at a great price.

Moreover, what is the range of his claims? Is Zelditch suggesting that participant observation in some sense is *always* more efficient and more informationally adequate in obtaining information about incidents in the community? What sort of incidents are these? If the incident occurred 40 years ago, far from participant observation in some sense being the best method, it is not possible to use it at all. An old reliable informant like Billy Williams may be an anthropologist's only hope. But this may be true even with a contemporary incident. For instance, if Beattie had suddenly discovered that an important and rare ceremony was to occur in a Bunyoro village close to the Bunyoro village he was a participant observer in, it would have been impossible for him to have obtained information about this incident by his participant observation. He was not a participant observer in the neighboring village and in several relevant senses of 'participant observer' it would have been impossible under normal circumstances to have become one in the needed time. If Beattie needed this information to obtain an understanding of Bunyoro culture, some nonparticipant observer method would have been his only hope.

Of course Beattie might have hoped that this rare ceremony would occur in the village in which he was a participant observer given enough time. However, he only had a limited time to perform his research and in any case a long wait would be inefficient.

Moreover, there may be certain kinds of contemporary incidents occurring in a community in which an anthropologist is a participant observer which are not and cannot be observed by the anthropologist. For instance, there may be secret ceremonies to which the anthropologist has no access despite his having lived in the community for an extended period of time and despite his having developed a sense of social know-how. In this case also it would be necessary to use trusted informants if any are available. To be sure, the anthropologist may maintain that given more time and patience these ceremonies will be opened to him. But the time and energy needed to gain entrance into the inner circle may well make other methods more efficient.

Of course, it may be admitted that in certain cultures under certain conditions relative to certain types of information participant observation in some of the senses discussed may be more efficient than other methods. But this is an extremely weak claim; indeed, it would hardly by itself justify any widespread use of participant observation or justify the status participant observation seems to enjoy in the profession.

CONCLUSION

The main conclusions of this paper are as follows:

(1) Scientific understanding of a community consists in knowing certain facts about the community. What these facts are will depend on the context of inquiry.

(2) Scientific understanding of a community is logically independent of being understanding toward a community although there may be certain important practical connections between achieving scientific understanding of a community and being understanding toward that community.

(3) Being empathetic with a community in one important sense of the term (the adoption sense) is usually independent of both scientific understanding of the community and being understanding toward the community. Moreover, there are cases in which being empathetic is incompatible with understanding in both senses of understanding.

(4) Participant observation in the senses specified here is neither a sufficient condition nor a necessary condition for achieving scientific understanding of all communities.

(5) Moreover, participant observation is not in all cases the most efficient way of obtaining certain types of information.

(6) However, it may be true that in certain cases participant observation is a necessary condition for achieving scientific understanding of a community and in certain cases it may be the most efficient way of obtaining needed information. How far ranging these cases are is uncertain. If anthropologists put great stress on participant observation as has been often claimed and if the number of such cases is small, this emphasis can not be justified on the methodological grounds here considered.[32]

Several cautions or perhaps disclaimers should be added to these conclusions:

(1) It may be a myth that anthropologists do put great emphasis on participant observation as a method. If so, I am attacking a straw man, and I should get myself new 'native' informants since I have gathered my impressions about the status of participant observation from people in or close to the profession.

(2) Other methodological justifications might indeed be possible. Naturally, I have not exhausted all possible ones in my paper.

(3) Even if no methodological justification is possible it does not follow that participant observation should be deemphasized in the profession for some other sort of justification may be possible. Participant observation may give anthropologists great enjoyment, pleasure, adventure and a sense of esprit de corps, for example, and this perhaps is all the justification that a widespread use of participant observation may need in anthropology.

One final thought: I have evaluated the relevance of participant observation for achieving scientific understanding, but it may be a mistake to suppose that anthropologists are after this type of understanding alone when they investigate some particular society or culture. Kroeber has noted that there is an aesthetic dimension to anthropological thought and method.[33] There is a sense of 'understanding' distinct from scientific understanding that I have not yet mentioned. For example, when one speaks of understanding a poem or painting or a piece of music one may well have something different in mind than knowing certain facts about the work of art.[34] Thus aesthetic understanding may indeed be analogous to what the anthropologist is after in understanding a culture or society and participant observation may be much more relevant to achieving this sort of understanding than to achieving scientific understanding. But a consideration of this suggestion is another paper.

Boston University

REFERENCES

[1] Patrick Gallagher's review of Malinowski's *A Diary in the Strict Sense of the Term*, in *New Republic*, June 17, 1967, p. 14.

[2] I. C. Jarvie, *The Revolution In Anthropology*, The Humanities Press, New York, 1964, p. 29.

[3] See, for example, E. E. Evans-Pritchard's 'Social Anthropology: Past and Present', *Man* **50** (1950), 118–124; Conrad M. Arensberg, 'Anthropology as History', in K. Polanyi *et al.*, *Trade and Market in Early Empires.*

[4] It is important to note here we are discussing the notion of understanding something', i.e. expressions of the form *x* understands *y*. In these expressions 'understands' takes a noun or noun phrase as its grammatical object. We are not discussing here expressions where 'understands' takes some propositional clause as its grammatical object. For an analysis of these latter kinds of expressions see May Brodbeck, 'Meaning and Action', in *Readings in the Philosophy of Social Science* (ed. by May Brodbeck), Macmillan Co., New York, 1968.

[5] Cf. C. G. Hempel, *Aspects of Scientific Explanation*, The Free Press, New York 1965, p. 488. According to Hempel to understand a phenomena is to show that it fits' "into a nomic nexus". In our terms this would involve knowing that the phenomena fits into a nomic nexus. According to the present analysis Hempel's view would be a special case of scientific understanding.

[6] For a discussion of propositional knowledge see Israel Scheffler, *Conditions of Knowledge*, Scott, Foresman, Chicago, 1965.

[7] See John Beattie, *Understanding an African Community: Bunyoro*, Holt, Rinehart and Winston, New York, 1965.

[8] Cf. Leslie White, 'History, Evolutionism and Functionalism: Three Types of Interpretation of Culture', *SJA*, 1945, 221–248. One need not agree with White's evolutionism to agree that there are more than two ways to interpret cultural phenomena.

[9] These two notions of understanding were first explicitly separated and analysed by Jane Roland Martin in her doctoral thesis. See Jane Roland, *Understanding, Explaining and Teaching History*, unpublished Ph.D. dissertation, Radcliffe College, 1961.

[10] Bronislaw Malinowski, *A Diary in the Strict Sense of the Term*, Harcourt, Brace and World, New York, 1967.

[11] See for example Bronislaw Malinowski, *Coral Gardens and Their Magic*, Indiana University Press, Urban, 1967. See also Clifford Geertz's discussion of this work as well as of Malinowski's *Diary* in *The New York Review of Books*, Sept. 14, 1967.

[12] S. F. Nadel, *The Foundations of Social Anthropology*, Cohen & West, London, 1951, p. 18.

[13] Scheffler, *op. cit.*

[14] Howard S. Becker and Blanche Geer, 'Participant Observation' in R. N. Adams and J. J. Preiss, *Human Organization Research* pp. 267–289; for another attempt to distinguish different senses of participant observation see Buford H. Junker, *Field Work*, University of Chicago Press, Chicago, 1960.

[15] Sir James Frazer, 'Preface' to *Argonauts of the Western Pacific*, pp. vii–viii.

[16] Malinowski, *Argonauts of the Western Pacific*, p. 8.

[17] John Madge, *The Tools of Social Science*, Doubleday, New York, 1965, p. 137.

[18] Eric Fromm, 'Psychoanalysis and Zen Buddhism', in Suzuki, Fromm, and De Martino, *Zen Buddhism and Psychoanalysis*, Grove Press, New York, 1963, p. 112.

[19] M. Herskovits, *Life in a Haitian Valley*, New York 1937, p. 323.

[20] Beattie, *op. cit.*, p. 14.

[21] Florence Kluckhohn, 'The Participant Observer Technique in Small Communities', *AJS*, 1940, 331–343.

[22] Beattie, *op. cit.*, p. 47.

[23] John Beattie, *Other Cultures*, The Free Press, New York, 1964, p. 87.

[24] *Ibid.*, p. 82.

[25] George P. Murdock, 'British Social Anthropology' *AA* **53** (1951) 465–473.
[26] See Margaret Mead and Rhoda Metraux (ed.), *The Study of Culture at a Distance*, University of Chicago Press, Chicago, 1953, pp. 41–49.
[27] Cornelius Osgood, 'Informants', in *Yale University Publication in Anthropology*, 1940, pp. 50–55.
[28] See Margaret Mead, 'Native Language as Field Work Tools', *AA*, 1939, 189–205.
[29] R. Larco Hoyle, *Los Mochicas*, Museo R. Larco Herrera, Trujillo (Peru), 1945.
[30] Irving Rouse, 'The Strategy of Culture History', in *Anthropology Today* (ed. by A. L. Kroeber), The University of Chicago Press, Chicago, 1953, pp. 61–62.
[31] Morris Zelditch, 'Some Methodological Problems of Field Studies', *AJS*, 1962, 566–576.
[32] For a criticism of the thesis that all anthropologists should do field work, see I. C. Jarvie, 'On Theories of Fieldwork and the Scientific Character of Social Anthropology', *Philosophy of Science* **34** (1967) 223–242.
[33] A. L. Kroeber, *An Anthropologist Looks at History*, University of California Press, Berkeley, 1966, p. 124.
[34] See Carl R. Hausman, 'Intradiction: An Interpretation of Aesthetic Understanding', *Journal of Aesthetics*, 1963–64, 249–261.

JUDITH B. AGASSI

COMMENTS: THEORY AND PRACTICE OF PARTICIPANT-OBSERVATION

Prof. Martin has made a detailed semantic analysis of the concept of participant-observation-as-a-tool-for-understanding-societies, sorting out the different meanings of the concept of understanding and the concept of participation, discussing whether it is a sufficient and/or necessary condition of the former, and saying no to both (in all interpretations).

Since participant-observation was invented by the anthropologists Malinowski and Radcliffe-Brown as a means of inducing or verifying functional theories, and since Professor Martin hardly mentions such preoccupations, I do not know how helpful his semantic analysis may be to the working social scientist. Professor Martin does mention that some important anthropologists have stated that anthropology is an ideographic science, not aiming at discerning laws or theories but at understanding particular phenomena. Windelband's distinction between the nomothetic and the ideographic sciences is the same as between theoretical explanation and purely factual description. I do not know how understanding particular phenomena can be ideographic. Anyway, Professor Martin says whether the description of anthropology as ideographic is true is irrelevant to his analysis. He continues by stating that he needs not enter the following controversies: (a) whether anthropologists, although they may not aim to discover laws, nevertheless use laws in understanding communities; (b) whether anthropologists should aim at discovering laws – even if they do not do so. He further describes Beattie and other anthropologists as approaching communities either functionally or historically – therefore utilizing a theory. But again Professor Martin does not discuss the question of theory because "all these different ways of understanding the community..." came down to having what he calls propositional knowledge about the community. By propositional knowledge he means, presumably, any knowledge about somebody or something which is either factual or theoretical.

It seems to me that exactly here lies the main weakness of tonight's

Boston Studies in the Philosophy of Science, IV.

paper. The emphasis on the participant-observation method in anthropology and its peculiar techniques and mystiques as made by leaders in the field in the second quarter of the century, is only to be understood out of the theoretical orientation of anthropology during that period.

Defining the aim of anthropology (or sociology or political science or any other science or any other intellectual activity) as simply understanding something and therefore gaining *propositional knowledge*, and then examining the technique of participant-observation in relation to such a vague aim – is not very fruitful. Gaining propositional knowledge, for Professor Martin, includes such diverse activities as general undirected data-gathering, making inventories about anything in the community studied, or in the direction of the main preoccupations of anthropology only, or attempting to order them into some structural picture, or generalizing a theory out of such 'facts', or verifying a theory, very general or very specific, arrived at before setting out on the fieldwork, or perhaps even testing and refuting and reformulating theories through observing only such facts of the life of the community as are relevant for this purpose. With such a mixed bag of aims, Professor Martin's highly negative conclusion should not be surprising.

Gaining propositional knowledge includes for Professor Martin the activities of anthropologists who start out from problems and those who do not start out from problems, from practical social problems or theoretical problems, those who accept general theoretical frameworks such as functionalism or not, those who reject and those who accept the use of historical material as relevant because of their theoretical orientation, those who attempt to set out without any theory to verify or test and those who do not, those who see their chief duty as the gathering of as much data about the community as possible to increase the stock of anthropological knowledge, those who concentrate on major traditional interest areas of anthropology such as kinship, as well as those who set out looking for specific aspects of the community with the express purpose to verify or refute a theory.

Participant-observation or anthropological fieldwork was an all-or-nothing proposition rooted in a philosophy of a stock of data leading up to or verifying social theories of a holistic-functionalistic character. This is the framework within which participant observation has to be discussed first. If and when the framework is outgrown we have to take into account any new framework. The result will be then interesting to the working social scientist who contemplates endorsing either framework.

The initial framework of anthropology had been very different. The chief problems of 19th-century anthropologists were: what is the secret mechanism of the evolution of human societies, and why are there such great differences among them? Their answers were sweeping: there exists a scale of unilinear evolution along which all societies pass without skipping a stage. This scale they constructed out of the history of the literate peoples. At the lower end they filled in the lacunae by using the myths of, and making up a conjectural prehistory for, these historical peoples. In addition they drew freely on the miscellaneous information about primitive peoples from the tales of travellers, missionaries, and colonial administrators – and in turn they supplied the primitive peoples with a conjectural history of their own. They gave bold and conjectural answers to the question: what caused the backwardness of primitive societies? They raised extremely interesting problems, and produced naive though interesting solutions – often using rather dubious research methods. Their tools were mainly the arm-chair and the library. They used documents uncritically, and did not inspect the institutions of the primitive societies they used as examples for their theories; nor did they look for additional as yet unreported examples. Instead they relied on the observation of untrained others – mostly very much non-participant observers.

The generation of Malinowski and Radcliffe-Brown rebelled against this evolutionist-conjectural tradition, by instituting fieldwork or participant-observation as the prime method and technique of anthropology. Primitive peoples were no longer to be regarded as societies rather low on the ladder of evolution; indeed, the enquiry into their probable history or into their probable later evolution were at least temporarily banished from anthropology. Their beliefs, customs, and institutions, should no longer be regarded as quaint phenomena of one of the lower stages of human evolution, but should be considered worthy of serious observation and detailed enquiry in their own right, as functional parts of stable, well-functioning societies.

The current ethical orientation became not humanitarianism, but ethical relativism. It became the official taboo to express superiority, indignation, or disgust at the way of life of the primitives.

And, most of all, this generation of anthropologists rebelled against the conjectural method. Anthropology was to be a science proper – meaning primarily empirical and inductive. Primitive people being the subject matter of anthropology, the first duty of anthropologists was to go out into their field and gather as much as possible data as raw material for the storehouse of their science.

Data should be gathered first-hand by a qualified scientific observer and all care should be taken that the observer gets as near as humanly possible to the objects of his observation. Therefore, of course, he has to be in the field where his primitives live, and not just for short visits but for extended stays. To make his observation first hand, he also has to learn the language. Proper observation of a primitive society which is a functioning, living whole, can only be done from within that living whole. Therefore, the anthropologist has to live within the tribe. In order to observe the meaning that the activities of the primitives has for themselves, the anthropologist should participate in their activities. Thus he will become familiar with their way of thinking, their norms and values; he will be able to overcome the barrier of his own alien way of thinking and thus avoid the standard distortions of his predecessor the trader, missionary, and district officer.

By not using the material and moral baggage of his Western society, and by not approaching the primitive society as an active outsider such as a trader, missionary, administrator or teacher, he will avoid changing the very material he is attempting to observe.

Thus far goes the theoretical justification for fieldwork, or participant observation, as the essential method and technique of anthropologists.

This classical-functionalist framework of Malinowski and Radcliffe-Brown has since been transcended. As I. C. Jarvie has indicated in his *The Revolution in Anthropology*, in spite of their official hostility to speculation and thus to theorizing, they themselves not merely produced inventories but also concentrated on a few theoretical problems and produced a few major theories of primitive institutions. Soon the result of many an anthropologist's 'fieldwork' became the criticism of the theory produced by another anthropologist in his previous fieldwork in another corner of the primitive world. Soon also the rejection of any enquiry into the history of primitive peoples as conjectural was recognized as questionable, especially by those anthropologists who did research on primitive people who had in fact an undeniable history, sometimes even through their prolonged contact with literate outsiders who had recorded some of this history. Prominent among them are Heyerdahl and Beattie. And even more obvious became the evidence of ongoing and even rapid change of many of these societies under the impact of contact with the industrialized world. Some primitive societies disintegrated under the impact; but others adjusted, experiencing more or less internal strain in the process. The model of the ahistorical, self-sufficient, stationary primitive society became more and more unsatisfactory.

The classical model had been of a simple, organic, unstratified society. Recently more and more preliterate societies, especially in Africa, with a considerable and sometimes rigid social stratification have been described by anthropologists.

All these innovations came along with new techniques and even a new methodology. The traditional technique of pitching a tent or erecting a hut within the native village and thereafter awaiting patiently for something to happen, or somebody to look you up, is not suited to problem-oriented research. Instead of absorbing and recording in a leisurely fashion all the manifestations of native life, the anthropologist will want to find out as much as possible about his special field, such as the political system. He may use informants, interviews, and surveys. With beginning literacy he may even engage native research-assistants.

Emphasis on history will bring with it the study of documents and the files of a colonial administration. Here, again, growing literacy will permit the probing of the people's own image of their past through the writing of, for instance, prize essays.

When a primitive people turns out not to be homogeneous, but stratified, obviously 'entry' into, and participation in, the life of one social group may not help and may even hinder the anthropologist's rapport with members of other groups, as Professor Martin observes. John Beattie who set out to follow the classical model of participant observation was pushed by his new theoretical-methodological framework to add all the aforementioned techniques to his fieldwork. It seems obvious from his own account that he gave up his initial classical inductive-grass-roots-participant-observation technique because it limited and hindered his research which became increasingly both problem-oriented and historically minded, as well as increasingly less functionalist and more stratification-conscious.

However, the tradition and the mystique of fieldwork are strong. They seem to have hampered anthropologists' realization of the extent to which they themselves have in fact already changed the classical fieldwork.

It should also be pointed out that contemporary anthropology, especially in America, is using more and more the comparative method, attacking specific problems, building theories, and testing them, using the manifold second-hand observations of dozens of fieldworkers.

Fieldwork has a somewhat longer history in European sociology than in anthropology – although in the 1860's and 70's it was called 'social survey' and although people doing it did not call themselves sociologists. They were moved by the obvious misery of the new industrial working class and decided to find out its dimensions and causes. They did not use either census methods or questionnaires, but first and foremost went to see for themselves how the poor lived, observed and recorded copiously and usually busted some of the accepted notions on this topic (Booth, Le Play, Riehl, Eiler, Sundt).

This sociological tradition insisted on prolonged observation *in situ* but not on participation.

The second tradition of fieldwork in sociology was mainly anthropological. It originated in the 'Chicago School' in the twenties where 'social survey' and 'social anthropology' met under the auspices of Park's German-oriented philosophy. Sociologists substituted for the primitive tribe what they called 'the community', including rural or other small communities, but also neighborhoods of large cities, the ethnic-minority group, the deviant group and groups within walls of institutions, seeing all these as social wholes; patterns of behavior have meanings within wholes and so have to be observed from within.

Sociologists rather than anthropologists carried on the prolonged debate about degrees of 'participation' and about the role or roles that the observer could or should assume.

While for the anthropologist it was hardly possible to hide his identity as an outsider, for the sociologist who observed segments or fringe-groups of his own larger society, this was possible. He could pretend to aim at becoming a full and ordinary member of the group – settle in the small-town or in the New-Mexican village, put on the rags of the hobo, play the piano as a dance-musician, pretend to become a convert of a Californian apocalyptic sect, or even a mental patient. The sociologist could, in fact, completely disguise his role of observer (something which few anthropologists have attempted to do while they have varied in the explicitness of their explanation of their observer-role to the primitives). Sociologists tended to favor this course when either the group to be enquired into was known to be hostile to observers or when it was assumed that the known presence of an observer would considerably change the social reality to be observed. Several drawbacks of the disguised observer-technique were discovered: a 'non-observer' has difficulties to explain his wish to be present in many situations, his freedom of movement might be severely limited; in addition, his entire recording activities have to be clandestine; if discovered either while recording or later after publishing, he might be considered a spy or traitor. In the case of the deviant group the degree of his involvement proved a formidable problem. It also proved near-impossible to prevent any influence of the hidden observer in the group. It is obvious that three additional members – the hidden observers – who had to appear as seriously committed converts in order to be accepted, had a considerable reinforcing influence on the other members of an apocalyptic sect which at its height numbered only 24!

As to the overt observer he is ascribed a role by the observed which need not be the same. Thus W. F. Whyte told Cornervillers that he was a social historian and they declared him a novelist. Others adjusted to this tendency and introduced themselves as local historians, botanists, etc., as such observer roles were already locally familiar.

Choice of participant roles also became a refined art. Florence Kluckhohn assumed three different participant roles in one New-Mexican village: housewife, high-school girl, and lady-trader.

During the thirties and forties a great number of community-studies were undertaken by American sociologists. While all these community-researchers adhered to the ethnographic canon, most of them were intentionally or unintentionally, also concerned with some special aspects such as class-structure, migration and assimilation, power, race-relations, or the influence of industrialization on the countryside. Most used not only the classical, uncontrolled, unstructured techniques of observation, they also used structured interviews and questionnaires and, when available, even census and other statistical material as well as newspapers and some documents of local history.

And yet, many of these sociologists were severely limited by their theoretical framework. Functionalism had become the dominant theoretical framework in American sociology and most community researchers set out to see in their community a balanced and functional 'social system' in miniature, from which they hoped conveniently to learn about social phenomena in the larger social system. They tended to ignore the untypicalness of their chosen community as well as the influence of the larger society on it. Like the classical generation of anthropologists, many of this generation of American sociologists also ignored both the history of their communities and the ongoing change taking place within them. On these counts much criticism was heaped on the community researchers, unfortunately usually without analyzing or criticizing their theoretical background.

During the fifties a flood of criticism arose aimed at traditional sociological fieldwork not on account of its community centeredness but on account of its main traditional technique – the technique of participant-observation. Around the mid-fifties the ideal of maximum quantification gained acceptance in American sociological circles. Sociological data, just like data of natural science, should be quantitative, accurate and 'hard', in order to permit the working out of valid statistical relationships between different factors. The highly structured interview, and, even better, the questionnaire, could be impartially administered to hundreds and even thousands, considered samples representative of ensembles of millions, producing impressive quantities of hard computable data. The laboratory-type (or otherwise staged) social experiment became another preferred sociological technique, because it permits exact recording and measurement of many facets of behaviour of the members of a group under controlled conditions, as well as checking by the use of control groups. Participant observers were charged with producing impressionistic, qualitative, inaccurate, so-called 'soft' data – indeed with being old-fashioned and soft-headed.

While some partisans of participant-observation retorted defensively with accusations of shallowness and lack of 'empathy and feel' as characteristic of the questionnaire school, and artificiality as characteristic of the social-experiment school, others underwent considerable soul-searching and conceded sins of over-participation and some technical sloppiness.

Those who had conceded also now demanded of themselves more discipline in the reporting of the process of formulating hypotheses and of building models, as well as their subsequent scrapping and rebuilding during the course of fieldwork. They also devised techniques of measurement leading at least to quasi-statistics. Finally, they conceded that though pure participant observation was efficient, adequate, and preferable for obtaining some kind of information, i.e., about incidents and about unverbalized norms, interviewing informants and enumerations and sample-methods were preferable for other kinds of information such as frequency-distributions or institutionalized norms; they advised participant-observers to use the latter methods whenever appropriate.

Some advocates of participant-observation assigned the technique a special and somewhat limited place within the process of social research – where it was absolutely essential: to become acquainted with the life of the members of the community, organization, or institution under research was recommended before the formulation of working hypotheses – or even before the definition of specific problems to be tackled. Others recommended the technique as a step between the drawing-up of the interview or questionnaire and its pre-test – chiefly in order to acquire the appropriate special lingo of the persons about to be approached.

Apparently all this advice, perhaps useful on occasion, about the proper time and place for participant-observation, has not been widely taken by the younger sociologists; the technique of participant observation, and indeed of fieldwork in general, i.e., the observation of social behavior both *in vivo* and *in situ* – or 'on the hoof' as the Chicagoans used to say – has suffered an amazingly sharp decline during the last decade.

Unfortunately again the theoretical origins of the pretensions and exaggerations in the use of the technique of participant-observation in sociology have not been examined sufficiently: naive inductivism led to much undirected waste of time by field workers because of the taboo against early formulation of problems and explanatory theories. W. F. Whyte relates, "I was 18 months in the field, before I knew where my research was going." Naive inductivism caused the compilation of monstrous inventories of field-data. Community holism-functionalism led to the illusion that nation-wide social phenomena could be studied by participant-observation in small towns and without recourse to history. The functionalist-harmonious illusion about the possibility of participation in 'the' community led to much over-participation, which obscured a balanced approach to varied and sometimes antagonistic sections of the same population.

Eliminating these defects from the fieldwork technique and supplementing it and substituting for it as need be, would leave us with the valuable core which was unjustly condemned by the current fashion of the computer.

Massachusetts Institute of Technology

SIDNEY W. MINTZ

COMMENTS: PARTICIPANT-OBSERVATION AND THE COLLECTION OF DATA

Professor Martin's paper has raised many questions of interest to anthropologists, but I will attempt to discuss only one of them: the relevance of participant-observation to the collection of data.[1] Professor Martin asks whether participant-observation can be regarded as either a necessary or a sufficient condition for successful anthropological fieldwork. If by participant-observation is meant – at its simplest – living among the people one hopes to study and whose culture one aspires to understand, then some views of the methods and goals of anthropology would, almost by definition, include participation-observation. Conklin, for instance, writes of the field called 'ethnography' as follows:

> The data of cultural anthropology derive ultimately from *the direct observation of customary behavior* in particular societies. Making, reporting, and evaluating such observations are the tasks of ethnography. ... An ethnographer is an anthropologist who attempts – at least in part of his professional work – to record and describe the *culturally significant behaviors* of a particular society. Ideally, this description, an ethnography, requires *a long period of intimate study and residence* in a small, well-defined community, knowledge of the spoken language, and *the employment of a wide range of observational techniques including prolonged face-to-face contacts with members of the local group, direct participation in some of that group's activities*, and a greater emphasis on intensive work with informants than on the use of documentary or survey data. [Italics added, S.W.M.][2]

The stresses in this definition put upon lengthy stays in one's locus of study, prolonged face-to-face contacts with local people, and participation in local activities, attests to the importance anthropologists give to such procedures in collecting the information they desire. But there is yet another emphasis in this passage that deserves our attention: the concern with *observation of behavior*. Since Professor Martin asks whether the data anthropologists collect cannot be collected as well – or even more successfully – by methods other than participant-observation, I wish to remark particularly the concern with observing behavior, to which I will return shortly.

Boston Studies in the Philosophy of Science, IV.

Put crudely and imperfectly, anthropological study aims at the collection of four sorts of data: (1) data about things (and, sometimes, the collecting of the things themselves); (2) data about persons; (3) data about ideas; and finally, (4) data concerning, describing or revealing the relationships among things, persons, and ideas. One eminent anthropologist, Alfred Kroeber, suggested that anthropology's major concern is with culture – with, as he puts it, "... the way in which men relate themselves to one another by relating themselves to their cultural materials".[3] This phrasing puts heavy emphasis upon anthropological understanding of the social relationships among men. Many anthropologists – though perhaps not most, and certainly not all – would regard the study and analysis of social relationships as anthropology's most important undertaking.

Not all anthropological fieldwork, however, need be concerned with this last subject, nor have the understanding of social relationships as its primary objective. For example, studies of material culture – artifacts – can be carried out (if desired) without attention to the society and culture that has produced them. Again, archeological research, much of which is executed by anthropologists, may or may not be directed to the inference of social-organizational contexts. And reconstruction of the culture through intensive interviewing of single informants has long been a commonplace in American anthropology; Professor Martin, in referring to Osgood's work with a single Athapascan informant[4], provides an excellent example of solid anthropological research that did not involve – and hardly could have involved – participant-observation in the conventional sense. And one could provide other examples as well. Hence it is quite correct to argue that participant-observation is not essential to *all* kinds of anthropological research, and I know of no anthropologists who have argued seriously that it is.

But this assertion does not indicate whether participant-observation is essential to *any* kind of anthropological research – and this is a far more difficult question. If we return briefly to the quotation from Conklin, it will be noted that he writes of "the direct observation of customary behavior", and, at a later point, of the need to "record and describe the culturally significant behaviors of a particular society". The phrases "customary behavior" and "culturally significant behaviors" are doubly important. First, the stress on observing behavior must be noted. Verbal communication is, of course, one kind of behavior, and an anthropological fieldworker elicits much data from the verbal behavior of his informants – including their responses to his questions. Yet it should be clear that descriptions of behavior provided by the informant in answer to the fieldworker's questions, and descriptions based on the fieldworker's observations, are two different things – sometimes, two very different things. Second, the phrases employed by Conklin deserve careful notice because the terms "customary" and "culturally significant" strongly suggest certain theoretical preconceptions about culture: that culture is patterned, orderly, and – in some epistemologically significant ways – consistent. A culture is more than the sum of its parts and the 'carriers' of the culture function within some kind of social system. Individual participation in that system may be viewed in terms of statuses and roles, of sub-groups of various types and magnitudes (families, lineages, age-grades, clubs and societies, estates, castes and classes, etc.) and in other ways. But in any case, each individual's perception of the social system will be limited to some extent by the position he occupies within it.

I mention these matters as background for the question of the usefulness or necessity of participant-observation: is it *essential* for some kinds of anthropological research? In line with what has been said so far, it might be argued that one way to get at the culture would be to interview as many informants as possible, one by one, until a composite picture emerges from their responses. Such a procedure has much to recommend it; and it has already been suggested that, when the culture is in fact no longer viable or functioning, this is one of the few ways in which reconstruction can be attempted. But if there exist communities within which patterns of and for behavior can be comprehended by direct observation, it is difficult to see why the elicitation of verbal responses from individual informants, especially by interviews conducted outside the community, would be a preferable or equally efficient method.[5] In fact, reconstruction of the traditional culture by intensive interviewing of one or only a few informants is usually undertaken only because the culture is no longer really being 'carried' in such a way that the behaviors of its carriers can be observed directly.

But the argument can be advanced beyond questions of relative efficiency or completeness. I have noted the difference likely to appear between an informant's description of behavior, and a description written by the fieldworker, based on his direct observation of one or more cases of the behavior in question. Without intending to belabor this point, I suggest that one reflect on the difference between carefully observing a number of weddings, funerals, or political meetings, for instance, and asking a 'typical informant' to describe a wedding, a funeral, or a political meeting. Such reflection will make clear that 'the typical informant' must either provide a 'typical description', or else describe some particular such event – and in either case, he will probably omit certain information. Omissions can be caused by at least three different factors: (1) overfamiliarity with the pattern; (2) the confusion of the real with the ideal; and (3) the deliberate distortion or withholding of information. The first difficulty can be avoided – to some extent, anyway – if the fieldworker is able to observe the event. As a naive observer, the fieldworker can perceive much that the informant himself is likely to ignore in his description. The second difficulty – that of confusing what happens with what is supposed to happen – may also be partly corrected for, if the fieldworker can observe for himself. (Nonetheless, of course, verbal elicitation of information can be very important here, precisely because it helps the fieldworker to distinguish the real from the ideal; this, too, is an important datum.) And the third difficulty may also be in part reduced by direct observation. None of these compensations invalidates the utility of interview data; but they do suggest the advantages of participant-observation, when feasible.

At this point it is also necessary to stress that any anthropological fieldworker worth his salt will never forget that his own observations are much clouded by his cultural background, his values and preconceptions. Anthropology, like most of the social sciences, cannot escape methodologically from the fact that the observer's presence affects his findings, both because it introduces a new element into the social field, and because the observer's lens is flawed, as it were, by previous experience. An anthropological field report, in other words, is never entirely objective (in the way that a description of geological or astronomical phenomena, for instance, is objective); and such reports can only approach objectivity to the degree that the observer struggles successfully to identify and to weigh his own perceptual limitations. Nevertheless, one hopes that anthropological inquiry can aspire to objectivity, at least enough so that anthropologists from widely different cultural backgrounds can agree on what, indeed, has been seen, if not on the best way, or ways, to understand or to explain it.

In enumerating earlier the difficulties implicit in basing one's data collection solely on verbal elicitation. I noted three sources of distortion or incompleteness. To these, I would now like to add another. Not only do one's informants neglect to describe what they regard as obvious, confuse the real with the ideal, and sometimes deliberately withhold information – they also sometimes do not really know themselves why a certain pattern of behavior is practised or maintained, or how it relates to other behavioral patterns. It may seem arrogant indeed to propose that the anthropological fieldworker can discover some relationship among men, things and ideas in a culture foreign to him, that is not apparent to the carriers of that culture. But this is in fact the case; and it is highly unlikely that any amount of interviewing of informants can reveal such connections. Complete descriptions of these kinds of insights would be tedious; but Malinowski's analysis of the *kula*[6], incomplete and somewhat inconsistent though it is, doubtless reveals to Western readers more about the relationship of this institution to other Trobriand institutions than was known by any Trobriander. Again, Robert Lynd has pointed out that the automobile had already become firmly embedded in American life before any American recognized that it would revolutionize adolescent mores[7]; yet it seems more than probable that, had a Latin American anthropologist been studying our society at the time, this relationship would have become apparent to him very swiftly. In other words, the freshness and naivety of the fieldworker's view can contribute to (though it does not guarantee) insights not necessarily available to the native, if the fieldworker is able to observe behavior. This is far less likely to happen if questions are the only basis for data collection.

It might also be fair to argue that the study of certain problems – or the collection of certain bodies of information – requires some measure of mutual trust or respect most likely to be achieved through participant-observation by the fieldworker. At any rate, it is certainly true that some kinds of data can be collected with little or no personal contact or friendship between informant and investigator, while other kinds of data are probably more available if the basis for elicitation is a mutual, personal, high regard. Thus, to facilitate finding out about those aspects of life invested with great affect, rarely talked about, hedged by taboos, viewed as illegitimate in terms of local values, revered deeply or considered deeply shameful – it is probable that good rapport is useful, particularly in reducing the amount of deliberate distortion. In contrast, in order to find out about most things that anthropologists want to know – house types, fishing techniques, riddles, cuisine – a lengthy stay, cautious interviewing, or mutual friendship and sympathy may prove superfluous. However, the examples I have chosen are random; it is in fact not possible to know in advance whether a society has hedged some particular activity with taboos, or whether its members do or do not feel deep shame about some undertaking that would be completely free of affect in the anthropological fieldworker's own society. The point may really be that it is impossible to predict what activities or institutions are linked to deep feelings in some particular culture, such that direct questioning without the prior creation of good rapport may interdict responses, or increase the likelihood that the responses will be deliberately or unconsciously distorted.[8]

I would like to summarize the points I have attempted to make so far. I suggested first that lengthy residence in the community, and personal, face-to-face contact with one's informants, is regarded as essential to the anthropological (or ethnographic) method by many anthropologists. I went on to indicate that the study of social relationships is a very important task for anthropology, in the view of many of its practitioners. Yet much of what can be properly called anthropological research either cannot be carried out on the basis of participant-observation, or does not appear to require such a technique for the collection of reliable and satisfactory data. Nonetheless, anthropological information is usually collected from persons who participate as natives in a community or society with its own distinctive social system, and such informants occupy particular status and role-playing positions in that system. What people know about their own culture and society is partly a function of their positions in the society, and no society is entirely homogeneous; hence, information collected from any one informant is not likely to be complete, except in terms of his or her personal grasp of the relevant details.

Beyond this problem, there is the distinction between act and idea, between verbal descriptions of behavior and the behavior itself. I suggested several possible sources of distortion in this connection, since the informant describing some aspects of his culture must do so from the vantage-point of his or her own particular perceptions. While the difference between acts and descriptions of acts may itself be an important source of insight to the fieldworker, his understanding will be enhanced by the opportunity to compare what actually happens with what informants describe as happening. Here we must not lose sight of the fieldworker's own perceptual limitations; he must recognize that he is no camera, any more than are his informants. But it is better that he recognize this for himself, without assuming that his informants are necessarily better observers than he.

Further, I suggested that there may be linkages or interrelationships in the organization of culture and society that are not apparent to the 'natives', but that may become visible to the outside observer – if he can observe, rather than simply question, his informants. Such linkages are sometimes intuited by the observer, and take the form of hypotheses; it is sometimes possible to test these, and sometimes not. But they are more likely to become apparent through observation than through questioning, precisely because informants may not be aware of them. Finally, I raised the general question of the value of rapport. While it can be argued that good rapport may interfere in some ways with objectivity, there are certainly good reasons for arguing the opposite, particularly when the observer is in a position to check what people say against what they do.

I have tried to make clearer why I think that participant-observation is more than simply one way to collect anthropological data. For many problems, I would contend that participant-observation is the best way to collect such data; for any problems that may pose a contrast between verbal renderings and actual behavior, I would contend that it is probably the only readily available way. I do not think, however, that anthropology need be ashamed of its methodological tentativeness. Professor Martin has pointed out that there was a time when anthropologists did not engage in participant-observation, using Sir James Frazer as his case. Without meaning to impugn Sir James's greatness in the slightest, I think it fair to ask whether modern anthropology did not really begin with his student, Malinowski, and his contemporaries, who insisted that man's nature could not be deduced *ex nihilo*, but would eventually be understood only if his behavior were subjected to careful observation. It is possible to over-intellectualize one's criticisms of anthropological methods, I think. We go 'where the action is', and try to 'tell it like it is', because we like to, I readily admit – but also because we know of no other way better to find out what we want to know.

Yale University

REFERENCES

[1] The literature on participant-observation in the social sciences is considerable. Excellent bibliographies are provided in Harold C. Conklin, 'Ethnography', in *International Encyclopedia of the Social Sciences*, Vol. V, 1968, pp. 172–78; Hortense Powdermaker, 'Field Work', in *International Encyclopedia of the Social Sciences*, Vol. V, 1968, pp. 418–24; Rosalie Hankey Wax, 'Observation, Participant', in *International Encyclopedia of the Social Sciences*, Vol. II, 1968, pp. 238–41; Buford H. Junker, *Field Work*, 1960, pp. 159–207; Richard N. Adams and Jack J. Preiss (eds.), *Human Organization Research*, 1960, pp. 443–56; A. L. Kroeber (ed.), *Anthropology Today*, 1953, *passim*; and *The American Journal of Sociology* (special issue on problems of field study) **60**, No. 4 (January 1955) *passim*. Of a certain special interest with regard to the problems raised in anthropological fieldwork using participant-observation are: Hortense Powdermaker, *Stranger and Friend[2] the Way of an Anthropologist*, 1966; Gerald D. Berreman, *Behind Many Masks* (Society for Applied Anthropology, Monograph No. 4), 1962; Laura Bohannan, 'Shakespeare in the Bush', *Natural History* **7** (1966) 28–33; Elenore Smith Bowen (pseud.), *Return to Laughter*, 1954; Sidney W. Mintz, Review of G. D. Berreman, *Behind Many Masks*, in *American Anthropologist* **65**, No. 6 (December 1963) 1362–63.

[2] Harold C. Conklin, *op. cit.*, p. 172.

[3] Alfred L. Kroeber, *Anthropology*, 1948, p. 68.

[4] Cornelius Osgood, *Ingalik Material Culture* (Yale University Publications inAnthropology, No. 22), 1940.

[5] In the last 15 years, a new 'school' of ethnography has emerged, the theoretical and methodological implications of which may eventually affect significantly the importance anthropologists will attribute to fieldwork and to participant-observation. The 'New Ethnography' (discussed most usefully, I believe by W. Sturtevant, 'Studies in Ethnoscience', *American Anthropologist* **66**, No. 2 (June 1964) 99–131) views culture essentially as what one needs to know in order to behave appropriately under specifiable circumstances. (See Ward H. Goodenough, 'Cultural Anthropology and Linguistics', Report of the Seventh Annual Round Table Meeting on Linguistics and Language Study, in Paul L. Garvin (ed.), Monograph Series on Languages and Linguistics No. 9, Institute of Languages and Linguistics, Georgetown University, Washington, 1957, pp. 167–73.) If this view is theoretically sound, and if a particular culture can be validly derived from an aggregation of individual cognitive systems or 'maps' – both conditions I tend to mistrust – then presumably, elicitation from individual informants will produce or generate whatever the ethnographer needs to know in order to describe

that culture fully. At least, this is what I understand to be one of the implications of the methodology of the New Ethnography. However, since questions of this kind have not received much attention yet in the literature, I will not attempt to deal with them here. See the bibliography in Sturtevant, 1964, cited above, for more information on this school.

[6] Bronislaw Malinowski, 'Kula: the Circulating Exchange of Valuables in the Archipelagoes of Eastern New Guinea', *Man* **51** (1920); 'The Primitive Economics of the Trobriand Islanders', *The Economic Journal* **31** (1921) 1–16; *Argonauts of the Western Pacific*, London 1922.

[7] Robert S. Lynd and Helen Merrell Lynd, *Middletown*, New York 1929, p. 251, *et seq.*

[8] Though there is implicit in the view of some anthropologists the feeling that rapport may be an *obstacle* to good data collection, since mutual sympathy may excite an unobjective desire to please, on the part of both fieldworker and informant. Transference and counter-transference and, hence, deliberate omission and lying may result. This reservation is interesting, and a good fieldworker ought to be aware of it. Yet, in a curious way, there lies within this view the accompanying implication that the more one knows about one's informants, the closer one is to them, the less accurate one's data are likely to be – an interesting implication, indeed, for a philosophy of science. Though anthropologists will not admit it, there is within the profession a body of opinion about who is, or is not, 'a good fieldworker' – most of it gossip. Some of those anthropologists known for the objectivity of their field methods are also known for their inability to learn foreign languages well, or for their dislike of the field situation and the personal demands it makes upon them. The same may be true for teaching, psychoanalysis, etc.

ABRAHAM EDEL

PATTERNS OF USE OF SCIENCE IN ETHICS*

It is commonly recognised that the question of the relation of science to ethics has been a troubled one, especially in the 20th century. There is an interesting history to the question. Scientists have occasionally attempted to legislate for ethics. They thought to determine morality from their biological results – for example, their conclusions about human evolution. Or else, they offered lists of instincts or needs out of which they expected a conception of the good to be constructed. They framed concepts of mental health to serve as a basis for morals, or attempted to apply to moral inquiry the latest model from the latest science. Philosophers reacted with emphasis, often with impatience. They attempted to establish and maintain a concept of moral autonomy which would free them from an incipiently authoritarian science as it had freed them from an authoritarian religion. Sometimes they fell instead into the clutches of an authoritarian linguistics, and they confused autonomy with isolation. Many philosophers felt that the whole question had really been settled when G. E. Moore worked out the idea of the *naturalistic fallacy* in his *Principia Ethica* at the beginning of the century, and separated the moral sphere from the natural sphere; in one form or another, it has won assent in the analytic world. In more popular vein, and among perhaps most scientists, it has seemed enough to say 'Science can give us only means, not ends.'

My aim in this paper is not to give a historical survey, but rather to attempt an adequate formulation of the question. For in spite of – perhaps because of – the heat it has generated, formulations are loose and presuppositions are insufficiently analyzed. I find it significant that as late as 1959, so careful an analyst as R. B. Brandt, in his *Ethical Theory* should entitle a chapter 'Can Science Solve all Ethical Problems?' My leading idea – and it is no doubt an obvious one, or at least I hope it is – is that there is no single wholesale relation or wholesale absence of relation between science and ethics. There are numerous specific types of relations of specific results of specific sciences to specific parts of morality or ethical theory.

Boston Studies in the Philosophy of Science, IV.

The formulation I am suggesting begins with a recognition of morality as an on-going process in human life – with its own phenomena, concepts, methods, problems. There is no question of starting from science, and establishing a scientific morality from scratch on the basis of scientific data. But on the other hand, there is no question of isolating morality from scientific impact. There is no part of morality – whether concepts, rules, processes, or methods – which may not be refined, improved, or altered by the application of the knowledge arising in one or a number of the sciences. Similarly, there is no part of ethical theorizing – again, whether concepts, principles, processes or methods – which may not be amenable to refinement in the light of scientific knowledge.

It is very likely that you will all agree on this. And so we have reached a consensus, my argument is over – if I had one in the first place – and we can all go home. I go on for a very simple reason: I am suspicious of a too ready consensus. There has been so much argument on the question, so deep an embedding of the theory of the naturalistic fallacy in the last half century, so complete a disregard of the results of the growing science of man in the actual doing of philosophical ethics, that I find it difficult to believe that the controversies are over. I suspect rather that they will flare up again the moment we begin to apply the formulation suggested. The main aim of my paper is thus to test the consensus by beginning the kind of research program for which the formulation calls – the location of the precise points of relevance or points of entry of scientific materials into ethics.

There are two preliminary points to be noted about our procedure. We focus on the relation of science to ethics. Science and ethics are seen as two *fields* of human inquiry. We have no right to translate this directly into the relation of fact and value. The latter is a categorial reinterpretation of our problem. It is a proposed answer – and I might add, in many respects a vague one, and a dubious one. We shall reflect on it much later, but please do not start out by equating the problem of the relation of science and ethics with that of the relation of fact and value.

My second preliminary point is that we must choose whether to fix the science side or the ethics side in beginning the study of relations. We could, of course, fix both – for example, five sciences and three parts of ethics, and so study fifteen relations. Because our primary focus is on ethics, and because we expect the sciences of man to shift their borders,

I choose to start by fixing the ethics side and leaving the science side open. And on the ethics side I want to distinguish three parts. The first is moral judgment, usually called substantive morals or normative ethics. The second is the part of ethical theory that commonly goes in the texts under the heading 'the nature and tasks of morality', and is usually disposed of dogmatically in introductory chapters; it is the preliminary consideration of, as it were, the shape of the field. The third consists of theoretical ethics or meta-ethics – analysis of ethical concepts and their relations, justification procedures, and so forth. No doubt the field of ethics could be cut in many different ways, but this cutting will do well enough for our inquiry. We have then to look into the relations of science – that is, the field of the sciences of man as they exist today – to these three parts of ethics. I should like, however, to add both a prefatory section and a concluding one. The prefatory one is an unorthodox interpretation of Moore and the naturalistic fallacy in the light of our problems. The conclusion will reflect on the logical character of these issues such as the relation of science and ethics, of means and ends, of fact and value, and how far they are analytic issues or how far they are themselves scientific issues.

I. MOORE AND THE NATURALISTIC FALLACY

This is not an $n+1$th attempt to reckon with the naturalistic fallacy, but to suggest a mode of reckoning. The naturalistic fallacy has been so prominent as a theoretical barrier to relating science and ethics that it cannot be by-passed. I have too much respect for Moore to treat him as he has been treated by so many who purportedly agreed with him – that is, to use him to support their own views by saying that what they were saying was what he really meant. Look at *Principia Ethica* as a whole, then, not just at Chapter I.

Moore was rejecting a scientific basis for intrinsic value, as he rejected a metaphysical basis or a religious basis or a historical basis. His most outright statement is to be found in his review of Brentano's *The Origin of the Knowledge of Right and Wrong* when he says: "The great merit of this view over all except Sidgwick's is its recognition that all truths of the form 'This is good in itself' are logically independent of any truth about what exists. No ethical proposition of this form is such that, if a certain thing exists, it is true, whereas, if that thing does not exist, it is false.

All such ethical truths are true, *whatever the nature of the world may be*" (*International Journal of Ethics*, Oct. 1903, 116). In *Principia Ethica* he propounds the view that good is a simple non-natural quality, that the way to discover whether anything is intrinsically good is to isolate it from all extrinsic relations, contemplate it as a world by itself and simply *see* whether it has the property of intrinsic value. Ample and subtle illustration of his procedure is found throughout the book and especially in chapter VI on the Ideal, in which he reckons with the greater or higher values. The term 'naturalistic fallacy' is coined to condemn any view, like Utilitarianism, which offers an equivalence definition of good in terms of some descriptive property. (That good is non-natural means in this context, in terms of procedure, that the whole of which it is to be asserted or denied has *first* to be completely described as existing by itself before we look for the further property, good.) It is generally recognized that Moore's most successful argument for the naturalistic fallacy is the open-question argument – that it is always possible to ask, if you define good in terms of some specific content (e.g., pleasure, desire, what God wills) whether that content is itself good, and this question always makes sense. (Later writers applied it also to ought – e.g., if ought means what my community demands of me, I can always ask: Ought I to do what my community demands of me?)

A word on this argument's significance. Moore meant it to support the simplicity of good. But look at it this way: to be able to question something's goodness is to evaluate it. The open-question argument thus is the guarantee of the permanent possibility of evaluation for any described content. Moore's view of good as a simple non-natural quality guarantees the permanent possibility of evaluation. But so might other views about good. For example, we might even legislate in our theory of definition the right to change definitions if they did not explain our facts, and facts here includes the whole array of immediate moral reactions as well as existent principles. In fact, we might simply stipulate the permanent possibility of evaluation through a substantial concept of individual intellectual liberty to be worked into any satisfactory ethical theory. Thus what the open-question argument gives us can be achieved in many ways. It is not uniquely tied to Moore's rejection of science in dealing with intrinsic value.

Now it is important to note that Moore did not reject science directly in his theory of right and duty. In fact he incorporates its relevance in his very definitions of the terms. In Chapter V, on Ethics in Relation to Conduct, he insists that questions about right or duty are questions about the effects of our actions in making the whole world better in one way than in another. Thus while science is perfectly at liberty to come in and give us knowledge of the effects, Moore has set his requirements so high – the effect upon the totality of the world for all alternative courses of action – that he is not unreasonably pessimistic about acquiring the knowledge needed, and so he was led to acquiesce in an extreme conformism.

In sum, Moore's ethical theory represents the extreme rejection of science from ethics. Science is irrelevant in dealing with intrinsic value and helpless in dealing with duty. Moore's ethics thus gives us a very accurate and insightful picture of what an ethics without science is like – helpless conformity to tradition in matters of duty, intrinsic value reduced to isolated reactions in terms of attractive quality. So far from constituting a proof of the irrelevance of science to ethics, it seems rather to be an effective demonstration of the plight of ethics without science, and so to call for the rejection of its own conceptual structure for ethics because of the very results that ensue from its use in ethics.

Take one important example of such a theoretical result. It is rarely noticed that Moore's construal of intrinsic value as a quality apprehended when the object is held before one's mind as a total world by itself, made the judgment of comparative value impossible and almost meaningless. How could we judge whether beauty has a higher or lower intrinsic value than friendship? If I envisage beauty as a world alone, I apprehend directly its intrinsic value; similarly for friendship. But how compare them? If I put the two worlds together, I have the contradiction of one world which is two. If I put the objects together in one world, I get a single value-reading – say the value of a beautiful friendship, or beauty and friendship side by side. The only way out is to have the degree of value apprehended directly in each alone, and the readings compared – and Moore does speak of seeing a high or low value, as when he says that knowledge has a low intrinsic value though it adds greatly to the value of wholes in which it is combined with other things.[1] But if so, it is not simply intrinsic value that is a simple quality, but value in such-and-such a degree. And if degrees of goodness can be distinguished on a scale of *n*, we have *n* simple qualities. In the light of all that has been done on ordering of preferences by scientific attempts in different fields, Moore's concepts for ethical theory have simply not had the fruitfulness that warrants acceptance. When he said that good was as simple as yellow, it is a pity that philosophers did not go to the sciences that study color, and find out how complex yellow really is, what inner orderings are possible in it, how far it is either perceptible or meaningful without some context, today perhaps even what kind of op art is possible with yellow alone! If lessons emerged of the ways in which yellow depended on context, they might have been directly applied to good.

How strictly do the inadequacies of Moore's structure follow from its neglect of science? The difficulties about comparative value do seem to come directly from the isolation procedure that is a consequence of the way he construes intrinsic value in isolation from the relations that science would study. The element of conformity in duty may perhaps be altered by tinkering with Moore's definitions: for example, if we changed his definition of duty to refer to the effects within a limited part of the world over a limited stretch of time, or if we introduced qualifications specifying that duty be determined on the best available knowledge rather than ideally complete knowledge, perhaps the helplessness might be reduced. It may be suggested as a possibility that Moore's own belief about the lack of utility of scientific knowledge is itself a lesson of science, that he is using science itself to yield a kind of conservatism when he says that attempts to change rules are never warranted. If so, then the helplessness is a consequence of the specific scientific results reported, and might be altered if those results proved unreliable.

One further argument is worth noting because this style of query is likely to reappear at many points in our account. How, it will be said, do we justify a rejection of Moore's structure because it yields isolated attraction for intrinsic value and conformity for duty? Are we not simply assuming that these consequences are undesirable or bad? What if someone actually preferred a life of isolated responses about the good and strict tradition for duty?

Now it follows from what was said above concerning the permanent possibility of evaluation, that anything can be made the subject of evaluation. Hence we would expect that any orientation imposed upon us by accepting Moore's structure could be subject to evaluation. Such possibility of evaluation is a necessary consequence of our analysis rather than an objection to it. The crux is not then in the fact of such evaluation but in how it would be carried out. Could it be done satisfactorily by Moore's procedures – for example, judging a rule of conformity by a higher-order rule of conformity to conformity (as conservatism includes among its rules that one be conservative!)? Or would a more satisfactory evaluation of conformity require as an indispensable part a psychological study of the conformist attitude, and a historical study of the workings out of various forms of traditionalism? Of course any further values embedded in the idea of 'satisfactory' evaluation should be rendered explicit. The type of possible circularity here – of a non-vicious kind – is familiar from the parallel procedure concerning assumptions in scientific progress: they should be rendered explicit, and may in the context of a fresh investigation be examined for their worth, the fresh context admitting of other assumptions. All that is surrendered in such procedure is the guarantee of absoluteness.

I am not aware that Moore has been looked at in this way in 20th century ethics – as the supreme model of the plight of an ethical theory that rendered science either irrelevant or bound to helplessness. Therefore I have spoken of it as an unorthodox interpretation. But it seems to me so obvious a one that once we see the effects of the conceptual structure in ethics, so far from saying that Moore has proved science to have no role in ethics, we must look realistically for what role it has and what further role it can be given.

II. SCIENCE AND MORAL JUDGMENT

By 'moral judgment' I mean here the kinds of assertions that are made in substantive morality or normative ethics. The usual view is that science tells us about means to ends that we aim at, but cannot help us in the selection of ends. This is a difficult, often confusing formulation. For one thing, means and ends constitute only one categorial way in which the factors in moral judgment can be construed. The history of ethics, as well as ordinary use, shows us also ideals and their institutional embodiments, intrinsic experiences and the conditions of their occurrence, systematic wholes such as an organized whole mode of life and its constituents. When one attempts to stretch means and ends over these varied types, one often gets strange results. Examples from these alternative formulations also show a second important point – that it is often very difficult to separate out means and ends. Try to sort them out in, say, an ideal of liberty, or in a systematic notion of the good life. The paradoxical point is that it requires a scientific study of means and ends to learn how far they are separable! But of this later. Now let us adopt the formulation of the issue in the usual slogan, and see what science gives us when it gives us means, then ask how far it can muscle into the consideration of ends.

In the original and limited sense, a means to an end is a step which taken produces or helps produce the goal at which we are aiming. To discern the correct means thus involves knowledge of causes. We expect the science to give us better and more reliable as well as more extensive knowledge here. If we wish to curb inflation, what but economic science will tell us whether to regulate interest rates and on what principles? If we want to know whether certain types of threats against the people of North Vietnam will produce intended effects, surely the South-East Asian scholars who have taken an anthropological view of the character-structure of the people can make sounder predictions than one might have by introspecting on how we would respond. In short, it is obvious that science can help furnish better grounded beliefs about *instrumentalities for limited ends.*

In addition to helping us find effective means, science can help us trace the *collateral and subsequent consequences* of the use of means, beyond the fact that the end is achieved. The threats may secure the desired effect on the people of Vietnam but making them may have undesirable effects on the people of Europe. Nor can we stop with the effects of using the means. Even the effects of achieving the end have to be considered. Suppose someone offers taking LSD as a means to a certain kind of psychic experience. Who but the scientists of the mind and the body can tell us what fully goes on in that experience (including collateral happenings and their interpretation), and what the effects of the experience will continue to be?

If brief, when it is said that science gives us means, the package surely includes consequences as well, and consequences of achieving the end as well as consequences of pursuing the end by that means. We ought to add here unintended as well as intended consequences. In many cases, only an advanced science can discover these – for example, the unintended genetic effects of nuclear experiments carried on as instrumental to the end of, say national security.

Whether one will insist on assimilating the concept of consequences to that of means does not really concern us here. It could be done because the notion of means and ends has considerably ambiguity. Sometimes means has the more limited meaning of prior causal steps; here consequences are not prior and so not means. But sometimes an evaluative use of means emerges more sharply, and means becomes assimilated to extrinsic or instrumental value; hence consequences are part of the costs of the end, paid subsequently on an installment plan. If the view that science deals with means allows us – as it obviously must – to have science help discover the events which are costs, then the role of science is extended. For to investigate costs permits also the investigation of gains, that is, of further ends which are consequences. But, it must be noted, science is here still precluded from making the decision whether a given consequence is a cost or a gain; it discovers simply whether it occurs and leaves to others the labelling.

To have extended the scope of science up to this point has already, it should be noted, allowed it to deal with ends in the sense of exploring the consequences of *holding to or pursuing an end, or achieving an end.* Suppose the question is whether to hold to absolute victory and unconditional surrender as a goal in Vietnam; a team of scientists would have the task of working out the consequent increased destruction, prolongation of war, political and economic and moral or character effects, and so forth. The purist would add that an independent value-judgment is required to affirm the destruction of a whole people to be an evil. An alternative move would be to point out that victory is both end and means rolled into one; an end in the sense of furnishing a sense of triumph or achievement (the value of winning), and a means to securing an abiding peace and the safe-guarding of presumed national interests. (The scientist is given a green light on the means aspect.) We have here an important recognition that many ends are in fact instrumental as well, or else they are sub-ends in a hierarchical end scheme. But if so, they have a quality of partial means and are in that respect fit subject for scientific concern. Only pure and ultimate ends are beyond the pale. In a tightly hierarchical system, science would be unemployed in ethics only at the top.

But even the end-elements on the way up present interesting scientific questions. What, for example, is this sense of triumph which is the end-value in victory? Why do we want to win in a war, rather than stop the war – instrumental gains aside? Is there an end-value in sheer coming out on top? If so, does it lie in a sense of achievement, or in a humiliation of the other fellow, or in a reassurance of our own worth, or what? Note what is happening in such questioning. There is the assumption – or perhaps better seen as a major scientific discovery of modern psychology – that our holding of ends has a definite role within us, that it plays some organizing role in our internal economy, and that its quality in our consciousness is in part tied up with, and varies as, the role it plays. Dewey goes so far as to take the end-in-view to be itself a means for organizing and focusing our energies to secure their expression; the end is thus evaluated as means. But apart from this extreme formulation there is involved a basic recognition that the *holding of ends* by individuals – and similarly of social ends by groups such as nations – always takes place in the context of what Dewey calls a problem-situation of interests, needs, pressures, demands, claims, in which the achievement of the end would function as a mode of resolution of conflicts and difficulties and release of impediments. I state this in general terms so that it will admit of both psychological and social interpretation. Such a mode of analysis reflects a scientific discovery in the psychological and social sciences about the conditions of the occurrence of ends in particular individual and social life-histories. It need not be taken as derogation of the idea of intrinsic value – how can it for it has been declared neutral to that? Science can therefore be of use in ethics in helping to determine which of the proposed ends, if held to, would by its pursuit help solve what problems.

A further use lies just beyond this one. If science studies the functioning within of goals held by the individual or group, it is also in a position to understand how strongly they are held, whether they are peripheral and conditional, or central and integral to self or group identity. And so it can contribute to the old Socratic quest expressed in the injunction 'Know thyself'. In Sartre's example of the young man who wanted his advice on going to join de Gaulle vs. staying under the Nazi occupation to take care of his mother, Sartre asked whether the man's coming to him for advice did not already indicate that he wanted a particular answer. It is of course quite possible that resistance to the Nazis was the man's guiding value. But what if it was to get away from his mother? A recent press item says that Venezuelan guerrillas use psychiatrists to check on high school students who wish to join them; presumably part of the check-up is to see what they really want, as well as what they can bear.

Certainly we have come to expect from depth psychology understanding which will enable us to construct a scale from whole-hearted conscious desire at one end, to unconscious desires at the other. And similarly, we may expect from the social sciences, in their study of the functioning of group goals, and in their analysis of ideologies and utopias, some help in distinguishing an authentic national goal from one that serves some special often partial purpose. If knowing oneself is a central effort in the search for the good, it makes all the difference whether a man has a clarified stand on what he stands for, or just a feeling of internal pressure in a given direction. Some aims are given up when we know what we were looking for in them. Others are more effectively pursued. And some goods, however good in general, may not be good for particular persons. It is no easy task to find out what goods a given self might not be able to take.

Let us see what uses we have so far found. Science may help us in furnishing the means to an isolated end, the consequences of the means, the unintended consequences of the use of the means, the consequences of achieving the end, foreseen and unintended, the consequences of holding to an end and pursuing it, the role of the end within the person or group holding to it, the extent to which pursuit of the end will resolve problems to which the holding is instrumental, the depth and quality (or genuineness) of the holding itself.

This is a formidable array of uses, but we have only begun. So far we have spoken chiefly of a limited end; in fact men hold many ends at once, and serious moral issues arise from their conflict. Can science help here? In one very obvious sense it is indispensable. Opposition of ends is not a sheer logical problem like contradiction. Ends are incompatible because they cannot be achieved together, not because they cannot be desired together. For large-scale complex ends, only a scientific inquiry can determine compatibility or incompatibility. Is the goal of individual success in a profit economy compatible with the goal of cohesive social sympathy? Assuming a clarification of both goals, it is a question of psychology whether the institutional requirements of the former have an educative effect which militates against the latter. (It is often said that in our society the schools educate for one, actual life for the other.) Or again, is the ideal of liberty compatible with that of equality, or that of security? A full study in political science is required to see whether we have real incompatibilities or only conflicts of specific forms which these ideals have taken at particular historical periods.

In this area of apparent or real conflict of ends, knowledge of means plays a great part in the structuring of a moral problem – that is, whether one recognizes the presence of a problem, whether it is formulated one way or another, whether it breaks up into distinct and separate problems, and so on. Knowledge of a means may shut out concern with a moral issue by making it inapplicable. This is obvious in many non-scientific contexts. John Ladd gives a striking illustration of this when he reports his inability to make his Navaho informant face up to the moral question whether he should divide his water supply with a man he meets far out in the desert, when there is not enough to see both men through. A too ready 'yes' turned out to rest on knowledge of desert plants that could always be used to alleviate desperate thirst! Science and its application shelves many an issue by undercutting it and making it as irrelevant as whether to shout a warning 'fire' in a crowded theatre in a society where all theatres are fireproof.

Perhaps more morally significant are the effects of scientific knowledge in *restructuring* a moral problem so that it is seen as a quite different one. Economic knowledge in the Great Depression at last did this to the sense of individual responsibility for unemployment. A man stopped asking guiltily what was unworthy about himself that he could not provide for his family; the full picture of the economy and its functioning restructured the problem as the obligation of the society to maintain a reasonably high level of employment. Or again, what is felt as a single problem may be unravelled as a consequence of applied science into several problems – as the applied knowledge in techniques of birth control helps separate in the domain of sexual morality the issues of having children, sexual enjoyment and companionship, social family organization. (It also alters the pattern of sanctions, but this is a distinct matter from structuring of the moral problem. We shall not have time to look at this interesting problem, as where the introduction of knowledge of germs into a society that always thought of illness as punishment for immorality may force alteration in the sanction system.)

It is only fair to note that science may create fresh moral problems as well as shelve some, by the same procedure of increased control. A clear case is the moral problem likely to emerge when genetics offers means of determining to some degree the character of subsequent human stock – what features are to be encouraged as desirable?

In the light of the inroads that science is able to make in the concern with ends, what remains from which it is excluded? Is there any problem of ends that still remains beyond its reach – not, of course, the mere existence of the phenomenon of a man having a given end, but the deliberation about and the justification of ends?

There is, of course, the judgment of intrinsic value that Moore insisted on, the recognition that something is good in itself, when all else is excluded or detached from the object judged. Many a different ethical theory might say that it welcomed all the information on how science helps moral judgment and would welcome still more. But it would regard all this as (collectively) ancillary labor. The last word – about ultimate values – would remain beyond scientific reach. Intrinsic value, however, is an elusive quarry which moral philosophers have pursued without much success. Moore, it will be recalled, protected it with the logical gap laid down by the naturalistic fallacy, which had the effect of placing the intrinsic value (or in later modes, the ought) somehow beyond. Subsequent analytic ethics has worried a great deal about what the gap consists in and what constitutes the unique element on the far side; apart from Moore's intuitive grasping or vision, it has been interpreted as a morally charged definition of 'good', an emotive expression, an act of will or commitment, an acceptance of guidance for practice, a practical act of commending, and so on.

To be so elusive, intrinsic value (or its obligation counterpart) has to be either extremely simple or extremely complex. Neither course escapes the reach of science.

If it is some kind of simple, then its cash-value in moral judgment, no matter what its epistemic or ontologic status, turns out to be the criterion of *attractiveness* in evaluating an end taken by itself. This fits precisely the picture that Moore gives (or that phenomenological accounts have offered), including even the inability to give a justification other than seeing. But in that case, nothing hinders scientific study from considering the conditions of attractiveness for given classes of individuals, the nature and quality of attractiveness, projective elements in attraction, the role of learning in attraction, and so on; nor from experimenting on variations and determinants and the dependence of specific qualities on specific configurations. Bentham's attempt to establish intensity as a dimension of evaluating pleasures, or Ralph Barton Perry's treatment of intensity as degree of arousal of the organism, were philosophical attempts heading in the same direction.

If, on the other hand, intrinsic value is a complex concept, then it is perhaps best at this late stage in the history of ethics to confess that it admits of no easy decomposition or unpacking, that it is a highly theory-laden concept, serving as a coping-stone in diverse theoretical structures, and that it gets its meaning from the theory or model associated with it. This alternative in dealing with the concept seems to me much the more probable one, in the light of the history of ethics. It is because we have confused a highly theoretical term with a simple observational term that we have raised such difficult problems. This confusion is more significant in the long run than our inability to find pure intrinsic values of an ultimate Moorean sort.

On this analysis of the issue, there is again no simple answer to the question of the reach of science in dealing with intrinsic value. The path of investigation follows the path of analysis of the theories associated with the concept in its diverse uses.

III. SCIENTIFIC STUDY OF THE NATURE AND TASKS OF MORALITY

Every ethical theory includes somewhere or other some conception of the nature and tasks of morality. Our concern here is particularly with the tasks; the very concept of nature would take us too far afield. The conception of tasks is usually derived from common observation of the role that morality plays in human life. Although it does not at first sight appear to be a normative conception, it often involves or specifies criteria for the satisfactory functioning of moral regulation or moral discourse. For example, it is regarded as a truism in contemporary analytical ethics that the purpose of moral discourse is to guide conduct, and so no analysis of moral discourse will be satisfactory which does not show how conduct is being guided. I shall not here embark on a study of the variety of roles for which morality has been cast in the long history of ethical theorizing. My problem is rather the part that science may play in such questions. But let us remember that the two characteristics noted about statements of the tasks of morality are present even before science comes on the scene. The first is that to assign a task to morality is in some sense to find out what morality is up to – it is a statement about morality which is asserted as true or correct. Secondly, it becomes the basis for criticizing how a particular morality is being carried out, or again, how adequate a particular ethical theory may be in its analysis of the morality. When science attempts to study the functioning of morality, and to develop consequent criteria for evaluating morality, it is only extending a job that common observation and reflection have begun. Not only can it hope to do it more intensively, but it can – we shall see – even help make clearer what have been the normative and non-normative elements in the process.

The scientific study of morality – describing its phenomena, taking it to pieces, seeing how the parts work, what configurations they form – is thus a perfectly proper job of the bio-psychological and socio-cultural sciences. Perhaps anthropology has borne the brunt of the job. Considerable work has been done on it though not perhaps enough as yet. Its results, to repeat, are scientific propositions about morality, moral configurations, relations of moral codes and moralities to other aspects of human life. They are not moral propositions. And yet when the knowledge is accumulated – even when the outlines of the kinds of knowledge possible become distinct – this growing understanding of what enters into a morality, how it is organized, how it functions, is seen to be a *sine qua non* for formulating criteria for a regulative critique of moralities. Of course values enter into the process of formulation, but it is not as if the values came first fully developed and the knowledge simply found the means. The values themselves are systematically clarified and articulated in the process. The comparison with medicine is helpful. The initial values for medicine may be the desire to continue living, to avoid pain, and to continue functioning in some general sense. The knowledge furnished by the sciences of the body elaborates a comprehensive ideal of health, with varieties of prescriptions and hosts of indices. So too we develop the ideal of a well-functioning morality and articulate criteria for evaluating moralities in part and as wholes. In fact such a task is forced on us in the modern world by the variety of moralities in different peoples, the conflict of moral principles, and the pressure for choice and decision. Nor is our ordinary language without expressions that in some sense suggest that we are evaluating morality in part or whole. We sometimes speak of our sexual morality as being a failure, of such-and-such a morality as being one-sided or too rigid, and so on. What criteria concerning the nature and function of morality are we employing and what values are embedded in the criteria? Perhaps common-sense ones, such as that most people would find it impossible to live up to the morality, or that desirable institutions are militating against it, or that its concepts are unclear and confusing, or in some sense unrealistic, etc. Careful analysis may help sharpen some of these criteria and embedded values; growth of psychological and socio-cultural and socio-historical knowledge clarifies or stabilizes others. But even more basically, the whole outlook that is ready to think of morality as having psychological and social functions is a product of the growth of psychological and socio-historical knowledge. Once again, it need not necessarily be in conflict with other judgments of intrinsic worth in morality – though these have to be established on their own. It can be seen as the elaboration of additional dimensions for evaluating moralities and moral elements – for example the dimension of social utility in the light of social functioning for evaluating psychological processes, and psychological utility in the light of psychological functioning for evaluating social processes and institutions.

Many moral elements turn out to have this girding or supporting relation to social forms and institutions. It is clear in virtues – for example, the great stress on punctuality in an industrial society that requires precision, honesty in a scientific enterprise or credit institutions, hospitality in primitive societies that have tenuous external communications. Where the particular social utility is not required, we may find the virtue ignored, or even its opposite praised, as the cunning of the wily Odysseus is praised even by the gods in Homeric morality. Socio-historical studies of a whole moral pattern, such as Puritan ethics, in familiar writings such as those of Weber and Tawney, have made us conscious of the way in which the dominant shape of a morality – its particular set of rules, selective cluster of virtues, and so on – may be geared to the demands in economic and social problems of a given day. Sometimes this knowledge gets built into a critical concept of 'ideological' as a way of distinguishing the less authentic, more projective, from the authentic and more insightful and realistic. Sometimes the social role of the morality is made integral to its criteria of self-judgment and we may evaluate a morality in part by how it answers vital social issues.

Comparably, a psychological concept of maturity becomes increasingly useful with the growth of the psychology of personality. Since morality involves internal control of impulse, it follows that psychological study of the growth of impulse-control, its modes and patterns in relation to general psychic functioning is not without relevance to the understanding of moral processes. Of course, again, there are values embedded in the notion of effective psychic functioning, but these may be readily agreed on in terms of indices of ability to function without overriding anxiety, or break-down, and so forth. When the discovery of a developmental pattern and these broad basic values come together, it becomes possible to think in terms of successive pitfalls which a personality encounters in the course of its growth, and of successful outcomes in typical maturation. Specific criteria of maturity may thus be fashioned, fusing the knowledge of steps and dangers with the broad goals; examples of such criteria are the progress from unconscious to conscious, or from fear to security, or from dependence to self-determination, and so forth. The criteria are neither wholly physically compelled, nor sheerly evaluative (as, say, a desire for the novel as against the familiar, which is sometimes slipped in on the assumption that the desire for the familiar represents a kind of regressive clinging!); they represent rather an articulated scheme comparable to the articulated ideal of health referred to above.

The combination of psychological and social analysis, coupled with further scientific knowledge, occasionally makes it possible to undercut a whole value system and not to take it at its own estimation. The clearest illustration is the so-called Nazi ethics. Its rejection comes not from arbitrary disagreement with ultimate values of adventuresomeness and impulsive strength, but from taking it to pieces, seeing the falsity of many of its assumptions, seeing the distorted character of many of its psychological elements, and so on. For example, the well-known studies of the authoritarian personality, in spite of serious methodological issues and the influence of value elements in their shaping, show the kind of impact scientific knowledge may have on norms for evaluating moralities. Interestingly, Sartre's 'Portrait of an Anti-Semite' graps the same central point in an intuitive-phenomenological way, that anti-Semitism served not as just a particular set of beliefs or attitudes, but as a focal characterization of a personality governed by fear and self-hatred.

In general there is no reason why the scientific study of morality should not in the long run quite revolutionize our concept of morality, just as has happened to politics and law and religion and education. It does this by making us sensitive to the kind of jobs that the morality performs in a culture in relation to the whole of the life of the culture. It is not pertinent to the argument that this be a reductive functionalism which says morality 'is really just this and this alone'. That is a quite distinct issue, just as in the comparable developments of the scientific study of religion the existence of God is a quite separate question from the scientific determination of the projective elements that enter into traditional images of God, or the social analyses of particular uses of historical religions. So too in morality. The jobs turn out to be psychological: morality involves developing specific patterns of impulse control charged with specific types of emotion. Or the jobs turn out to be social: morality undertakes the burden of holding back agression in the society through the use of specific internalized sanctions; or it transmits through educational mechanisms specific patterns of desired character to a subsequent generation; or it guides the society to typical patterns of goals and activities; or it furnishes principles and modes of individual decision in certain classes of cases, and so on. So far from reducing the ethical, such scientific investigation of morality may widen the scope of normative ethics. For it brings into consciousness connections hitherto unconsidered, so that if they prove subject to control they now constitute something that can be changed, and the new question is whether they should be changed. This is obvious in ordinary matters: for example, if we discover that some of our moral rules are supported by subtle threats of punishment, we may want instead to shift to the promise of rewards (or demonstrations where virtue is its own reward). But it enters increasingly into larger theoretical issues. Compare the older treatment of conscience with that found in recent psychological writings on ethics. Conscience used to be a faculty of determining with epistemic certainty what was right and what was wrong, expressing a ruling element in our nature. The chief problem was to hear its voice distinctly in the clamor of the passions. With modern inquiries into its origin and development and the realization that it is constituted by a patterning of specific emotional development, questions are raised whether there may not be different types of conscience – e.g., rigid authoritarian and democratic humanistic, each with its own psychological etiology. The implication is that with sufficient understanding men may build one rather than another type in subsequent generations – hence a broadened scope of decision. And what holds in the domain of conscience holds equally in the kinds of prescriptions a culture builds into its morality – inflexible guilt-laden, or flexible utilitarian.

In all such cases, a scientific study, going beyond common reflection, is able to see what is capable of alteration and thus is able to pinpoint where normative decision is required, even if only to maintain the same path that has previously been trodden. In this way, the discovery of causal conditions may open the way to need for moral judgment where none was seen or even feasible before. This may take place even with regard to basic assertions of the functions of morality, for it becomes possible *thereafter* to ask whether fresh functions should be taken on or some older ones abandoned. The logical character of such a process which starts with discovery of existent functions, elaborates in terms of them criteria for the assessment of the morality, and ends with a call for normative evaluation of the very bases from which the criteria were developed[2], is obviously an extremely complex one. It does injustice to it to dismiss it as mere psychology or sociology, just because it begins with psychology or sociology; or to dismiss it as arbitrary value judgment just because it ends with a normative assessment. In between lie processes of criteria-formation which as we saw in the case of the criteria of maturity, carefully fuse broad values with responsible knowledge. There are also logical processes of concept-formation, into which the criteria are built. There are empirical elements entering into the establishment of criteria of effectiveness for the functions accepted. And finally, there are normative processes of the broadest sort, joining major values with growing knowledge if criticism is directed on some initial task of morality itself – as when, for example, some functions assigned to morality are shifted to medicine, or other functions assigned to law are shifted to morality. Ethical theory has suffered much in the past from the failure to grasp the great complexity of its own processes and from an over-dogmatism in reducing them to simple value-commitments as contrasted with empirical knowledge.

IV. USES OF SCIENCE IN META-ETHICS OR THEORETICAL ETHICS

There remains the question of theoretical ethics or meta-ethics – analysis of ethical concepts, their relations, justification procedures, and so forth. I do not propose to consider the whole image that meta-ethics has had of itself as neutral to values and pure of empirical elements. My concern is with the latter only since I am tracking down the use of science in this area.

Let us take the kind of items just listed, beginning with analysis of ethical concepts. Consider, for example, analyses of the good. Most conceptions of it tie it to some human goal, property or process, either by identification or some looser logical bond – whether it be happiness, pleasure, desire, purposive striving, or guiding and commending. It requires little insight to realize that the analysis of the concept tends to follow the structure of the materials to which it is related. A clear instance can be seen in Bentham: he assumes that all goods are comparable because he assumes that all pleasures are homogeneous. But how are such concepts as pleasure (or purposive activity, etc.) themselves to be understood? Either in terms of an older introspective psychology, or in terms of a more developed psychology which interrelates physiological, introspective, depth accounts and functional studies. (By functional, I mean such studies as, for example, assign to pain the function of signalling illness.) There is no understanding of the concepts without any psychology. You do not escape probing for underlying psychology by simply following linguistic usage. If you distinguish meanings of 'good' as good-of-its-kind, and good-for, and specify the logical properties of each, so far so good; but to get a deeper understanding you need to study men's activities in fashioning or determining kinds and in using instruments. The same holds for guiding and commending.

For an example of the relation of concepts, take the familiar controversies over the relation of means and ends? If intrinsic good is identified with ultimate end, the theory of the good will take shape from the study of ultimate ends and the kinds of ultimacy to be found in human thought and action. Thus Aristotle has ends as the separate objects of wish, with means the object of choice, and ultimacy belongs to the end in the series that occupies the last point and has no instrumentality whatsoever. Dewey extends the domain of choice to cover ends and sees a thorough fusion of means and ends. Fundamentally, it is because his basic psychological picture of man is different. He thinks of human motivation in terms of an integrated reconstruction of a situation in which conflicting tendencies have arisen, and he refuses to separate the cognitive, volitional, and emotive into distinct entities. Whatever the complexities of doctrine involved, the theory of ultimate ends will reflect the theory of psychological processes. Ultimate ends for Dewey could only mean those large and permeating goals which gain their strength from the multitude of aims they support and organize and by which they are continually supported. An ultimate end could not be a sheer desire or isolated will-act.

Nor is the outcome any different if we consider the problem of method. Every general methodological claim has its underlying scientific picture. Bentham's claim for an inductive methodology in ethics rests on the view that there is a clearly specified end – pleasure – sought in all human actions. Therefore moral judgments report reliable means for achievement of this end. Emotive theory rules out deductive and inductive method in ethics because in its theory of the emotions the relation of belief and attitude is taken to be contingent. Emotions are treated as attitudes that may be affected by beliefs but not in a dependable way; ordinary life examples are invoked as evidence, based on roughly an introspective psychology. If the psychological theory of the emotions is altered, the methodological view requires reconsideration.

Perhaps the most striking illustrations of the reach of science in theoretical ethics are those in which we are not merely forced to give different answers to the ethical questions when scientific results are changed, but in which we are forced to change the questions themselves. This is clearly seen in the history of the discussions of egoism in ethical theory. Arguments in older formulations varied from attempts to show that rational egoism yielded altruism to G. E. Moore's one-paragraph proof that egoism is logically self-contradictory! When advancing knowledge turned on the question of what an ego is like, how it develops, what shapes its pattern, the form of the question changed. There were experiments in task-completion, for example, which showed the existing differences between people who are task-oriented and are satisfied if someone else completes the task on which they have been interrupted, and people who are not satisfied unless they complete it themselves. Or again, psychological studies differentiated personality types and their etiology – for example, people who have a passive-receptive orientation, and people who show an active consideration for others. Social studies that began with contrasts of co-operation and competition found a proliferation of concepts necessary according to modes of work, types of motives, and so forth. In general, then, the question whether man is inherently egoistic or altruistic gives way, through a realization of the complex psychological and social determinants of character, to the question of the causes for development of specific types of character; and the question whether egoism is a justified ethics gives way to the normative question of deciding desirable types – for example, whether to encourage individualistic self-assertive attitudes or socially cooperative ones, and so on.

Many more questions of ethical theory than we are prone to think may admit of such refinement and reconstruction. There are some signs that analytic approaches may be moving in such a direction. So at least I interpret Rawls' illuminating treatment of promising as a practice – in effect an institution – in his effort to cut through the intuitionists' claim of self-evidence for the rightness of keeping promises. Why have promising? This means why have such institutional procedures, and is answered by a utilitarian reckoning of the good consequences of having such a practice. Why do I have an obligation to keep this promise? To promise is ipso facto to take on an obligation; therefore the question can only mean whether I have proper excuses for dropping it. And so on. What the analytic philosopher works through to with great effort and at the end of an epoch in which it is realized that language itself is a system of practices, was long known to the anthropologist and to philosophers with a sense of history. Nietzsche, for example, in touching promises, gets directly to the question under what historical conditions promising makes sense; it presupposes that a man can know what he is going to do in the future, therefore sufficient regularity in his life. Hence it can arise as a practice only with some degree, he says, of the domestication of man! But then Nietzsche always did complain that philosophers lacked a sense of history, and no man could understand moral philosophy who had no realization that there had been many moralities.

Let us try to formulate more systematically the sort of relations between science and meta-ethics or theoretical ethics that we have been considering. We are obviously not dealing with so simple a relation as that of scientific premise to ethical conclusion. It is rather that of a scientific presupposition involved in the very formulation of questions and statements in ethical theory. But a general reference to scientific presuppositions will not do – we have to find the precise points of entry of scientific materials into statements of ethical theory. Perhaps the simplest way to begin would be to collect the usual non-ethical terms that appear in the discourse of ethical theory – e.g. 'aims', 'ends', 'purposes', 'feelings', 'desire', 'choice', 'want', 'will', 'law', 'self', etc. These terms function in diverse ways: sometimes in definitions of ethical terms, sometimes as interpretations of ethical terms or as indices for their application, sometimes locating the phenomena or situations in which morality is applied, sometimes as constituents in the statements about human life that are offered as justification for theses about the nature and tasks of morality, and so on. There is no uniform or standard way in which they function; it varies with different ethical theories. Now most of these are terms of psychology – whether individual or social – and sometimes of biology or sociology. And while any given term may be dispensed with in a particular ethical theory, any theory that had no such terms at all would rob itself of relevance to human life – as Moore's theory in fact did. If then we more or less formalized an ethical theory, we could specify which variables had such terms as their values. These variables might then be regarded as placeholders to be filled in with terms whose explication was essentially scientific. There is no avoiding such entry of scientific materials. The question is whether it is good science or bad science, yesterday's science or today's science, the best available science or hangovers from common sense precursors of science.

That the non-ethical terms have an influential role in ethical theories can be readily seen by inspection. Take, again, the theoretical consequences of the concept of pleasure in Benthamite ethical theory in the light of the particular psychological approach taken to the concept itself. If pleasure is taken by an introspective psychology to be limited to individual consciousness, the ethics is individualistic, and the idea of a common good has only an aggregative character. Again, since the good is identified with pleasure and pleasure lies in individual consciousness, and since pleasure so construed is subjective, the individual is his own final judge of his good, without appeal beyond. Since Bentham regards pleasure as homogeneous, he regards goods as summable. If, on the other hand, the psychological approach to pleasure is behavioristic, preferential acts become the units of inquiry, the theory of value measurement is revised correspondingly, the concept of group preference is an easier construct than that of group pleasure. Again, if pleasure is regarded primarily as a signal of functioning well, a more biologically oriented conception may take over, and the ethics go in for ideals of health rather than maximal accumulation. These suggestions are cursory, but I hope they indicate the way in which an ethical theory is shaped by the scientific materials that enter through the variables we have pinpointed.

The aim in such an undertaking is not merely to find the relevant scientific commitments of the ethics, but to learn how changes in the ethics take place as changes come about in the underlying assumptions. In becoming conscious in this way of the role of science in ethics we are enabled to develop a regulative critique of our ethics. As scientific results change – which in the present growth of science in general and especially the anticipated break-through in the sciences of man, we may almost take for granted – a developed critique would enable us to plug in the new results, as it were, and look for the changes in ethical theory that are desirable.

Perhaps this type of formulation is not essentially new. Obviously a great part of the naturalistic tradition in the history of ethics recognized that there were psychological and social commitments of ethical theory, and often began inquiry by setting them forth – witness Hobbes or Hume, Adam Smith or Bentham. And in spite of different commitments they all shared the hope that once men saw the role of the underlying psychological processes in the operations of the morality, morality would be carried on more smoothly, more consciously, and get into fewer difficulties. They thus hoped that knowledge of man's nature, revealed by the relevant sciences, would help provide a critique for moral judgment in the obvious sense that a man who is conscious of what he is doing can see how to do it better. Because they assumed a permanent structure to human life and endeavor, they seldom raised the question whether a man who became conscious of what he was doing and saw how he might do it better, might want to do something else instead! The issue, of course, is whether he is unavoidably bent on that undertaking – or whether it is changeable. I do not wish to create a gap in principle between this earlier naturalism and post-evolutionary naturalism with its greater sense of change and its greater openness to varied reactions. How permanent a structure there is in human endeavor is itself a scientific question, and the character of men's reaction to the consciousness of what they discover is also a scientific matter. The idea of turning the consciousness of human processes into principles of a critique of those processes is central to both the older and the present inquiries. But we must have a keener sense of the values that are built into this transition and their justification. It is a fitting paradox that the growth of a scientific study of ethics and a search for the scientific commitments of ethical theory should itself turn out by its results to promote a sounder conception of the *autonomy* of morals. But autonomy means here not isolation, but an actual discovery of the wide limits within which human choice is capable of determining outcomes; this in turn generates a demand for a concept of rationality in terms of which decisions may be criticized and evaluated.

V. CONCLUDING REFLECTIONS

In these concluding reflections I want to push ahead beyond what I have tried to do so far. I tried to set the terms in which the multiple relations of science to ethics could be explored and to show how the program of such exploration could be initiated. I said at the outset that this paper was programmatic, but I hope I have shown that this particular program has the advantage that it can be put to work, not merely argued about, and that it renders obsolete much of the general controversy about science and ethics that has held the center of the stage. But I want to go further and suggest that the carrying out of the program will extend a breakthrough in the treatment of some of the general categories and general issues that have confined work in ethical theory. We have already noted the result in considering means and ends – the recognition that the utility of the means-ends categories, the extent to which they can be sharply separated, and their scope in ethics as compared to alternative categories, are all issues in which fruitful analysis requires a scientific study of the human processes in purposive behavior. It is the results of this study which show what theoretical use may be made of these categories. Similarly, in the paradox mentioned above concerning moral autonomy, it is the results of the scientific study of man which determine the scope of possible autonomy and the domains of significant choice. I should like to extend this type of analysis to two important theoretical problems. One is the question whether scientific method can be used in ethics. The other is the question of the scope and utility of the fact-value distinction in ethics.

The question whether scientific method can be used in moral judgment is sometimes made to seem the sum and substance of the question of science and ethics. Actually, it is a derivative question. In one sense, nothing can stop someone from trying to use scientific method in ethics. But trying is not equivalent to succeeding. Certain conditions have to be satisfied, such as a minimal stability of the phenomena, some degree of generalizability, and so forth. Of course, the best way to show the method can be used is to use it successfully. I think that a good case can be made for a Deweyan type of use, at least in many types of problems. But it is very easy to conceive of a possible discovery about man which should render the use of the method inapplicable: suppose we found that man was basically irrational whenever he approached serious questions of interpersonal relations – just as Hobbes thought divisive property interests, if they cut across, would affect thinking even about Euclid! It might then be a discovery of science that scientific method could not be used in ethics. I do not wish to be deprecating the use of scientific method in ethics by hypothetical considerations; I want simply to make it clear that this is a dependent not a primary issue in the relation of science to ethics.

Now on the question of the fact-value distinction. I stressed at the outset that this was not equivalent to but an interpretation of the distinction of fields – of science and ethics; it assumes that science gives us facts and ethics gives us values, without questioning seriously whether both fields contain both facts and values. The distinction of fact and value has, as we all know, been one of the primary handles for keeping science and ethics apart. And this has troubled many philosophers. They counter with different moves. Some will see science itself as a value enterprise, or at least as committed to some values. C. I. Lewis finds the problem of right decision pervasive and therefore that there are prescriptions in all domains of knowledge, not just in ethics. Dewey gives an instrumental view of knowledge that is equally applicable to ethics. Some recent analytic writers look to the domain of institutional facts with built-in values as an intermediate zone. But I wonder whether we really need general moves to solve our present problem. We can instead recognize that a distinction of the fact-value type is a fashioned categorial one oriented to the solution of certain problems; it may be useful in some contexts and obstructive in others. It is useful in maintaining the permanent possibility of evaluation and in preventing values from being smuggled in and imposed on us. But if used indiscriminately it creates gaps where they are not needed. Perhaps we ought to explore its functioning more carefully in a manner parallel to the manner in which we examined the means-end distinction, which bears some resemblance to it. And in ethics the notion of fact does behave suspiciously. The facts of the case prior to evaluation may include all past value judgments – everything gets factualized. On a Sartrean analysis every decision is free, so all my past value judgments now are simply facts about my past decisions. I now have to decide whether to stick to my past trends of decision. In effect, what Sartre is doing is replacing the fact-value cutting of the conceptual pie with a past-future cutting, with the present as moment of decision. Still other categorial selections are possible. Much of current analytic ethics uses the distinction of observing and acting, or the contemplative and the practical attitudes, where the earlier positivist analysis used the fact-value or the indicative-imperative, or the knowledge-volition distinction. I prefer instead the attempt to experiment with a fourfold distinction of enterprises – describing, analyzing, explaining, evaluating – while insisting that each of them can cover all the materials as well as the processes of any of the others. Hence there is no partition of field or data among them.[3] In any case, the fact-value distinction has not proved helpful in much of ethics, except to guarantee the permanent possibility of evaluation, a consequence which can probably be achieved in other ways. Both concepts – fact and value – are too stretchy and results of application not always happy. We need a fuller investigation of both to see what has gone into their stretching – why all scientific propositions whether of theory or generalization or particular observation got cast within a single rubric, and why all selective tendencies from sheer desire to complex ideal and prescription have been construed in another. It is too late here even to offer suggestions in this inquiry. For our present purposes it is enough to note that if the integrity and permanence of evaluation are safely established in other ways, we can happily experiment with other categories for ethical theory. At the very least, the fact-value dichotomy can be removed from its position as obstacle to the fruitful relation of science and ethics. In some sense our present formulation of the relations of science and ethics has been carried out without involving the fact-value categories. This should add to the conviction that the fact-value distinction is not an absolute one but is context-bound. We have not, of course, even begun the investigation into criteria for determining its proper contexts and proper limits.

Finally, what about the logical status of the inquiry into the relations of science and ethics themselves? It is certainly a philosophical inquiry, but that designation does not settle the extent of conceptual-analytic, scientific, and even ethical-normative elements that may go into the solution. Conceptual-analytical elements are clearly involved, for the meanings of 'science' and 'ethics' and their distinction into components are matters that involve careful analysis. Ethical-normative elements are involved in a definite but perhaps peripheral way: it is desirable to find relations of science and ethics so that knowledge of the world and its ways can act as good reasons for courses of conduct. But the crucial question is whether the world indeed has the requisite type of order. And this is pre-eminently a question to be settled by the results of scientific inquiry. Thus in a fundamental sense the issue of the relations of science and ethics is itself a scientific issue. And this is so, whatever the outcome. Suppose for a moment that the verdict were that an ethical theory needs no change whatsoever, no matter what the psychological theory of human striving and purpose, of human choice and emotion proved to be. This would not have the status of a metaphysical presupposition or an analysis of the logic of linguistic use. It would be the scientific hypothesis of a constancy or an invariance in ethical theory for all transformations in psychological theory. But in fact such a verdict of invariance is far from the emerging picture today. The analysis of the diverse patterns in which science and ethics are related is thus of basic importance for the reconstruction of the enterprise of ethical theorizing and for fashioning an adequate ethical theory.

The City College of the City University of New York

REFERENCES

* In revising this paper, the writer is generally indebted for critical suggestions to Professors John Ladd and Ruth Putnam. The latter's comments on the treatment of G. E. Moore were especially helpful.

1 Moore's principle of organic unities denies that we can go directly from the value of the parts of a whole – each evaluated separately (that is, as a whole) – to the value of the whole with them as parts. In short, every complex has to undergo the isolation test itself.

2 It is worth stressing again that given the permanent possibility of evaluation, there is no limit to which this questioning of the functions of morals may not go. In theory,

even the whole of morality as a human institution may be put on the scales and found wanting, if all the functions suggested for it can be carried out better in other ways or through other human agencies. (Morality might then cling to a place in the criteria of 'better' in such investigation – apart from the separate questions of intrinsic value.)
[3] Cf. Abraham Edel, *Method in Ethical Theory*, Bobbs-Merrill, Indianapolis, 1963.

RUTH ANNA PUTNAM

COMMENTS*

Professor Edel has presented a view concerning ethics – or rather an application of a view – with which I agree on the whole. Roughly, the view may be summarized by saying that (a) there is moral knowledge (or opinion), i.e. there are not merely attitudes, or decisions, or whatever your favorite version of emotivism says there are. And (b) this knowledge does not differ radically in kind from the sort of knowledge we hope we are gathering in the sciences. Consequently, just as there are no *a priori* reasons for denying the relevance of one science to another so there is no *a priori* reason for denying the relevance of the sciences to morality and ethical theory. I take it that Professor Edel's paper constitutes an application of this view in the sense that he actually shows us some ways in which knowledge gained in the sciences is relevant to morality and ethical theory. In short, he goes beyond the merely negative claim: there are no *a priori* reasons for *denying* the relevance of science to morality and ethical theory, to the positive claim: there are some, indeed many, *a posteriori* reasons for *affirming* the relevance of science to morality and ethical theory.

The question which I should now like to examine a bit further is, simply: How persuasive are these reasons? Fortunately, inspite of the title of Professor Edel's paper, it is not necessary for us to look at 27 reasons; we need to look at only four kinds of reasons. The first set of reasons may be reduced to the following argument: A careful examination of Moore's *Principia Ethica* shows us that an ethical theory which maintains that science is irrelevant to the determination of the good and helpless with respect to the determination of duty leads to consequences which are mostly bad. Therefore, we should attempt to develop an ethical theory which has room for science.

Let us agree for the moment that Moore's theory leads indeed to the consequences that Professor Edel has uncovered. There remains the question whther these consequences follow inevitably from a rejection of science. I shall ignore the good consequence that such a theory leaves room for the permanent possibility of re-evaluation, for I agree with Edel that this desirable condition can be obtained in other ways.

The two bad consequences which Professor Edel discovers are (a) that judgments of what is good are reduced to isolated reactions, and (b) that with respect to duty Moore's theory issues in conformism. The first is supposed to be a result simply of the claim that good is a simple non-natural quality coupled with Moore's methodology for determining what is good, viz. examine the thing to be judged in isolation. But I think Professor Edel's argument here depends at least in part on ignoring Moore's doctrine of organic unities and in part on a too strict interpretation of Moore's methodological advice. It seems to me to be quite in keeping with the spirit of Moore to imagine, for example, three worlds, one full of friendship, one full of beauty and a third full of both beauty and friendship – which is not the same, I suppose, as to imagine a beautiful friendship – and to compare them while holding all three before one's mind. The principle of organic unities warns us not to assume that the value of a world full of beauty and full of friendship is simply the sum of the values of the first two worlds – and also not to assume that its value is that of a world full of beautiful friendships. I must confess that I find it impossible to do what Moore tells me to do, to imagine worlds that are so very simple. But if I could do that, then I suppose there would be no great difficulty in imagining several such simple worlds at once and hence to 'see' not only their values but their comparative values.

If good is a simple quality, like yellow, it may nevertheless be the case that some complex wholes are more intensely good than others (just as some yellow patches are more intensely yellow than others). In any case, Moore obviously made this assumption, otherwise he would not have emphasized the principle of organic unities. Thus I am not convinced that on Moore's theory judgments of what is good "reduce to isolated reactions".

Let us then examine the second claim that Moore's theory leads to conformism with repect to duty. I share Professor Edel's distress on reading 'Ethics in Relation to Conduct'. Somehow one expected new and exciting insights when, instead, one is presented with a consistent act-utilitarianism (Rawl's term) coupled with an unwillingness to make concessions to a human frailty which is perhaps magnified beyond necessity. It is after all our total inability to foresee all the consequences of our actions and of all their alternatives which leads Moore to argue for what Professor Edel calls conformism. Here it is, I think, important to notice that any uncompromising act-utilitarianism would find itself in exactly the same position. Any act utilitarian-would have to confess that one can never know, in some strict sense of 'know', what one ought to do. But no moral philosopher is willing to stop at this point; for to stop here would be tantamount to saying not that science is irelevant to ethics but that ethics is irrelevant to conduct.

Thus the question arises, must one turn to conformism and conservatism in order to mitigate this moral scepticism. I submit that it is by using science – a perhaps primitive sociology – that Moore is led to his conservative views. For does he not argue in part from the fact that most attempts at moral innovation are doomed to be ineffectual? Secondly, I should like to ask whether it is not possible to argue that a thoroughly intuitionist account of duty may lead to a radically new set of *prima facie* obligations? In other words, granted that Moore reduces science to helplessness by demanding too much of it, does it follow that a dynamic and changing moral code can be developed only by making room for science?

Sofar I have denied that Moore's ethical theory is as convincing an example of the evils of excluding science from ethics as Professor Edel claims it to be. I have denied it on the grounds that excluding science from ethics is merely one element in Moore's theory and perhaps not the decisive one.

Let me now raise a further difficulty. Even if we grant that rejecting science leads to incomparable value judgments and to conformism and conservatism with repect to duty, we may now ask: but are these consequences bad or undesirable? Do not misunderstand me, I fully agree that they are bad or undesirable. Surely, one *raison d'être* for an ethical theory is that it should enable one to choose rationally between competing values, and to provide for a morality that can change with a changing world. But, presumably, Moore did not see matters in this light and so would not have been persuaded by Professor Edel's argument.

Let us now turn to Professor Edel's second set of reasons. The argument here consists in showing by enumeration the many uses of science in morality even for those who hold that "science can provide only means". For precisely the sort of information for which we look to science when we want to know how best to achieve an end will also be of value when we have to choose between competing ends, and it may even lead us to abandon an end by showing us that it is unattainable or attainable only at a price that we are not willing to pay. Thus this set of reasons amounts to saying: look how much you gain if you admit that science has relevance for morality. So this constitutes an invitation to develop an ethical theory which leaves room for science.

With this set of reasons I have no quarrel whatsoever. For even an intuitionist should be willing to accept this invitation. After all, having constructed an ethical theory which leaves room for science, he is still free to intuit whether it is better or worse than an alternative which does not. In short, this is only an invitation to perform a *Gedanken* experiment. Its virtue consists in preparing the ground for the experiment with great care.

But how shall one evaluate the outcome of the experiment? This question leads back to my last point concerning Moore and forward to Professor Edel's third set of reasons. As members of a community we look for a morality to guide our conduct; as philosophers, specifically as ethical theorists, we look for criteria on the basis of which we can evaluate competing and conflicting moralities. And now it seems that we are caught in a vicious circle – too judge whether a morality is good we must already know what values it ought to uphold. But we cannot know that unless we have already a set of values, i.e. a morality. Professor Edel suggests that here once more science can come to our rescue. For, he says, let us look at many existing moralities; underlying all their conflicting prescriptions and exhotations we can discern certain common aims. One purpose, if not the only purpose, is to provide rules by which men can live, rules which are obeyed by most people most of the time so that the members of a community are able to predict each other's behavior and plan accordingly. This common aim leads to a set of criteria for evaluating every morality which transcend any given morality, such criteria as clarity, consistency, possibility of being lived up to. It would seem that these criteria could be disputed only by one who fails to see that any morality attempts to satisfy a vital social need. But such a person is not concerned to evaluate conflicting moralities but to argue against all moralities. I am inclined to say that such a person has no business being an ethical theorist. And Professor Edel might add that historical sociology would show the rest of us that such persons are sufficiently rare so that we need not bother arguing against their view.

Thus the value of science to ethical theory consists in the fact that it provides a neutral starting-point. But Professor Edel lists further criteria, at least by implication. A morality should not be one-sided, rigid, inflexible. These criteria do not follow immediately from the fucntion of any morality. However, if science could show us that societies or individuals who subscribe to one-sided, rigid, inflexible moralities find themselves suddenly without any morality because of changed circumstances, then science would lend support also to these further criteria. Whether it does is a question that I am not competent to answer.

Professor Edel's discussion of uses of science in meta-ethics seems to me to belong with his second set of reasons – the uses of science in morality. Here, as there, he shows us with great care and most interestingly how much more science can do for us than we might have expected.

What I have said so far boils down to this. There are two types of arguments which Professor Edel has advanced for the thesis: science is relevant to ethics. One type, his second and fourth, consists in demonstrating uses which anyone would grant are uses. Even Moore would have no quarral with the second section of the paper under discussion; he would simply doubt that science can supply us with enough information. And I have suggested before that Moore himself uses science – although perhaps a too primitive science – in the manner suggested in the fourth section.

My qualms concern only the second type of arguments, those advanced in Sections I and III, and here my difficulty is most easily stated by saying that Professor Edel's value judgments seem to be essential premisses in his argument. I hesitate to use this terminology, for I agree that the fact-value distinction is merely one way to cut the ethical pie. What is more, I doubt that the pie can be cut successfully in this way. Finally, even if fact-value is sometimes a useful set of categories, like any other set of categories it should be abandoned when it is a hindrance rather than a help. And my argument may be interpreted as showing just that. The fact-value distinction gets in our way when our task is the evaluation of values.

Why then did I draw attention to this difficulty? I did so primarily because Professor Edel did not. Indeed, he seems to claim that his argument does not involve the fact-value dichotomy. Undoubtedly it does not from his point of view. But from the other point of view it must look as if he were guilty of 'smuggling values in', a charge against which, according to Professor Edel, the fact-value distinction guards us. My own inclination in such a situation is to stand up and be counted, i.e. to admit explicitly that I hold a certain set of values on the basis of which I can argue against conformist, conservative, rigid, one-sided, inflexible, guilt-laden, etc. moralities and for dynamic, flexible, humanitarian moralities. For to admit that one holds a set of values is not tantamount to admitting that values are not facts (whatever that may mean). How one regards the fact-value categories will emerge only if one is then challenged to defend the values one advocates. Fortunately, this is not my task here; my task here has been only that of a devil's advocate.

Wellesley College

NOTE

* Strictly speaking, these comments apply to an earlier version of Professor Edel's paper, the version which he read on February 7, 1967, rather than to the version now published.

JOHN LADD

COMMENTS ON ABRAHAM EDEL'S 'PATTERNS OF USE OF SCIENCE IN ETHICS'

The relationship between science and ethics is a complicated one, as Mr. Edel points out, and there is a variety of different ways in which science has a bearing on ethics, or, perhaps we should say, in which parts of science have a bearing on parts of ethics. We are not, I assume, interested in the relations that have in the past subsisted between these two disciplines, that is, we are not interested in historical influences. Rather, we are interested in how 'scientific materials' can be utilized in ethics, that is, we are interested in what Mr. Edel candidly admits to be a program for future investigations. The question is: does science have a use in ethics and, if so, what is it?

Before we can answer this question, we must first have a clearer conception of what is meant by the term 'use'. I shall therefore preface my remarks with a few observations concerning the meaning of the term 'use'. To begin with, we must distinguish between the *use* of a thing, how it is or could be *used*, and its being *useful*. 'Use' carries with it the connotation of propriety, i.e. proper use, and as such contrasts with abuse or misuse, i.e. improper use. Not every way a thing is used reflects its (proper) use; for it may be used improperly, i.e. be misused, or it may be used neither properly nor improperly. Finally, a thing may be useful for certain purposes quite apart from whether or not using it for such purposes reflects its use; e.g. a telephone book may be used to sit on, but that is not its (proper) use.

It is clear that science has often been used improperly in ethical discussions, that is, it has been misused or abused. The more notorious cases come quickly to mind: Herbert Spencer's using 'scientific materials' borrowed from the theory of evolution to back up his own peculiar blend of evolutionary ethics and laissez-faire political views, and Plato's using geometry and astronomy to back up his totalitarian political philosophy and idealistic ethics. Even today there are still many scientists who venture pretentiously into the field of ethics assuming that they, as successful scientists, are equipped with the philosopher's stone that can turn all the old lead in ethics into pure scientific gold. And in addition to physicists and biologists with their philosopher's stones, we now have a large and influential group of mathematicians and social scientists who believe that they have found in the mathematical theory of games and decision theory the key to the solution of the problems of ethics. The net result of using science in these ways, of these misuses and abuses, is to sacrifice the ethics of human dignity, justice and liberty at the altar of the sacred cow, science.

Science has, of course, also been useful for ethics in a more modest way, namely, as a source for new ideas and new ways of thinking in ethics. But in this sense of usefulness, the usefulness of science for ethics is on a par with the usefulness of many other subjects, e.g. literature, history and law; for a study of any of these subjects may be of great help to anyone doing ethics.

To speak of the *use* of science implies, however, that scientific materials are related to ethics in a more specific way, namely, they are not only useful but also relevant. For by the proper use, in this connection, we mean the relevant use, just as by an improper use we mean an irrelevant use parading as a relevant one. By 'relevance' I do not mean simply a psychological (causal) connection, that is, the psychological influence something may, as a matter of fact, have on the way a person thinks. Rather, I mean by it a *logical* connection, that is, a connection that ought to determine the way a person thinks about the subject in question. In calling it a 'logical' connection I am using this term in the broad sense in which it is used by ordinary language philosophers. That is, what is logical in this sense embraces the 'informal logic of concepts', presuppositions, categorical distinctions and logical functions as well as the relation of premises to conclusions.

In its interesting and important form, therefore, the question concerning the use of science in ethics becomes the question of what scientific materials are logically relevant to ethics and how they are relevant. This question itself is an ethical question, I submit, and not a scientific question, although Mr. Edel appears to regard it, at least in part, as a scientific question. Thus, towards the end of his paper, he writes: "Thus in a fundamental sense the issue of the relations of science and ethics is itself a scientific issue."

The fact that the relevance or irrelevance of a scientific proposition to ethics can be determined only on ethical grounds comes out clearly when we observe that the relevance of a scientific fact is always open to challenge and that the only acceptable reply to such a challenge is to provide ethical reasons for regarding that fact as relevant. In a larger sense, ethics itself might indeed be regarded as a body of principles for the determination of the relevance or irrelevance of factual considerations to our actions. For instance, the fact that a certain kind of bomb kills people more quickly and cheaply than other means of killing that are available is, according to one's ethical principles, rejected as irrelevant to the question of whether or not it should be used against people who get in our way, or, again, the fact that it would be to our personal disadvantage to fulfill a promise is excluded as irrelevant when it comes to deciding whether or not to keep it.

It follows that to assume that certain scientific propositions or theories are automatically relevant to ethics is to beg the ethical question. This, I think, is precisely what is done in a great deal of Mr. Edel's paper. Specifically, there are three points that are made in the paper where, as I shall try to show, the scientific materials mentioned are logically irrelevant to ethics and they are irrelevant for ethical reasons. (That is not to say, of course, that such materials may not be quite useful in the broader sense mentioned earlier.) The three points with which I disagree are set forth in the three middle sections of the paper and represent the chief, although not the only points made in the respective sections.

(1) In the discussion of the relevance of science to normative ethics in Section II, it is taken for granted that the principal concern of ethics is with a person's ends, goals or ideals, that is, with what are often called his 'values'. The use of scientific materials, then, is to assist him in 'refining', 'structuring', 'organizing', and integrating his ends, etc. In Dewey's terms, we may say that science is used in the "construction of good (or value)".

Now, the term 'value', as we all know, is quite ambiguous: thus, 'to be a value' may mean simply to be a thing that has value, or, on the other hand, 'to be a value' may mean to be something that someone holds to be valuable. In the second sense, 'values' refer to value-conceptions and they are always somebody's values: thus, we speak of Smith's values, or of middle-class American values, or of Navaho values. Furthermore, values in the sense of value-conception may be characterized as enlightened, mature, stable, significant, traditional, narrow, etc., or the opposite of any of these. Ethics, if I understand Mr. Edel's view correctly, is concerned with the promotion of meaningful values in the conception sense of 'value', and for this purpose science is useful because of its causal influence in structuring our values (=conceptions).

Let us skip over the question of whether or not there is any warrant for believing, as Dewey and Edel do, that the causal influence of scientific knowledge on the formation of our values (=value-conceptions) will, under the right conditions, (always, generally?) be benign. The question I want to ask instead is: *what has this to do with ethics?* Ethics, I take it, is concerned with right and wrong, duty and obligation, and, on a social plane, with justice and human dignity. But none of these is dependent upon what people conceive them to be and much less on their conceptions of what is valuable. They depend, rather, on the nature of the acts themselves and the situations in which they take place. The fact that a person has certain goals (=value-conceptions) is logically irrelevant to the question of whether or not it is his duty, e.g. to tell the truth. A person's value-conceptions, regardless of how refined or mature they may be, cannot make or unmake an action of his a duty. For this reason, it seems to me that the discussion of ends in Section II of Mr. Edel's paper is quite beside the point. The scientific materials he mentions are not relevant to ethics in the way he suggests simply because the subject-matter to which they are supposed to be related is not really ethics at all.

(2) My criticism of the proposals in Section III of Mr. Edel's paper are somewhat in the same vein, namely, that 'morality' as conceived by Mr. Edel does not properly belong to ethics but, instead, to anthropology or one of the other social sciences. For by 'morality' it is clear that Mr. Edel means what Austin called 'positive morality', i.e. the sum of a person's or group's operative conceptions about what constitutes right and wrong conduct, etc. I have elsewhere called this a person's 'moral code'. The scientific investigation of 'morality' in this sense belongs to a discipline called 'descriptive ethics'.

In the positive sense of morality, there are as many different moralities as there are different sets of conceptions of what is right and wrong: there is a middle-class American morality, a Navaho morality, Greek morality, etc. And with regard to moralities in this sense it is quite proper to ask what are their psychological and social functions. Such an inquiry might be called the 'sociology of morality'.

Most moral philosophers, however, use the term 'morality' in a quite different sense, namely, to refer to the principles of right and wrong conduct that *ought* to determine what we do. Morality in this sense, i.e. true morality, is part of normative ethics, if not the whole of it, and as such it is logically independent of people's opinions of what is right and wrong, i.e. their positive morality (= 'morality' in Mr. Edel's sense). It is as absurd to hold that the social and psychological functions of people's positive morality have anything to do with what is actually right or wrong (i.e. true morality) as it is to hold that the same kind of functions have anything to do with what is actually the case scientifically. The functions of people's opinions, whether they be ethical or scientific, cannot determine the truth or validity of those opinions; logically they are simply irrelevant. Hence, we must conclude that any scientific materials relating to the function of a person's positive morality are logically irrelevant to true morality, i.e. morality in the sense in which it is part of ethics, even though, as I have already said, consideration of such materials may turn out to be helpful to us for other reasons.

Morality, in Mr. Edel's sense, may also be viewed as a social institution like law or etiquette and, as such, it is again perfectly in order for one to inquire into its functions. By the same token, we must be able to evaluate a morality as we do, e.g. a legal system, and say of it that it is good or bad, enlightened or archaic, etc. But for this purpose there must be some sort of standard for the evaluation of 'moralities'. This standard, I take it, is provided by the theory of values presented in Section II. In effect, then, in Mr. Edel's paper morality is politely shown out of the door in order to make place for a theory of values.

(3) Finally, in Section IV, in his discussion of the relation of science to meta-ethics, that is, the analysis of ethical concepts, scientific materials are referred to that are logically irrelevant to meta-ethics as such. Mr. Edel's argument is that many concepts used in ethical discourse are actually psychological or sociological concepts, and so we should look to psychology and sociology for clarification and refinement of these concepts. We are all aware of the fact that, for theoretical purposes, scientific theory often assigns new, technical meanings to old, familiar expressions and that these new meanings frequently diverge quite considerably from the original ordinary meaning of these expressions. Therefore, from the fact that ethics and psychology sometimes employ the same expressions we are not entitled to infer that these expressions are used with the

same meaning; indeed, they may represent entirely different concepts.

In fact, I think that it can be shown that most of the ethically relevant concepts on Mr. Edel's list are not the same concepts as the technical concepts of psychology at all, but are, instead, concepts of ordinary language. As such they are inexact, fuzzy and multi-functional, for, in Wittgenstein's terms, they are intended for an entirely different language-game from the language-game of psychology and the social sciences. Unlike the latter, the language-game in which they are used is that of justifying, guiding and planning action. The term 'want', which is one of those on Mr. Edel's list, provides a good example of an ordinary concept of this kind; since in ethics and ordinary language it performs certain functions that it does not perform in psychology, namely, the special logical function of justifying action as well as a performative function.

In conclusion, I want to reiterate once more that nothing I have said should be taken to derogate from the benefits for ethics that may be expected from a knowledge of science. There is much to be gained by way of 'cross-fertilization', as the jargon goes, and moral philosophers should always seize the opportunity to broaden their horizon. But I believe that I am safe in assuming that we are interested here in a stronger, logical relation of science to ethics, a relation such that, if a certain scientific proposition is accepted, then one is logically compelled to accept or reject some ethical proposition or at least to modify or qualify it to take the scientific proposition into account. This is what I mean when I say that the one proposition is logically relevant to the other.

All the cases that involve the logical relevance of scientific propositions to ethics come under an ethical principle that I call the Reality Principle. This principle forbids the adoption of false beliefs into one's ethical thinking and asserts that, if a conclusion is founded on false beliefs, then it must be rejected as invalid and unacceptable. (Note that this principle permits us to reject any moral assertion that does not have a valid and true foundation. But the position I am stating here must be elaborated elsewhere.) The Reality Principle may be summarized by saying that *ethics must be realistic*.

The Reality Principle, it should be observed, is at once more general and more restricted than any of the principles mentioned by Mr. Edel. It is more general because it involves not only facts about means and consequences but also facts about acts and situations, as well as facts involved in the various presuppositions and assumptions implied in ethical thinking. On the other hand, the Reality Principle is more restricted, for in order to invoke it and to recognize scientific propositions as logically relevant, it is necessary to show how and why these particular scientific propositions are implied in the ethical discussion in question, say, either as premises or as presuppositions. If this cannot be shown, then we must conclude that the scientific propositions are logically irrelevant to any ethical ones.

How the Reality Principle operates may best be seen from an example. Let us consider the controversy about the morality of capital punishment. The question of whether or not capital punishment acts as a deterrent is one, we all agree, for the social scientists to answer. Suppose that they come up with the answer: No, it is not an effective deterrent. Then any ethical argument that cites the deterrent value of capital punishment to justify it is by the Reality Principle invalid and unacceptable. On the other hand, suppose that the answer is: Yes, it is an effective deterrent. Then the scientific proposition to that effect might still be irrelevant according to the restricted character of the Reality Principle; for it would first have to be shown that deterrence provides a valid, moral justification for punishment. If that cannot be shown, then any use of the sociological facts about deterrence are strictly speaking irrelevant.

There are other and more exciting facets to the Reality Principle, e.g. its application to ethical presuppositions, but they cannot be discussed here. It is my belief and hope that this principle will be able to account for all the cases in which science has a use in ethics, i.e. is logically relevant to it.

Brown University

RUTH ANNA PUTNAM

ON EMPIRICAL KNOWLEDGE

> A philosopher constantly afraid of stepping into another domain runs the danger of asserting empty generalities, because his philosophy is not sufficiently anchored in the specific sciences.
> Hans Reichenbach, *The Philosophy of Space and Time.*

Tom Lehrer has a song which begins "Whatever Happened to Hubert?". I should like to change that to "whatever happened to empirical knowledge?". By empirical knowledge I mean, to borrow phrases of David Hume, knowledge of matters of fact based on experience. Hume, as you know, declared that all knowledge of matters of fact is based on experience, and I am inclined to believe that almost all philosophers hold the converse, namely that all knowledge based on experience is knowledge of matters of fact. Yet a perusal of the recent literature leads one to wonder whether either of these generalizations can be maintained, and whether there is indeed any interesting empirical knowledge. Where by interesting empirical knowledge I mean what Hume called knowledge of matters of fact "beyond the present testimony of our senses, or the records of our memory".[1] Let me hasten to assure you, however, that despite these references to Hume my concern is not with scepticism.

I just said that one might wonder whether there is any interesting empirical knowledge at all. Surely such doubts must seem outrageous in Boston. For Boston is the home of Quine, and we all know that in 'Two Dogmas of Empiricism' Quine argued persuasively for the opposite point of view, namely that there is hardly any interesting knowledge which is not empirical knowledge. Is it not at the very least presumptuous and perhaps as futile as flogging a dead horse to try to defend again, in Boston, what is essentially a Quinean point of view? Perhaps so. And yet, truth, philosophical or otherwise, needs not only its prophets but also its minor preachers.

I

"All our knowledge of matters of fact is based on experience." For years

I took this to be a quite unambiguous, though certainly vague, claim. Then I read Peter Unger's recent paper 'Experience and Factual Knowledge'[2] in which he argued that this claim implies

> For any sentential value of *p*, where it is not necessarily the case that *p*, a man knows that *p* only if that man has had or presently has some experience

and further undertook to show that

> Counterexamples to these doctrines can be found; consequently we can show that they embody no sound philosophy.[3]

Let us ignore for the moment the last part concerning sound philosophy and pay attention only to the supposed implication and the alleged counterexamples. What strikes one is that the qualifying clause "where it is not necessarily the case that *p*", appears to be idle. Has any philosopher, other than Unger, maintained in recent years that any man may know anything at all without any experience? This is a view which Locke mendaciously ascribed to the defenders of the so-called doctrine of innate ideas; but that surely is ancient history. At any rate, since Kant, we have known that all our knowledge begins with experience. This, of course, is not to say that we should ignore Unger; rather it shows us that either "based on experience" or "knowledge" is curiously ambiguous. Hume says[4]:

> We learn the events of former ages from history; but then we must peruse the volumes in which this instruction is contained, and thence carry up our inferences from one testimony to another, till we arrive at the eyewitnesses and spectators of these distant events.

What then makes my belief that an important battle was fought on the Lexington Commons in 1775 knowledge based on experience? What is crucial? My experience of reading on a historical marker in Lexington that this event occurred, or the experience of the surviving participants and spectators, who first wrote about it? Clearly, both are crucial. Had I not read about the battle of Lexington (or been told about it), I should not know that such a battle was fought. But if no battle had been fought, or if, though a battle was fought, no one lived to tell of it, and by some miracle all traces of it had vanished as well, then again I should not know of the battle of Lexington be there ever so many 'historical markers' to be read about it. That this is so on the first alternative (no battle was fought) is obvious. Although, upon the evidence of historical markers I may be *entitled to say* that I know, I cannot in fact know what is not so. The second alternative is more complicated and goes to the heart of the difficulty.

Clearly, if no battle was fought, then no one had the kind of experience which someone must have had in order that anyone may know that a battle was fought. But, on the hypothesis that no battle was fought we did not have to refer to these non-existent experiences to establish the claim that I could not know that a battle was fought. Now the situation is different: a battle was fought, the participants had the kind of experience which someone must have had in order that anyone may know that a battle was fought. Yet, I say that if none of the participants lived to tell of it, and if a miracle removed all trace of the battle before anyone came to the Lexington Commons who might have seen them, then again we cannot know. But did I not say that there was a historical marker? To be sure, and how it came to be there is a mystery – someone put it up as a hoax, someone had a divine revelation, *you* make up a story, only note that when we trace the causes of the marker's being there we do not find as the first cause that someone had an experience which makes it not an accident that what the marker says is true.

Now I claim that inspite of the fact that almost everything Hume says makes it appear that my reading of the marker is *the* crucial experience on which my knowledge is based, the quotation I read refers not by accident finally to the eyewitnesses of the event. And I claim further that this must be so. The experience of the participants and spectators must be the experience on which our knowledge (yours as well as mine) that a battle was fought on the Lexington Commons in 1775 is based, in that sense of "based on experience" which makes "all our knowledge of matters of fact is based on experience" interesting and important.

This I must now defend. First, in the context of Hume's philosophy and psychology there can be no knowledge at all which is not based on experience if the relevant experience is my experience. Consider e.g. my knowledge that $2+3=5$. This is a paradigm of what Hume called relations of ideas, a "proposition discoverable by the mere operation of thought".[5] However, thought must operate upon something, namely ideas, and for Hume all ideas have their origin finally, although perhaps quite indirectly, in experience. Thus, had I never had any experience, I should not now have the ideas, or concepts, necessary to formulate and entertain the proposition that $2+3=5$.

Even if you maintain, what is surely contrary to fact, that once I entertain the proposition I shall inevitably believe it no matter how I came to entertain it, it is still true that $2+3=5$ is based on experience in the sense just outlined. And it is further true, although I cannot document that Hume would agree with me, that my entertaining the proposition that $2+3=5$ was in the first instance, and indeed in many later ones, prompted by experiences, the sort of experiences which every child has when it first learns arithmetic. Just as my entertaining the proposition that a battle was fought at Lexington was prompted by my experience of reading the historical marker.

Let me then summarize the burden of my argument so far. What I have tried to show is this. "All our knowledge of matters of fact is based on experience" has two readings using the paraphrases suggested by Unger, the Ungerian reading

(U) For any sentential value of *p*, where it is not necessarily the case that *p*, a man may know that *p* only if that man has had or presently has some experience on which that knowledge is based, i.e., an experience which prompts the man to entertain *p* and accept it as true,

and what, for want of a better label, I shall call the Putnamian reading[5a]

(P) For any sentential value of *p*, where it is not necessarily the case that *p*, a man may know that *p* only if some man has had or presently has some experience on which that knowledge is based, i.e. an experience which is such that a) if no man could have had that experience, then *p* would not be true, and b) that experience is a near or remote cause of the belief that *p* of the man who knows that *p*.

Note, please, that this condition does not deny that a sentence *p* might be true although no man could have had an experience which is such that *p* would not be true if no man could have had that experience; the condition denies only that such a sentence could be known. And note also that both (U) and (P) state merely necessary not sufficient conditions.

Let me continue my summary. I have claimed further that insofar as empiricists have wanted to use *being based on experience* as a criterion by which they could distinguish knowledge of matters of fact from relations of ideas on the one hand and what Hume called "sophistry and illusion" [6] on the other, they must have meant it in the sense of (P).

Now, what Unger does in the article which prompted these musings is to refute not (U) but the weaker claim that (U) is what he calls a conceptual truth. I shall now present Unger's argument that a man may know the population of China is greater than that of India without having had, or indeed ever having, any experience.

Consider Unger as he was just before he typed the following words (which occur in his article) on a piece of paper in his typewriter [7]

I confidently type onto a piece of paper what I know to be so: that there are more people in China than there are in India.

Suppose that clever scientists construct an exact duplicate of Unger-at-that-moment, call him Unger*; suppose that Unger* is kept under total anaesthesia until he is finished, that he is placed before Unger's typewriter and that the typewriter contains a piece of paper with whatever words were already on Unger's paper before he began to type the sentence just quoted. All conditions are exactly as they were just before Unger began to type that sentence. Unger* is now awakened to instant full consciousness. We may suppose then that Unger* will instantly type the sentence, "I confidently type onto a piece of paper what I know to be so: that there are more people in China than there are in India." Unger argues that Unger* is a man who has knowledge of a matter of fact but no prior experience (and if we had killed Unger* before awakening him, he would have been a man who knew a matter of fact without ever having any experience).

To be sure, the mere fact that Unger* types that sentence does not show that Unger* knows that the population of China exceeds that of India. We could have constructed a very simple automaton which types the quoted sentence whenever placed in front of any typewriter with any piece of paper in it regardless of any other conditions. Such an automaton would not, of course, know anything. It seems essential to Unger's argument that Unger* be an exact duplicate of Unger, and that he be such not by accident but because clever scientists not making any mistakes produced him to be an exact duplicate of Unger as Unger was at that crucial moment.

Why must Unger* be an exact duplicate? At this point one is inclined to argue as follows, and indeed Unger invites us to argue in this way. He invites us to argue in this way because he believes that the argument, if correct, would be fatal to his claim that Unger* has knowledge, and so, he wants to refute that argument. But one cannot refute unproduced arguments any more than one can slay non-existent dragons. One wants to argue: Unger, or Unger*, knows that there are more people in China than in India if and only if

(a) there are more people in China than in India,

(b) Unger, or Unger*, believes that there are more people in China than in India, and

(c) Unger, or Unger*, is epistemically reasonable in believing this.

Now, for Unger* to believe that there are more people in China than there are in India, he need not be an exact duplicate of Unger but merely one which believes as Unger believes. But Unger is epistemically reasonable in believing that the population of China is greater than that of India because his belief is grounded in certain others, e.g., that he has read certain statistics in a certain book and that that book is reliable, and these grounding beliefs are true. Now Unger* must have similar grounding beliefs, and of this we can be sure most easily if we let him be an exact duplicate of Unger, for then he too will believe that he read the book in question and that that book is reliable. Of course, Unger*'s belief that he read that book is false, and so this argument would in fact refute Unger. But Unger argues that a man may have knowledge even though he has no true grounding beliefs, and even though he is not epistemically reasonable in believing what he knows. I shall return to his arguments for these contentions later. For the moment we seem to be left without any adequate answer to our question, why must Unger* be an exact duplicate of Unger. Unger's answer is this: "A man's belief that *p* is an instance of knowledge only if it is not an accident that the man's belief is true."[8] And it is not an accident that Unger*'s belief that there are more people in China than in India is true since Unger* was specifically constructed to have this true belief.

All this seems very puzzling. Let us retrace our steps. Surely, the first and most obvious reason for constructing Unger* to be an exact duplicate of Unger is this: Since Unger* is an exact duplicate of Unger, there is no conceivable test which anyone could apply which would show that Unger* did not know that the population of China exceeds that of India unless that test would also show that Unger's claim to know this is false. E.g., if we discovered that the population of China was smaller than that of India, we should have to withdraw the claim that Unger knows as well as the claim that Unger* knows. If, on questioning Unger, or Unger*, we are told that he concluded that there must be more people in China than in India because he knew that the total land area of China was greater than that of India, we should be inclined to say he did not know it. Again, if either of them vehemently denied, or placed a large bet against the proposition that more people live in China than in India, we would suspect that either of them lied when he wrote: "I confidently type..." But since they are exact duplicates, one of them will do this if and only if the other does. In other words, there must be no way for us to find out that Unger* does not know what Unger knows except by finding out that Unger* is not a normal man but a constructed duplicate man.

Now, having discovered that Unger* is a constructed duplicate we find out that Unger* never read any book, hence his belief that he read population statistics in a reliable book is false. Is this sufficient to show that he does not know? Clearly, by itself it is not. Unger's belief that he read the book may also be false. Actually Unger learned about the populations of China and India by reading a reliable newspaper, or by listening to a reliable informant, or even by reading the facts in so many sources that he is justified in his belief even though no one of his sources is always reliable. Nevertheless, we are inclined to say that Unger knows. I suspect that this is so, (a) because the information concerning the populations of China and India is the sort of thing which someone could know (many people do know it), and (b) Unger, being a normal man, is likely to have come by this information in some way which would entitle him to say that he knows it. Thus Unger is quite right when he claims that a man may have knowledge even though he has no true grounding beliefs. I know many things concerning which I no longer know how I came to know them. Can any of you remember how you came to know that John Adams was the second President of the United States? Yet only some philosophers would deny that you know this.

Nevertheless, and contrary to Unger, one feels disinclined to grant that Unger* knows. That is so because there is no plausible account of how Unger* might have come to know. But here the fact that Unger* is an exact duplicate of Unger comes to Unger's rescue. (Here again, as in the just preceding paragraph, my argument differs from Unger's own argument.) Unger*, we may suppose, does not remain an exact duplicate of Unger. In particular, Unger* may be told what he is and how he came into being, and this story would provide him *inter alia* with true grounding beliefs for his belief that the population of China exceeds that of India although no mention was made of this fact when Unger* was made aware of his true nature. Thus, would it not be exceedingly odd to say: "By learning that he was a deliberately constructed duplicate of a man who knew many things Unger* came to know many things"? That suggests that we told Unger* explicitly everything that Unger knows. But that is precisely what we did not do and did not have to do. Unger* already knew everything that Unger knew, what he learns from our story is merely that Unger knows all these things too.

You see then that I am inclined to agree with Unger that Unger* knows many things. And so I agree with Unger that (U) is not a conceptual truth. But showing that (U) is not a conceptual truth has no bearing on the question whether (P) is true; indeed it has no bearing on the question whether (U) is true either. (U) may very well be, and I confidently affirm that it is, true of all normal men. But that is not very interesting to me. Unlike Unger I do not demand that sound philosophy consist only of conceptual truths for reasons which I shall mention below. The truth of (U) does not interest me because I have shown above that (U) cuts no philosophical ice. The clause, "where it is not necessarily the case that *p*" should be stricken from (U); for even without that phrase (U) is not a conceptual truth, and even without that phrase (U) is true of all normal men. To be sure, there may be no philosophical ice to cut; neither the analytic-synthetic distinction nor the meaningful-meaningless dichotomy are really solid. But that is still not a sufficient reason for finding (U) interesting.

(U) is not interesting; but (P) is interesting, and Unger's argument fails altogether to refute (P). Indeed Unger's insistence that Unger* be a deliberately constructed duplicate rather than one which came into being through some cosmic accident is based on his belief, mentioned above, that

For any sentential value of *p*, a man's belief that *p* is an instance of knowledge only if it is not an accident that the man's belief is true.

What Unger means by this is precisely that the man's belief that *p* must be caused by the event, or fact, in virtue of which *p* is true. To be sure, it does not follow from this that this fact, or some effect of this fact, has to be *experienced* by someone; but Unger's own construction does not challenge this further condition. That is to say, everything Unger says is compatible with the view that if (P) states a necessary condition for Unger's knowing that the population of China exceeds that of India, then (P) also states a necessary condition for Unger*'s knowing this.

I have not, of course, proven (P). I have merely shown that the most recent attempt to refute (P) has failed. (P) asserts, let me remind you,

For any sentential value of *p*, where it is not necessarily the case that *p*, a man may know that *p* only if some man has had, or presently has, some experience *e* which is such that *p* would not be true if no man could have, or could have had, *e*, and *e* is a near or remote cause of the belief that *p* of the man who knows that *p*.

Leaving aside qualms about "necessarily", is this principle true? We were led to its formulation by considering knowledge of historical events, i.e., the sort of event which someone experiences, or which leaves traces (fossils, dead bodies, ruins, etc.) that someone experiences. And for these examples the truth of (P) seems very evident. But what makes one hesitate to embrace (P) wholeheartedly is the fact that it looks too much like yet another "empiricist criterion of meaning", one fears that (P), just like every other such criterion known to man, will be found to be inadequate, that on the basis of (P) one might conclude that a man may know something which no empiricist philosopher would regard as knowable, or not know something which every empiricist philosopher would regard as knowable. (P) might then turn out to be a variant of the second dogma of empiricism castigated by Quine, the domga of reductionism. Or, at the very least, (P) might come to play a role in philosophy analogous to that of the second dogma, i.e. it might tempt one to look for better formulations of (P) instead of looking without prejudice at knowledge, or science, itself.

Let me illustrate these fears very briefly. Consider the following instance of (P):

> A man may know that no particle moves faster than a photon in vacuo only if some man has had or now has some experience which is a remote or near cause of the belief that no particle moves faster than a photon in vacuo of the man who knows that no particle moves faster than a photon in vacuo, and which is such that no man could have that experience only if some particle moved faster than a photon in vacuo.

And now someone might ask, "What is that experience?", and then he might be tempted to look for a statement describing that experience, a statement which would be, somehow, a deductive consequence of "no particle moves faster than a photon in vacuo". And when he fails to find such a statement, and he will fail, he might then try to formulate some more generous condition which would allow additional premisses, and he might try to find ways of limiting these premisses to premisses of a certain kind, and after he had in effect repeated that portion of the recent history of philosophy that dealt with the search for an empiricist criterion of meaningfulness, he might give up in despair. If there is even the slightest danger that this might happen, then I think that in the interest of sound and fruitful philosophy (P) should be allowed to die tonight.

But in fact the man just envisaged did not understand (P) at all. What he should do, if he wanted to find the answer to the question "What is that experience?" is to study physics and the history of physics. And what he would discover would be that a number of physicists and astronomers had a number of experiences which jointly led to a crisis in physics, which crisis led to what Kuhn calls a revolution in physics and a new Kuhnian paradigm[9], and that anyone who accepts that paradigm, accepts and may be said to know that no particle moves faster than a photon in vacuo.

What I have just said is to be taken as a gloss on (P). (P) itself is not important, it is only a slogan. But, if I may paraphrase Wittgenstein, in philosophy there should be no slogans, only glosses. If we understand (P) in the sense of this gloss, then I do not believe that (P) can do any philosophical harm. If we understand it in this way, then it is not, and does not claim to be, a conceptual truth. And yet, contrary to Unger, I am inclined to claim that it is sound philosophy because, as a French philosopher of science has said in *Lire Le Capital*, "Philosophy, then, is nothing other than the knowledge of the history of the sciences. Philosophers ... practice the history of the theories, and *at the same time* the theory of this history ... to engage in Philosophy is to study *in what conditions*, and *on what conditions* scientific problems are posed."[10] Moreover, if one takes it that the force of Quine's rejection of the analytic-synthetic distinction and of his holistic view of knowledge is that there are almost no, and at any rate no interesting, analytic truths, then to restrict sound philosophy to conceptual truths is tantamount to killing philosophy. But I for one do not wish to bury philosophy yet.

II

Let me now turn to the converse of the principle (P), namely, the claim that all knowledge based on experience in the sense of (P) is knowledge of matters of fact. Since to my knowledge no philosopher has openly embraced the analytic *a posteriori*, since even those philosophers whom I shall soon accuse of having done so implicitly would recoil in horror from this specter, I shall assume that all of you agree that the converse of (P) is true. But, as before, what is important is not the principle but the gloss, i.e., what we do with it.

The usual gloss seems to be this: a certain proposition is claimed to be analytic, and it is then confidently asserted that therefore it is not based on experience. It will be no surprise to you that I shall argue, rather, that discovering that a proposition is based on experience should lead one to reject the claim that it is analytic.

But let us look for a moment at what can be said if one accepts the analytic-synthetic dichotomy and draws the line in the traditional empiricist way. Being told that a bachelor is a man who has never been married is an experience which prompts one to entertain the proposition that all bachelors are unmarried; but it is also the type of experience which makes one realize that there is no experience which is such that if no man could have had it, then some bachelor would be married. In fact, of course, one realizes that there is no experience that any one could have which would be either a confirmatory or a disconfirmatory experience. Again, so the argument runs, while it is experience with counting objects which prompts one to entertain the proposition that $2+3=5$, since it is 'analytic', anyone who 'fully understands' the proposition will also see that it is not based on experience. In general, then, while experience prompts one to entertain an analytic proposition, and while seeing that it is true and analytic may be an experience – e.g., the exciting experience of finding a proof – there is no experience, no complex of experiences, on which knowledge of this proposition, or of its analyticity, is based in the sense of (P).

If, then, a philosopher of science holds views from which it follows that knowledge of analytic propositions, or knowledge that a proposition is analytic, or knowledge that a proposition said to be analytic is not analytic, is based on experience, those views are, to say the least, suspect. And if such views are held by an important philosopher of science, it behoves us to ask how he might have been led to these views. I shall now try to show that there are such false views and how they might have arisen.

I alluded above briefly to the attempts by various philosophers to make precise an empiricist criterion of meaning. Of these attempts the most impressive is that given by Carnap in his article 'The Methodological Character of Empirical Concepts'.[11] As you know, in that article Carnap assumes that the language of science can be regarded as a dual language system consisting of an observation language and a theoretical language. The descriptive vocabularies of these languages are said to be disjoint, and the question of meaningfulness is raised with respect to the theoretical terms. Carnap supposes further that in the theoretical language we have theoretical postulates, i.e. the postulates of a scientific theory, and in the total language we have correspondence postulates, i.e. statements containing both theoretical and observation terms. The theoretical and correspondence postulates jointly provide a 'partial interpretation' for the theoretical terms. I shall not bother you with the details. I merely wanted to remind you of Carnap's work so that the following two quotations would be intelligible. Carnap says:

> The definition of meaningfulness must be relative to a theory *T*, because the same term may be meaningful with respect to one theory and meaningless with respect to another.[12]

and later:

> ... the theory *T* which is here presupposed in the examination of the significance of a term, contains only the postulates, that is the fundamental laws of science ...

the class of the terms of L_T admitted as significant ... will generally be changed only when a radical revolution in the system of science is made.[13]

But, of course, radical revolutions do occur; I have argued above, and Carnap would undoubtedly agree, that such revolutions are based on experience. It follows, therefore, that on Carnap's views a term can change its meaning because some men have certain experiences. Presumably such a change of meaning would occassion a change in the class of sentences regarded as analytic. Thus, finally, on the basis of experience in the sense of (P), not merely in the sense of (U) one may come to know a new analytic truth. Moreover, no one could have come to know that analytic truth unless someone had had those experiences, or others which would be similarly related to the scientific revolution in question. All this, of course, assumes the truth of (P); but we have every reason to believe that Carnap accepts (P).

Lest you accuse me of willful and mischievous misconstrual of Carnap's views, let me remind you of his work on analyticity in theoretical languages. Again he supposes that a theory is given by theoretical postulates T and correspondence postulates C. Next, following a suggestion due to Ramsey, we are to replace all theoretical terms in this conjunction by variables of the appropriate types, finally we are to form the existential closure of the resulting open sentence. What we now have is the so-called Ramsey sentence of the theory TC, call it R_{TC}. Now form the conditional $R_{TC} \supset TC$, let me call it the Carnap conditional. This, Carnap tells us, is analytic, and therefore has only analytic consequences. He says[14]:

It can easily be shown that this sentence is factually empty. It tells nothing about the world. All the factual content is in the ... Ramsey sentence R_{TC}. The [Carnap conditional] simply asserts that if the Ramsey sentence is true, we must then understand the theoretical terms in such a way that the entire theory is true. It is a purely analytic sentence, because its semantic truth is based on the meanings intended for the theoretical terms.

Many years ago I suggested that on certain highly plausible assumptions concerning the vocabularies of different scientific theories, it might be impossible to maintain simultaneously the analyticity of all the Carnap conditionals for various theories TC. In particular, suppose that we have two theories TC and $T'C'$ the vocabularies of which overlap and which are such that the conjunction $TCT'C'$ is inconsistent. Then the conjunction of the two supposedly analytic conditionals $R_{TC} \supset TC$ and $R_{T'C'} \supset T'C'$ will imply the denial of the conjunction of the two Ramsey sentences; but this denial is in general not a truth of logic, nor is there any reason to suppose that it would be analytic in some wider sense. Thus, one would have to conclude that at least one of the two so-called analytic conditionals was not (or was no longer?) analytic.

But how might the kind of situation envisaged abstractly arise in a concrete case? Well, TC might be the pre-Copernican geo-centric theory of the solar system, and $T'C'$ might be the Copernican helio-centric theory. Clearly, rejecting the former and substituting the latter was based on experience in the sense of (P), although it was a very complex event with many and varied causes. All I wish to claim here is that after both the pre-Copernican and the Copernican theories existed, one could no longer maintain the analyticity of both the associated Carnap conditionals.

My own inclination is to say at this point that neither Carnap conditional is now, nor ever was, analytic; indeed I wish to say that no Carnap conditional is analytic, since knowledge of a Carnap conditional is not merely prompted by the experiences on which the theory TC is based, but is itself based on those experiences. For, the Carnap conditional of a false theory may be false, and thus to know that the Carnap conditional of a theory is true, and therefore (if we retain as much of Carnap's proposal as we can) analytic is based on knowledge of the theory, and Carnap would agree with me that the latter knowledge is based on experience in the sense of (P).

Obviously, someone could claim that both Carnap conditionals are really and timelessly analytic by maintaining that the alleged self-contradictoriness of $TCT'C'$ is merely apparent, that TC cannot contradict $T'C'$ because, although the same words occur in both theories, these words do not have the same meaning in both theories. This view, which Carnap would surely endorse, is held for example by Kuhn precisely with respect to the example just cited. He says[15]:

... the Copernicans who denied its traditional title 'planet' to the sun were not only learning what 'planet' meant or what the sun was. Instead, they were changing the meaning of 'planet' ...

and later[16]:

Consider, for another example, the men who called Copernicus mad because he proclaimed that the earth moved. They were not either just wrong or quite wrong. Part of what they meant by 'earth' was fixed position.
Their earth, at least, could not be moved. Correspondingly, Copernicus' innovation was not simply to move the earth. Rather, it was a whole new way of regarding the problems of physics and astronomy, one that necessarily changed the meaning of both 'earth' and 'motion'.

This defense amounts to saying that while our acceptance of the Copernican theory was based on experience in the sense of (P), it merely prompted our entertaining the associated conditional; i.e. our knowledge of that analytic truth was based on experience only in the sense of (U). But to say that the terms 'sun', 'earth', 'planet' and 'motion' changed their meaning *seems* to say that by coming to know the Copernican theory we came to know the new meanings of these terms. And since the Carnap conditional of the Copernican theory can be recognized as true only by someone who has come to know these new meanings, knowledge that this conditional is an analytic truth seems after all to be knowledge based on experience in the sense of (P).

But this will not do. I have spoken just now as if, in some strange way, words had meanings as people wear clothes, and that words could change their meanings as people change their clothes. And that view is surely absurd. *We* can change the meaning of a word, and what experience teaches us is that it may be a good idea to do so. When we do so, then, of course, we change the set of analytic truths. It is wrong to regard this series of changes brought about by us as a coming to know something. Changing the meaning of a word is like changing gears in your car; you do it because you have come to know something (that the sun is stationary, or that you are going too slow for third gear) but doing it is not knowing, or coming to know, something.

This line of reasoning is appealing. It is, however, false to the facts in the particular case under consideration. I do not think that what the Copernicans did was *inter alia* to change the meanings of any words. I do not think that the mere fact that a word occurs in a radically new theory is proof that it has been given a new meaning. But although the argument implied in the last paragraph is false to the facts of the Copernican revolution, it is less false then the view that words have meanings and can change their meanings so-to-speak without human help. For, needless to say, people sometimes change the meanings of words. Speakers of English, for example, no longer mean by 'knave' what they used to mean.

But, what in the light of all of this can we finally say about the Carnapian views on analyticity in theoretical languages? If you agree with me all the way, i.e., if you agree that the Copernicans did not change the meanings of the crucial words 'planet', 'earth', 'sun' and 'motion', then my original criticism of Carnap stands unaltered. The Copernican theory is incompatible with the pre-Copernican theory and the truth of a given Carnap conditional can be learned only *a posteriori*. But since it is impossible for an *a posteriori* truth to be analytic, Carnap's view must be false.

If, however, you agree with me only to the extent of saying that words do not change their meanings, we change their meanings, i.e., 'the scientists' change the meanings of words when they discover, or accept, a new theory, then you can maintain that the Carnap conditionals of various theories, even of seemingly incompatible theories, are all analytic. And since Carnap is my one and only example of a philosopher whose views might be said to imply the existence of analytic *a posteriori* truths (Kuhn does not say enough to be definitely charged with this philosophical misdemeanor), you may say that I have failed to establish my case.[17] But that is all to the good. For I am not here to villify one of my favorite teachers but to defend empirical knowledge. And if the Carnapian views are less of a threat to empirical knowledge than I thought they were, then empirical knowledge is that much easier to defend.

Permit me, however, to be wary and resume where I left off, i.e. with the view that my original criticism of Carnap is tenable. What we must then ask is why anyone, and in particular why an empiricist might be tempted to adopt this position, a position which, like Unger's position discussed above, and like the position of the sceptics of yester-year, leads one to hold that much which passes for empirical knowledge is not empirical knowledge. To be sure there are differences between the sceptic and the Carnapian and Unger. The sceptic denies that we have knowledge, the Carnapian denies that it is of matters of fact, and Unger denies that it is based on experience. We have dealt with Unger, we have chosen to ignore the sceptic, we should like to understand the Carnapians.

What might lead one to adopt the view that the same term might have different meanings in different theories? There are, perhaps, two reasons. First, sometimes a scientist deliberately and explicitly changes the meaning of a word ('acid' is an example of this), and the obvious method for discovering this is to read the scientist's work. One is then tempted to think that by reading any scientific treatise one will discover what its author, and the people who agree with him, meant, by any crucial term used in that treatise. So far so good; but from the mere fact that the Copernicans said different things about the sun than the pre-Copernicans we cannot infer that they meant something different by 'sun'. If I maintain that elephants are found only in India, while everyone else holds that they are found both in India and in Africa, you are not entitled to infer that 'elephant' has a meaning in my idiolect different from its meaning in ordinary English. What you should infer is that I am either ill informed or quite mad.

But, of course, the philosophical reason for maintaining that the same word might have different meanings in different scientific theories is quite different. The reason is that this is the view to which one is almost inevitably led if one begins with some modern linguistic version of Hume's views on the origin of ideas. It is also the view to which one is almost inevitably led if one wants to use "all our knowledge of matters of fact is based on experience" as a criterion for rooting out "sophistry and illusion", in short it is a view to which an empiricist philosopher might easily be led. The details of the journey from straight reductionism to the views of Carnap to which I alluded above does not concern us here. What concerns us is that traditionally an empiricist philosopher has owed allegiance to two principles, one a principle concerning meaning the other a principle concerning knowledge. And I should like to persuade you that allegiance to the former principle may have, and in a sense has had, the effect of ignoring the second. For while it does not lead to a denial of the principle (P) it might lead to a denial of its converse. And even if it does not have this drastic effect, it tends to have the effect of limiting the extent to which we are said to have empirical knowledge. Time does not permit me to demonstrate this, let me suggest to you only that operationism, instrumentalism, and Kuhn's rhetorical question[18]

Does it really help to imagine that there is some one full, objective, true account of nature and that the proper measure of scientific achievement is the extent to which it brings us closer to that ultimate goal?

may all be regarded as in some way saying that we know less than we seem to know.

If, on the other hand, one accepts Quine's view that reductionism is an unempirical dogma of empiricism, and follows his admonition to eschew talk of meaning, at least talk of meaning in the traditional way, one is then free to accept both the empiricist principle concerning knowledge and its converse. And if one does that, one will discover once again the importance of the history of science for philosophy.

The correct gloss on "all knowledge based on experience is knowledge of matters of fact" is *not*: find out what knowledge is not knowledge of matters of fact and then deny that it is based on experience. The correct gloss is rather: find out what knowledge is based on experience, for all such knowledge is knowledge of matters of fact. But in order to find out what knowledge is based on experience one must once again study knowledge, i.e. science, and its origins, i.e. the history of science. And if one does this, then one will discover, as has been shown in detail in several papers by Hilary Putnam, that almost everything we know is knowledge of matters of fact based on experience, that is to say, empirical knowledge.

Thus I conclude that nothing happened to empirical knowledge. Mankind has much empiciral knowledge, and our task as philosophers is both to share it and to add to it.

Wellesley College

NOTES

[1] David Hume, *An Enquiry Concerning Human Understanding*, section IV, part I.

[2] Peter Unger 'Experience and Factual Knowledge', *Journal of Philosophy* **64** (1967) 152–173.

[3] *Ibid.*, 153.

[4] Hume, *op. cit.*, section V, part I.

[5] *Ibid.*, section IV, part I.

[5a] Since the paper was read, it has been pointed out to me that the reading (P) is too strong and that the following weaker reading is in fact compatible with the gloss on p. 140:

(P′) For any sentential value of *p*, where it is not necessarily the case that *p*, a man may know that *p* only if some man has had or presently has some experience on which that knowledge is based, i.e. an experience which belongs to every class C such that no man could know that *p* unless some experience belonging to C is a near remote cause of his belief that *p*.

[6] *Ibid.*, last sentence.

[7] Unger, *op. cit.*, 157.
[8] *Ibid.*, 172.
[9] Thomas S. Kuhn, *The Structure of Scientific Revolutions*, Chicago 1962, passim.
[10] Louis Althusser *et al.*, *Lire Le Capital*, Paris 1965, vol. I, p. 216.
[11] Rudolf Carnap, 'The Methodological Character of Empirical Concepts', in *Minnesota Studies in the Philosophy of Science* (ed. by Herbert Feigl and Michael Scrive), vol. I, Minneapolis 1956, pp. 38–76.
[12] *Ibid.*, p. 48.
[13] *Ibid.*, p. 51.
[14] Rudolf Carnap, *Philosophical Foundations of Physics*, New York 1966, p. 270.
[15] Kuhn, *op. cit.*, p. 127.
[16] *Ibid.*, pp. 148–149.
[17] To say that does not, of course, commit one to acceptance of Carnap's views on analyticity. These views can be challenged on grounds quite different from those presented here.
[18] Kuhn, *op. cit.*, p. 170.

JOHN COMPTON

COMMENTS ON 'ON EMPIRICAL KNOWLEDGE'

I admire Professor Putnam's paper very much. It is a subtle and intricate argument which rests, somewhat in the fashion of the exposed portion of an iceberg, upon the enormous history of empiricist analysis. And it ramifies in many directions beyond itself. The result, however, is that I have had some trouble in finding its own proper thesis and in finding the natural lines of its own special argument. I hope I do not exert too heavy a hand in my effort to get before us for discussion what I find to be its central concerns.

The two parts of the paper are intended to do distinct but complementary jobs. The first – in interaction with Peter Unger – formulates and glosses an improved empiricist principle of knowledge. The second – in interaction with Carnap and Kuhn – aims to show that what has been alleged to be analytical or definitional in scientific knowledge is empirical and, therefore, genuinely synthetic knowledge after all. I am really uncertain how to take these two claims and what I will do in my comments is simply to put these uncertainties before you in the form of queries.

In regard to the first section of the paper, the temptation is almost overwhelming to indulge in what Professor Putnam most urgently wishes us to avoid, namely a rehearsal of the whole tiresome dialectic of the empiricist meaning criterion. Her description of a necessary condition for 'someone's knowing that *p*' is certainly more interesting, because more specific, than Unger's is but, because the terms are not explicated in detail, it is extremely difficult to evaluate it. It may be true that all knowledge of matters of fact is knowledge by experience in the sense that a man may 'know that *p*' only if he has had or presently has an experience *e* which is both a cause of his belief that *p*, and which is one he genuinely could have if *p* is true. But one wants to know what sorts of experiences *e* will be admissible, in what sense of 'could have' it must be true that a man could have them, and in what sense the experience is to be said to be the cause of the belief that *p*. My impression is that this statement of the principle is a most generous one if, for example, Copernicus can be assumed to be one of those who has come to know some *p*, namely his theory, and if this knowledge is knowledge by experience in a way which satisfies the Putnamian conditions, as she later says it does. Experience will not then be restricted in any Humean way to the impressions of any sense, but will include experiences of aesthetic, geometric, and metaphysical pattern among the facts. I quite enthusiastically welcome this move, but it would be helpful in an essay on empirical knowledge if we were to learn rather more clearly what empirical knowledge is to be taken to be, than we have from the statement of the principle alone. But I want to be true to Putnam's intent and not become sidetracked by such issues of detail.

The question is, however, if we are not to take the stated principle as a *dogma* to be refined and refuted in the manner of the recent past, how then *are* we to take it? This is really the first issue I wish to raise. She does not believe that it represents a *conceptual* truth that a man may know *p* only if the stated conditions are satisfied; she seems to accept Unger's proof of this. Parenthetically, for myself, I would not accept his ingenious proof, on the grounds that to say that a man knows that *p* is to say something of a person, and to say something of a person is to say something of a being with a history, and therefore that some account of how he came to know *is* requisite, as Putnam at one point suggests it might be. The result is that Unger*, having had no personal history, could not be said to know that *p*, even if he were in all other respects identical to Unger. For the sake of argument, though, let us say that this empiricist principle is not a conceptual truth. Putnam goes on to acknowledge that she has not proved this principle; I do not think she has even refuted an attempt to disprove it by showing that Unger's argument fails against it, because, as she points out, his argument is simply irrelevant to whether or not the principle is true *per se*. The only evidence given for its acceptance, so far as I can determine, is that we are led to it from a consideration of and an extrapolation from an example of historical knowledge; to wit, our knowing that the battle of Lexington took place. Perhaps, then, the principle should be construed as an *empirical generalization*, but then it is a very weak one because it is stated in much more general terms than the data warrant. Perhaps, since it arises from an attempt to analyze the notion of knowing, we should call it an analytic statement which is yet based upon a consideration of empirical examples: would it be too perverse to call it an analytic *a posteriori* claim?

The way the principle functions in Professor Putnam's argument is likely to be a better clue; she wishes us to use it as a heuristic, as a *regulative principle* of inquiry, and it is in this light that its rootage in historical examples becomes important. For what we should do if we take this principle seriously, she says, is to study the history of the sciences, and she quotes with favor the statement that "Philosophy, then, is nothing other than the knowledge of the history of the sciences". This is a fascinating and extremely important contention, the truth and meaning of which should be made out in detail, although I really do not see how it could be said to follow from the principle as such. There is considerable discussion in part II of Putnam's paper that bears on this new claim, and I want to turn to it as a central concern in what follows. The sum of my objection at this point, however, is simply that we need not have gone through all the trouble of reformulating an empiricist principle of knowledge in order to yield this interesting recommendation when, in effect, the recommendation could have been offered by itself with as much justification.

In part II of her paper, Professor Putnam wishes to challenge what she thinks is a prevalent view to the effect that we know less in science than we seem to know, a view which she believes to arise because of a neglect of the converse of the empiricist principle of knowledge, namely that all knowledge by experience is knowledge of matters of fact. She wishes to argue, as against Carnap and Kuhn, that scientific knowledge contains far fewer analytical or definitional components than they suspect. The problem behind this dispute – as I see it, at least – is how science can count as genuine knowledge, if large parts of it are taken, as by operationists, instrumentalists, and some empiricists, to be generated from initially adopted, implicit or explicit, definitions of basic terms. Even if empirically tested to a suitable level of corroboration, the theory may be construed simply as a conceptually convenient symbolism or algorithm for arranging and predicting phenomena. On this view, a modification in theory is too easily conceived as a mere change, or in radical cases an *ex*change, of conceptual structures, a change or exchange in the meanings of words. Conflicting theories, then, would not be said to be logically incompatible, but rather simply to be different from one another; the notion of clean, rational confirmation of one and the disconfirmation of the others in a group of incompatible theories, then gives way to the idea of a more pragmatic evaluation of competing alternatives; the accumulative character of scientific knowledge tends to be replaced by a model which emphasizes conceptual revolution; and the objective aim of scientific inquiry, namely to articulate the structures of natural events, comes to be replaced by a conception of science which sees its theoretical constructs in the context of shifting world-views and, as a quite logical extension, productive of a changing physical reality. Carnap appears to give some comfort to this point of view in his doctrine of analyticity in theoretical languages and Kuhn explicitly advocates it as an outcome of an historical analysis of science.

Now there *is* a crucial problem posed by these views – largely, it is important to note, because they are so nearly true. It is the deepest problem I see in the philosophy of science today, because it involves rectifying the self-understanding of science in such a way as to give its products genuine intelligibility, that is philosophical or ontological value, in the face not only of the philosophical currents to which Putnam refers but of the persistent trivializing of their own work which characterizes the professional and technical attitudes of scientists themselves. I very much agree that in a confusing situation such as this we need to re-examine our basic concepts such as those of 'empirical knowledge' and 'matters of fact', but I question whether redrawing the lines between the synthetic and analytic, or the *a priori* and *a posteriori* is of any use in this enterprise. The problem *does* seem to me to be that which Professor Putnam elected to avoid, namely scepticism or, more precisely, phenomenalism and historicism. What is needed is a re-examination of the whole idea of experience in order to take it quite beyond its Humean roots. We need to understand the forms of experience and their genetic constitution through our interactions with the world, through our language, and through our science. This can hardly be done in anything short of a fresh systematic effort; it requires more than re-allocating territory to the empirical which had at one point been called analytic.

Because I think this is so and because I think this is the fundamental problem which is at stake in Professor Putnam's discussion, even though she spends more time with Carnap and the analytic *a posteriori*, I will spend more time on Kuhn and the history of science. I do not think Putnam has faced Kuhn head on. In fact it seems to me that her implicit criticism of his view has the ring of paradox to it.

For consider what happens if we follow the recommendation of Part I of her paper. There we are urged to construe the empiricist principle of knowledge as an injunction to ask just what experiences, in what historical context, posed what problems for physics and what ones, in what context, led to the resolution of those problems. It would be the task of philosophy of science to pursue this historical inquiry. The paradox is that it is just such historical inquiry which has led Kuhn to affirm that some major changes in scientific thinking such as that from the pre-Copernican to the Copernican astronomy are well described as including shifts in the meanings of fundamental terms and that, therefore, we must recognize incommensurate cognitive matrices at the heart of at least some scientific controversies. Yet it is just such a conclusion, which her own injunction to do history renders plausible, which Professor Putnam wishes to reject at all costs!

The controversial doctrine is that the same term may have different meanings in different theories. Putnam can think of only two reasons why this view might be held: first, a hasty inference from a few cases in which scientists explicitly and deliberately change the meanings of words, as in the use of the word 'acid', to the conclusion that when, for example, the Copernicans said different things about the 'sun' from what the pre-Copernicans said, that they meant something different by the word 'sun'. And second, a philosophical reason, deriving from the empiricist principle that knowledge is based on experience, when this principle is used in an overzealous way to root out 'sophistry and illusion'. But there is another reason, an empirical one, for coming to this doctrine of meaning-shift. This is the reason why an historian would propose it as an hypothesis to account for certain characteristics of scientific change. And it seems to me that this empirical reason must be taken seriously.

The meaning-shift doctrine is not introduced just because certain different things are said about the sun by the Copernicans; there are perfectly respectable disagreements within Kuhnian paradigms which betoken no differences of meanings but rather require shared ones – for example differences over the position of Venus, the date of an eclipse, or spots on the sun. But the meaning-shift view is introduced in order to account for an irresolvable or enduring disagreement, or to describe a break, such as Kepler's, with a long tradition. The reasonability of this kind of hypothesis is precisely indicated by the *difference* between these historical situations and the example envisioned by Professor Putnam to the effect that, when we differ about where elephants are to be found, what one should infer is that one of us is either misinformed or mad. This is just the conclusion which an historian must not tolerate; neither the Copernicans nor the pre-Copernicans were simply ignorant or mad; to say either of those things is to put an end to the attempt to understand the dimensions of that revolution. They dwelt in different 'worlds'; they regarded all the phenomena with which they dealt in different ways – at least for scientific purposes; so much so that, as Kuhn points out, a variety of rhetorical devices and metaphysical and aesthetic arguments were of use simply to get those dwellers in the pre-Copernican world to 'see' what the newer world was like.

Now in the face of this sort of historical argument, Professor Putnam says that she does not think that the Copernicans did change the meanings of any words. The only reason I can find for this denial is not an empirical one based on a more faithful account of the historical events, but a philosophical objection to the effect that the notion of a meaning-shift is misleading because it risks the monstrosity of an analytic *a posteriori*. Thus, in order to protect the sanctity of certain categories, we should deny that certain controversies in science involve different meanings given to the same terms or that we can come to know anything when we change the meanings we give to some of our basic terms! The last thing I wish to do is to defend the view that there *are* any truths which are analytic in some strict sense and at the same time *a posteriori* in some strict sense; it does sound monstrous; but whether there are or are not such truths is not the point. I do not even know what these *terms* mean anymore. Nor do I know anyone else who does. And I do find it extraordinary that someone following Quine's program should be brought to the sort of conclusion Putnam is brought to. All that I want to urge is that if we come to the point of rejecting reasonable empirical hypotheses in order to protect our logical categories, then we had better look again at our categories.

Of course, it *is* a mistake to press a comparison between the shift of the meanings of 'earth', 'sun', 'planet', and the like in the Copernican revolution and a deliberate choice to change the meanings of words. The meaning-shift about which Kuhn speaks is largely independent of personal choice; it is bound up with a change in perceptual Gestalt and in the whole conceptual matrix within which a scientist works; it often comes about gradually, however momentarily it may be recognized; and it is only secondarily expressed in words. The scientist does not change the meanings of these words so much as he acknowledges in his words the changes in the meaning of his world. Surely, it is not absurd to regard this kind of meaning-change as a coming to know something. And the something he comes to know is the world of that theory and, through it, the world (period). No doubt we need better language to describe this process, but I see no reason for not trying to develop it. It might be better to call these shifts in meaning changes in very broad empirical descriptions of things. But this is misleading too, for these changes are not simply changes in descriptions or explanations of fact in accord with some commonly accepted, neutral language. They are more fundamental than this and seem to infect the whole fabric of experimenting and theorizing, the envisionment of problems and the suggestion of solutions. Such a fundamental shift merits a title something like a shift in our language. At any rate, the phenomenon must be saved, however we may domesticate it within our logical categories.

The problem I see as fundamental in this entire discussion can only be raised once the essential truth of this historical perspective is accepted. It is, once again, the problem of scepticism, of the relation of the world of tested theory to the world of everyday and to the world (period), of the relation of historical reason to reality, the problem of rationality in its ultimate sense. I devoutly wish I could find some light on this sort of question from Professor Putnam's point of view.

My only conclusion will be an attempt to summarize the questions which occur to me after a reading of Professor Putnam's paper. She has beautifully combined niceness of analysis with largeness of purpose in a most constructive way. But constructive of *what*?, that is the rub. What is our revised empiricist principle of knowledge to do for us? What kind of a principle is it and, more particularly, how do we get from it to the conclusion that history of science is the proper task of philosophy of science? And 'history of science' in what sense – in some sense that will do without the hypothesis of shifts in meaning from one theoretical world to another? And what sort of history would this be? I wonder, finally, whether, in the light of these many questions, it can reasonably be maintained that history of science alone can exhaust the province of philosophy of science – after all, to assert that it is, and to justify that assertion, and to explain its meaning, are all themselves philosophical questions in the old, traditional sense, and not historical ones. The best we could say, I suppose, is that philosophy of science should become philosophy of the history of science, and I think there is very much to be said for *that* position indeed.

Vanderbilt University

WILLIAM RUDDICK

CAUSAL CONNECTION

Philosophers often attend to our common judgments, however casual or erroneous. The language we commonly use and the errors we commonly make are supposed to clarify the principles of sound judgment. That supposition has, I think, been justified by the results of attending closely to our common judgments in matters of perception, morals and logic. But curiously, philosophers have tended to neglect our common causal judgments. They have been causal snobs preoccupied with 'the right causes', those endorsed by the scientific élite. There is concern only with long-established causes such as billiard balls, freezing water, flipped switches, and indigestible dinners. The popular causes of the common man – Fate, charms, conspiracies, and various divinities – are beneath consideration and contempt. Even causes once favored by scientists themselves are neglected: the planet Vulcan attracts no more philosophic attention than its ancient namesake.

This penchant for the scientist's current causes has been, I think, both a cause and an effect of unduly restrictive analyses of causation, causal error, and causal explanation. If the only phenomena considered are those subject to laboratory isolation and variation, it becomes tempting to identify the concept of cause with rules for finding *true* causes, causal error with misuse of those rules, and causal explanation with scientific laws and conditions that permit accurate prediction. Conversely, these identifications once made will restrict attention to the scientist's causes and explanations. Who but a scientist knows the rules for discovering "the whole of the contingencies of every description, which being realized, the consequent invariably follows", or even "an *insufficient* but *necessary* part of a condition which is itself *unnecessary* but *sufficient* for the result"?[1] If that is what it takes to be *the*, or even *a* cause of an event, who but physicists and chemists know of causes? By the standards set by such definitions, our daily causal judgments are bound to seem crude, hasty and capricious, and our causal errors are bound to seem naive and easily correctible. And if causal explanation is, or requires deduction from laws and conditions, our ordinary explanations will be as rough and sketchy as Professor Hempel would have us believe. Given these definitions and analyses, philosophers are bound to think common causal judgments a waste of time.

I am not, however, inclined to accept these identifications, and I do not find common causal judgments a waste of time. In the hope of breaking the spell of the scientist's causes and the restrictions of the positivist doctrine of causation, I wish to examine some of our common causal errors and common causal statements. This examination will reveal a special relation between some causes and their effects, that of *congruity*. In forming causal hypotheses, in choosing which causal hypothesis to test, and in retaining causal hypotheses even against the evidence, we tend to be guided by considerations of congruity. And, likewise, congruity with a given effect is often the basis for our choice of one from among several equally necessary causal factors as *the* cause of the effect we want explained. Such choices reflect the fact that congruity is a main source of explanatory power.

I do not wish to claim that congruity is either a necessary or sufficient condition for an event to explain another. Either claim would require arguments which, in its present state of definition, this notion of congruity cannot provide. Accordingly, I do not expect to worry or satisfy adherents of the positivist account of causation and causal explanation. But I hope at least to suggest that the positivist attempt to give a formal, syntactic analysis of causal explanation ignores the fact that our descriptive language contains explanatory connections. My talk of congruity is a way of attracting attention to this fact, for congruity is a *semantic*, or *intensional* connection. A cause and an effect are congruous if they are describable by terms that are congruous, and congruous terms are semantically connected in ways I shall illustrate even if not completely characterize.

A catalogue of common causal misjudgments will help to introduce and justify this notion of congruity.

I. MISJUDGED CAUSES

We are inclined to think some putative causes are better than they prove to be, and others worse. Of those we overvalue, there are the *Fictitious Cause*, the *Spurious Cause*, and the *Vacuous Cause*. Of those we undervalue, there are the *Paradoxical Cause*, the *Incommensurable Cause*, the *Chance Cause*, and the *Heterogeneous Cause*.

The *Fictitious Cause* is the biggest and most hopeless liar. It is a double deceiver, being an alleged object (event, property, or process) which is (or was) falsely believed to exist *and* falsely believed to have certain effects. Such totally false causes abound in folk and children's tales, as well as in the histories of both genuine and pseudo-sciences. Supernatural decrees, ghosts, both Vulcans (god and planet), entelechies, and the *horror vacui* are examples. They are hopeless causes, for eventual rejection as frauds is their only fate.

A *Spurious Cause* tells fewer lies. While fictitious causes lie about both their existence and their efficacy, spurious causes lie only about their efficacy: they exist, but do not do what they are taken to do. Perhaps for this reason, spurious causes command a wider audience, even among the practical-minded. Being perceptible, they demand less faith than do fictitious causes, and can fool even taxi-drivers, historians, and doctors (whether amateur or professional). I have in mind such spurious causes as travellers' charms, 'the great men' of history, purgings and vitamins. Like fictitious causes, they face eventual exposure.

The *Vacuous Cause* that Moliere taught us to ridicule has, by contrast, more hope. To begin with, vacuous causes are not totally empty. When in *Le Malade imaginaire*, the mock candidate for the Faculty of Medicine tells his mock examiners that opium makes one sleep because there is in it a soporific quality whose nature it is to dull the senses, he is saying more than is usually allowed. By assigning opium a quality or power, he has committed it to having a *general* effect, that is, an effect on most men in most circumstances. And by locating that quality in opium, he has excluded the possibilities that opium is a mere *placebo* dependent for its action on the false beliefs of patients, or that opium is a mere pharmaceutical base or incidental medium. Moreover, unless sleep is defined as a dulling of the senses, the candidate has even suggested a way in which opium exercises its *virtus dormitiva*, namely by dulling the senses.

His future patients would not be wrong in thinking that Argan knew something.[2] But, no doubt, they would take him to know more than what was implied by the mere naming and locating of the cause. Uncritically, they might suppose him to know details of structure and physiological action that were not known for centuries.[3] The deceit of a (near-)vacuous causal diagnosis lies not in its seeming to say *something*, but in its seeming to say *more* than current knowledge entitles it to say or imply. Some near-vacuous causes, however, live to see their pretensions made good by subsequent discoveries. Accordingly, of the three overvalued causes, vacuous causes are least deceitful and most hopeful.

But why, we should wonder, are we prey to any of these deceits? Perhaps the vices most often blamed for causal error – hasty generalization and uncured anthropomorphism – are at fault here. Logic texts say that superstitions are often the result of the jump from *post hoc* to *propter hoc*, from chance conjunction to lawful connection, and perhaps beliefs in *spurious* causes have such an origin. Belief in *fictitious* and *vacuous* causes, on the other hand, might seem to derive from the desire for humanoid agents throughout nature. According to various students of our past, such a desire appears in our earliest selves – the primitive tribesman, the child and the Greek philosopher. And according to philosophers, our concept of power or force – deriving does from our experience of moving limbs and chairs – shows that we still seek to satisfy this desire. Are not fate, ghosts and entelechies but further satisfactions?

Perhaps. But haste and self-projection alone cannot account for many of these misbeliefs. Superstition conjoins certain events which have probably never been observed together, and hence have never given hasty generalizers a chance. (Have seven years of bad luck ever dogged a mirror-breaker? Has a child, having stepped on a sidewalk crack, ever found his mother at home with a newly broken back?) Conversely, most chance conjunctions give rise to no false generalizations. Superstition seems choosey.

Since Frazer, anthropologists have noted a pattern among primitive causal practices. Magicians tend to be homeopaths, following a principle of 'sympathy' (or 'direct symbolism', or 'similarity'). A cause must resemble its effect. Thus, to induce rain, the magician scatters water from a tree, or carries an undiapered infant through the fields. To insure fertile fields, pregnant women are set to sowing at full moon or high tide. To cure slow speech, a diet of lizard tongues is prescribed. If chance conjunctions had any part in the origin of these practices, it was the resemblance of the conjuncts, not their conjunction that invited causal inferences.

Moreover, many causal misbeliefs are too tenacious to arise from chance conjunction and hasty generalization. Were this their origin, they should be short-lived. The first failure of the chance conjunction would expose them. But as we learn from missionaries, anthropologists and our own vain attempts at persuasion, causal misbeliefs can withstand not just one, but constant failure of co-occurrence.

The other common diagnosis of causal error, anthropomorphism, is less easily dismissed. Theorists differ as to which of our various aspects we are most anxious to find among natural phenomena. According to most philosophers, we tend to project ourselves as *physical* agents: choosing to move our limbs or our chairs and doing so is the paradigm for natural causation. According to other theorists, however, our paradigm is *social* action. According to Kelsen, the Greeks took what we now call the natural order to be a moral order; for them, an effect was a retributive punishment which a cause, as a disturbance of a just balance, inescapably called down upon itself.[4] According to Piaget, a child attributes life and consciousness to natural objects so that they may be capable of responding to directions and commands, especially those issued by the child himself.[5] In this, children resemble Lévy-Bruhl's primitives: cooperation, or joint participation is the aspect of human life projected into nature.[6]

Let us suppose (contrary to subsequent critics) that these claims are well-supported, and let us suppose further that we are still Greek, child-like, and primitive in our causal desires. Nonetheless, before anthropomorphism can be an adequate explanation of causal mistakes, we must investigate the appeal of these paradigms of physical and social actions. Why is the alleged desire to have nature be like us – retributive, responsive to command and cooperative – strong enough to make men blind to evidence of causal mistake? I shall suggest an answer, but not until a very different kind of mistake is catalogued.

We not only overvalue alleged causes; we undervalue them. That is, we not only form and retain causal hypotheses independently of evidence; we also resist causal hypotheses independently of evidence. We hesitate to allow as causes what subsequent research proves to be a cause, or at least a causal factor. Upon hearing of such alleged causes, we react, 'But how could *that* be the cause?' This is no request for details of production, but an incredulous challenge. Nonetheless, certain details of production can on occasion remove our resistance.

The most repugnant of such undervalued causes is the *Paradoxical Cause*. Our repugnance and suspicion, even when there is statistical support for such a cause, are understandable, for a paradoxical cause is one whose description is logically incompatible with that of its alleged effect. For example, several years ago it was reported that highly excitable school-children had been tranquilized by a certain stimulant. 'But how', people reacted, 'could a stimulant tranquilize anyone, let alone a school-child?'

I do not know how such resistance was in fact overcome, but I know how it could be. Suppose the children's hyperexcitability proved to be due to a *hypo*active inhibitory center in the midbrain, and suppose further that the paradoxical drug was shown to stimulate that center. The clash between stimulant and tranquilization would be resolved by this third, mediating factor. There is nothing paradoxical about a stimulant stimulating an inhibitory center, or about an inhibitory center reducing activity. The initial paradox has been dissolved by finding a suitable 'immediate' cause of the effect and thereby reducing the initially paradoxical cause to the status of a 'remote' cause.

The case is not unique. Scientists often find puzzling correlations, and often know in advance what would remove the puzzlement. Indeed, such knowledge guides the formation and choice of hypotheses to be tested. The initial puzzlement and repugnance need not, of course, arise from an apparent paradox. An apparent *Incommensurable Cause* can be almost as distressing. Admittedly, we talk easily of the straw that broke the camel's back, but we do not mean what we say. By 'the straw', we mean the last straw, or the last of many straws and thereby refer to what we all take to be the cause, namely, a load (or overload) of straw. Should anyone insist that he meant what he said, that the single straw was the cause of the camel's collapse, we would object, 'But how could a *single straw* break the back of a beast of burden?' We would allow, of course, that a single straw, or rather a small number of straws might break the back of a camel suffering from osteoporesis, or some other decalcifying disease. But then there is nothing incommensurable about a small load and the broken, fragile back of a sick camel no longer fit to be a beast of burden.

We tend to hold with the Scholastic principle of commensurability of cause and effect, a principle with several historical roots. Like tribal magicians, Aristotle held that "agent and patient must be 'like' [i.e., identical] in kind", and such similarity, of course, would include magnitude.[7] It is doubtful Aristotle was still laboring under the juridical view of things Kelsen ascribes to earlier Greeks. But there are other obvious charms to this principle. It allows one to think of causal influence in a literal fashion, as a flowing of cause into effect. And it favors theistic accounts of creation. If all causation is *literally* causal influence, then a cause must be at least as great as its effects, for a donor cannot give what it does not first possess. And what is as 'great' as the universe if not God?

Whatever our reasons, we do tend to demand that an incommensurable causal diagnosis be revised, either by enlarging the cause or diminishing the effect. Such causal enlargements reduce the initially alleged cause to the status of causal factor, or of part-cause; while reductions of effect qualify the original alleged effect. In either case, the initial diagnosis is replaced, but not totally rejected.

Some instances of a third repugnant cause, the *Chance Cause*, resemble incommensurable causes. Indeed, on Poincaré's analysis we say that an event is due to chance in cases in which "a very small cause which escapes our notice determines a considerable effect that we cannot fail to see".[8] For example, whether I win or lose at roulette is due to chance, because an effect which is for me 'of the highest importance' is decided by a 'variation of a thousandth or a two-thousandth in the impulse' which sets the wheel turning.

Not all chance events, however, have incommensurably small causes. On Aristotle's account, an event is said to be due to chance when it has an 'incidental' cause.[9] For example, a man goes for a certain purpose to a place he rarely visits, and while there accomplishes something totally unrelated to his intended mission. His going there is the chance cause of his, say, selling a ticket to a feast, for 'it was only incidentally that he got his money by going there'. Chance, for Aristotle, is "an incidental cause in the sphere of those actions for the sake of something which involve purpose".

Being 'incidental' to its effect is a misfortune that any cause can suffer during temporary misdescription. Most causes, however, can be redescribed in a way that makes them 'proper' to their effects, that is, makes them proper answers to specific causal questions. But causes of effects due to chance cannot, Aristotle thought, be so redeemed. They are forever incidental to their effects. The gap between, say, going-in-order-to-do-*X* and doing-*Y* is irreducible.

Likewise, there is a gap between cause and effect in Poincaré's roulette case. The initial increment of torque that makes the difference between winning and losing differs from the effect not only in magnitude, but in category. It is a physicist's event, while the effect is a gambler's event. We can remove these disparities of size and category if we redescribe the effect in physicists' terms. Rather than the psychologically loaded phrase 'my winning number', let us use 'small spatial interval *R*' for describing the wheel's final resting place. There is nothing 'considerable' about the difference between small interval *R* and its neighbor, interval *B*. Hence there is no longer a disparity in size between a small increment of torque and the small increment of motion from *B* to *R*. And, of course, torque and motion are at home together.

If, however, we retain the gambler's description of the outcome, there may be no redescription or supplementation of the initially cited cause which will remove the disparity between cause and effect. For example, beefing up the decisive increment of torque with the rest of the torque and other forces will not work. Gambling outcomes and physical forces, of whatever magnitude, remain incommensurable in category.

In this respect, the torque which determines a chance success at roulette is also an instance of a fourth kind of repugnant cause, the *Heterogeneous Cause*. Just as philosophers have argued that chance causation is impossible, so they have claimed that psycho-physical interaction and divine interference – paradigms of heterogeneous causation – are impossible. How could something *immaterial* have material results? An idea might produce another idea, but how could an idea produce neural stimulation?

The disparity between mind and body, or between God and the world, as they are usually conceived, has inspired materialism and idealism. To reduce mind to body or body to mind is to remove an obstacle to their interaction. Just as the roulette case could be redeemed by 'reducing' the gambler's winning number to the physical space on which his stake happened to rest, so psycho-physical interaction can be made acceptable by 'reducing' desires to neuro-muscular events. Just as torque and motion are homogeneous, so too are contractions and movements of approach or avoidance. Or, alternatively, one may try to identify limbs and goal-objects with alleged mental surrogates. There is nothing incongruous about desires producing appropriate images, or, more generally, with ideas causing ideas (within limits set by, say, traditional laws of association).[10] Similar reductive remedies of pantheism and idealism help overcome resistance to the possibility of divine interaction with the world.

The four causes we undervalue – Paradoxical, Incommensurable, Chance, and Heterogeneous – and their remedies are obviously alike, indeed sufficiently alike to count them as species of a single kind, the *Incongruous Cause*. All four lack a certain property which their various remedies supply, thereby removing their incongruity and our initial resistance. Let us call that property, *congruity*. It is only when a causal diagnosis provides a cause *congruous* with the effect in question that we are likely to give any *supporting* evidence its full weight.

And, I suggest, that it is because the causes we overvalue – Fictitious, Spurious, and Vacuous – have this property of congruity that we are inclined to dismiss *subverting* evidence. They are *Congruous Causes* par excellence, and as such are allowed lower standards of evidence, or none at all.

But why should congruity and incongruity have such influence? The answer, I think, is this. We overvalue congruous causes because, although there may be little or no evidence that they produce certain effects, they do explain those effects. And we undervalue incongruous causes because, although there may be evidence that they produce certain effects, they do not explain those effects.

This answer will be plausible only after congruity has been characterized (in Section II) and shown to be a source of explanatory power (in Section III).

II. CONGRUITY

From our overvalued causes and the remedies for undervalued causes, let us compile a partial table of congruous causes and effects:

Cause	*Effect*
command: Do *X*	doing *X*
desire for *X*	getting *X*
doing *X*-like action	occurrence of *X*
exercise of *X*-ive power, or force	production of *X*
taking a stimulant	increased activity
firing of inhibitory center	reduced activity
overloading	break (or breakdown)
application of torque	motion
muscle-stimulation	movement

There are two features of these pairs of terms worthy of comment:

(1) one term includes the other, in its meaning, referred to either as a normal cause or a normal effect;

(2) certain conjunctions formed with these pairs imply causal statements.

For example, take 'overload' and 'break'. Neither term includes the other explicitly, but an ample dictionary might well contain the following entry:

> **overload**, *n.* A burden greater than something can long bear under normal conditions without breaking, malfunction or failure of function.

The entry for 'break', however, would not mention overloading: the terms are *not* mutually definable.

Moreover, this pair of terms in conjunction has causal implications. Consider the following conjunctions:

(1) The wire was overpriced and broke.
(2) Something was placed on the wire and the wire broke.
(3) The wire was overloaded and broke.

Were any of these three asserted, we would suppose that the first event mentioned bore, or was believed to bear some relevance to the second. For the third conjunction, that relevance is clearly causal; for the first, it is not causal; and for the second, it may be causal. That is, assertion of

(p) The wire was overloaded and broke.

implies

(q) The overload caused the wire to break.

and perhaps

(q′) The overload snapped the wire.

I mention (q′), because a pair of cause and effect terms of the sort in question (i.e., 'overload' and 'break') often have one or more associated transitive verbs (here, 'snap'). Depending on the objects, or events in question, the conjunction (upon assertion) may either imply (as here) or be implied by a causal judgment containing one of these associated verbs. But these associations are of less importance than the nature of the implication of (q) or (q′) by (p).

The implication of causation is not as strong as logical entailment. If (p) entails (q), (p) must be false if (q) is false. But it may be false that an overload caused the wire to break and yet true that the wire was overloaded and broke. If I say 'The wire was overloaded and broke, but the overload was not the cause', what I report may call for an explanation and perhaps I must have one in hand if I am justified in saying what I said. But whatever my remark invites or needs for justification, it makes perfectly clear sense.

H. P. Grice has introduced a notion of implication weaker than entailment which is useful here.[11] It is not the conjunction which implies a causal statement, but rather the *assertion* of the conjunction which implies (or 'implicates') a causal statement. As we have already seen, this implication (or 'implicature') is subject to 'cancellation'. That is, the speaker can explicitly cancel the causal implication of his assertion of the conjunction without thereby cancelling, or withdrawing his assertion. Were the conjunction itself, rather than its assertion, the vehicle of implication, such cancellation would not be possible.

These implications of causal connection also have what Grice calls a high degree of 'nondetachability'. That is, most other ways of saying the same thing carry the same causal implication. For example, the causal implication of asserting 'The wire was overloaded and broke' is not detached by asserting instead, 'The wire was overloaded, then broke', or, 'Overloaded, the wire broke', or 'The wire broke: it was overloaded'.

This last sentence is not to be confused with 'The wire broke and was overloaded'. Although the sentences are superficially equivalent by truth-functional criteria, asserting 'The wire broke: it was overloaded' implies that overloading was the cause of the break, while asserting 'The wire broke and was overloaded' implies that the break was an occasion, or reason for overloading the wire.

This last conjunction reverses the original order of the conjuncts. Since by truth-functional criteria, conjunctions of the forms '(p) and (q)' and '(q) and (p)' say the same thing, reversing the conjuncts is at least one way of detaching the causal implication of the original conjunction. Another method of detachment is the truth-functionally innocuous replacement of 'and' by 'but'. Asserting 'The wire was overloaded but broke' implies (if anything) that the wire broke in spite of its overload. That this is implied, however, is doubtful, for its very sense is doubtful. That an overload would tend to prevent, rather than to produce breakdown verges on nonsense.

Of course, 'but' or some other contrastive ('yet', 'nevertheless') *is* permissible when an overload is *not* accompanied by a break. Indeed, a contrastive is needed if we do not wish to surprise and mystify our hearers. A contrastive signals a discontinuity: it warns the hearer that what is to follow is not what he has been led to expect. Asserting 'The wire was overloaded' raises expectations of malfunction. If those expectations are to be dashed, then a speaker who wishes to make his remarks unarresting had best introduce surprising news of normal functioning by 'but'. He will thereby prepare his hearer and show that he realizes that a conjunction as odd as that of overload and normal function cannot pass without explanation.

It may be supposed that these several expectations, oddities, and implications derive from common knowledge of first- or second-hand experienced regularities. We have learned directly or indirectly that overloaded wires break: hence the facts about assertable and unassertable conjunctions and contrastives noted above.

But glib appeal to learning and experience is not as convincing as it once was. Nor should it be here. Few philosophers, even few philosophers of science have had great experience of wires. And yet we are certain of some general truths connecting overloaded wires and breaks. But what exactly are those truths? I, for one, do not know for certain that all or most overloaded wires break. It may well be the case that many wire-users are sufficiently knowledgeable and careful to take compensating precautions whenever they overload a wire. They may relieve other strains on an overloaded wire, or allow an overload for only short periods. But without knowing whether or not they do, I know that overloading increases the probability of breaking, and hence makes compensations desirable or imperative. And I know for certain, whatever the actual frequency of breaks occurring or allowed, that overloads cause breaks.

It may be said that my scant experience of wires is amply supplemented by knowledge drawn from experience of similar things like ropes or threads. As a reader of philosophy of science, I have for example been told on the highest authority that there is a well-tested 'hypothesis of the character of a natural law' to the effect that,

> Whenever a thread is loaded with a weight exceeding that which characterizes the tensile strength of the thread, then it will break.[12]

But if such a law is a source of my confidence in my general claims about wires, empiricists can take little satisfaction. Since tensile strength is definable as 'the greatest longitudinal stress a substance can bear without tearing asunder', the 'strict universality' (and truth) of this 'hypothesis' is derivable from meanings, *as well as* from cases.

My initial, less general claims about overloaded wires have a similar linguistic, or semantic defense. Should someone seem to question either 'Overloading increases the probability of a wire breaking' or 'Overloading causes wires to break', we could suspect him of ignorance of at least some of the relevant English words and would very likely cite something like the ample definition of 'overload' given above.

These claims, however, are not simple tautologies. If we could take a challenge to such platitudes seriously, we could defend them by performing a simple demonstration with a wire and a weight. Although in no sense an experiment, such a demonstration is an appeal to experience, and as such would provide an empirical defense for claims that would more usually be defended semantically. These causal and probability claims, then, are *doubly* defensible.

Verificationists would, of course, object that claims so variously defensible must be ambiguous. If they seem to have both empirical and semantic defenses, that can only be due to their harboring both tautologies and generalizations under a single verbal cover.

I grant that suitable qualifications can drive unqualified causal and probability statements into tautologies, thereby depriving them of empirical support, and other suitable qualifications can turn them into generalizations, thereby denying them of semantic, or conceptual support. For example, adding features of my definition of 'overload' produces the tautologies, '*Under normal conditions*, uncompensated overloads *eventually* cause breaks or breakdown', and '*Under normal conditions*, uncompensated overloads *increase the probability of eventual* breaking or breakdown'. Likewise, adding frequency qualifications will produce the frank empirical generalizations, 'Overloading *often* causes breaks or breakdown' and 'Overloads *usually* increase the probability of breaks or breakdown'.

But the results of such additions do not show the ambiguity of the initial, unqualified claims. Nor do they show that those claims are vague or imprecise: unqualified generality is not identical with vagueness, nor qualifying with making precise. Nor do these derivative tautologies and generalizations commit me to counting the parent propositions as hovering between, let alone falling into, both the Analytic and the Synthetic baskets. I have avoided these disputed labels, for they have at least temporarily lost their philosophic usefulness. But if forced to use the labels, I would count the unqualified causal and probability claims as synthetic – on two conditions: namely, that

(1) 'synthetic' is taken to mean no more than 'non-analytic'; and

(2) 'analytic' is taken to mean (*à la* Frege) 'defensible *only* by meanings alone'.

Calling them synthetic is, however, not merely useless. It is misleading, for it is the semantic support for such claims that is usually neglected. Even philosophers most committed to Hume and extensional logic give as paradigms of empirical laws truisms whose appeal is not simply the familiarity of supporting correlations.[13]

Why I have dwelt on these 'causal generalities' will become clearer in my subsequent remarks on explanation. Their immediate relevance is simply this: they add to the evidence that congruity is a semantic relation. The pairs of terms listed at the outset of this section (p. 427) are such that one member may be used to define, or at least explain the other. In addition we see that lying behind the causal implications, expectations, and oddities associated with conjunctions formed of these pairs are causal generalities which have a linguistic or semantic defense.

That congruity is a semantic relation is of some importance if, as I believe, congruity is a principal source of the explanatory power of events. But before examining congruity and explanation, I shall mention a few more characteristics and examples of congruity. Congruity is a dyadic relation between descriptions of events, objects, states, or processes. And it is a symmetric, but ordered relation. The first member of the pair often marks an event beginning earlier than that marked by the second, as with the congruous pairs, 'asks for' and 'is given', 'needs' and or 'receives'. But only at the momentary outset does pushing against, or pulling at an object precede its motion. When 'pushing against' and 'moving' are used of concurrent events, there is no question as to which is the first, or causal term.

Congruity is not a biunique relation. ('Moves', for example, is congruous with various verbs of force.) Nor is it a transitive relation. ('Press upon' is congruous with 'crush'; 'crush', with 'break' and 'break' with 'make useless'. But 'press upon' is not congruous with 'make useless'.)

Nor is congruity a reflexive relation: a term is not congruous with itself. This follows from the facts that *only* one of two congruous terms is definable or explicable with the help of the other and no term can explain itself. Nor can a grammatically active term explain its passive cognate. To understand either 'pushing' or 'being pushed', a person must understand the other. Hence, neither term is available for explaining the other. Likewise, for correlative terms like 'giving' and 'receiving', or 'searching' and 'finding'. Unlike these correlative pairs, one term of a congruous pair may be understood without the other and thus will be available for explaining the other.

I do not pretend to have given a list of properties which can be used for identifying further congruous pairs. The notion of one term helping to define or explain the other is too vague to allow this list such a use. My concern here is merely to sketch a relation which holds between what are cause- and effect-terms when they describe two events or objects, and to show that relation to be semantic. So much is enough to give sense to the thesis of the next section, namely, congruity, a semantic relation between descriptive terms, is a principal source of explanatory power.

III. EXPLANOTORY POWER

What is the source of an event's explanatory power? The usual answer is: covering laws of appropriate form. Laws of the forms $(x)\,(Fx \supset Gx)$ and $\text{prob}(G|F) = n$ (where n is near 1) confer on an event whose description has the form *Fa* the power to explain an event whose description has the form *Ga*. Various counter-examples have shown that this formal, or syntactical condition does not insure explanatory power[14]. Perhaps further conditions, formal and otherwise, can dismiss these counter-examples.

But, whether sufficient with or without emendation, coverage by laws of the above standard forms does not seem necessary for explanatory power. If any general proposition is needed to enable one event to explain another, it need not, I think, be either a universal generalization or a probability statement. Something 'looser' will do, namely, what in the last section I called 'generalities'. Examples are the causal generality 'Overloads cause breakdown' and the related probability generality 'Overloads increase the probability of breakdown'. Since the latter states an *increased* probability, not a *high* probability, there is no temptation to make it fit the form, prob$(G|F)=n$, where n is close to 1. The causal generality, however, might seem to be elliptical for 'Overloads always cause breakdown', and thereby lend itself to representation by $(x)\ (Fx \supset Gx)$.

Even if a universal causal claim were adequately representable by this formula, the causal generality cannot be so represented for, as we have seen, it is not a universal claim. It is true that overloads cause breakdown, but false that they always, or even usually do.

Why has it seemed plausible that universal truths are necessary for explanatory power? Perhaps one set of connected assumptions is this: only a cause explains its effect; a cause is a sufficient condition for the occurrence of its effect; only laws can insure sufficiency; hence, laws are necessary if an effect is to be explained.

John Stuart Mill, for one, held such a view. To insure that a cause was sufficient for its effect, he had to make a cause indefinitely complex:

> The cause ... philosophically speaking, is the sum total of the conditions positive and negative taken together; the whole of the contingencies of every description, which being realized, the consequent invariably follows.[15]

He noted, however, certain vulgar prejudices which kept us from admitting this complexity. We are so struck by a condition which is 'productive', or which immediately precedes the effect, that we single it out from among the innumerably many conditions which compose the cause (philosophically speaking). We especially favor immediately preceding conditions as 'the cause' (commonly speaking). Thus, we take as the cause of a camp being attacked by surprise the absence of the sentinel – a mere 'absence of preventing causes' – instead of a 'producing cause' such as a bribe, or the fatigue which put the soldiers to sleep, or the events that led to mobilization of the attacking force. And rather than the solid, indisputable earth, we take as the cause of falling bodies the 'logical fiction' of a force, or attraction exerted by the earth. These 'abstractions', unlike the earth itself, can be represented as 'exhausted by each effort, and therefore constituting at each successive instant a fresh fact, simultaneous with or only immediately preceding the effect'.

Mill has our preferences right, but our reasons wrong. It is not a demand for a temporally proximate cause that leads to the 'invention' of forces of attraction. As a fact of theory, forces are not conceived as a succession of 'exhausted efforts' each being simultaneous with or only immediately preceding an effect. But even if they were so conceived, it would be their congruity with their effects, rather than their temporal proximity that would make them preferable to the earth *tout court* as the cause of the falling of bodies toward the earth. As vacuous causes show, the relation between a force and its appropriate effect is an especially appealing case of congruity.

Our preference for the sentinel's absence is also due to congruity with, not to its temporal proximity to, the surprise capture. Even if one of the 'productive events' that put the soldiers to sleep (say, their fatigue) had overcome them just before the attack, we should still prefer as *the* cause the sentinel's absence. As conditions *sine quibus non*, the fatigue and the sentinel's absence are on a par. The camp would not have been taken by surprise if either the soldiers were awake and alert or the sentinel had been at his post. But as candidate for *the* cause of the surprise capture, the sentinel's absence is superior. The soldiers' fatigue causes, or creates the need for a sentinel to stand guard while the soldiers sleep. It is the sentinel's absence, that is, his failure to stand guard that causes, or allows the camp to be taken by surprise.

It may seem that the sentinel's absence is preferred because it, rather than the fatigue, is abnormal or is a deviation from routine. But suppose there had been no war and no apparent threat or war prior to the attack, and suppose that sentinels were often off their posts. In what sense was the sentinel's absence on the fatal night abnormal? Only, I think, in the sense that the sentinel was not doing what *by definition* he was meant to do, namely, guarding against the very kind of surprise that took place. Or, in my jargon, the pair 'sentinel' and 'guarding from surprise', and hence the pair 'absent sentinel' and 'taken by surprise' are congruous pairs. By contrast, 'fatigue' and 'taken by surprise' are not. Our preference is for the causal factor congruous with the effect.

And we can recognize that factor without knowledge of 'strictly universal' laws, or indeed any laws at all. In their concern to exclude untestable, non-empirical explanations, empiricists have made explanation seem an unachievable ideal. To explain fully, say, the cracking of a car radiator, Hempel would require us to complete a long list of scientific laws which he has only begun.[16] But, I suggest, only one general truth (if any) is needed to explain fully the famous case, namely, "Internal pressure bursts inflexible containers." Of course, for this generality to apply, it must also be true that the metal of which the radiator is made be inflexible and that the fluid with which it is filled be able to exert internal pressure. And to know why the radiator cracked when it did (on a cold night), we would have to know that the radiator contents expand, and thereby exert internal pressure, upon freezing. But this additional information merely justifies subsuming the case in question under the causal generality, "Internal pressure bursts inflexible containers". It is the generality so applied that confers explanatory power on the freezing water. And ultimately it is the semantic connection between 'expansive pressure' and 'bursting' that gives this generality the power it confers, for this generality *expresses* that connection. It does not *establish* it.

The ability to understand the cracking of the radiator is, I claim, enjoyed by anyone who knows the relevant bits of English (or some equivalent). The ability to predict the cracking is, of course, not so widely shared. Since Hume, predictability and explicability have been closely associated. Hume's demands on any impression which would legitimate the concept of power are demands for both *inevitability* (hence predictability) and *intelligibility*. For him, an impression of power (or, necessary connection) would have to reveal

(1) what effect that power will always produce,

(2) why it has that effect and no other, and

(3) why it *must* have that effect and *can* have no other.

Hempel's thesis that explanation and prediction are 'symmetrical', that is, that the source of explanatory power *is* the source of predictive power, is the refined version of Hume's conflation of intelligibility and inevitability (or conditional necessity).

But to know, given certain conditions, what event will occur and when is not the same as knowing, given those same conditions, why that event and no other has occurred. To say what the latter does involve is, of course, not easy. But certain metaphors and analogies are helpful, especially in the light of what has been said about commensurability and congruity. In expressing dissatisfaction with psycho-physical interaction, Sherrington speaks of the absence of 'a hint' of the cause in the alleged effect:

> ... there follows on, or attends, the stage of brain-cortex reaction an event or set of events quite inexplicable to us, which both as to themselves and as to the causal tie between them and what preceded them science does not help us;

> a set of events seemingly incommensurable with any of the events leading up to it. The self 'sees' the sun; ... in the sequence of events a step is reached where a physical situation in the brain leads to a psychical, which however contains no hint of the brain or of any other bodily part.[17]

And Santayana put equally well what we want. People expect

> to have an insight into causal connections, such as they have in action, or in answer to a prayer, or obedience to an order given or the execution of a plan.[18]

Or, at least they did so expect, Santayana thought, until Hume freed them from such vain hopes.

I think we still have such expectations, and that they are both legitimate and satisfiable. That insight which involves the 'hint' of the cause in the effect is found when cause and effect are congruous.

At least some of Hume's demands for a notion of causal connections can be met, namely, those for clear examples of *intelligible* causal connections. And a congruity does not satisfy Hume's demands for examples of *inevitable* causal connections, congruity does give rise to certain qualified inferences about future and past unknown cases. For example, if something is overloaded and no compensating precautions are taken, then failure of function is more probable and will eventually occur. And if something has broken while overloaded, then the overload is a probable cause.

It was Hume's phenomenalism and his preoccupation with visual properties that insured his failure to find anything resembling the causal connections he pretended to seek. Whenever Hume wished to show that nothing could be inferred about the character of an effect from its cause, he tended to describe the cause solely in visual terms. Thus, the cause of a second billiard ball moving from rest was said to be the 'approach and touch', or 'contact' of another ball.[19] And for Hume's Adam, an animal falling into water and drowning is an animal falling into "a fluid, transparent substance"[20] (water = air?), and a flame consuming paper is a mere "bright, flickering" object (fire = firelight?).

It is not surprising Hume despaired of finding causal connections which were nothing more than our projected expectations of familiar consequents. Visual properties do not tend to explain other physical properties, or as I put it, visual terms are not congruous with other physical terms.[21] At times, however, Hume did count as the cause of a ball's moving from rest the *shock*, or *impact* of another ball. Is it obvious that nothing about the effect, or likely effect can be inferred from such dynamical, non-visual descriptions of the cause? Only someone indifferent, or insensitive to causal language could treat 'impact' and 'touch' as equivalent descriptions of the same event.

Except for Aristotle and certain Idealists[22], philosophers have been indifferent and insensitive to our causal language. In my efforts to correct these lapses, I have no doubt given the impression that I take *what we say*, rather than *what happens* to be the explanation of natural and social phenomena. I do not wish to deny a distinction between events and their descriptions, however vexed that distinction may be. But causal explanation is closely tied, as Aristotle saw, to the answering of questions. And it can be argued (but not here) that questions, by virtue of their meaning, delimit what will count as an answer and what will count as a good answer.[23] Accordingly, it is neither paradoxical nor merely modish to claim that causal explanation involves congruity, a *meaning*-relation between the terms which describe a cause and the effects it explains.

It is not our expectations, but our language that is the source of whatever 'causal connections' we project into nature. And far from being the mysterious connections empiricists allege, those causal connections are as intelligible as the meaning-connections they reflect.

The notion of congruity is not intended to restore the possibility of arm-chair science. Reflection is not the way to discover what causes what. Causal knowledge arises from the investigation of co-occurrences which are independently describable and identifiable. Although such co-occurrence need not be constant, causal claims are implicitly general to the extent that they are open to challenge from parallel cases in which alleged cause and effect fail to co-occur. And since an event may fail (in special circumstances) to have its usual effect (or usual cause), causally paired events must be describable in conceptually distinct ways that allow such failures to be detected. If events are separable, they must be distinguishable.

Hume, so understood, was right about causation. But he mistakenly thought that truth was the only issue. We value causes for their explanatory power, and such power is largely independent of truth. We are all facile explainers, able to produce (possible) explanations for novel phenomena without concern for tests or testability. We know what will explain what well before we know what causes what. And accordingly, we tend to jump to, and cling to false causal hypotheses in the absence of evidence, or even in the presence of counter-evidence. Superstition is not the result of hasty generalization, but of hasty explanation.

Experience, of course, feeds this explanatory fertility. Causal claims once singly defensible by observed co-occurrences become doubly defensible. Our concepts and terms incorporate our experience, in ways perhaps best understood by Kant (in his discussion of the transition from 'judgments of perception' [when the sun shines on the stone, it grows warm] to 'judgments of experience' [the sun warms the stone] in the *Prolegomena*, Second Part, secs. 20–23). What was once a matter for discovery becomes a matter of meaning, but given its origin, it does not thereby become trivial. By such transitions, conceptual connections develop which render certain causal claims self-explanatory.

Accordingly scientists can often provide both conceptually *independent* and conceptually *dependent* descriptions of a cause and effect. The relation of these two descriptions can be diagrammed as follows:

$$\begin{array}{ccc} D_c & - & D_e \\ \cdot & & \cdot \\ \cdot & & \cdot \\ d_1 & \ldots & d_2 \\ \cdot & & \cdot \\ \cdot & & \cdot \\ \mathrm{Ph}_1 & \ldots & \mathrm{Ph}_2 \end{array}$$

Only terms D_c, D_e are conceptually connected; there is no kind of necessity between phenomena Ph_1 and Ph_2 nor between the phenomena and either pair of descriptions, nor between the pair of descriptions d_1, d_2 and the congruous pair, D_c, D_e. Indeed, the pair d_1, d_2 give the contingent criteria for applying the congruous pair. My thesis is that *testability* of causal claims requires conceptually *independent* descriptions (d_1, d_2), while *explicability* requires, or at least derives from conceptually *dependent* descriptions of the sort illustrated by congruous terms.

New York University

REFERENCES

[1] Mill's definition of 'the cause' (*A System of Logic*, III, v. 3) and J. L. Mackie's definition of 'a cause' (*Amer. Phil. Quart.* **2** (1965) 245), respectively.
[2] At the outset of the play, Molière himself shows some sympathy for *virtus dormitiva.* "Un julep hépatique, soporatif et somnifère, composé pour faire dormir Monsieur" is the only item on the apothecary's bill which Argan pays for in full and without complaint.
[3] Details of the sort that the woodcutter concocts when forced in *Le Médecin malgré lui* to give a cause of the impediment to the action of the tongue by which he has initially explained Lucinde's muteness (Act II, scene 6).
[4] 'Causality and Retribution', *Philosophy of Science* **8** (1941) 533–556; also *Society and Nature*, London 1946.
[5] *The Child's Conception of Causality.*
[6] On one possible reading of the term *participation* in *Primitive Mentality.*
[7] *On Generation and Corruption*, 323^b33ff.
[8] *Science and Method*, ch. IV.
[9] *Physics*, Bk. II, ch. 5 (196^b10ff.).
[10] Berkeley found ideas essentially passive, hence incapable of producing any other ideas. But most empiricists have allowed an idea associated by 'contiguity' or 'resemblance' with another idea to be its effortless cause.
[11] 'The Causal Theory of Perception', *Aristotelian Society Supplementary Volume*, XXXV (1961), Sec. 3, pp. 121–152.
[12] Karl Popper, *The Logic of Scientific Discovery*, p. 61. For this law, Popper substitutes two non-tautologous laws in a footnote. But the explanatory power is not thereby increased. That power resides in the notion of an exceeded breaking point. But I am anticipating Section III below.
[13] Consider these 'empirical generalizations':

'Blows cause injuries.' (Hart and Honore)

'Fragile things tend to break when struck hard enough, other conditions being right'. (D. Davidson)

'Flames are hot and cause melting'. (Hume)

'When hard particles of steel are in immediate contact with soft particles of bone, the latter give way'. (Schlick)

'Insults cause anger'. (Nagel)

'Whenever the sun (or some other luminous body) is above the horizon, and its rays travel in straight lines, and there is no opaque body in a straight line between that luminous body and the part of the earth where we are situated, then it is day there' (J. S. Mill).

[14] For counter-examples to the Popper-Hempel models of explanation, see, for example, S. Bromberger, 'Why-Questions', in *Mind and Cosmos* (ed. by R. G. Colodny), sec. V, pp. 91–95. For additional conditions to meet these and other counter-examples, see forthcoming papers by J. Kim.
[15] *A System of Logic*, III, v. 3.
[16] 'The Function of General Laws in History', *J. Phil.* **39** (1942), sec. 2.1.
[17] *The Integrative Action of the Nervous System* (1947 ed.), pp. xvi–xvii.
[18] In a letter quoted by C. J. Ducasse in 'Causality, Creation and Ecstasy', *The Philosophical Forum* **9** (1953), 12.

[19] *An Abstract of A Treatise of Human Nature*; the *Treatise*, I.iii.2

[20] *An Enquiry concerning Human Understanding*, sec. 4, Pt. I.

[21] The non-explanatory laws cited against Hempel's model of explanation often have visual terms as antecedents (e.g., 'angle of light-ray incidence', 'apparent stellar brightness'). See Bromberger, *op. cit.* (note 14 above).

[22] See *Physica* II, 3; J. McT. E. McTaggart, *Philosophical Studies* 7, pp. 156–178.

[23] See J. J. Katz, 'The Logic of Questions', in *Proceedings of the Third International Congress for Logic, Methodology and the Philosophy of Science* (forthcoming).

M. M. SCHUSTER

SOME COMMENTS TO 'CAUSAL CONNECTION'

I am in basic sympathy with Prof. Ruddick's heretical position and I agree that the explanation of causality lies along some such line as that drawn by his notion of congruity. His argument represents for me a movement away from the contingent world of the Humean tradition with its externally related objects unnecessarily connected, where the actual order of things might logically have been the reverse of what it is or even turned inside out so that the middle is at the beginning, the beginning at the end, and the end at the middle. In that world, nothing entails anything else, nothing stands in need of anything else; and whatever meaning may be given to causal connection, it will always be the case that any connection between a particular cause and a particular effect might be other than it is. Now I happen to be quite dissatisfied with this view of the nature of things. I think it explains a world, but not the world that we set out to explain; hence, if we must accept it, we should do so with the utmost reluctance and only as a last resort.

These remarks should explain why I cannot be fundamentally critical of a theory of causal connection that smacks of internal relations. Still, I do not think that the kind of defense I seek can be provided by the type of argument Prof. Ruddick employs. What is more to the point here, I do not think that what he is arguing for can be adequately defended on the level on which he works, for he deals with descriptions rather than things, but it would seem that a prior argument is required which, perhaps, deals with things instead of descriptions. At any rate, I shall explore this possibility, first, by analyzing the nature of his argument and the conclusion he actually reaches; second, by examining the status of his doubly defensible causal generalities; and third, by examining the conclusion which it seems he would like to reach.

Prof. Ruddick embarks upon a kind of Freudian undertaking when he attempts to show that the apparently irrational and disorderly realm of causal misbeliefs has in fact a unity, order, and rationale which is the same as that governing the sphere of genuine causal judgments. His approach – working from deceptive causes to causal connection in general – is indeed novel; however, though the study of deceptive causes may have been the source of Prof. Ruddick's insight, I fear that it is also the source of some confusion. Let me explain. Ruddick's belief that the investigation of erroneous causes reveals congruity in operation leads him to be highly critical of traditional empiricists for having devoted themselves almost exclusively to the study of 'right' or 'true' causes. Those who have thus confined themselves would reply that the important thing is to determine the nature of legitimate causal judgments and not what some people mistakenly consider causal judgments, and they would add that anyone might have any reason or anything in mind for making false judgments. It is precisely because of this view that deceptive causes have not been of philosophic interest to the traditionalist; and it is because of this lack of interest that the traditionalist has not attempted to render a philosophic account of them. The explanations he has offered, e.g., "anthropomorphism" or "the desire for humanoid agents throughout nature", "hasty generalization" or sloppy thinking, may with some justification be called psychological. Prof. Ruddick fails to perceive this. He fails to perceive that his own explanation is of a different – a linguistic – order, that he is concerned with the analysis of *'what we say'*, the causal judgments that we actually make, rather than with their psychological origins. I am exaggerating, of course. Prof. Ruddick does see this to some extent, as witness the last quotation and his praise of "Aristotle and certain Idealists" for their sensitivity "to our causal language". But this only points up the confusion. His criticism of the conventional psychological accounts is irrelevant to his purpose and incapable of being substantiated by his linguistic argument. It may well be the case both that an erroneous causal judgment exhibits congruity and that its utterer was directed by vanity, sloth, or one of a multitude of other possible psychological reasons. But whatever the case is, a mere analysis of statements can no more inform us of its psychological aspects than a psychological analysis can inform us of its purely linguistic aspects.

It should be clear, then, that Prof. Ruddick only clouds the issue when he pits his notion of congruity against anthropomorphism and hasty generalization, thereby putting them on the same level and treating them as mutually exclusive alternatives; and when, in the same vein, he seeks to explain "our causal desires". For he is not examining the reasons or motives of those who make the judgments, as we have already seen; instead, he is examining the judgments themselves, and what he finds is that they conform to his concept of congruity.

Here I come to a difficulty. Is he saying that all causal judgments – true as well as false – are instances of congruity? If so, then this is a matter to be settled by summoning up case after case in order to discover possible exceptions. I have neither the patience nor the facility for such a task. At any rate, if this is his point, then it should be considered a hypothesis which can be more or less probable. But I am quite sure that it is not, for although Ruddick is not very clear on the subject, he does introduce the whole category of *incongruous* causes. He tells us that any particular incongruous cause may have *"supporting evidence"*, that "subsequent research" may prove that it is a true cause, that it may have "statistical support", that it may be 'redeemed' by the use of an appropriate remedy. In short, incongruous causes are thought to be, and often are, genuine causes, and judgments referring to them are causal judgments. But they are *not* instances of congruity. This is almost paradoxical. One of Prof. Ruddick's criticisms of the conventionalists is that they fail adequately to explain some kinds of erroneous causes, e.g., fictitious and spurious ones – and even this criticism we saw was unwarranted – but it now appears that there is a category of deceptive but remediable causes which cannot be explained by congruity.

Prof. Ruddick says that his concept of congruity at least explains why we feel uneasy about incongruous causes. (Note that this is to resort to a psychological explanation, namely, our preference for the congruous. It dispenses with the analysis of statements that are actually made; it even disregards what is in the mind of the person making a statement in order to get into the minds of the people who are confronted by it.) The conventionalist will counter with the claim that he can set forth rules which explain why we bother to make such causal judgments at all. These rules are implicit in Ruddick's admission that incongruous causes may have experimental support. But we have not yet touched the basic point. Prof. Ruddick might retort: fictitious and spurious causes will never be adequately explained by the conventional account of causality, but incongruous causes will serve as instances of congruity when properly remedied. That may be. Yet such judgments are incongruous when they are made because we do *not* have knowledge of the congruity of cause and effect and because congruity is consequently *not* reflected in the meanings of the terms involved. Thus it seems to me that Prof. Ruddick is not setting forth a hypothesis, as we thought might be the case, but that he is prescribing a rule for the analysis of cause and effect. What he seems to be saying is that we ought to look at all causal connections in this way, we ought to define causes and effects such that they will be congruous, and any judgments which cannot be made to fit the rule would not then be actual (true or false) causal judgments. (Perhaps, then, Prof. Ruddick should not include incongruous causal judgments as causal judgments at all. They would become causal judgments only upon ceasing to be incongruous.)

If I am correct and he is merely proposing a rule, then I think it likely that a conventional empiricist would reject it. Ruddick might say that his view is more plausible, though if he means 'plausible' in his sense then he begs the question: one cannot defend one's position by an argument presupposing it. In fact, as I said at the outset, I do not think that the question – whether to choose Ruddick's rule or the conventionalist's – can be settled without moving to an argument of an entirely different sort and on an entirely different level.

Let me turn now to my second point: the status of doubly defensible causal generalities. Ruddick says that he would label them synthetic rather than analytic if he were obliged to use that terminology. In fact, however, the analytic-synthetic distinction is inappropriate, inasmuch as causal generalities are supposed to be a third type of statement which may be made analytic or synthetic by "suitable qualifications". Ruddick is apparently conscious of the traditional interpretation of the rationalists – that the truth of such statements is for them a matter of definition – and particularly conscious of the traditional criticism of the rationalists – that such statements consequently tell us about the meaning of words but not about the existence of things. His third category of statement enables him to gain the advantages of rationalism without its disadvantages. Despite this ploy, I think it is prefectly reasonable to consider the charge that if we do not classify, e.g., "overloads cause breakdown", as analytic or synthetic it is only because the statement is not just a platitude, which Ruddick admits it is, but something vague and ambiguous, which Ruddick emphatically denies. That is, the meaning of "overloads cause breakdown" awaits the filling in of detail, not to redeem it, but to determine whether, when it is used, it is either analytic or synthetic.

I am aware of Prof. Ruddick's assertion that the analytic-synthetic distinction has been at least temporarily deprived of its philosophic usefulness because of the dispute which surrounds it. My point is not particularly to defend the distinction but to show that Prof. Ruddick's argument does not provide an adequate defense of his own position, and I plan to accomplish this by showing that he offers no basis for choosing his categorization instead of the analytic-synthetic, or some other conceivable, one.

We are offered the following definition of 'overload':

A burden greater than something can long bear under normal conditions without breaking, malfunction or failure of function.

Ruddick maintains that this definition does not make overload the cause of breakdown, and one of his points here is that we may have overload without breakdown. I wonder whether this is so. Because it seems to me that a beyond-normal burden which did not after a period of time and under normal conditions result in incapacitating malfunction would, by his definition, not be an overload. The introduction of "compensating precautions" to prevent malfunction immediately alters the situation and deprives us of the 'normal conditions' specified in the definition. Thus it seems that "overloads cause breakdown" is true by virtue of the meaning of words alone, for the definition of overload, although worded differently, actually boils down to this: that overload is a beyond-normal burden that causes breakdown.

To object that the generality cannot be analytic because it is defensible by appeal to experience as well as by appeal to meaning is to beg the question, because that is precisely the point at issue. Nonetheless, explanations of *apparent* double defensibility are not lacking. Consider, for example, Ayer's "criticism of the rationalist interpretation of causal laws".[1] Ayer concedes to the rationalist that "it is possible, logically, that a system that had a purely fanciful origin might subsequently turn out to be realized in fact. But", he continues, "in the case where the derived propositions are 'the laws of dynamics', is the origin of the system really so fanciful? Surely those who construct it already have empirical knowledge of what it would be useful for its premises to entail? And is not their selection of the premises to a large extent governed by this?" He concludes: "It is only when the theory has been very strongly substantiated by empirical observations that the tendency arises to make it wholly analytic."[2] In short, Prof. Ruddick's causal generalities might be analytic and still appear to derive support from experiences, because in reality they would only be synthetic propositions in analytic clothing. That is, their origin, their explanatory power, and their defensibility would ultimately reside in observation rather than in meaning connections. In contrast to Ruddick, then, who claims that explanatory possibilities are provided by language, it would be argued that experience provides the possibilities which later become established as congruities.

Another alternative is that the generality, when made precise, is synthetic. On this view, the definition of 'overload' would be, let us say, a beyond-normal stress. The judgment "overloads cause breakdown" would then be an empirical generalization to the effect that in most cases, or frequently, overloads are followed by breakdown. There are other ways to doctor the definition and view the situation. We could introduce a third factor, condition *C*, which could then be placed between overload and breakdown. The connection between overload and *C* and breakdown might then be entirely a matter of definition or partly – on one side – a matter of definition and partly – on the other side – a matter of empirical generalization.

These are all possibilities and I do not see that Prof. Ruddick has sufficient reason for rejecting them and accepting his. And it is not a problem that can be settled by appealing to a mass of specific causal judgments. We have, on the one hand, a two-category, analytic-synthetic rule. When we use it and come across such troublesome generalizations as "overloads cause breakdown", we say that they are ambiguous – they do not fit our categories – and thus by the rule so alter them that they do fit our categories. That is what the rule is for. On the other hand we have a three-category rule. It is important for Ruddick to keep his causal generalities in his third category, for in that way he avoids the pitfalls of tautological rationalism and arbitrary conjunctionism. And this is a worthy objective, but it does reveal that his whole approach to the subject is oriented toward the traditional view from which he deviates. He treats of internal relations and yet is hesitant to make a full commitment to them. I would expect that if the objects in the world were really internally related then our definitions of them would reflect that fact, since internal relations supposedly "infect the nature of their terms".[3] This, I think, would of necessity make causal judgments analytic. There would then be a sense in which the truth of such judgments would be merely a matter of definition, but in another sense – that they are true of the world – something more than meaning alone would be involved. Clearly, in order to determine whether this view is valid or not, we cannot simply cry analytic and synthetic, but some other sort of argument – an argument on a different level – must be brought in. And without that other argument I do not see how it is possible to decide among the alternatives, and therefore I do not see how it is possible to judge the validity of Prof. Ruddick's position.

I come, then, to my final point, to the conclusion that I believe Prof. Ruddick would like to reach. I have argued thus far that his heresy consists of nothing more than the proposal that we adopt certain unconventional rules, that it is unlikely that a confirmed conventionalist would follow his suggestion, and that we are consequently confronted by alternative rules and by the problem of making a choice. Now it seems to me that none can be made as long as we confine ourselves to 'our causal language' and its analysis. Prof. Ruddick's flirtation with psychological reasons was shown to be both improper and unsuccessful. Some other argument is needed, an argument that will enable us to determine which are the appropriate rules to use when dealing with things and events. And such an argument, I would think, must be about things and events themselves.

What I find extremely interesting, and extremely odd, is the fact that Prof. Ruddick is apparently of the opinion that he is telling us about the world and not just about our use of language. I refer in particular to his closing paragraph. He does recognize a distinction between events and their description, between *'what happens'* and *'what we say'*. Nevertheless, he adds a 'but' which I interpret to mean that his account of language is also an account of events. He therefore seems to be guilty of a procedure which Richard M. Gale not only terms "highly suspect" but also a "crime", namely, the procedure of drawing "positive metaphysical implications from a piece of linguistic analysis...".[4] Surely, though, one cannot treat the difference between the study of *events* with the aid of language and the study of *language* with the aid of language as something negligible without first providing a sound defense.

Moreover, it is not altogether clear exactly what Ruddick's view of the world is. Perhaps I was wrong in thinking of it as anti-Humean, because his linguistic argument leaves open the most divergent metaphysical possibilities. On the one hand, he might mean that events are internally related in the sense that cause and effect terms are; on the other hand, he might mean that they are not. For example, he states that "it was Hume's phenomenalism and his preoccupation with visual properties that insured his failure to find anything resembling the causal connections he pretended to seek"; and again, he stresses that "visual properties do not tend to explain other physical properties, or as I put it, visual terms are not congruous with other physical terms". This is to say that causal connections are not to be found in experience but in our descriptions of experience, i.e., in language. Ruddick therefore at least appears to be a Humean – a modified one but a Humean nonetheless. Of course he might have introduced a brief statement clarifying his own position about the actual connections between events. My point is, however, that such a statement would be nothing more than the expression of an opinion rather than the conclusion of an argument, for the argument, taken by itself, does not settle the question. Even if we grant the notion of congruity and the third category of doubly defensible generalities, the argument alone provides no basis for determining whether the connections Prof. Ruddick underscores are merely linguistic or involved in events themselves. The problem is metaphysical and seems to demand a metaphysical solution. Yet that solution, as we have already seen, appears to be the basis for deciding whether or not to grant Prof. Ruddick's proposals.

Boston University

REFERENCES

[1] Alfred J. Ayer, *The Foundations of Empirical Knowledge*, Macmillan & Co., London, 1963, pp. 199ff.
[2] *Ibid.*, p. 206.
[3] Paul Weiss, *Reality*, Peter Smith, New York, 1949, p. 172.
[4] Richard M. Gale, 'A Reply to Smart, Mayo and Thalberg on "Tensed Statements"', *The Philosophical Quarterly* **13**, 352.

EDWARD H. MADDEN

CAUSALITY AND THE NOTION OF NECESSITY*

I

What is the main contention that Hume and contemporary Humeans share about inductive inference? It is essentially that no conclusions about the future or unobserved follow with certainty from propositions about what we have experienced in the past. The non-demonstrative nature of induction, it seems, follows inevitably from the truism that induction cannot qualify as deductive inference. The assertion of C and not E is never self-contradictory; hence one cannot say that E *must* occur. Necessity is not ascribable to causal propositions, except (and here the divergence begins) in a psychological sense (Hume) or in a 'nomological' sense (contemporary Humeans). There are, of course, other crucial differences among Humeans. For Hume, conclusions about the future do not even follow with any probability from propositions about past experience, a conclusion which Carnap, for example, tried to avoid by claiming that certain types of probability statements cannot be refuted by subsequent experience. But then the question arises of how such propositions can be construed as relevant in any sense to the future. In any case, we shall not pursue this question but will confine ourselves to the weaker and *prima facie* legitimate contention, shared by all Humeans, that there is no irreducible sense of natural necessity that is involved in causal judgments. My claim will be that this contention is only *prima facie* legitimate and is in reality false. I will develop the notion of a definitional 'must' implicit in the concept of cause itself, and in all concepts which have a causal component, and argue that this notion, far from being conventional or arbitrary, reflects certain non-linguistic aspects of that universe which we in fact have.

II

Before we can get to the heart of our problem, we need certain preliminary clarifications of causal terms and examples of what are to count as causal propositions. Consider what might be called the cause (C) of a forest fire (E) at time t_0. A forest fire occurred, one might say, because these conditions existed: (f) some means of ignition (lightning struck a tree); (g) dry underbrush; (h) a prevailing wind; (i) oxygen in the air, and so on. Each condition is necessary in the sense that without it the effect, the forest fire, would not occur. It is admittedly very difficult to find the whole set of necessary conditions, but this whole set, whatever it may be, is the condition sufficient, or able, to produce the effect and thus is the cause of it. We have, in effect, this string of definitions: 'cause' = *df.* 'sufficient condition'; 'sufficient condition' = 'complete set of necessary conditions'; 'necessary condition' = 'condition without which the effect would not occur'.

The crucial question for us is what we would say if at t_1, f, g, h, and i occurred but not E. We would clearly not say then, or ever, that while ordinarily C is sufficient to produce E, this time it was not; rather we would say that somehow we were mistaken in thinking that f, g, h, and i at t_1 were identical with f, g, h, and i at t_0. Since we would withdraw the use of C at t_1 and would never admit that f, g, h, and i at any t, if genuine instances of f, g, h, and i, would not produce E, we are clearly using C in such a way that actually producing E is part of its meaning. That the same conditions, under the same circumstances, would necessarily behave in the same way is, strictly speaking, an implication of the principle of identity; what it *means* to suppose the conditions to be strictly the same (except, of course, for the difference in t) is that they will behave in the same way.

The implication of the above seems to be that any breakdown in supposed C-E relations is always the result of being mistaken and that I am thus taking for granted the uniformity of nature. I want to avoid this implication; I need to show that it is, in fact, not really an implication at all. I agree that asserting C and denying E is not self-contradictory since there is no contradiction in supposing a change in nature. Causal relations themselves might conceivably change, that we cannot doubt. But granting such a possibility, what would we say were it to occur? Presumably we would say that C used to be the cause of E but now (depending upon how the world changed) E seems to have no cause at all (and so is no E any longer) or is caused by C', D, or (if the world has become very restless) Z. Since such talk still implies that whenever you have a cause it must produce its effect, the concept of causal necessity is still essentially present. This concept embraces both the notion of being mistaken about an instance of C and of the world changing so there is no longer that C. In either case, the use of the notion of cause would be withdrawn.

It is tempting to infer from the definitional necessity of 'f, g, h, and i is the cause of E' (asserted to be the case at t_0 and withdrawn upon any subsequent instance of not E) that we have discovered necessarily true propositions about matters of fact. Such an inference, however, would be totally unwarranted. All propositions about factual matters remain in principle corrigible on my analysis because while we know that 'f, g, h, and i at t_1 is the cause of E', if true, is necessarily true, we do not know that ft_1, gt_1, ht_1, it_1 are genuine instances of f, g, h, and i. We may be mistaken in thinking that ft_0 and ft_1, gt_0 and gt_1, ht_0 and ht_1, and it_0 and it_1 are identical except for the difference in time. The definitional necessity shows up when we deny they are identical when E does not occur. But we cannot know beforehand given what we take to be genuine instances of f, g, h, and i that E will necessarily occur. There is, in short, a crucial difference between the *meaning* of a causal proposition – which conceptually does involve the notion of necessity – and *knowing* that any given causal proposition is true – which conceptually involves the notion of corrigibility. 'Causal necessity', as one should expect, is of no use in practice for the discovery of causal relations. That the same conditions, under the same circumstances, would necessarily behave in the same way is an implication of the principle of identity; i.e., that is what it *means* to suppose that something is strictly the same. But in fact ft_0 and ft_1, etc., are not usually identical except for the time lapse; they are only similar in various respects and degrees. What we have to discover therefore is how much similarity, and in what particular respects, events and their circumstances must have in order that an event of a specified kind (E) must result.

III

It might be objected that the analysis of 'cause' as 'set of necessary conditions' is faulty and hence that my analysis of definitional necessity in causal judgments is mistaken. So I need to show that the latter is not dependent on the 'set of necessary conditions' analysis but arises also on the alternative analysis of 'cause'. Considering the notion of causal necessity in this second context also provides me the opportunity of sharpening still more the concept itself.

Consider again what might be called the cause (C) of a forest fire (E) at t_0. A forest fire occurred, one might say, because these standing conditions existed: (g) dry underbrush, (h) a prevailing wind, (i) oxygen in the air, and so on, to which was added one change that was sufficient, i.e., just enough, to produce E, namely, (f) dropping a lighted match into the underbrush. While (f) is sufficient to E, it is not necessary since some other change in the same standing conditions could also produce E, namely, the striking of a tree by lightning. And while g, h, and i are necessary conditions for the occurrence of E, they are not sufficient to it – some change in the standing conditions must occur to produce E.

'Causal necessity' can best be explicated at this point by considering what the converse of 'sufficient condition' is upon the present analysis. The converse of '(f) is sufficient to E' is 'E is necessitated by (f)'. Assume we have a balance with a five-pound weight in one pan. We add seven apples to the other pan but nothing happens. They are 'not enough', we say, they 'do not suffice', to raise the other pan. We add the eighth apple and the weighted pan moves up and after a bit of oscillation settles level with the apples. Adding the eighth apple was 'just enough', we say, it 'sufficed', to balance the pans. Another way of saying the same thing is that the balancing of the pans was necessitated by adding the eighth apple – *that is, adding it was just enough to make the pans balance.* And this 'necessitated by' is reflected in our notion of causal necessity. If in the future we had seven apples of precisely the same weight, the identical pan balance, and added an eighth apple of precisely the same weight, and the pans did not balance, we would not declare that this time (f) was not sufficient to E but rather that we mistakenly thought the balance was identically the same as before (except for the time lapse), or the apples were the same weight as before, and so on. If we kept trying assiduously to control these factors and had good evidence to suppose we had but never got the same results again, then we would conclude that adding the eighth apple *had* been the cause of balancing heretofore but no longer was. Presumably nature had altered its course and we must withdraw the old causal claims.

IV

So far I have shown how the concept of causality is necessarily kept intact. The crucial question now facing us is whether this conceptual necessity reflects any sort of necessity in the world itself. I believe that there is genuine necessity in the physical world (and the psychological too, no doubt), although it is not the 'ontological tie that binds' which many philosophers have traditionally cherished. The reason the concept of causality is necessarily kept intact is that the constitution of the world, or of any possible world revealed by ordinary and scientific methods of inquiry, requires it.

Consider the case of the farmer who removed what he took to be a superfluous beam only to have his barn collapse. What was the cause of the collapse? Different analyses, of course, can be given. One might argue that since the collapse was an *event*, where 'event' means 'change in a state of affairs', the cause of this event must have been another event, namely, the removal of the beam. A cause and effect, after all, must be of the same logical type. Or one might argue that the whole set of necessary conditions – gravity, the condition of the wood in the barn, the removal of the beam, etc. – was the cause of the collapse (thus denying that *C* and *E* must be of the same logical type). Or one might argue that the removal of the beam made it possible for gravity to produce its effect, and hence that it was the releasing occasion for the operation of the cause itself. 'Cause', on this view, must always refer to some pervasive feature of our particular universe. There is, however, a common factor in all three analyses which suggests that these alternatives are simply different conceptual ways of 'slicing' the same physical facts for whatever practical considerations one has in mind. The common factor is the notion that unsupported objects *must* fall, not simply that they always do. If an object did not fall, we would infer that we were mistaken in thinking it unsupported. This intractability reflects our belief that the nature of physical reality requires that unsupported objects fall. We know that they must fall because of the nature of gravity when this physical characteristic is understood as a local feature of space itself. We do not know that objects *will always fall* because the local features of the universe might conceivably change, and frequently do, and even the fundamental features might change; but the point is that whatever physical system might take its place would require that something else must occur unless or until displaced. The point in saying that x is the cause of y is that given x, y physically (or psychologically) must occur, and we have good evidence for thinking x is the case. That it is always possible the 'good evidence' was misleading is why it is possible to say that y must occur even though the assertion of x and the denial of y is not self-contradictory. We do not, of course, know why the universe must have any fundamental feature x. A given x at t_0 provides the framework in terms of which 'must' judgments make sense and hence cannot themselves be shown to be necessary. And admittedly the framework of necessity may be different at t_1. But again the point is that if we have good reason for accepting x' at t_1 and know 'if x', y' must occur', then we have good reason for saying that y' must occur. And this is compatible with insisting that the assertion of x' and the denial of y' is not self-contradictory.

The relationship of the use of 'cause' and the nature of physical systems is a complex one indeed and needs to be seen from different angles to be fully understood. Consider again the case of the forest fire. One might argue that the lightning striking the tree was the cause, since the forest fire is an event and hence the cause of it must be an event also. Or one might argue that the means of ignition, dry underbrush, prevailing wind, etc., altogether was the cause of the fire. Or one might argue that the striking of the tree made it possible for the standing conditions to produce their effect, and hence that it was the releasing occasion – among several possible ones – for the operation of the cause itself. The multiple ways of using the word cause, each useful for some particular purpose, again suggests that such alternatives are simply different conceptual ways of slicing the same physical facts. The common feature that is reflected in all such talk is that some event or some combination of conditions and events was sufficient to produce or able to bring about the forest fire. The converse of 'x is sufficient to y' or 'x is able to bring about y' is 'y is necessitated by x', where 'necessitated by', it should be clear, is quite distinct from 'necessary to'. Physical necessity, provided by some physical system, is precisely this relationship of sufficiency between events and states of systems and is not an ontological tie that binds, hidden behind events and states of systems. What is sufficient to what, and what necessitated by what, is discovered by the experiments of science and the experience of daily life, and the necessity involved is the necessity provided by the systems themselves and is nothing in addition. The experiments and experience that constitute the sole evidence for ascribing 'sufficient to' and 'necessitated by' may be faulty and the ascription of these terms thus erroneous. But if we have no reason to think them faulty, we are justified in saying, in whatever case is involved, that *x was sufficient to y and y necessitated by x*. In a subsequent case of $(x \cdot \sim y)$ we infer that we are mistaken in thinking that this is a genuine case of *x*. Upon continued cases of $(x \cdot \sim y)$ we might well infer that the world had altered. But again such a conclusion neither counts against the fact that *x was* sufficient to *y* and *y was* necessitated by *x* nor the fact that whatever new system prevails the same intractability of physical necessity and the concept of cause will be present. That the concept of cause functions the way it does in our language is clearly not simply a matter of convention but reflects the fact that the universe has been sufficiently stable to make it a tenable conceptual tool for investigating that universe. But the concept of cause is powerless to insure anything about the future. The notion of cause *is* intractable; a cause *must* produce its effect. This is part of the meaning of 'cause' and the nature of any particular physical system. But this does not insure anything about the future facts of the universe. Particular causes and effects might come and go; only the concept of causation and the necessity of each physical system remain constant.

V

So far I have claimed that the concept of cause itself is an intimate mixture of definitional and natural necessity, the former ultimately reflecting the latter. (The redefinition of scientific terms by virtue of subsequent theoretical 'discoveries' is well known, but the significance of it for our problem has never been sufficiently understood.) This intimate mixture is further evident in any concept whatever that has a causal component and careful examination of several such examples should throw further light on our general theme.

Consider the following analysis of the meaning of 'copper'. By this term we mean something having the properties malleability, fusibility, ductility, electric conductivity, atomic weight 63.54, m.p. 1083 °C, and density 8.92 (properties both dispositional and explanatory in nature, be it noted). Perhaps this is not a perfect definition of 'copper', but let us agree for the sake of advancing the argument that having these properties is required for any piece of metal to be called copper and that any further properties used to refer to some particular piece of copper are part of the description of that piece of metal. Now what does it mean to say that just these properties are 'required' to *x*'s being a piece of copper and that all other properties are part of a specific description? Say that a person (*P*) at time t_0 calls a certain object (*O*) copper (*C*) and at time t_1 *C* turns out not to be ductile. One of two things might be said, either that *P* was mistaken at t_0 in calling *O* '*C*' (it looked like copper, but the light was bad) or that by the time t_1 came around *O* had changed its character and was no longer an instance of *C*. In either case, ductility is taken as a 'required' character of *C* because this dispositional property is interwoven closely with atomic structure and hence with every other dispositional property and so is part of a 'web of definition', any part of which if missing would render the application of the concept pointless. It would simply be silly to say that this piece of copper has an unusual atomic structure and therefore is not ductile. Hence it follows that the ascription of *C* to *O* at any given time necessarily implies the presence of a cluster of properties. It is true that the cluster is not absolutely rigid. There are variants which still qualify as referents of 'copper' since copper has two isotopes. The notion of isotope is extremely interesting here and might be called theory-laden variation – a notion, thus, that reinforces rather than destroys the ineluctable use and withholding of concepts that have a causal base. Scientific terms are not ineluctably used, of course, in the sense that they are insensitive to empirical results. On the contrary, they depend wholly for their ineluctable use on a well-established scientific theory.

This analysis, of course, is not limited to scientific concepts but applies as well to any ordinary ones that have a causal component. Consider the simple case of striking a match. I rub *O* against *S* where *O* has the properties of being round, wooden, and tipped with phosphorus sulfide while *S* is the rough surface of a matchbox, but the head of *O* crumbles instead of igniting. Round is certainly not a required characteristic, wooden is crucial insofar as it refers to something combustible, and igniting certainly seems central to the concept of a match. And yet it is unlikely that we would say in this case that *O* was not a match but rather that it was a damp one, a soft one, or faulty in some yet unspecified way. One would, however, never say of the Lucifer Match Company that it manufactures strange matches because while they look and smell like other matches they never ignite when rubbed over a rough surface. One would rather conclude that the Lucifer Match Company was not really manufacturing matches at all but was playing epistemological games. In the last analysis, igniting is conceptually necessary to the concept of a match because the physical system of striking a match in a certain way and its igniting is a causal one and exhibits the natural necessity of any system, be it physical or psychological. The reason that we are willing to speak in a concrete case of a 'faulty match' is that we know what the maker was trying to do and hence that something *must* have gone awry. So conceptual necessity is present whether we say in the particular case 'This was a faulty match' or 'This was not a match after all'.

VI

According to Hume, the major reason for denying any 'physical necessity' is that the denial of *E* and the assertion of *C* is not self-contradictory even though they have always occurred together in the past. The major reason for this view, in turn, is that "no contradiction is involved in supposing a change in the course of nature" (*Treatise*, p. 89). The first claim is a major insight. The sort of necessity that does exist in the use of causal terms is not at all the relationship of entailment between propositions. The second claim is also true but likely to send one off in the wrong direction. Indeed, there is no contradiction in supposing a change in the course of nature, but there is a contradiction in supposing that when *x* was the cause of *y* it was not a necessary relationship. And the same notion of causal necessity would hold for whatever was discovered about the subsequent course of nature. We would sort out terms to fit the new occasion: C' is now the sufficient condition of *E* and *C* is the sufficient condition of E', or whatever turned out to be the case, and the same conceptual necessity of 'cause' would prevail here. What would be constant in all the change would be the *meaning* of the concept cause, namely, that it is sufficient, of necessity, to produce its effect. The course of nature, it is true, might become so fluid that the notion of causality would no longer be applicable; but even this would not affect the meaning of 'cause'. And, one might add, realizing the involvement of causal notions in those of 'object', 'property', 'quality', and so on, it is difficult to see just what concepts *would* be applicable in such a world.

Hume thought that what seems to be physical necessity in causal contexts is really 'psychological necessity'. According to him, "necessity is something that exists in the mind, not in Objects" (*Treatise*, p. 165). Necessity is not a relation between objects but designates only the propensity which custom produces to pass from an object to the idea of its usual attendant. However, this definition of 'psychological necessity' already contains, as C. J. Ducasse points out, three terms of causality, namely, 'custom', 'produces', and 'propensity', which, in Hume's own terms, can be explicated only as regularity of sequence. Hence it turns out that Hume cannot really talk about psychological necessity any more than physical necessity. It is instructive to see how thoroughly blended into even philosophical talk is the notion of natural necessity (a concept embracing both physical and psychological) even when it has been supposedly exorcised.

VII

In cases of volition, we are told, necessary relations are directly experienced. In willing to move my arm I directly experience the volitional force or energy that *makes* the movement of my arm occur. In the physical realm, of course, we do not have any direct experience of the causal connection between *x*'s and *y*'s, but not experiencing such a tie is no evidence that it does not exist. We must postulate the existence of some unobservable force or ontological tie which binds together the cause and effect so that we can say the events are connected and not simply conjoined.

There are a number of confusions in this activity talk which, once seen, quickly suggest that there is no justice that needs to be done to the concepts of 'causal force' and 'connecting ties'. While the following three confusions are not exhaustive, they are the most flagrant ones.

(i) In saying that the cause *forces* or *makes* the effect occur, the activity theorist has confused prescriptive and causal laws. A prescriptive law is the expression of a norm; that is, what must be done if one is to avoid sanctions. It makes perfectly good sense to say that the sanctions of a prescriptive law force a person to accept it or make him follow it, since he might well prefer on other grounds not to accept or follow it. However, the notion of forcing or making is inappropriate in the context of causal laws, where 'causal law' means a generalization of concrete cases of sufficiency each one of which is experimentally established.

(ii) It is true to say that not experiencing the causal tie is no evidence of its non-existence. But it is not a very helpful truth. What we need is some experimental evidence that it does exist, and there seems to be none whatever for this.

(iii) What, after all, do we mean when we say events *A* and *B* are *connected*? We do not mean that there is a tie between them because the terminus of the tie on each end would need to be tied, and so on *ad infinitum*. Rather we mean that *A* and *B* are connected in the sense that there is a chain of intermediate events so that *A-a-b-B* is perfectly continuous in space and time. Since there is no gap in space and time between *b* and *B*, there is no further connection for these events to exhibit. Consider the example of a doctor experimenting with a new drug. There are two patients one of whom is given the drug and the other not, all other factors presumably being held constant. The former is cured but not the latter. Clearly the doctor has not yet succeeded in finding out why the drug was sufficient to cure the illness – why the disappearance was necessitated by administering the drug – and will not know this until he succeeds in making *A* and *B* contiguous in space and time – that is, connecting them. He would know how the events are connected and why *A* is sufficient to *B* if he knew, e.g., that the drug after being injected into the veins comes into immediate contact with blood particles which undergo a chemical change and travel through the body until they come into contact with some organ, thereby changing it in a particular way. This sort of connected sufficiency *is* physical necessity and is always provided by some particular physical system. There is no causal tie behind every physical system. Rather the system itself provides its own particular sort of sufficiency and necessity. The trouble with the activity theorist is that he begins with a mistaken assumption about what is present in the world. Not finding it taken into account in other views of causality, he always finds them lacking something. We must avoid the temptation of false expectations in order to avoid unnecessary disappointment.

VIII

It is really silly to maintain that unsupported objects (unsupported in any way, that is, including air currents, motors, etc.) may or may not fall to the ground. If the falling of unsupported objects were not explained in any way, then it would be an amazing coincidence that these two aspects of experience should be invariably conjoined. A. C. Ewing asks, in view of such considerations, what sort of explanation could there be "except that the nature of the bodies or the nature of the physical universe as a whole somehow entailed their moving in that way?" (*The Fundamental Questions of Philosophy*, p. 164). To be sure, he says, it is false that inductive inference is identical with logical entailment, but it must be similar or analogous to it in order to make sense out of what no one really doubts, namely, that it is not a coincidence that unsupported objects always fall. It is not identical with entailment because the assertion of *C* and the denial of *E* is not self-contradictory. The reason for this, however, is not that there is no necessary connection between cause and effect but rather that a cause is a whole set of necessary conditions only some of which in any given case are known. Hence causal judgments may be mistaken without destroying the claim that the causal relation, if genuinely present, is necessary.

Ewing is quite right in insisting that it would be silly to maintain that what we *mean* by causation is constant conjunction. If 'cause' meant only 'de facto conjunction', then all *C-E* relationships would entail 'going together' but not 'belonging together'. But certainly in our ordinary concrete causal judgments we intend to say that two events belong together. Their association is not a cosmic coincidence. He is wrong, however, in trying to explicate this concept of belongingness by the concept of entailment. He never explains how the causal relation can *possibly* be analogous to entailment since one holds between events and the other between propositions. Moreover, since the truth of 'The speaker has a vivid imagination' entails the truth of 'Either the speaker has a vivid imagination or he ate a peanut butter sandwich for lunch', it seems clear that any extensional explication of the concept of causal necessity is useless.

Ewing, it seems to me, is mistaken in still another way. He believes that his theory avoids the mischievous problem of induction. If the meaning of 'cause' is taken to be 'de facto correlation', he says, then Hume's strong form of skepticism is justified: experienced correlations do not even make causal judgments probably true. On the other hand, if the meaning of 'cause' involves the notion of entailment, then causal judgments can be extrapolated justifiably to new cases. However, this reassurance is worthless, since the problem of induction now becomes one of being sure that you have a genuine instance of some *C*. Ewing, as we have seen, recognizes this problem, for he writes that a cause is a whole set of necessary conditions only some of which in a given case are known. Hence causal judgments may be mistaken and that is why the denials of causal propositions are not self-contradictory. But if we can never be sure that we have a genuine instance of *A* we can never be sure of any given extrapolation and hence it is unclear how the entailment theory helps in the slightest in avoiding the problem of induction. It simply has a new version of it. And it should be clear that my analysis of causal necessity also encounters this new version of the problem of induction. If I encounter a case of $(C \cdot \sim E)$ I will infer not that *C* malfunctioned this time but that I really did not have a genuine instance of *C*. The problem of induction now becomes, of course, how can I ever legitimately claim beforehand that I have a genuine instance of *C*?

IX

There is, in sum, more to causality than Hume thought, though he was right in saying that inductive inference is never demonstrative. The entailment theorist is right in thinking that there is something more in causality than *de facto* conjunction but wrong in thinking it is entailment. The activity theorist is right in thinking events are connected but wrong in thinking it is an ontological tie that binds. There is no necessity for ontological rubber bands when language and physical (or psychological) systems provide their own sort of subtly interwoven necessity.

State University of New York at Buffalo

NOTE

* Revised version of a paper read at the Boston Colloquium for the Philosophy of Science on October 24, 1967. I am indebted to C. J. Ducasse for several points and examples and to Kenneth Barber, Robert Greenberg, and John Woods for helpful criticisms.

JOSEPH AGASSI

UNITY AND DIVERSITY IN SCIENCE*

> ... the value of science as metaphysic belongs ... with religion and art and love, with the pursuit of the beatific vision, with the Promethean madness that leads the greatest men to strive to become gods. Perhaps the only ultimate value of human life is to be found in this Promethean madness. But it is a value that is religious, not political, or even moral.
>
> (Russell, *The Scientific Outlook*, ch. iv, 'Scientific Metaphysics')

> He who has science and art has religion.
>
> (Goethe)

TABLE OF CONTENTS:

ABSTRACT. The idea of the unity of science is the historically very important idea of total rationality and objectivity.

This idea is a utopian dream, and a rather dangerous one.

The Popperian view of rationality as a goal directed, i.e., as problem-solving, method of trial and error is a better view or rationality.

Solutions to problems offer the element of unification, and their criticisms offer the element of diversification.

The idea of the unity of science is a utopian dream, which is appealing because it has so many aspects – emotional, moral, social, methodological, epistemological, and ontological – especially, as we shall see, ontological. The idea has seldom been articulated, and almost never been examined – even superficially. On the face of it, at least, it may look as if the idea of the unity of science clashes with the idea of diversity of unique individual things; which latter idea has a great appeal too. Therefore, still on the face of it, there is much room for discussion. Both the idea of unity and the idea of diversity are by no means confined to science. People prefer to apply the idea of unity to science because they take science

Boston Studies in the Philosophy of Science, IV.

as synonymous with, or at least as a paradigm of, rationality. Moreover, behind the general idea of unity itself stands the idea of rationality. This idea of rationality, in ethics as well as in epistemology, was the idea of universalizability – of all individual phenomena in all their aspects. Furthermore, behind the idea of universalizability stands the idea of rationality again: namely, that universal laws not only cover *everything*, they can be proved by arguments that all rational *men* must accept. Still further, behind the idea of proofs acceptable to all rational men stands the idea of the rational unity of mankind.

Now this idea of utter universalizability, as it is used in the history of thought, is rather dangerous. I would contend that the dangerous element in it is precisely the theory of rationality which stands behind it: behind the idea of universalizability stands the idea that rationality is comprehensibility and comprehensibility is universalizability; hence everything is rational (i.e. universalizable). In opposition to this confident promise of complete rationality, I suggest that science is propelled not by aiming single-mindedly at universality, and not by aiming at diversity either, but by the constant conflict between these two tendencies. I further suggest that this thesis can be maintained only at the cost of replacing the old-fashioned idea of rationality as universalizability by means of proof, with the Socratic idea of rationality as a process of conflict between universality and specificity, often rooted in conflict of ideas about universality, to wit rationality as Socratic dialectic. And insomuch as science may be viewed as a paradigm of rationality – and (considering Plato's early dialogues seriously in this light) certainly not *the* paradigm – its history has to be reconstructed as that of a dialectical procedure.

Thus far the dialectic view of rationality presented here is that expressed by Popper in Chapter 24 of his *Open Society*; I shall now present an elaboration on it, which may be attributed collectively to Popper and his school.[1] In the *Open Society* Popper identifies the principle of rationality with the principle of criticism of putative views, proposals, criteria, desiderata, etc.; here I shall present rationality of thought as a special case of rational action in general. Rational action is (traditionally) viewed as conduct conducive to given ends – i.e. goal directed. In intellectual discourse, the ends are called desiderata; we begin with desiderata. Hence, rational thought is goal directed, and hence rational. (For example, a discussion within the traditional field of astronomy will at times begin with the description of phenomena to be explained, at times with difficulties stemming from empirical observations to be ironed out, at times with the aim of modifying an existing theory which is at odds with other existing theories, at times with the aim of inventing new tests of a given theory.) Whatever the situation is, we may begin with stating the agenda, the desiderata. These may be critically debated, or ideas allegedly conducive to them may be critically debated. This raises a problem: in order to criticize our standards of *x*, we need standards of criticism (of *x*); when *x* stands for criticism, things (seemingly, at least) get out of hand. How then, can our ideals of rational thinking – as conducive to our desiderata – be subject to critical debate? Criticism is not more fundamental than rationality, since criticism is impossible without reference to desiderata – in this case desiderata for rationality. Nor do the desiderata stand above criticism.

Let us leave this difficulty now, and go back to my historical sketch.

I. AMBIVALENCE TOWARDS UNITY: AN IMPRESSION

Let us examine impressionistically a few intellectual clichés and their transformations – remembering that in the present case, at least, clichés are not necessarily of the vulgar. 17th- and 18th-century philosophy entertained an optimistic mood that was largely[2] shared by 19th-century philosophy, though it hardly agrees with out present apprehensive mood. We are now as a matter of course worried about tasks which may be beyond human capacity. H. G. Wells stand on the watershed, as the author of *The Shape of Things to Come*, who foresaw the possibility of a return to the Middle Ages, and as the author of *Joan and Peter* where he puts in the mouth of God the declaration, "There isn't a thing in the whole of this concern of mine, that man cannot control if only he chooses to control it. It's arranged like that. ... But ... men ... are too lazy...." It's arranged like that, he says. Today we turn to apprehensive pre-Enlightenment clichés suggestive of an inflated intellect and a deflated morality; Russell echoes Rabelais' saying "Science without conscience is damnation." Yesterday the cliché was Spinoza's: the rational man knows that it is the best policy to react to hatred with love; or Kant's: the sense of respect for the law, the desire to universalize, is what makes one moral. For Kant, even love and friendship and so on, were either derivative of rationality, or peripheral to morality since, *eo ipso*, the problem of ethics is, how should I, as a rational being, act in the present circumstances. How far this is from the Christian disquisitions on ethics, where love was central and all else peripheral![3]

The optimistic doctrine of rationality as a doctrine of total proof – in principle of all truths – of total control – in principle of all circumstances – and of total propriety – in principle of all conduct – is not easier but harder to grasp than the old problematic idea of an omniscient, omnipotent omnibenevolent deity. Yet the idea of man-God is what I am trying now to present as historically powerful, as an idea advocated by Descartes and Spinoza, by Kant and Laplace, by Comte and Marx, as the most significant regulative idea in the post-medieval period.

Admittedly, this idea was seldom discussed fully and openly. Yet its pervasiveness is symbolized in the overpowering optimism of Condorcet and of Madame Roland even when persecuted by the French Revolution itself. The optimism Condorcet displayed was hardly a matter of temperament – it was a case of an ideology which overrode any temperament. A young man by the name of Berthollet, known merely because his father was a famous chemical philosopher, was so depressive that he was driven to suicide. He sealed himself off hermetically, retaining a candle, pen, and paper, and resolved to record his own observations of his dying self. His writing, incidentally, soon became illegible; as an act of faith it strikes us as remarkable, perhaps at the time it was less so.[4]

The optimism, then, of the idea of the unifying supremacy of reason was not in the psychological domain but in the intellectual, as was the reluctance to discuss it; indeed, both factors still persist. Even nowadays, when the idea is so much more tempered than it used to be, it is seldom clearly and lengthily discussed; it is often alluded to, but no more. We still are very ambivalent about it. In the opening paragraphs of *The Identity of Man*, Jacob Bronowski hits it exactly. The fundamental wish man has to be at home in the world, the intuitive assumption so-to-speak that man is a part of nature, he says, has a strong intuitive appeal and also clashes head-on with the equally strong intuition that man is unique, in a sense above inanimate matter and even above all other creation.

This ambivalence can be further illustrated by Freud's personal attitude. The feeling man has of being part of nature he called the oceanic feeling, and this feeling plays quite a significant role in his metapsychology. Yet in the very beginning of *Civilization and its Discontents* he expresses an ambivalence towards it. His ambivalence at least in part derives from his fear that there might be a kernel of truth in the idea that the oceanic feeling is associated with the need for religion. When he comes to technology he exults – but he first creates a barrier; he identifies oceanic feelings with narcissism and religious needs with a sense of helplessness, the need for mother's womb, for father's protection. These needs, he tells us, give rise to the religious ideas of omnipotence and omniscience. And these same ideas come impressively near to realization, he adds, in modern technology which has rendered man almost Godlike. Yet the connection between narcissism and a sense of helplessness could not escape Freud. "There may be something behind that", he admits, "but for the present it is wrapped in obscurity." "For the present" means when he discusses oceanic feelings; elsewhere he easily connects helplessness with narcissism. Freud usually allows for the fusion of psychological extremes, which may replace each other in a variety of mental mechanisms; and the same goes for the fusion of their symbols. Freud's theories of all dreams, of many neuroses, and of schizophrenia, are instances. We have the axis impotence-omnipotence, and its symbols demonology-religion; we have the axis of insecurity-security or alienation-oneness, and its variety of symbols of self-hate and of narcissism. If you allow impotence, omnipotence, demonology, and religion, to mix together, you might just as well allow impotence and insecurity to mix together, especially since, according to Freud, a sense of insecurity is what gives rise to a sense of impotence. Indeed, this mixing is the core idea of Freud's theory of (paranoid) schizophrenia (where insecurity leads to the deluded sense of omnipotence rather than of absolute security).

Freud cannot tell us why a sense of impotence, in its transformation to its opposite, may lead one person to a religious vision of omnipotence and another to a technological vision of power. Nor does his doctrine enable us to tell why a sense of insecurity and alienation, in its transformation to its opposite, to the sense of security and integration, may drive one person to a mystic dissipation of his soul, through rituals of purification, into unity with the soul of nature and another person to a technological dissipation of his body, through a control panel, with a panorama of bulldozers devouring the landscape and transforming it into a human artifact.

Once we are alerted, through Freud, to the ability of insecurity or alienation to transform itself into security and oneness, we may notice this transformation occurs fairly frequently. The connection is all too obvious even from a seemingly casual example which Wittgenstein carefully analyzes in his lectures on ethics. Consider a remark, such as, "I feel so safe, as if no harm will happen to me no matter whatever else may happen." Wittgenstein analyzes the verbal formulation of such statements, and shows that the above statement should read, "I feel so safe, as if nothing can happen which may harm me." He does not explain why he first formulates it in the objectionable manner except to say it is common. In the light of Wittgenstein's and Freud's analyses we can say, it is common because it expresses – adequately – both the oceanic feeling of oneness, and insecurity or alienation. Wittgenstein himself, however, centered on the religious aspect of the matter alone.[5]

My stress on ambivalence here is not a digression to psychology but a running explanatory theme. I wish to explain why the idea of unity which, as I claim, is so common, is so rarely expressed and never analyzed in the cold light of reason.

Let me give an example from recent philosophy before going to the history of the treatment of the topic. In the first monograph of the *Encyclopedia of Unified Science* Carnap briefly formulates the thesis of the unity of science which is substantially almost identical with Bacon's. This is a remarkable fact, particularly in view of the absence of literature on the topic – even Bacon's presentation is all too brief. Now Carnap admits that his thesis cannot be proven, but he consoles himself by saying that at least it cannot be refuted. He adds that a precondition of the idea of unity of science is the idea of the unity of the language of science and he is thus able to conclude his discussion in a few sentences. He does not even notice that his own philosophy makes it possible to have diverse languages of science developing into a unified language as they go along. All this is amazing. Even the slightest acquaintance with Carnap's philosophy, or of the philosophy of the Vienna Circle, suffices to raise an alarm: ideas which cannot be proven or refuted surely sound dangerously metaphysical and hence meaningless. The alarm may turn out to be a false alarm; my point is that Carnap does not even notice it! He really wishes to take the unity of science so much for granted as to have to say about it the barest minimum.[6]

Bacon's conduct is almost identical. For Bacon metaphysics is evil not in itself but in its methods. The proper method is inductive. From facts we conclude inductively some laws; from the laws the more general, and hence fewer, laws – the *axiomata media*; from these we conclude inductively the few laws of metaphysics; and from the few laws of metaphysics we conclude, by induction again, the single law of natural theology. This natural metaphysics and natural theology are proper, as they are based on induction and thus scientific. But they belong to the future. Traditional metaphysics is bad due to its speculative methods, due to its being based not on induction but on speculation! And yet, the very idea of the unity of science just outlined is highly speculative. And Bacon freely admits this. "And therefore the speculation was excellent in Parmenides and Plato", he says unabashed, "although but a speculation in them, that all things by scale did ascend to unity."[7]

If the idea of unity had met with a single critic, surely his task would not have been hard to perform, especially since in the Western tradition of methodology the dominant school was inductive, not deductive.[8] Yet there was hardly a critic around.

The deductive version of the idea of unity was more consistent – especially in the work of Spinoza; but it was even more optimistic in that version. To believe that already now we possess in essence all knowledge, and that knowledge is a sufficient basis for ethics, already makes us uncomfortably Godlike – in reality: in principle, Spinoza claims, he possesses the knowledge of the essence of all things. In all the literature against the Cartesian apriorist school, however, this point is never taken up: the apriorism of the school was attacked, and specific scientific views – yet what it shares with Bacon's speculations was curiously left unmentioned; on the contrary, one feels that the quarrel between inductivists and apriorists was a family quarrel just because of so much unmentioned common ground and mutual understanding.

(I should like to mention that this explains a quaint historiographic fact. However Baconian the French thinkers of the Enlightenment have declared themselves, again and again certain historians of ideas cannot avoid declaring them Cartesian. Indeed, the further away from the Age of Reason we get, the harder it becomes to distinguish between Bacon and Descartes, and the latter's private letter to Mersenne in which he declares almost full agreement with the former sounds commonplace; yet when we meet this communication immediately after the standard sophomore course in the history of philosophy from Descartes to Kant we feel numbed as though we had had intellectual shock. Incidently, Descartes does not explain his agreement with Bacon: as I have said, high pitched rationalistic optimism was seldom fully articulated.[9])

II. THE ETHICS OF SCIENCE AS A UNIFIER OF SCIENCE

Thus far I have discussed the strange fact that the unity of science was seldom fully presented and seldom criticized. I have said, it is often alluded to and then always highly favourably, and it was always conceived as a sacred point of science – but there is almost no criticism of it. Let me outline some of its appealing aspects.

I have already mentioned the enormous emotional appeal, when discussing the oceanic feeling. This, really, is only a variant of the overflowing sweetness allegedly felt by the mystic when uniting with the universe. And there is little doubt in my mind that if one traces the idea of the unity of science back to the 16th century one arrives at the mystic doctrine of the time, now called light-metaphysics, then called Pythagoreanism, and previously called Cabbalism – which is a mutant of neo-Platonism. And, if, with Bacon, one goes back to ancient Greece, one finds the same mysticism there, in the Pre-Socratics, particularly Parmenides, and even in Socrates, his skepticism notwithstanding.[10]

One need hardly say that the mystic doctrine is, as Bacon noticed, not only emotional but also, even chiefly, ontological. And rationalism, whether inductive or apriorist, always rested on the thesis of the utter comprehensibility of the universe by reason. Hence the hostility of science to religion. Hence, likewise, the call to return to religion, which stemmed from the aftermath of the crisis in physics of the turn of our century; once one claims that reason may be limited, the old conflict is gone, once and for all. This is how Maimonides put it; and this is how things still stand. Though Maimonides believed in unity and in comprehensibility, he stressed the limits of reason, as his chief quarrel with the rational unbelieving philosophers.[11]

In order to show that the utter comprehensibility of the universe drives religion out, one has to place morality within the domain of comprehensibility. This move looks incredible; at the very least it looks like the error known as naturalism, or perhaps as the naturalistic fallacy (not quite in the sense of G. E. Moore, inventor of the phrase), namely of (invalidly) deriving prescriptions from descriptions. Not necessarily so; Kant, at least, knew better, and yet he did subsume morality under rationality. There is one obvious naturalistic element in Kant, namely, the fact that we are rational (he strangely never called this highly questionable view into question); yet he did not commit a fallacy, since the link is in his claim to observe a conscience – in the form of a sense of respect for the law – which, he claims, obliges us to universalize, to take others as we take ourselves, which is the golden rule.

How is this egalitarian universalistic sentiment, one may wonder, integrated within the optimistic tradition?

This is a point much more articulated in popular philosophy and cheap science-fiction than in more serious philosophy. Its epitome may well be the mad scientist from the marvellous cheap chiller-movies of the thirties, sipping his claret, and airily, half-pensively, mumbling in perfect Oxford diction, 'Murder, you say? My dear fellow, this is a scientific experiment!' The mad scientist does not recognize morality and, in effect he claims, he has no duty to morality as long as he is fulfilling his duty to science: as long as he is a good scientist, he implies, he is good. Being a good scientist first was conditioned on being a good citizen; soon being a good scientist became a touchstone for being a good citizen (show me a good scientist and I will show you a good citizen); the last step, the perversion, is the conclusion that *eo ipso*, being a good scientist implies being a good citizen. In the beginning of this line of reasoning, morality is taken for granted, and science is declared subject to its laws; in the end of this line of reasoning, the roles have subtly been reversed (in a manner, incidentally, quite common in schizophrenia: the mad scientist is clinically nearer to the truth than the whole of Ibsen's lunatic asylum).

It ought to be stressed, if confusion is to be avoided, that the idea of the separateness of science and morality is 20th-century; traditionally we have pious science and evil science and nothing in between. Also, one must notice, the idea of evil science is as old as the idea of pious science. It is a serious mistake to think that fears of evil science were invented by Mary Shelley, or, more generally, by the 19th-century antiscientific Romantic movement. The idea is as old as the myth of the proximity of the gates of Hell to the gates of Heaven; the practical Cabbalist and the alchemist shared it, and their monster of Frankenstein was the Golem of Prague. Francis Bacon had it too. His idea that the scientist must be pure is not accidental to his philosophy, even though, as C. W. Lemmi has shown in his *Classical Deities in Bacon* of 1933, Bacon borrowed it from an alchemist by the name of Natalis Comes.

Although Bacon often spoke of the inductive method of demonstration, he never explained it; much to the frustration of his chief commentator Robert Leslie Ellis. His real view of the matter, however, had nothing to do with demonstration, but with the idea of grace as understood by the practical cabbalist and the alchemist. Speculation is putting Mother Nature in chains – we force her to conform to our blueprint. Induction is not so much collecting facts for the purpose of demonstration as the act of worshipping Mother Nature, of showing her humble obedience and respect. And, obedience is the best method of enlisting Her services; for She will reward us by revealing to us all Her secrets. You may be sure of that, since it is the prerogative of God alone to be hidden forever.[12]

Abstract Speculation, says Bacon, is tyranny over Nature as well as the conquering of kingdoms of the mind, which are philosophical schools. The speculant is impatient and wishing to achieve fame he arrives at his theories by spurious methods, attaining the goods by force instead of by honest work. Most attempts at rational reconstruction of Bacon's philosophy are directed at finding some technical (inductive) rationale for his demand for hard work and attention to detail. I do agree that Bacon does provide snippets of such technical justifications of his demands to work hard and amass details without end (e.g., theories are squeezed out of facts like wine out of grapes); that he even promises much more to come. Quite apart from the fact that he rejected the most common rationale – we collect facts in order to generalize them – he is definitely baffling on this point. Justus von Liebig claims that Bacon's promises are as vain as those of his patron James I. Robert Leslie Ellis was more sympathetic and he tried hard to make sense of Bacon's diverse requirement – but only to admit defeat. Quite possibly, and in my opinion more than possibly, the large collection of facts was less the material for input into the induction-making machine which Bacon promises, and more a matter of ritual labor which, like all ordeals, cleanses its bearer and makes him worthy of success. The rationale for hard work, then, is not technical but moral. There is an element of the so-called Protestant ethic in science – nowadays still so popular under the impact of Weber and Merton. With the exception that it is not Protestant but Medieval – alchemist-cabbalist – as I have already noted.

We have three traditional characters worthy of examination, equally at home in popular literature, in traditional philosophy, and in scholarship of all sorts. We have the good scientist who is a good man, a good citizen, a pious person – a man answering all descriptions of your personal ideals. He is the hero of most so-called science-fiction literature; he is the hero of most of the trash and nonsense and willful distortions and embarrassingly obvious self-deceptions of the so-called biographies of such impossibly difficult individuals as Newton and Laplace, heavily made-up as placid, warm, and worldly-wise. Also, we have the extreme opposites of such ideals – the enemies of science, with no redeeming features – stupid and brutal all at one. The two are reminiscent of Dr. Jekyll and Mr. Hyde. (Stevenson tells us that Jekyll is the usual mixture of good and bad, who succeeded in isolating only the bad – the Hyde – but not the good. He describes Jekyll, however, as an angel, perhaps because he is a scientist.) These two are the only utterly unproblematic and straightforward stereotypes. The mad scientist is the freak, the only somewhat problematic stereotype: he is the person whose mind is bright but whose heart is dark; Dr. Faustus.

Whereas the only three traditional archetypes are Dr. Jekyll, Mr. Hyde, and Dr. Faustus, the 20th century has created a new one, best presented in the biographical movie on the life of Dr. Wernher von Braun. Whether the movie tells the truth about its hero's scientific or moral career is not of much significance here. The movie does, it is well-known, represent its hero as he himself has chosen to appear to the public: able and concerned as far as science is concerned, but rather indifferent to moral and political questions, or at least one who puts scientific considerations first and moral ones second.

Dr. Faustus is not at all Goethe's Faust. Goethe's hero is a really unified person – he loves science, youth, love, happiness, righteousness and, finally, action. Unlike Faust, Dr. Faustus comes from the same folklore to which Bacon's philosophy belongs: he is an occultist whose mind got twisted and who became evil, poor fellow. He is not a refutation of Bacon's doctrine: he is a fallen angel, a freak. He was a good scientist and hence a good citizen, but he lost his soul to the devil. Like his heir, the mad scientist, he lost his heart but not his brain; to be precise, not the storage part of his brain. It is von Braun who breaks away from the Baconian tradition. We are still ambivalent about him – emotionally, morally, philosophically. We cannot decide, on each of these three levels, whether we prefer a mad scientist to a heartless one. Emotionally, the mad scientist repels us violently, we fear empathizing too much with him; whereas the heartless scientist arouses only revulsion. We answer the heartless scientist with heartlessness, yet we can empathize with the very same hateful, sadistic mad scientist – especially when invited to empathize with his psychiatrist. Indeed, when we meet Jules Verne's mass-murderer Captain Nemo and see how unjustly and severely he had been tortured, how kind and clever he still is, we tend to forget and forgive his aberration as a bad dream. Morally, too, we vacillate; in *The Devil's Disciple*, George Bernard Shaw places opposition and hate nearer to love than cold and deep-seated indifference; Shaw speaks the same language as Buber's *I and Thou*. Buber offers the same quasi-Christian message philosophically (there is more life of dialogue in disagreement, says Buber, than in ignoring an opponent). Of course, the situation can get completely distorted and revert to the worst versions of Christianity where indifference equals hostility. This, however, is not forced on us because there is a difference between our attitude to those with whom we happen to have no ties of sympathy and our attitude to those with whom in principle we can have no ties of sympathy. It is the latter class that, morally, we do not quite know where to place. Those who have put themselves beyond morality, have made us unable to feel anything for them, and thus they make us doubt whether our morality applies to them at all. Over this ground much contemporary moral philosophy wrestles. Much of the interest in and concern about the Eichmann trial may have stemmed from this, rather than from empathy or antipathy towards him. But our ambivalence is also philosophical, and it reflects a reluctance to give up the philosophy of the unity of science.

To illustrate this, let me refer to a strange similarity between the views of the leader of the Vienna Circle, Otto Neurath, and those expressed by the great scientist and Nobel Prize Laureate Werner Heisenberg. In his contribution to the *International Encyclopedia of Unified Science* of 1936 Neurath stresses the traditional hope that unified science will bring its practitioners together from different disciplines, from different walks of life, and from different countries. There is nothing objectionable in

Neurath's view except perhaps that, especially for the mid-thirties, it is somewhat naive – and who does not regret this! The same thesis is to be found in a lecture given by Heisenberg in Göttingen University in 1946 and reprinted a few times under the impressive title 'Science as a Means of International Understanding'.[13]

The problem aired in that paper was, understandably, very much in the air where it was delivered: what should a German science-student do to remedy the moral and political situation of post-war Germany? There is perhaps a lot to be said on this question, e.g., regarding the need for expiation and the like. I shall say no more than that Heisenberg barely faces it. Yet he does give an answer which is the same perversion of Baconianism that we have already met. He argues, to use my words, that since a good scientist must be a good citizen, you should now forget as much as is reasonably possible about all your civic duties so as to concentrate maximally on your duties as science-students. Become good scientists, and *eo ipso* you will also be good citizens. So much for my words; now I shall give a critical summary of the paper in its own words, so that you may judge whether I was fair.

"It has often been said ... repeatedly been stressed, with justification" that science is international, interracial, interreligious. "At this particular time [1946] it is important that we should not make things too easy for ourselves. We must also discuss the opposite thesis, which is still fresh in our ears." Why must we?

It is no doubt the case that we need not make things too easy for ourselves, that we should listen to critics, pay serious attention to opposite opinions, be ready to learn from opponents and change our minds. But there is a limit to all this. We are not obliged to take seriously every opponent. Heisenberg goes on: "I should like [!] to discuss which of these two views", he says, namely the unity of science and the national, racial exclusive view of science (i.e. Nazism), "is correct and what are the relative merits of the arguments that can be produced in their favour." I hope the reader suspects that there is a distortion here. Fortunately, indeed, there is: Heisenberg does not quite mean what I have quoted him to say. Yet the quotation is rather thought-provoking.

"To gain clarity on this question", he says, he wants to show how science works; and he tells the story which he can tell from personal experience: how international and unified the Copenhagen school in physics was when he joined it in the twenties. He learned in Copenhagen, Heisenberg says, that race and nationality are of no use in science; he received later "further proof of the 'objectivity' of science and its independence of language, race or belief". The thesis of the unity of science, then, is proven. The argument, opened on the first page of the lecture, is over on its third page; before the opponent had the chance of a hearing, it seems.

But Heisenberg goes on, nonetheless. A digression to the story of the internationalism of science in the 17th century, including a quotation from Dilthey, and we are ready to slip into a difficulty. "It has always been considered self-evident that adherence to such an international circle would not prevent the individual scientist from devotedly serving his own people and feeling himself one of them. On the contrary", internationalism enhances one's sense of love and indebtedness to one's own country.

Is there a problem lurking here? Indeed, it soon appears. How is it, asks Heisenberg, that all these cases of international scientific collaboration "seemingly do so little in preventing animosity and war?" Note the word "seemingly". The nation of Gauss and Helmholtz and Kirchhoff and Fechner and Wundt and Koch and Planck, the nation of Kant and Goethe and Beethoven, to throw in an additional bit of international collaboration, only "seemingly" benefitted "so little" from that volume of collaboration in its attempts at "preventing animosity and war".

But I misread: the words "seemingly" and "so little" merely serve to make an understatement. Heisenberg clearly states that science is politically impotent; and for a simple reason: there are very few scientists around, and the forces of politics are much too strong, resting, as they do, on masses of people, their economic conditions, and on "a few privileged groups favoured by tradition". The Junkers and their like plus mass unemployment, in case you miss the hint. "The political influence of science has always been very small, and this is understandable enough. It does, however, place the scientist in a position which is in some ways more difficult." We remember that it is not a question of divided loyalty: the scientist loves his country as much as anyone else. The reason for the difficulty, says Heisenberg, is that politics largely depends on applied science. "Thus, the action of an individual scientist often carries far more weight than he would wish and he frequently has to decide, according to his own conscience, whether a cause is good or bad. When the difference between two nations cannot be reconciled he is therefore often faced with the painful decision" of whether to work with his colleagues or countrymen.

Not that the unity of science has been disproved, mind you: it has been proven less than two pages ago. Not that there is a conflict of loyalties either – this was established less than a page ago. Yet, on the fifth page of Heisenberg's lecture all of a sudden the poor scientist is torn by inner conflict and outer strife.

The lecture goes on. There was a profound recent change in the relations between scientists and governments. In World War I, scientists adhered to their national loyalties. Not so in World War II, when "scientists claimed the right to judge the policies of their governments independently and without ideological bias. ... Eventually scientists were sometimes even treated like prisoners in their own country and their international relations considered even immoral." In 1946 Heisenberg spoke of governments in general, as if the status of scientist in one part of the world was comparable to that of another. Historians of post-war Germany may care to take notice of the fact that in Göttingen such words were spoken in 1946.

Heisenberg's outline of the history of international collaboration is not clear to me. In the 17th century and in the twenties of the 20th, science was international. In World War I it was not; yet 'a profound change' took place in the thirties! I ask the reader to forgive me if I do not discuss this puzzlement, however, and if I now become a bit more cursory in my summary. The next paragraph speaks of the martyrs Bruno and Galileo. The next to it speaks of the bomb, chemical warfare, and the like. The paragraphs which follow speak of the moral burden this puts on the individual scientist. "Can science really contribute to" international understanding in the present circumstances? After a brief digression to the Middle Ages, we come to the history of nihilism (which "we find ... in many parts of the world today" (not only in Germany?) even in "its most unpleasant form" (they too have concentration camps?) "disguised by illusion and self-deception" (they too dream of world domination?)), and to the possibility of its reinforcement by science, if not more than that. I am again at a loss: was science ever influential? If not, then, since nihilism is influential "in many parts of the world", the accusation that science contributed to nihilism fizzles out. If yes, how can we maintain the belief in the unity of science in Germany in 1946? I am at a loss. Why does he not refute this charge?

Heisenberg's discussion is nearly over: against his critics he must reaffirm his faith. The fact is that science is the way to "the centre", which is often called "spirit", or "God", or by other names. As the highway to God, then, science, in principle, is accessible to all. And the more people learn science, the greater the numbers of people who learn science, the nearer we shall come to God, and hence to live in a happier and nicer world.

And so Werner Heisenberg begins and ends with the profession of the same traditional faith as Otto Neurath. On the way he somehow succeeds in adumbrating certain severe criticisms of the thesis of the unity of science (e.g., in an unexplained manner it has led to nihilism), and he even speaks in a vein, totally alien to the traditional spirit of the thesis of the unity of science, that above and beyond his duties to science a scientist has to consult his conscience about the ethics and politics of science. But these 'above and beyond' are transient, I understand Heisenberg to suggest; soon science shall triumph and one's duties to science and one's other duties will merge again. Hence, the advancement and spreading of science are of great importance because a scientifically orientated world is one in which science can indeed help foster international cooperation and thus minimize hostility and war. Science, you remember, is in Heisenberg's view the highway to God.[13]

Heisenberg is only an extreme example of a less extravagant but rather popular view. We are all willing to agree that the world is not black-and-white, and so our archetypes are not quite true; but they are, by and large, not too remote from attainment and so can serve as reasonable ideals. C. P. Snow, in his celebrated *The Two Cultures*, blames artists for being Luddites and commends scientists as progressivists. For the sake of balance he admits that men of science all too often have no use for Dickens, even for books sometimes; but he sees in this only an aberration. He is, by and large, committed to progress, to science, to rounded scientific education, to an art – especially literature – which is committed to both science and progress. Literature is committed to progress, science equals progress, humanism equals progress, progress is a free man's worship – these ideas are common to the tradition of the unity of science. In his dream Descartes sees a poetry book and a dictionary, and in his dream he interprets the dictionary to be science and the poetry book to praise wisdom. So felt Dryden, so felt the Moderns of the late seventeenth century England, so feels Sir Charles Snow.

Sir Charles will admit that in fact things are not so simple; he will admit, in particular, that all too often men of science ignore the arts. Indeed, the great novelty and the importance of his work lies precisely in his recognition, which is his starting point, that scientists have no more commerce with artists than artists with scientists. Yet, finally, he is mildly critical of the scientists but he comes down hard on the artists. His bias cannot be overcome by the mere observation of symmetry.

What I have tried to illustrate with these two examples – Heisenberg and Snow – is, first, that empirical criticisms of the thesis of the unity of science can be easily dismissed, and they often are. Second, that the thesis may be employed to defend positions alien to its spirit.

The idea, then, that scientists must be pure and disinterested etc., is in many new variants still very popular; what I have tried to show previously is that this is an essential element in Bacon's philosophy, just as much as in Spinoza's or Kant's or Russell's. The comprehensibility of the world is not sufficient – there must also be the acceptance of the task or duty to comprehend it, the intellectual love of God. The fault of the Middle Ages was not that man had less capacity to comprehend, not that the world was then less comprehensible, but that man had degraded himself by intellectual laziness which made him give up his intellectual freedom of choice, just as Sartre's modern morally lazy man sells out his moral freedom of choice: in both cases the condemnation is total.

It is perhaps hard to articulate fully the moral-religious aspect of rationalism as I have presented it thus far – as the task of universalizing and comprehending universality, of moral and intellectual universalism, of the moral duty of high optimism. Let me leave all this now, however, and show how it ties in with the idea of rationality as proof.

III. PROOF AS THE UNIFIER OF SCIENCE

Bacon saw the existence of contending schools as an unmitigated evil – similar to divisions between warring empires. Had any school possessed the truth, he said, there would be neither need nor room for schools. Truth speaks for itself, truth shines, truth appears naked.

This doctrine, which Popper labels the doctrine of manifest truth or of truth as manifest, may sound as if it were a version of naive realism and thus in contradiction with the theory that Nature loves to hide. Far from it: nature does love to hide, but not forever; under some conditions, She even has to reveal Herself; but whereas when She hides She merely drops hints and clues as to Her whereabouts, when She does appear, Truth appears in her stark nakedness and dazzling beauty; and then no one need nor can doubt Her; the Naked Truth does not carry a handbag containing identification, corroboration, or letters of recommendation.

Taking what I have just said somewhat seriously may suggest that there is never any need for proof; truth is just known as truth. To this one might retort, but we must be able to distinguish the feeling of certainty from certainty proper; to use Sir John Herschel's terminology, we must differentiate psychological certitude from mathematical certitude. Indeed. But if we take seriously the idea that truth is manifest, then this aspect of truth is covered as well. If we could see the truth clearly yet doubt that the clarity is in fact and not in our minds, then there would be at least this lack of clarity in the picture. You may say, perhaps, that if we demand such a high degree of clarity from ourselves before we can claim knowledge of the truth, then we can never claim knowledge. Perhaps; it is harder however, to deny that once we do clearly see the truth, we can claim knowledge. And so, it looks as if it were analytic or near-analytic to declare that we do not need proof when we see the truth. This is what we mean when we say, it is so obvious, I do not know how to put it; it is so clear, I do not know how to explain it. To put it more congenially, knowing a truth is knowing a proof inside out – at a glance, as we say – seeing all steps at once. This idea is shared by many thinkers up to, and including, Wittgenstein – of the *Tractatus* in which truth '*shows* itself' as well as of the *Remarks on The Foundations of Mathematics*. Once you accept this idea, you see that when we speak of a truth in need of a proof, we either speak of a hypothesis which only *may* be true, of a truth revealed to *me* needing proving to *you* so that you can see it as well. So the role of proof, of proving, is inevitably that of introducing the newcomer to the truth. And the mode of introduction is, in my opinion, what differentiates – at least traditionally – science from mysticism. The mystic too – be he Oriental or European Mediaeval – introduces novices to the truth; but his introduction is only highly preliminary: most of the pilgrimage on the road to truth must be done alone, and he who has travelled it successfully is an initiate. Not so in science. This is a point endorsed even in the twentieth century by those who address themselves to the problem: it is agreed by most writers that it is the public character of science (to use Popper's expression of a common sentiment), as opposed to the esoteric character of mysticism which makes us prefer science to mysticism. And so, again, we see the universalist ethic behind science.

To approach matters differently: Russell says there must be a difference between the views of Einstein and the ravings of a madman. The difference lies precisely in the universality of the one, through rational justification, as against the caprice and idiosyncrasy of the other. What Bartley calls "the justificationist doctrine of rationality" is the view that what distinguishes the rational from the mad is the former's appeal to universally valid proof. Take as a conspicuous instance Parmenides' doctrine that the world is full, round, immobile, unalterable, etc. As a doctrine it may be declared a variant of mysticism, not much different from some Oriental or Occidental doctrines. As a doctrine it may be declared madness – as many if not all mystic doctrines were viewed by rationalists through the ages. But Parmenides presented proofs of his doctrine – indeed one may view him as the father of Western justificationism, as Popper does, and as the father of Western mathematics as Lakatos does.

The justificationist approach seems essentially different from that of truth as manifest; rationality is identified with proof rather than self-evidence. But not at all: for, self-evidence and proof are one: proof is what makes one see the objective truth, and acceptance of objective truth is rational as opposed to subjective, idiosyncratic, arbitrary. Rationality is universality, i.e. objectivity, i.e. truth; proof is the universalist method of revealing the truth. Indeed, were proof a matter for the initiate, then our whole edifice would collapse. Or, were proof of falsehood possible, then our whole edifice would collapse.

Consider the discoveries of Lakatos in this light. What has happened to Parmenides' proofs? They have been refuted, of course, and his view is not a proven theorem, of course. For, were the proof valid, we would be able to prove falsehoods. But how can one claim to have a proof without having it? How is error or deception at all possible if proof is the same as revealing a truth in all its clarity? There is no answer; such cases are freaks, they are rare, their proper place is therefore in the origins of mathematics or of a mathematical theory. But look at the history of mathematics, says Lakatos. There is almost no acceptable proof in Euclid, almost no acceptable possible proof in the calculus before the mid-19th century; there is, in brief, no proof a century old or more that an ordinary mathematician today cannot knock out.

What, then, is the role of proof? For Lakatos it is a challenge, an invitation to refute a proven theorem and a guideline for a possible refutation. This is exactly the same, I contend, as the role of every kind of proof in science, whether *a priori*, a *gedanken experiment*, or empirical support. Just as Lakatos views the world of mathematics as a system of ever increasing possible alternatives, so I view science as a system of ever increasing possible alternatives. And so, I think, every proof indicates something – and is thus quite informative – of the prover's shortcoming, of the limitations of his (intellectual) imagination.

I do not invite the reader to accept such views but to consider the difference between viewing proofs in this way from viewing proofs as establishing the truth in a manner open to all to follow and be convinced by. Indeed, once justification is rejected, the problem of objectivity arises full force all over again. Popper has stressed this point in connexion with science in his *Logic of Scientific Discovery*, and he has substituted for the universality of proof the universality of empirical criticism. In his *Open Society* he raised the problem a little more generally, but there he neither gave examples of criticisms outside the domain of empirical science, with the exception of an all-too-sketchy outline of criticism in ethics, nor did he claim that criticism pertains to all intellectual activity: he explicitly claimed, for instance, that interpretations in history, historical points of view, are immune to criticism.

The broadest presentation of the problem, how can we develop a non-justificationist philosophy of objectivity, equally applicable to science, religion, and morality, was forcefully presented by Bartley, in his *Retreat to Commitment* and elsewhere. He feels that the *tu quoque* argument of the esotericists of our century, the claim that rationalism too rests on esoteric foundations, must be answered, not to the satisfaction of the irrationalist – whose judgments may be too arbitrary anyway – but to the satisfaction of that rationalist who, though a non-justificationist, wishes to carry over from justificationist philosophy the vestiges of objectivism and universalism that can be salvaged.

For my part, although I do not agree with Bartley's solution (comprehensively critical rationalism), I find his presentation of the problem admirable. Indeed, if universalism and objectivism is to be retained, I feel, it can be retained not by a doctrine, be it Kant's or Bartley's, but by a perennial struggle towards objectivity and universality. On this point I am in full agreement with the doctrine which Gurwitsch ascribes (rightly or not, I cannot judge) to Husserl (who was, nonetheless, a justificationist). According to Gurwitsch, the most rational thing to do in the absence of a theory of rationality is to search for one (rather than sink into irrationalist despair).[14] This point is, doubtlessly, a minimum thesis that all non-justificationists would accept; it is in intention close to the attitude of the Age of Reason, but in points of technicality it is its extreme opposite. For Bacon and his followers, disagreements and splits between and within schools are anathema, the symptom of faulty justification; for Popper and his followers the opposite should obviously hold. That it does not always hold, only shows that non-justificationists can hold nothing as manifest, not even non-justificationism, not even to the arch-non-justificationist. I should therefore state tentatively, that it is always the case that we do not know that we have a satisfactory theory of rationality, so we may always try to either criticize existing doctrines, or create alternatives to them, or both. We may or may not be successful in this, but we may always try.

This point, too, is no novelty. Sextus Empiricus already notes that the consistent skeptic (etymologically, searcher) exercises a skeptical attitude even towards his own skepticism. This, he adds, leads to peace of mind. But, he adds a rider to this rider: this claim for peace of mind is not a dogma but a (psychological) hypothesis which, historically, came to his school as an after-thought.[15]

IV. MANIFEST TRUTH AS THE UNIFIER OF SCIENCE

The standard criticism of justificationism can be found very early in Sextus Empiricus' *Elements of Pyrrhonism*: all demonstration uses criteria which are either arbitrary or in need of further demonstration etc. *ad infinitum*. There are two reasons for not taking this criticism seriously; one traditional, one modern; one from science, one from commonsense. Both modern science and traditional commonsense tell us that obviously we do justify all the time, and hence that there must be a mistake in the argument to the contrary. Hence, though we should try to locate and exhibit the fault in this argument, the argument need not be taken seriously even in the meantime.

The appeal to commonsense justification is not quite up to standard; at least it is not as strong as the appeal to scientific justification. It is not up to standard for two reasons. First, one may take it as a part of commonsense that, if any truth is obvious then the fault in a criticism of it is obvious, and that, if a fault is obvious it can be detected after a relatively brief and straightforward inspection. Now, if truth is manifest, then criticisms of it, including the criticism from infinite regress, contain serious faults. And if it is obvious that there is a fault in the argument from infinite regress, then the period allotted for finding it should be brief. Yet the fact is that since the days of Hume attempts to locate the fault have failed dismally. If, however, the fault is not obvious, our whole tradition of justification may be resting on an error and hence on an arbitrary or accidental development. Hence, we must view the fault as obvious, and hence Whitehead was right in viewing our inability to locate and exhibit it – our inability to solve the problem of rational justification, which, for empiricists, is the problem of induction – as no less than 'the scandal of philosophy'.

The second reason for the weakness of the commonsense argument against Sextus' criticism of justificationism is a bit less obvious, not because of intrinsic difficulty but because it has in the past been (mis)-directed against non-justificationism. It has been claimed again and again that since in ordinary circumstances we justify our opinions and act on them, then the very fact that a non-justificationist acts proves him a justificationist. But this is only one reading of the situation, and a reading which leads one – in vain, as we shall soon see – to a search for justifications in ordinary life. An alternative reading is this. Justificationists maintain that we must justify our opinions in order to act, and hence conclude that they, as actors, do justify; but the very existence of non-justificationist actors proves that unjustified action is possible. In particular, they cannot justify their action in proclaiming themselves non-justificationists (since, say the justificationists, non-justificationism cannot be justified). Hence, the existence of non-justificationists in our midst refutes justificationism. It is only commonsense that even in cases when we all agree to justify our opinions – say in courts of law – we only agree to justify them relative to accepted fundamental propositions – say, the constitution – which we do not allow to challenge and which we shall not justify. The argument from infinite regress is commonsensically so powerful, that even if you reduce its infinity to, say, ten steps, it is still valid; in any ordinary circumstance there comes a point beyond which we bluntly refuse to justify our views and call anyone who persists in challenging us to do so a pest and a skeptic and a highfalutin philosopher. It looks as if the persistent challenger is a non-justificationist, but a moment's thought will reveal him as a justificationist.

The one extraordinary circumstance where the regress is allowed to be explored beyond what is commonly accepted as reasonable is in intellectual discourse, especially scientific debate. So let us consider, then, the argument from science against Sextus. Do we justify in modern science?

Take Descartes' justification of science, that is his theory of the natural light; as a criterion of truth it is subject to the criticism from infinite regress which is applicable to all criteria. Take the natural light to be a fact, however, i.e. that Truth imposes Herself on us, and you see that we arrive at a stalemate: the skeptic can doubt the veracity of even the most imposing ideas, whereas, Descartes will say, after having performed my duty (Protestant ethic) and having been forced by Nature to believe this or that, if even then I am not right, then God is a liar. It is not an accident that the skeptic Boyle attacked just this point of Descartes' epistemology: he said Descartes still blames God for his own errors; meaning, Descartes was in fact not forced to believe in Cartesian physics.[16]

The same holds for all the philosophers of the Age of Reason: to be able to take minimal sufficient account of Sextus they had to claim that in fact Truth has imposed Herself on us (to call this imposition revelation is in line with the religious doctrine of revelation proper: God's word imposes itself on us). Hence the enormous importance of Newton, so stressed by E. A. Burtt in his by now classical *Metaphysical Foundations of Modern Physical Science* (which is, not accidentally, 20th-century); if Truth has revealed herself we should be able to point at her, and Newton's Theory is a (the only, really) paradigm of Truth Manifest.

All this may explain a puzzling fact. There is a difficulty, in the traditional literature, of distinguishing between methodology and epistemology. Methodology and epistemology are theories concerning the road to knowledge and theories concerning existing knowledge respectively. They seem to be as easily distinguishable from each other as a road-map to Rome is distinguishable from a street-map of Rome. Of course, in each of these two cases the distinction need not be hard and fast, but they are unproblematic. Now, one would expect the distinction between epistemology and methodology to be equally unproblematic; yet this is not the case. Whereas Descartes wrote on method, not on knowledge, he counts as an epistemologist of the first order of significance, but hardly at all as a methodologist, to take but one example. Why is the distinction between methodology and epistemology so hard to make? Perhaps it is because it looks as if proof belongs to epistemology, since it supports our claim for knowledge, whereas this is not so: as long as we need proof we are groping. And so proof belongs to methodology, whereas epistemology can say nothing except that truth *shows* itself. So did Helmholtz understand matters when he observed that investigation is empirical yet knowledge is *a priori*. Helmholtz compares this to the difference between climbing uphill while struggling through the bush and not seeing any road to the peak, and the standing on the hill-top and being able to see with ease all the shortcuts. Perhaps the best metaphor is Wittgenstein's ladder which one has no need for after having arrived at the rooftop. This is why throughout the modern age the empiricists criticized apriorists on methodological – moral, I should say – grounds, expressing contempt at their lazy armchair method. In epistemology, empiricists and apriorists do not differ.[17]

To illustrate all this consider the following passages from Victor Weisskopf's essay on Niels Bohr in the *New York Review of Books* of April 20, 1967. Perhaps casual passages are not the best illustrations of serious and important and deep-seated convictions, but since I have already invoked the name of Freud, the arch reader of deep meanings into casual passages, allow me to go on for awhile. Let me add, however, that I am quoting what is evidently the central paragraph of the essay, that the essay appears in a not so unimportant a periodical, and that Weisskopf usually does not take lightly his duty as an advocate of science. Well, then; let me quote.

> During this great period of physics [the formation of the new quantum mechanics] Bohr and his small group of scientists touched the nerve of the universe. Man's eyes were opened to the inner workings of Nature which before had been obscure. Once the fundamental tenets of atomic mechanics were established, it was possible to understand and calculate almost every phenomenon in the world of atoms, including atomic radiation, the chemical bond, the structure of crystals, the metallic state. Before that time, the environment was composed of unknown forces: electric, adhesive, chemical, and elastic. All these were reduced to one – the electromagnetic force....

So much for the quotation, which I find quite fascinating. The claim that quantum theory is not a revolution but only a further insight into the electromagnetic force of a century ago is in line with the philosophy Weisskopf accepts from Rutherford and expounds early in the essay: in science there are only insights, no revolutions. The claim that the new quantum theory can explain so much and unify so much is highly questionable. No doubt it has grounding in fact and rings a familiar bell to the physicist. No doubt it also gives a misleading impression to the layman. More important, it enables Weisskopf to touch the nerve of his own faith in science: Bohr and his followers, let me repeat, touched the nerve of the universe and saw the working of nature. Once all this "was established, it was possible to understand and calculate". Not, quantum theory enabled us to understand and calculate and was thereby established, but once it was established, we could understand and calculate. Not, our calculations justify our theory, but, our justified theory enables us to calculate. Perhaps I am making too much of a piece of eulogistic writing. Let it be so; let us see what nerves do eulogies touch, what chords resonate more freely in deep sympathy.

To stress this point, I would refer to a less euphoric part of Weisskopf's essay. "Nuclear structure presents impressive evidence for the general validity of quantum mechanics", he says in a later part. After quantum mechanics "was established" and "Man's eyes were opened to the inner workings..." etc. one more success merely "presents impressive evidence"! Evidently this additional success resolves doubts which only it raises. Indeed, Weisskopf explains: "The weight of Bohr's personality was so great that [under his influence] for a decade the typical differences between nuclear and atomic states were at the center of interest, whereas the similarities between these states were [erroneously] pushed aside." What Weisskopf finds hard to admit is that the present state of nuclear physics is still not half as satisfactory as that of atomic physics – in spite of the latest advances which he is alluding to. Hence, clearly, "impressive evidence" is only Nature's hint that our eyes are turning towards the right "center of interest". Thus, there is a difference between turning our eyes toward the right "center of interest" (methodology) and seeing the "inner Workings of Nature" (epistemology). Though in methodology a sworn empiricist who has to turn his eyes in the right direction, in epistemology Weisskopf endorses the views of Giordano Bruno, Galileo, and Descartes that Truth reveals herself. Again, the barriers between apriorist and empiricist epistemology break down: empiricists cannot escape a touch of apriorism; only the difference in methodology remains, the empiricist methodology being inductive – hard labour and humble efforts and purity of heart – or, as Robert Merton calls it (approvingly) "Protestant".

Whether all this is true or false, it is meant to explain why the skeptical criticism from infinite regress which is so obvious and clinching has failed to dent the optimistic attitude of the classical scientists which assured them that they could find the unity of nature. For my part I prefer to combat the classical view not by the criticism from infinite regress or by any variant of it – as Popper frequently does – out by the argument from diversity which seems to me equally as appealing from any vantage viewpoint as the argument from unity. Let me show, then, that the classical view of rationality imposes on reasonable people an unreasonable degree of unanimity.

V. UNITY OF SCIENCE AS A DICTATOR OF UNANIMITY ON ALL QUESTIONS

When we distinguish the rational from the capricious, from the whimsical – even merely from the private and personal – we scarcely do more than accept the minimum program for rationality, the desideratum for a doctrine of rationality accepted by everyone. To claim that taste is entirely personal, or subjective, or arbitrary, is also to claim that taste is entirely lacking in objectivity and rationality. Hence rationality is linked with objectivity and hence with truth. From this to justificationism is but a small and reasonable step – with one most unreasonable consequence.

Justificationism, I claim, totally polarizes the subjective from the objective, the non-rational from the rational. It leaves no room for any middle ground of partial rationality of opinion, of partial reasonableness, of degrees of rationality; similarly, it leaves no middle ground for any degree of rationality of a decision, in the realm of morality and responsibility, in the realm of daring and courage and boldness, in the realm of putting political and social proposals to critical debate and attempted implementation; there is no room within justificationism for the partial justification even of a scientific project, be it a theoretical exploration or a design of an experiment or a sociological survey or anthropological fieldwork.

This contention of mine is open to a number of criticisms which I shall soon expound and deflate. First let me show how very unreasonable the polarization is, especially in that it makes rationality so uniform that no room is left for diversity. Both the Middle Ages and the 18th century, at opposite poles, suffered from this.

One of the central tenets of Mediaeval philosophy was that man is irrational; for all men sin, and the wages of sin are infinitesimal in comparison with both the punishment sin entails and the rewards of righteousness. The theory of rationality of the 18th century pointed at the very opposite. Granted that some men are beset by prejudices and are thus not rational, there is no irrationality that a person can commit once he has given up his prejudices. His ends are given, his circumstances are given, his knowledge is solid as far as it goes, and it goes no further (i.e. he is unprejudiced); how else can he act but in the light of all these? Still, the eighteenth century philosophers admitted not only the possibility of prejudice but even the fact that the multitude is prejudiced. Nowadays even this concession to irrationality has been tacitly omitted.[18]

Nowadays the leading theory of rational conduct is that of Bruno de Finetti, the theory of subjective probability and of revealed preferences. The theory is very simple. It grants, first, that there is no universal scale of assessing the probability of a given prediction in the light of given experiences; it assumes, second, that everyone adopts his own scale and everyone acts on it. If any person claims to have no judgment of the likelihood of a prediction, we may in principle show he conceals his opinion by challenging him to a wager. This will call him to act on his assessment, and since he will act rationally he will reveal it. As long as we are not concerned with the cause of the concealment, be it deception or self-deception, the theory seems to cover all possible cases – in principle, that is.

This view of de Finetti is now endorsed by Rudolf Carnap and many other leading thinkers.[19] It represents the pole that we are all rational. Contrary to it, and contrary to the Mediaeval idea, as well as the 18th-century idea, I suggest, as a matter of sheer commonsense, that we are all partially rational, that there are degrees of rationality, that there are cases of better and poorer rationale, that there are degrees of responsibility, that we are sometimes more, sometimes less reasonable, responsible, deliberate. In particular, we may be in two minds, both reasonable, or belong to controversial groups differing from each other yet respecting each other as rational, reasonable, adult, responsible. I suggest, further, that this should be treated as a hard fact (in as much as there are such things) to be contended with even if we have no theory as yet to account for it.

We have already met the opposite idea, explicitly stated by Bacon, that competing schools, doubts, controversies, are symptoms of lack of reason. Bacon also stated in the preface to his *Great Instauration*, you can't criticize a man and yet respect him. This is in character, yet fantastic. Adam Smith, in his obituary of his friend David Hume says, of his philosophy I shall say nothing since those who agree with it admire it, and those who do not, despise it. I think today most of us respect and yet dissent from Hume. Benjamin Franklin exhibits in his *Autobiography* a curious ambivalence to debate. He enjoyed debate in his youth, but took pride in not having argued with his scientific opponents who lost ground even without a battle. Much as we disagree with Adam Smith, we usually agree with Franklin. At least we have a problem here.

In a well-known paper [20] Stephen Toulmin rejects the thesis that Priestley was foolish to reject Lavoisier's doctrines and seeks to show that he was justified. The criterion of choice of a hypothesis, says Toulmin, is that of simplicity; and which of the available doctrines was simpler was an open question for quite a while.

This is a nice point. Toulmin agrees with Bacon and his followers about clear cut decisions and the inadmissibility of dissenting from them once they were computed; he only claims that the interim period of computation may be long. But Toulmin is even historically mistaken: it was Kirwan, not Priestley, who held the criterion of simplicity, and applying the criterion he soon gave up his phlogistonist views and joined Lavoisier; Priestley had different criteria, which are too complicated to present here.[21] Whatever the criteria within science, even when one accepts them within science, and even when their application is unproblematic, dissent may still be sustained, say within metaphysics.

Examples: Faraday agreed that Newtonian mechanics had been as well confirmed as possible, yet he rejected it on metaphysical grounds: he rejected action at a distance on the basis of the idea of conservation of force. Einstein agreed that quantum theory is highly satisfactory as a consistent explanation, yet his determinism made him reject it. Were these people unreasonable? By implication, Toulmin's paper comes dangerously close to saying, yes. I think it is not too much to suggest, intuitively if you like, that they were both reasonable, though most of us would side with Faraday more easily than with Einstein.

The doctrine of proof requires that all arguments be quickly terminable. The verification principle is but a variant of it. To claim that instead of verification we have only probability looks like partial rationality, but this is not so: we all have to endorse the most probable hypothesis, and to the degree that it has been rendered probable. The question is, of course: What hypothesis, among competing ones, is the most probable? To this there is no generally accepted answer, but since inductive philosophers are still struggling to find such a universal criterion, claiming to be able to find it in the relatively near future, we should not press this question now; suffice it that they claim that in science unanimity is achieved over any specific case of choice of hypothesis. Of course, the unanimity thus reported is a myth too, but let us not press this point either. Rather, we can ask: What does the endorsement of a hypothesis amount to? Endorsing the most probable hypothesis, most inductive philosophers claim, is the bare recognition of the (logical) fact that a given hypothesis is the most probable among the class of competing hypotheses in the light of experience and thus the recognition of its rational preferability.[22] Does this solve any problem? Does Einstein's recognition of the merits of quantum theory amount to endorsement of that theory? In the accepted sense it does; in fact it does not. The stress on unanimous endorsement of the most probable hypothesis is so heavy in the current literature, that there is almost no explanation of what endorsement means in terms of 'cash value'; in practical terms, that is. And the few explanations that exist have not been put to any critical examination.

But perhaps we can find two equally probable hypotheses? Will not this, at least, justify diversity of opinion?

We barely allow for diversity this way and the choice looks to us dangerously arbitrary. So we add a criterion of choice: of two equiprobable hypotheses choose the simplest. And farewell to diversity.

Ambivalence plays its tricks here; diversity looks arbitrary and hence irrational since we have contrasted caprice and idiosyncrasy with objectivity and universality and rationality. Once we have done that we cannot ameliorate matters. We have to start afresh: we must identify caprice and lack of rationality but not conclude from that that rationality is compulsive to all rational men. How can this be done is a question which I shall come to in the next paragraph. First I wish to state that – paradoxical as it sounds – we do in common experience identify caprice with lack of rationality yet refuse to view all diversity as caprice; hence we refuse to see rationality and caprice as contrasts though we identify caprice with lack of rationality.

The paradox, in other words, is a logical error; identifying the minimal points of two scales is not identifying the two scales. Even if each is composed of two points each may have only one point in common with the other. It may happen that all good money is genuine, and that all forged money is bad, yet from this it does not follow that all bad money is forgery! It *a fortiori* does not follow from all irrationality is sheer caprice that all rationality is totally binding. We may introduce leeway, and it is commonsense that we should do so.

Even on a question on which we have utterly no evidence, the classical doctrine of rationality allows no diversity! Some say, have utterly no judgement on matters concerning which you have no proof, and with no judgement engage no action. Others, notably Descartes, require conventional conduct on matters on which there is no rational judgement. Others, still, suggest consulting *a priori* probabilities. Strange as it sounds, there is disagreement on the criteria to be applied, but not on the doctrine that the criteria should eradicate all disagreement. The doctrine of induction is based on the transcendental argument that there must exist a principle of induction since scientists are in perfect agreement, and surely not on a religious authority or some such silly criterion. Yet the debate over the principle of induction indicates as wide a disagreement between inductive philosophers as can be. Why cannot they use the methods of science and determine empirically what is the principle scientists employ (as they say scientists do) and eradicate this shameful disagreement?

The intuitive clash between universality and diversity was felt in different ways by different thinkers. Let me mention one. William James felt it would be a bigotry to consider his own brand of Christianity the only rational one or to say that choice of religion is totally arbitrary. He developed pragmatism in order to permit everyone to choose a religion, that is to say, to offer more than one religion as rational choices. He ended up by a criterion which, if it were operative at all, would compel each one to choose one and only one religion – the most useful one. I suppose it is the religion of his environment, but I am not sure.[23]

The same may be said of Carnap's principle of tolerance. Carnap first claimed that metaphysical problems – and solutions to them – are meaningless; he then claimed that they may be rendered meaningful by translating them to questions concerning the language-systems we may employ. Now, it seems, the question of choice between competing metaphysical systems gets translated or transformed into questions concerning the choice of language-systems. This question Carnap has answered with a principle he called "the principle of tolerance": everyone has the complete right to choose his own language-system. Now this principle may be a principle of indifference, in which case there is little value in the tolerance employed: we do not have to display much tolerance to allow people to choose between alternatives of equal merit. If, on the contrary, one language-system is preferable to another, tolerance (which we all take for granted anyway – otherwise there would be no problem of choice) is no solution to the problem of choice: we need criteria and rules of rational discussion to facilitate rational choice – i.e. the choice of the best language-system. We are caught here in the same net. Either there is no rational choice of a framework – the framework is arbitrary whether it be a religion or a language-system (a metaphysics) – or there is a criterion of rational choice and it imposes on us a unanimity; all who share it are rational and all dissenters are irrational.

One final example, an intriguing one, and with a note of depair. R. G. Collingwood, *Speculum Mentis or the Map of Knowledge* of 1924; chapter on History, section on the World of Fact as Absolute Object. I quote:

In science the attempt is made to bridge the gulf between God and the world. Instead of God we find the concept or law, which because it is the law of the world no longer stands outside the world. Here what was transcendent has become immanent. But in this very success, failure is revealed; for it becomes plain that the immanence is a false immanence [for the following reason: in science the fiction is held that] the world is not a world of individuals but a world of particulars, and because the concept [i.e. the law] is indifferent to the various particulars in which it is embodied, their diversity remains meaningless and the world is to that extent a chaos ... the failure of religion is repeated and the quest for immanence ends in transcendence, but an abstract transcendence more intolerable than the concrete transcendence of religion. Thus [art, religion, and science] ... are attempts ... to reach ... organized individuality. ... Art comes nearest to success. ...

This conclusion of Collingwood is very impressive, but it leaves us with the problem open. And the problem, how to have unity and diversity combined, hinges on the problem of rationality. And, anyway, the problem of rationality is significant in its own right.

Anyway you look at it, there is still a significant and challenging difference between the choice of the madman and the choice of the reasonable man; only the reasonable man has a number of reasonable alternatives, some of them equally reasonable, some more than others. This should be the desired corollary to any doctrine of rationality that may claim superiority over the traditional doctrines of rationality.

VI. A THEORY OF RATIONAL DISAGREEMENT

Einstein agreed with the majority of physicists that quantum theory satisfied certain desiderata. His contention, however, was that these desiderata are not sufficient. In other words, we need not specify *a priori* the desiderata of rationality, but leave them open to rational debate. Also, given the desiderata, we may leave open to rational debate the question of which alternative satisfies them best, and if few are adequate, leave them to arbitrary choice without declaring this arbitrariness irrational.

This seems straightforward enough, and it applies to views as well as to conduct.

There are also two moves that seem widely acceptable (at least they have been proposed by Kant, and were endorsed by Neurath, Carnap, and practically all members of the Vienna Circle at one time or another), but which *within justificationism* do nothing to relieve the difficulty; within the above framework of rational debate are quite helpful. They are the idea of presenting universality as a methodological desideratum rather than as an ontological claim and the replacement of objectivity by intersubjectivity.

Let me explain.

It is agreed by practically all philosophers and historians of science that something similar to Bacon's pyramid of levels of generality takes place in science; the classical instances are Newton's unification of Kepler's and Galileo's theory, electrodynamics as a unification of theories of electricity and magneticism, and electromagnetic field theory as a unification of electrodynamics and physical optics. The question is, what is the status of this pyramid? Does the pyramid in science correspond to a factual one, to one in the real world? Or is the pyramid something to which there can be no correspondence, is it a characteristic similar to the linguistic structure of science? For justificationists this question makes little difference: they have to endorse each law of science, independently of the question on which level of the pyramid it happens to be. It is a secondary matter to them that laws are ordered in a pyramid, the primary matter being that the laws are true. This, however, raises an old difficulty: does not science rest on the *a priori* assumption that the pyramid does exist, that the next level of generality will be discovered, that laws will fit in a pyramid? In reply to this it might be claimed that whether or not the pyramid reflects the true state of affairs in the cosmos, science may try to find it, and if the pyramid is there science may even be successful. Thus, the pyramid belongs to methodology. For justificationists, who claim that science proves its doctrines, this move is of no avail. For, when we prove a higher level theory we show *a fortiori* that one more level of the pyramid is filled by true laws. If we replace proof with probability, (meaning when we say that a scientific theory is probable, that most scientific doctrines are true) then the previous sentence will have to be modified, but not in a way prejudicial to the present discussion. And so, justificationists must accept the pyramid not only *a posteriori*, but also as the *a posteriori* corollary to any *future* development of science. This means that it is a transcendental part of all science, that for science to be possible God must have created parallel pyramids, one in nature and one in the intellect or in language or in some human quality. This is not as primitive a thesis as Wittgenstein's picture theory of every statement in language, but it is identical with Bacon's claim that God has created Nature and the Intellect on a par, so as to make science possible.[24]

If, however, we claim that the lower-level theories are often refuted, or simply clash with the theories they replace even when the former are not refuted, then we can deontologize or methodologize the claim for universality in science. Moreover, in such a case, tests and crucial experiments are the particularizing effect that keeps the universalizing effect going, that keeps us dissatisfied with existing universal theories and pushes us towards further generalities.

To take an example. William Whewell claimed that Newtonian mechanics cannot be further deduced from a more general law. Why? The tendency, he said, was to explain both existing laws and the as yet unexplained facts, by newer laws which are thus more universal. But in the case of mechanics there were no such facts. Simple. But where do such facts come from? Whewell stressed that there are no uninterpreted facts, and that interpreted facts follow from the theories they verify. So there can be no theory-isolated facts to increase the generality of a theory! To my knowledge this simple criticism of Whewell has never been noticed. To forestall dismissal of Whewell as an isolated case, let me point out that on this point Whewell's view – of explaining theories and isolated facts by higher-level theories – was endorsed by Duhem and Meyerson.[25]

The reason for the general significance of the difficulty implicit in this view – namely, that isolated or uninterpreted facts are unavailable – is that more and more philosophers tend to share it. They increasingly tend to endorse the view that there are no uninterpreted or isolated observations of fact. This means, for almost all philosophers of science, that every new observation of fact confirms some known theory, which is a bit incredible; and that a new observation of fact cannot refute a theory except if we first invent a new alternative to that theory, which is even more incredible. There is little doubt that any critical examination of the current theory of interpretation, which (adumbrated by Kant, Johannes Mueller and Helmholtz) is due to Whewell and Duhem, is no satisfactory substitute for the traditionally most popular (Baconian-Lockean) doctrine of science as based on uninterpreted data. But there is an alternative theory of observation.

Both traditional theories of observation derive from their originators' views of the role of observation in theoretical science. Bacon viewed facts as the raw material which is somehow processed into theories; hence his insistence on raw data. Whewell viewed facts as verifications; hence his view of interpretations as predictions. Popper views new facts as refutations, and so his theory of interpretation allows for both predicted and counter-predicted observations, the latter constituting novelties relative to the theories which they refute.

The result of this new view of interpretation is the methodological aspect of scientific theory. A theory is designed for explanation. It is thus the refuted theory, especially one which had once served a given purpose, which cries for a replacement which may better fulfill the same purpose. And here Whewell's doctrine is rectified, as well as Duhem's – into Popper's.

Thus, I have argued, methodologizing or deontologizing universality can be effective only in a non-justificationist system. The same may be said of replacing objectivity with intersubjectivity.

Indeed, Kant replaced objectivity by intersubjectivity in order to deontologize universality, as should be obvious upon a simple reflection. But his deontology allowed no diversity, nor was it intended to. Nor did it remove the polarization of all human thought and action into the rational and the irrational. When Kant argued for the limits of reason and concluded that beyond these limits scientific intersubjectivity breaks down and can only be replaced by arbitrariness, he plainly and assuredly concluded from this that attempts to transcend the limits of reason must be irrational. This attitude of Kant's should have made him view rationality in the narrow fashion common to his time, and practice it with consistency uncommon at any time. To an extent he was, indeed, quite impatient: not in the least glad to suffer fools and quite willing to consider all opponents as fools. Yet, if there is one aspect of Kant's conduct which was grossly out of character, it was his ambivalence towards differences of opinion. I do not mean his table-talk: light-hearted controversies, even serious yet admittedly preliminary controversies, were tolerated as a matter of course. But Kant was willing to admire even opposition to the last as a symptom of a keen mind concerned with truth. It is perhaps amazing that such a prolific writer, who was so willing and able to elaborate a point, was so brief on some most crucial points of controversy, to the effect that we still know too little about his change from Leibnizianism to his own, critical, version of Newtonianism. No doubt, ambivalence often is the source of reticence – certainly not any lack of frankness in this case. It is also quite possible to see the intellectual reason for the ambivalence: Kant's critical doctrine of intersubjectivity seems to be a liberating doctrine, an idea enabling us to divorce rationality from unanimity; but it fails to do that: rather than establish possible understanding between individuals, it clinched understanding to the point that it became full consent.

The newer theory of the intersubjectivity of tests of theories likewise does not help diversify thought and depolarize rationality and irrationality. For, tests are confirmatory and, we are told, the acceptance of the confirmed theory is declared imperative. The same idea does, however, work if tests are attempted criticisms; they are then aimed at particulars and when successful they diversify – they open new areas of choice which are free for all.

In concluding this paper, which is intended more to open up issues than to close them, I wish to state three points.

First, if, as I suggest, instead of laying down desiderata we allow different contenders to state their own desiderata and debate them, then this may lead to an infinite regress. Indeed a regress there is here, just as there is the regress from one explanation to a higher level one; both these regresses may be infinite, yet they are not futile – not necessarily so; each step in the regress may offer its own interests and challenges.

Second, the desiderata may be solutions (perhaps of given characteristics) to given problems. Popper's orientation has always been towards problems, and in his post-war work he has stressed that problems are the starting-point. I hope I have introduced this idea of his in a somewhat integrated fashion. To mention one corollary: Popper speaks of the objectivity of problems and perhaps also of their relative significance; but he says nothing about the objective choice of a problem to study – sometimes he even suggests one must fall in love with a problem, which smacks (quite erroneously, of course) of subjectivism. Once we suggest that problems are chosen by more general criteria, which themselves are open to criticism, we render matters of choice of problems inter-subjective to some degree.

Finally, the matter of choice of problems seems to me to be the crux of the philosophy of the future. Problems abound; yet most of them are not interesting, or not important, at least in comparison between the expected investment in, and the expected return from, the forthcoming solution. But all this is essentially highly speculative and vague. (Mössbauer was assured by his superior that he was wasting time on a useless problem, very much as E. O. Lawrence before him, and as Planck even before.) And so this may lead to honest disagreement and even some debate. More important, the sense of significance is thus brought into research in a new way. In the classical doctrine it was brought in in the abstract, in the sense that science is important and requires dedication. Here it is introduced particularly, as a particular problem pertaining differently to any different research project. It thus leads anew to some connection between the prescriptive and the descriptive and again *via* rationality, but with a built-in diversity.

Boston University

REFERENCES

* Earlier versions of this paper were read on April 4, 1967, in the philosophy colloquium of Simon Frazer University, Barnaby, B.C., and on May 22, 1967 in the Continuing Science Discussion Group, Boston. I am grateful to Robert S. Cohen for his investment of much work and his detailed critical comments, and to I. C. Jarvie for his correction of the final draft.

[1] The development of a new theory of rationality, following the lead of Chapter 24 of Popper's *The Open Society*, seems to me to be a collective effort, but Bartley has the lion's share. The following is a partial list of the efforts in this direction.

K. R. Popper, *Logic of Scientific Discovery*, London 1959, Preface to The English Edition, 1958, and new footnotes pp. 44, 98, 206 (and compare with old note on p. 55), and new appendixes ix (introductory part), and xi. *The Open Society*, 4th ed., London and New York (Paperback), 1962, Vol. 2, Addendum. *Conjectures and Refutations*, London and New York 1963, 1965, Preface to 1st and 2nd ed., Chapters 4 (originally 1948), 5, 8, (2), 10 (v, vi). See also his 'Naturgesetze und theoretische Systeme', in *Gesetz und Wirklichkeit* (ed. by S. Moser), Innsbruck 1949, and 'The Aim of Science', in *Ratio* (Oxford) **1**, No. 1 (1957) 24–35.

J. O. Wisdom, *Foundation of Inference in Natural Science*, London 1951, final chapter. 'Respect for Persons, the Pleasure Principle, and Obligation' in *Atti del XII Congress Internationale di Filosofia*, Vol. VII, G. C. Sansoni, Florence, 1958–9. 'The Refutability of "Irrefutable" Laws', *Brit. J. Phil. Sci.* **13** (1963).

J. W. N. Watkins, 'Confirmable and Influential Metaphysics', *Mind* **67** (1958), 'The Haunted Universe', *The Listener* **57** (1957), 'Epistemology and Politics', in *Proceedings of the Aristotelian Society*, London 1957.

W. W. Bartley, 'I Call Myself a Protestant', *Harper's*, May, 1959, reprinted in *Essays of Our Time* (ed. by L. Hamalian and E. L. Volpe), New York 1960. *The Retreat to Commitment*, Knopf, New York, 1962, 'How is the House of Science Built?', *Architectural Association Journal*, February 1965. 'Rationality vs. Theories of Rationality', in *The Critical Approach to Science and Philosophy. Essays in Honor of K. R. Popper* (ed. by M. Bunge), Macmillan, London, 1964. 'Theories of Demarcation and the History of the Philosophy of Science, in *Problems in the Philosophy of Science, Proceedings of the International Colloquium in the Philosophy of Science, London*, 1965 (ed. by I. Lakatos and A. Musgrave), Amsterdam 1968.

I. Lakatos, 'Proofs and Refutations (in four parts)', *Brit. J. Phil. Sci.*, 1963–4. 'Infinite Regress', *Proceedings of the Aristotelian Society* (Supplementary Volume 36, London 1962).

P. K. Feyerabend, *Knowledge Without Foundation*, Oberlin 1961. 'How to be a Good Empiricist – A Plea for Tolerance in Matters Epistemological', in *Philosophy of Science*, The Delaware Seminar (ed. by Bernard Baumrin), New York 1963. 'Problems of Empiricism', in *Beyond the Edge of Certainty* (ed. by R. G. Colodny), Prentice Hall, 1965. See also his contributions to Vols. II and III of *Boston Studies in the Philosophy of Science*.

I. C. Jarvie, *The Revolution in Anthropology*, London 1964 (see Index, art. 'Rationality'). 'The Objectivity of Criticism of the Arts', *Ratio* **9** (1967).

Jarvie and Agassi, 'The Problem of Rationality of Magic', *Brit. J. of Soc.* **18** (1967).

M. Bunge, *The Myth of Simplicity*, Prentice Hall 1963.

S. Anderson, 'Planning for Fullness', in *Planning for Diversity and Choice* (ed. by S. Anderson), MIT Press, Cambridge, Mass., 1968.

J. Agassi, 'Epistemology as an Aid to Science', *Brit. J. Phil. Sci.*, 1958. 'Science in Flux: Footnotes to Popper', *Boston Studies*, Vol. III, Reidel, Dordrecht, 1967.

[2] The pessimistic mood which crept into 19th-century philosophy is a fascinating topic which I am not qualified to discuss. Largely, I should only say in parentheses, it is the irrationalists' tribute to reason that their despair of reason led to total despair – even against their will. Walter Kaufmann is right in viewing Hegel, for example, as by and large a pessimist (*Philosophical Review* **60**, 1951) – quite contrary to Hegel's own pretended optimism and progressivism (and rationality of sorts), one should add. The Malthusian view was rational and, in at least one aspect, pessimistic; yet, the evil predicted by Malthus was a matter to overcome by a technicality; and when a technical suggestion to overcome the evil was made, it was advocated with incredible zest and great zeal. Darwinism, too, had an aspect which could be viewed pessimistically, and the advocates of German nationalism did stress it; but the rationalists saw in Darwinism the strongest support for progressivism. The first collapse of the equation of the two dichotomies – rationalist–irrationalist and optimist–pessimist – takes place with Freud's world-view, since he was both a staunch rationalist and a confirmed pessimist. This is, perhaps, one of the most traumatic aspects of Freudianism.

Norman O. Brown, in *Life against Death, the Psychoanalytical Meaning of History* (Wesleyan University, 1959, Part 5, Chapter XV, section: 'Rationality and Irrationality') says: "That the instinct of psychoanalysis – for it too has instincts which it represses – makes it want to attack the rationality of prudential calculation and quantitative science is an indubitable but not widely advertised fact. It is concealed by the use of a quite naive and traditional (therefore unpsychoanalytical) notion of the 'reality-principle' and 'reality-thinking'. Behind this naive notion of 'reality-thinking' is Freud's unquestioning (he could not question everything) attitude to science, that Comtian attitude which saw man passing through the stages of magic and religion till it finally arrives at the scientific stage, where he is at last mature – i.e., where he has abandoned the pleasure-principle, has adapted himself to reality, and has learned to direct his libido toward real objects in the outer world. Behind this scientistic pose of the psychoanalyst lies the repressed problem of the psychoanalysis of psychoanalysis itself."

Here we see the ideal of rationality in the double sense of (a) objective comprehension and (b) it being used successfully in purposive action. This is optimism regarding both the descriptive and the prescriptive aspects of reason – welded so well that Brown cannot but see a conflict between Freud's rationalism and pessimism.

Kleist sounds like a precursor of Freud and Adler, not, as some presume, because of his intense obsessive personality (such abound in history), but because he despaired of reason (after reading Kant, recognizing his excellence, and not being satisfied with him), yet remained crystal clear and rational in style as well as in mode or reasoning – psychological or otherwise. Thomas Mann's claim that he was a romantic (Preface to English ed. of *The Marquise of O-----*) is based on most incredibly weak evidence (such as Kleist's making the midwife reaffirm the view that with the exception of the Holy Virgin no woman ever conceived without sexual intercourse; in the described circumstances, what else can a Christian midwife say?).

Mann's claim that Kleist is a romantic (meaning somewhat mystic, meaning irrationalist) is a deduction from Kleist's pessimism and Mann's equation of the two dichotomies – rationalist–irrationalist and optimist–pessimist. There are other rationalistic pessimists, of course, prior to Freud, the most notable of which is a certain literary school in Russia, culminating with Chekhov (and excluding pessimistic irrationalists like Dostoevsky and optimists like Tolstoy). Mann's essay on Chekhov is an attempt to explain away Chekhov the rationalistic pessimist. It looks as if Mann held on to the equation of the two dichotomies in order to convert his commitment to rationalism into a much-desired but never-quite-felt optimism (see also note 15 below). See his ambivalent attitude to Freud whose views are used optimistically in *Tonio Kröger*, but not everywhere else. Indeed, here lies Mann's literary effort, in his struggle through reason towards optimism, the 20th-century fallen Faust. See Henry Hatfield, 'Religion in Thomas Mann's *Joseph and His Brothers*', *Boston University Graduate Journal* **15** (Fall 1967).

To conclude, rationalism or irrationalism can, *apriori*, be optimistic or pessimistic, depending on many factors. That irrationalism should lead to pessimism is, however, quite reasonable (in view of the high value of reason), though not very pleasant, of course. But then, pessimistic rationalism is the least pleasant choice (*Ecclesiastes*). This, however, is not to say that optimistic rationalism is easy – see the quotation from Basil Willey in the next note. The easiest, obviously, is optimistic irrationalism, also known as fools'-paradise. This exhausts the possibilities, unless we become more detailed and cautious in describing rationalism and optimism.

[3] See Spinoza, *Ethics* (trans. A. Boyle), Everyman's 1910, Pt. IV, *Prop. XLV* to *Prop. LII*. The following extracts are sufficiently indicative.

"*Prop. XLV.* Hatred can never be good.

Prop. XLVI. He who lives under the guidance of reason endeavors as much as possible to repay his fellow's hatred, rage, contempt, etc. with love and nobleness.

Prop. XLVII. The emotions of hope and fear cannot be in themselves good.

Proof. – The emotions of hope and fear are not given without pain.

Note. – To this must be added that these emotions indicate a want of knowledge and weakness of mind ...

Prop. XLVIII. The emotions of partiality and disparagement are always bad.

Prop. XLIX. Partiality easily renders the man who is overestimated proud.

Prop. LII. Self-complacency can arise from reason, and that self-complacency which arises from reason alone is the greatest.

Note. – Self-complacency is the greatest good we expect.

Prop. LXI. Desire which arises from reason can have no excess."

Spinoza is not likely to have been influenced by Bacon, yet the similarity between the above passage about the deficiency of hope and the following, from Bacon's *Meditationes Sacrae*, 'On Earthly Hope' (*Works*, ed. by J. Spedding, R. L. Ellis, and D. D. Heath, new ed., London 1870, vii, p. 248), is neither accidental nor insignificant. "Certainly ... to keep the mind tranquil and steadfast ... I hold to be the chief firmament of human life; but such tranquility as depends on hope I reject, as light and unsure." See Aristotle, *Metaphysics*, XII, 1972b-1073a; Moses Maimonides, *Guide for the Perplexed*, Shlomo Pines' translation, Chicago 1963, part III, Chapters 51 and 52; Gershom G. Scholem, *Major Trends in Jewish Mysticism*, paperback edition, New York 1961, pp. 131–132; cp. pp. 56–59, 345. Consult also works referred to in note 10 below.

It is hard for me to comprehend how Kant's affinity with Spinoza has been so persistently overlooked. In his *Critique of Pure Reason* Pt. II, Ch. II: 'The Canons of Pure Reason', Section 2: 'The Ideal of the Highest Good, as a Determining Ground of the Ultimate End of Pure Reason', quotes Leibniz (A812B840) but makes no reference to Spinoza. Yet, clearly, its sentiment (A816B844) is Spinozist, its conclusion (A814B842) is Spinozist, and its critique of previous attempts to arrive at the same conclusion (A815B843) is obviously a critique of Spinoza's system. But the greatest compliment in that section, also not explicitly directed at anyone in particular (A817B845), is perhaps the most obvious allusion to the *Ethics* of Spinoza: 'Accordingly we find, in the history of human reason, that until moral concepts were sufficiently purified and determined, and until the systematic unity of their ends [i.e. the intellectual love of God] was understood in accordance with these concepts and from necessary principles, the knowledge of nature, and even a quite considerable development of reason in many [?] other sciences, could give rise only to crude and incoherent concepts of the Deity, or ... resulted in an astonishing indifference ...." (The observation of the indifference, I suppose, is in an allusion to Descartes.) See also Ch. III, 'The Architectonic of Pure Reason' (A840B868): "The legislation of human reason (philosophy) has two objects, nature and freedom, and therefore contains not only the law of nature, but also the moral law, presenting them at first in two distinct systems but ultimately in one single philosophical system."

Note also that the principle of "architectonic of pure reason" (Ch. III) itself provides a whole system of metaphysics culminating with rational theology. Kant asserts that too much was claimed for metaphysics once, thus leading it to disrepute; but "we shall always return to metaphysics as to a beloved one with whom we have had a quarrel." ... "Mathematics, natural science ... have a high value [but merely] as means ... to ends that are necessary and essential to humanity. ... Metaphysics is the full and complete development of human reason ... it is an indispensable discipline ... [T] hat, as mere speculation, it serves rather to prevent error than to extend knowledge, does not detract from the value. On the contrary this gives it dignity and authority, through that censorship which secures general order and harmony, and indeed the well-being of the scientific commonwealth, preventing those who labour courageously and fruitfully on its behalf from losing sight of the supreme end, the happiness of all mankind."

Schiller criticised Kant's moral philosophy (*Über Anmuth und Wurde*, 1793) as too austere – "carrying with [the sense of duty] a monastic cast of mind", as Kant puts it in his reply to it in his *Religion Within the Limits of Reason Alone* (note (in the second edition) to the first *Observation* in Book I, p. 18 of *Harper Torchbook* ed., transl. by T. M. Greene and H. H. Hudson, New York 1960). "The majesty of the moral law (as of the law on Sinai) instils awe (not dread, which repels, nor yet charm, which invites familiarity)", replies Kant. (The 'Anmuth' in Schiller's title is the charm Kant refers to.) See also F. Überweg, *A History of Philosophy* (transl. by G. S. Morris), 4th ed., London 1885, Vol. 2, p. 198.

The same view has been stated by Sir Leslie Stephen, in his *History of English Thought in the Eighteenth Century*, 3rd ed., New York 1902, 1927, Vol. I, pp. 19–20: "Newton laid down mathematical doctrines which were speedily accepted by all mathematicians. To study Newton is therefore to study the history of mathematical investigation of the time. The difference between his views and those of other inquirers is simply a difference of extent, not of substance. One thinker has more knowledge and a wider intellectual horizon; but all thinkers agree so far as their knowledge goes. If the same statement held true in philosophy, we should simply have to expound the views of Locke and Hume, and to show how those views were developed by later inquirers. The thoughts of the greatest man would include those of the less, and afford a starting point for his successors. In fact, however, we have to consider a complex process of antagonistic theorizing, where every position is in turn assumed and abandoned, instead of a simple evolution of thought. ... Men have been arguing metaphysical questions for many centuries without deciding them. Why are these studies, so apparently fruitless, so perennially fascinating? ... What is this world in which we live? What are the ultimate limits of our knowledge? How can it be increased? ... What are the rules to be deduced for the conduct of life? If we could answer [metaphysical] questions, we could satisfy the demand of the intellect for a firm basis of knowledge and a systematic coordination of all discoverable truth. But ... the true theory is reduced by blundering into every possible error ..."

See also Basil Willey, *The Eighteenth Century Background, Studies on the Idea of Nature in the Thought of the Period*, London 1940, p. 44: "Optimism at this level – the level at which Spinoza could declare that *omnis existentia est perfectio* – so far from being a facile or complacent creed, is admittedly almost impossibly hard to attain, and can never be long sustained by flesh and blood. ... With a Spinoza or a Leibniz, unquestionably, it represents the conclusion of long and arduous metaphysical reflection. But with Pangloss ... Thoreau ... or Browning ..., it generally seems to denote contentment. ... In the early and middle years of the eighteenth century the wealthy and the educated of Europe must have enjoyed almost the nearest approach to earthly felicity ever known to man. Centuries of superstition, error, and strife, lay behind. ... 'The vulgar', not yet indoctrinated with the Rights of Man, were contented with their lot. ... The universe had been explained. ..."

For further details of the 18th-century unitary views see next note.

For the difference between Christian ethics and 'scientific' ethics see Paul Hazard, *European Thought in the Eighteenth Century from Montesquieu to Lessing*, New Haven 1954, Chapters IV–VI, esp. p. 64ff.

The references to H. G. Wells and to Russell (*The Scientific Outlook*) I have borrowed from A. E. Baker, Canon of York, *Science, Christianity, and Truth*, London 1943, which I consider a charming and informative period piece, pp. 109, 104–5. See also Russell's *Religion and Science*, London 1935, p. 175: "When Canon Streeter says that 'science is not enough', he is, in one sense, uttering a truism. Science does not include art, or friendship, or various other elements in life. But of course more than that is meant." I think when Russell says this truth is a truism (i.e. widely accepted) he overlooks the fact that in the rationalist tradition of the 17th and 18th centuries (which the Canon was fighting in 1930) this was anything but a truism. Indeed, not only an apologetic Canon, but even an ally to Russell such as Schrödinger, had to repeat the truism and even at a much later day; and a truth, however obvious to Russell, which is repeated by such a polished and laconic writer as Schrödinger, is not quite a truism. See E. Schrödinger, 'On the Peculiarity of the Scientific World-View' and 'The Spirit of Science', both reprinted in *What is Life and Other Scientific Essays*, Doubleday Anchor, New York, 1956, *Mind and Matter*, Cambridge 1958, pp. 44–7: "Dear reader or, better still, dear lady reader, recall the bright, joyful eyes with which your child beams ... and then let the physicist tell you that in reality nothing emerges from these eyes. ... In reality! A strange reality! Something seems to be missing in it."

[4] For the clichés of the period see Peter Gay, *The Enlightenment: An Interpretation*, New York 1967, pp. 34–6; see also John T. Mertz, *A History of European Thought in the Nineteenth Century*, Dover, New York, 1965, I, 147.

For Condorcet's martyrdom see Henry Ellis, *The Centenary of Condorcet, an Address*, London 1894, and J. S. Schapiro, *Condorcet and the Rise of Liberalism*, New York 1934, pp. 105–9: "... a modern instance comparable to Socrates ... calmly writing on the perfectability of mankind under the shadow of the guillotine. [His] true greatness ... was ... evident ... in his attitude towards the Revolution which now threatens to destroy him. 'I have the good fortune' he declared, 'to write in a country in which neither fear, ... nor respect for national prejudice has the power to suppress or to veil any universal truth'."

For Madame Roland's martyrdom see Carl L. Becker, *The Heavenly City of the Eighteenth Century Philosophers*, New Haven 1932, Chapter 3, Section III (paperback ed., 1959, pp. 151 ff.).

For Berthollet's son's suicide see Dr. Thomas Thomson, *History of Chemistry*, London 1830, II, p. 151.

Kleist's suicide was not in spite of optimism, but due to despair – he could "neither learn nor gain anything" by staying alive, he said. Even in the deepest despair, his Enlightenment background (see note 2 above) was showing. So with Freud, who was in constant terror lest he lost his originality and thus be driven to suicide. See my 'Revolutions in Science, Occasional or Permanent?', *Organon* (Warsaw) **3** (1966).

[5] L. Wittgenstein, 'A lecture on Ethics', *Philosophical Review*, January 1965, 8–9: "I will mention another experience straight away which I also know and which others of you might be acquainted with: it is, what one might call, the experience of feeling *absolutely* safe. I mean the state of mind in which one is inclined to say 'I am safe, nothing can injure me whatever happens.' ... The first thing I have to say is, that the verbal expression which we give to these experiences is nonsense! ... We all know what it means in ordinary life to be safe. ... To be safe essentially [*sic*] means that it is physically impossible that certain things should happen to me and therefore it's nonsense to say that I am safe *whatever* happens. ... This is a misuse of the word 'safe'. ... I want to impress on you that a certain characteristic misuse of our language runs through *all* ethical and religious expression. ..."

For the connection between the senses of absolute security and of omnipotence, see, e.g., Freud, 'On Narcissism: An Introduction' (1914), *Collected Papers*, Vol. IV, London 1946, p. 98, and Thomas Freeman, John L. Cameron, and Andrew McGhie, *Chronic Schizophrenia*, Tavistock Publication, London, 1958, pp. 26–42. Cf. Norman Malcolm, *Ludwig Wittgenstein, A Memoir* (with a 'Biographical Sketch' by G. H. Von Wright), O.U.P., London, 1958, pp. 70–71 for the close link Wittgenstein saw between the remarks on absolute safety and religion. See *loc. cit.*, pp. 3, 10–11, 20–21, and 32 for Wittgenstein's own character. See also L. Wittgenstein, *Remarks on the Foundations of Mathematics*, Oxford 1956 (ed. by G. H. Von Wright, R. Rhees, and G. E. M. Anscombe, transl. by G. E. M. Anscombe), Pt. IV, Section 53, p. 157: "The philosopher is a man who has to cure himself of many sicknesses of the understanding before he can arrive at the notion of the sound human understanding. If in the midst of life we are in death, so in sanity we are surrounded by madness." The customary reference to Wittgenstein's use of psychotherapeutic terminology as if it were metaphorical, I suppose, is myopic or more likely mythical. See also the quotation from Wittgenstein in note 15 below.

[6] Robert McRae, *The Problem of the Unity of Sciences: Bacon to Kant*, Toronto 1961, claims (Preface) that it is not even clear what exactly is the thesis of the unity of science: "The differences in the conception of unity" he adds (viii), "are in some cases so great that one may wonder indeed whether they belong within the discussion of the same

subject." Therefore, his book is mainly an attempt "to bring them together at the onset systematically and unhistorically, to make it clear that there is in fact a common subject".

One might expect the modern followers of the new movement for the unity of science to be more explicit. The impression one gets from browsing in the literature, however, is disappointing. There is little on the topic, and that little contains complaints, such as the one launched by P. Oppenheim and H. Putnam in their 'Unity of Science as a Working Hypothesis', in *Minnesota Studies in the Philosophy of Science*, vol. II (ed. by H. Feigl, M. Scriven, and G. Maxwell), Minneapolis 1958, opening sentence: "The expression 'Unity of Science' is often encountered, but its precise content is difficult to specify in a satisfactory manner." "A concern with Unity of Science", they add, "hardly needs justification", and they justify it by the need for "counterbalancing specialization".

In their conclusion, Oppenheim and Putnam admit that the Unity of Science has thus far been adumbrated "without very deepgoing justification". "It has been our aim", in the paper here cited, they add, "first to provide precise definitions for the crucial concepts involved, and, second, reply to the frequently made accusations that belief in the attainability of unitary science is 'a mere act of faith'."

The definition of Oppenheim and Putnam, incidentally, is the theory of the reduction of all science to microphysics in a hierarchy of steps reminiscent of Comte's hierarchy of sciences. I am not clear about it all, since they do not refer to Comte and they do not explain how to reduce, say, general relativity to microphysics. Their 'justification' of the unity of science is inductive: it has been impressively successful in the past, etc.

Perhaps Herbert Feigl sketches ('Unity of Science and Unitary Science', in *Readings in the Philosophy of Science* (ed. by H. Feigl and M. Brodbeck), New York 1953) the same view – it is hard to say on account of his brevity.

The modern place for the thesis of the unity of science seems to me to be, perhaps, Wittgenstein's *Tractatus*, which, of course, nobody claims to be in full comprehension of (see note 14 below). The clearest expression of the modern view seems to me to be that of Rudolf Carnap. See his *The Logical Structure of the World and Pseudoproblems in Philosophy* (transl. by Rolf A. George), Berkeley, 1967, Introduction, p. 9: "*there is only one domain of objects and therefore only one science* ...", (p. 10) "in construction theory we sometimes speak of constructed objects, sometimes of constructed concepts, without differentiating." This is neutral monism, cf. p. 284: "177. *Construction Theory Contradicts Neither Realism, Idealism, nor Phenomenalism.*"

See also p. 290: "180. *About the Limitations of Scientific Knowledge. ... There is no question whose answer is in principle unattainable by science* ... [there exist] 'mere technical obstacle[s]', not an 'obstacle insurmountable in principle'."

Nonetheless, Carnap's view is not as classical as it sounds. See pp. 292–3: "181. *Faith and Knowledge.* ... we do not here wish to make either a negative or a positive value judgment about faith and intuition (in the nonrational sense). They are areas of life just like poetry and love ... as far as their content is concerned, they are altogether different from science. Those nonrational areas ... and science ... can neither confirm nor disprove one another." See also pp. 296–7: "183. *Rationalism*? ... For us there is no '*Ignoramibus*'; nevertheless, there are perhaps unsolvable riddles of life. This is not a contradiction. *Ignoramibus* would mean: there are questions to which it is in principle impossible to find answers. *However, the 'riddles of life' are not questions but practical situations.* The 'riddle of death' ... has nothing to do with questions about death. ... These questions can be answered by biology, but these answers are of no help to a grieved person. ... Rather, the riddle consists in the task of 'getting over' this life situation. ..."

Contrast this with Spinoza's solution of the problem of death: *Ethics*, IV, *Prop. LXVII*:

"A free man thinks of nothing less than of death, and his wisdom is a meditation not of death but of life.

Proof. – A free man, that is, one who lives according to the dictate of reason alone ..."

Spinoza says the extraordinary and Carnap says what is widely accepted and expressed in a multiform of manners and styles. In *The Story of Gilgamesh* Gilgamesh does not change his view of life when he loses his alter ego Enkidu; he simply loses the taste for life. In his suicide note, likewise, Stephan Zweig implies clearly that in his view life is still worth living and fighting for – if one has the strength; yet he simply lacked the strength to start all over again past middle age. All this well illustrates Carnap's view. By immense contrast, Spinoza says, science proves that life is worth living to such a high degree and in such a poignant manner, that he who sees the proof can have no emotional problem facing life, no matter under what conditions. This is the optimism which, Basil Willey observes (see note 3 above), is so hard to sustain emotionally for a thinking person with strong feelings, and yet so easy to sustain for a shallow person like Pangloss.

It is easy, then, to overcome the emotional problems of death if one does not suffer them strongly. Otherwise all philosophy is not enough of a consolation. Philosophy is not enough of a consolation, may I add, because the optimism to the extent which both Spinoza and Carnap advocate is not very convincing: perhaps the intellectual problem of death cannot exist, but the optimism with which they both dismiss it is doubtful. So Carnap's view of the emotional problem of death is true; yet it tallies less well with his view of the intellectual problem of death, than Spinoza's view of the emotional problem tallies with the same view of the intellectual problem: Spinoza's optimism is more of one cloth than Carnap's.

To conclude, it may be said that much as the thesis of the unity of science does exist and writers on it do have much in common, it is hard to make the thesis so specific as to declare that it does or does not contain reductionism of this or that sort. The unity may be of approach, of method, of language, or of subject-matter. The subject-matter may be mother-nature or atomic facts or physics. The unity may be the indifference to various existing diversities of opinions, e.g., concerning the mind-body problems. It is even hard to say whether Oppenheim and Putnam are reductionists although doubtless, reductionism of some sort is what they advocate as the thesis of the unity of science. Similarly, when Schrödinger attacks reductionism he is attacking the traditional unity thesis, just as Russell does; but unlike Russell, Schrödinger is advocating a deeper doctrine of unity.

[7] F. Bacon, *The Advancement of Learning*, Book II: Human Learning, III: Philosophy, 2: Natural, (1) science, (2) metaphysical. *Works* (ed. by J. Spedding, R. L. Ellis, and D. D. Heath), new ed., London 1870, pp. 361–2; see also *ibid.*, p. 321. Note that Bacon uses both images of the pyramid and of the ladder. See also notes 12, 17, and 24 below.

[8] I do not mean to say that the inductive method is used in science, of course, but rather that scientists often said (and believed) that they used that method regularly. And sometimes, no doubt, after sufficiently idealizing experiments, and adding to them some powerful thought-experiments, they could indeed apply the inductive method to them, and perhaps they did. Funnily enough, it is because this practice was recommended by both apriorists and inductivists that it is not easy to say where they differ.

They differ, of course, mainly in their claims for the final ground for knowledge, and this difference is somehow reflected in their different views on method, but it is not so easy to say how.

Laplace, in his *System of the World*, Book 5, reports that the victory of Newtonianism over Cartesianism signified the victory of inductivism over deductivism. For my part, I think the story is somewhat more complicated, and linked with the success of the Royal Society, the success of British politics, its link with Locke's theories of toleration and of balance of power, and with Locke's friendship with Newton – as well, of course, as the inductivism of Newton, Locke, and the Royal Society. See also P. Duhem, *The Aim and Structure of Physical Theory*, Princeton 1964, Part I, Chapter IV, Section 8: 'The Diffusion of the English Methods'. See also Paul Hazard, *European Thought, etc.* and H. B. Acton's 'The Philosophy of Language in Revolutionary France', *Proc. British Academy*, 1960, reprinted in *Studies in Philosophy* (ed. by J. N. Findlay), Oxford Paperbacks, 1966.

Nevertheless, apriorism was by no means dead, and the traditions of Euler, Kant, Oersted, and Helmholtz, kept it alive and very significant indeed yet, doubtless, it was more than somewhat disreputable to be an apriorist in the eighteenth and nineteenth centuries. And it is this last point that is pertinent to the text above, since the unity thesis obviously squares less easily with inductivism than with apriorism.

[9] Descartes, *Correspondence* (ed. by Ch. Adam and G. Milhaud), Paris 1936, Vol. I, to Mersenne, 23 December 1630 (p. 184): "About this [making useful experiment] I have nothing to say after what Verulam [Bacon] has written about it, namely, without being too curious to research into all the small particulars touching on a matter, mainly one must make general collections of all those things that are most common and which are very certain and which can be known without expenditure: such as, that all colchea are rotated in the same direction, and to know if this is the same after the equinox; that the bodies of all animals are divided into three parts, *caput*, *pectus*, and *ventrem*, and also other examples; since these are those which serve infallibly in the search for truth. For more detailed items, it is impossible that one does not get many superfluous ones, and even false ones, because one does not know the truth of things before one makes them [the experiments]."

Op. cit., to Mersenne, 10 May 1632 (p. 226), a follow-up on the previous letter: "You have informed me elsewhere that you know people who are pleased to work for the advancement of the sciences, even of their desire to make all arts of experiments at their own expense. If any one of this disposition would want to undertake to write a history of celestial appearances, according to the method of Verulam [Bacon], and if without putting forward neither any reason nor any hypothesis, he would describe for us exactly the heavens as it appears now, which position each fixed star is in respect to its neighbors, which difference, either of size or of color or of clarity or of more or less twinkling, etc.; *item*, if this corresponds to what the ancient astronomers have written of it and what difference he has found in it (since I do not at all doubt that the stars do not ever change their relative positions because one deems them fixed)." And he goes on about comets and their orbs, and about ecliptics and apogees of planets, "very useful to the public ... will relieve me of a lot of labour, but I do not hope that someone else will do it. ..." (My translation.)

See also Martha Ornstein, *The Role of Scientific Societies in the Seventeenth Century*, Chicago 1928, pp. 44–6, for an exposition of Descartes' Baconianism.

See also Leibniz's praise of Bacon, and its roots in the influence of Bacon's doctrine of prejudice on him, in Ellis's preface to *Thoughts on the Nature of Things* (*Works, op. cit.*, Vol. 3, pp. 70 ff.). "We do well to think highly of Verulam", i.e. Bacon, says Leibniz, "for his hard sayings have deep meaning in them." And the "hard sayings" are criticisms of the scholastics, whom Leibniz admired all his life, alone in the world of radicalists whose contempt for everything mediaeval was constantly mounting.

Spinoza, however, dismissed Bacon as confused and dogmatic, especially regarding his doctrine of prejudice (of the cause of error); see his first letter to Oldenburg, *Works*, Dover, New York, 1951, Vol. II, p. 278. Nevertheless, one should note that he dismisses both Bacon and Descartes in the same letter, and expressly at Oldenburg's invitation. His final verdict of Bacon is, that whatever valuable he has said, has since been better said by Descartes.

Concerning the dispute regarding the French philosophers and their following Bacon or Descartes, see C. C. Gillispie's and L. Pearce Williams' contributions to *Critical Problems in the History of Science* (ed. by M. Clagett), Madison 1959, and comments there; see also R. Emerson's 'Peter Gay and the Heavenly City', in *Journal Hist. Ideas* **28** (1967), and references there.

There is, of course, the idea that science is both inductive and deductive, or both compositive and resolutive, or both synthetic and analytic (these are three sets of more-or-less synonymous, though always intolerably vague, terms). This idea may be methodological, and it may be epistemological. See J. H. Randall Jr., *The School of Padua*, Padova 1961, and comments on it in J. W. N. Watkins, *Hobbes' System of Ideas*, London 1965, Section 9. See also Justus von Liebig, *Induktion und Deduktion*, 1865, and F. Engels, *Anti-Dühring*. There is a paper by G. Buchdahl on 'Descartes' Anticipation of a Logic of Scientific Discovery', in *Scientific Change* (ed. by A. C. Crombie), New York 1963, and comments on it by N. R. Hanson there. I find Buchdahl's summary of his problem in his second paragraph (p. 399) incomprehensible, as he raises the question, how shall we handle the question of scientific truth, which second question he leaves unformulated. Still, I suppose Hanson is right when he says (p. 461) Buchdahl has rendered us a service "in revealing as myth-eaten the picture of Descartes as a naive Cartesian rationalist".

To say that Descartes was not quite a Cartesian, as Hanson rightly says, is not the same as to say that Marx was not a Marxist. For, what Marx meant was that he was at liberty to change his mind, whereas what Hanson means is much more interesting. When one has a major theory which one assumes to be true, one may accept only conclusions from that major theory, or also other truths which ought to follow from, or at least harmonize with, that major theory, even though we may not know how. And, of course, two truths must harmonize; but one's major theory may turn out to be false after all. In this sense, we may say, our false major theory did not prevent us from accepting truths which contradict it: had we known these truths contradict the major theory we would simply have given up the major theory.

And so, what Hanson says to Buchdahl is that quoting instances from Descartes which are, say, inductive-deductive in character, does not prove that Cartesianism is inductive-deductive, but merely that Descartes could think out of character while believing that he was thinking in character.

Buchdahl's error is part of a thesis of A. C. Crombie's *Robert Grosseteste and the Origins of Experimental Science 1100–1700*, Oxford 1953. There Grosseteste's philosophy is linked with the view of the school of Padua, and the view of that school is declared to be the one both widely preached and widely practiced in the seventeenth century. If this were so, then the book's lengthy title would indeed read, 'Robert Grosseteste as the Origin of Experimental Science'. The thesis is expressed at the end of the Introduction (p. 13): "Grosseteste took the double, inductive-deductive procedure described by Aristotle in the *Posterior Analytics*", he says. "He found the inductive side illustrated by the writings of the medical school and the deductive side illustrated by the writings of Euclid, Ptolemy, and others. ... Grosseteste was favoured ... by unusual opportunities to make his influence felt. ... With Grosseteste, Oxford became the first centre of the methodological revolution with which modern science began. On the continent, Grosseteste's influence may be traced with certainty in several writers, and there is evidence to suggest that the methodology of the Oxford school exerted a decisive influence on European science as a whole." Paris, he admits, was also influential, and was somewhat independent of Oxford. But "Oxford took the lead. ... There is no doubt that from the time of Grosseteste the experimental science ... began to appear in centre after centre...."

In his Conclusion Crombie follows Randall in claiming that Padua was the most important 16th-century methodological centre, adding in a footnote (p. 297): "According to Randall ... the Paduan teacher Paulus Venetus was sent by his Order to Oxford in 1390 and remained there for three years ... after which he taught for two more years in Paris", and this, we are led to believe, was the world's most significant, world shaking, travel-grant. "In his various encyclopedic writings he fully though critically expounded Oxford ideas on logic and dynamics", says Crombie in a manner of proof.

Those who need criticism of such a 'proof', can consult N. Gilbert, 'Galileo and the School of Padua', *Journal of the History of Philosophy* **1** (1963) 223–31, and Benjamin Nelson's 'The Modern Revolution in Science and Philosophy', *Boston Studies*, vol. III, especially p. 10.

So much for Crombie's claim of Grosseteste's influence on Padua. As to Padua's influence on posterity, he tells us that, for instance, (p. 301), "Francis Bacon's method of discovering the form was precisely the 'double procedure' worked out during the preceding four centuries". And (p. 302), "His method was mentioned and used by more than one seventeenth century scientist, particularly in England. For example, Harvey [this is no slip of Crombie's pen but a point supported by a very scholarly reference], Hooke, and Boyle all referred to it in the midst of their investigations." Not only "Bacon's method ... was precisely the 'double procedure'", Galileo's was too, it seems (p. 303). "The originality of Galileo's method lay precisely in his effective combination of mathematics with experiment." "To connect the observation with a theory" Crombie adds (p. 305), "Galileo described precisely the double procedure of resolution and composition which his predecessors in Oxford and Padua had made familiar." Precisely Bacon, Galileo, everybody, is precisely Grosseteste.

Crombie's thesis is no less than that all important methodologists preached and all important scientists practiced the method of induction-deduction, or resolution-composition, or analysis-synthesis – including Bacon and Galileo. It is amazing to me that in the mid-20th century such a thesis should be advocated, and that the volume which contains this as part of a major thesis (the thesis being, all this goes back to Grosseteste) should be viewed as a scholarly contribution merely because it is (doubtlessly) very scholarly.

Nevertheless, incredible as the thesis that Galileo and Bacon preached the same view, there is a certain amount of truth in it. In retrospect we may differ so much from both that we may see their differences as a family quarrel, much as they would not like our viewpoint. This can be said by one who, like myself, rejects both the inductive and the deductive view (in favour of the skeptic view). But Crombie agrees with both Bacon and Galileo, which is by no means an easy feat.

In defense of Crombie one can say, it is in epistemology that Bacon and Galileo differ, not in methodology. In my view, however, the opposite is the case. The methodological differences between the inductivist and the apriorist are much easier to delineate than the epistemological ones. In method the empiricists insist that experience plays a role much more crucial than intuition, and apriorists hold the converse; in epistemology it is hard to maintain the same distinction with equal clarity and sharpness. Already Kant, in the last chapter of his *Critique of Pure Reason*, on the history of pure reason, makes the observation that though the distinction between intuitionism and sensationalism (a priorism and inductivism) is subtle, these two positions are represented by schools which go back uninterrupted to the very beginning of ancient philosophy. Unfortunately, Kant does not elaborate here. We shall return to this point in notes 17 and 24 below.

Note also in Crombie (ed.), *op. cit.*, Henry Guerlac's 'Some Historical Assumptions of the History of Science', especially Section III, on historians' failure to treat science as a unity, let alone as a part of the intellectual unity of our tradition. See, however, Koyré's comments on Guerlac there and Koyré's subtle move from unity proper to interconnectedness and its problems.

[10] See H. A. Wolfson, *Philo*, Cambridge, Mass., 1947, especially the first and final chapters, and *Crescas' Critique of Aristotle*, Cambridge, Mass., 1929, especially the first chapter. See also his 'Extradeical and Intradeical Interpretations of Platonic Ideas', in *Ideas in Cultural Perspective* (ed. by P. P. Wiener and A. Noland), New Brunswick 1962. See also J. L. Blau, *The Christian Interpretation of the Cabala in the Renaissance*, New York 1948, and 1965; Frances A. Yates, *Giordano Bruno*, Chicago 1964; P. P. Wiener, 'Problems and Methods in the History of Ideas', in Wiener and Noland, *op. cit.*, esp. p. 28.

[11] Moses Maimonides, *The Guide of the Perplexed*, Shlomo Pines' translation, University of Chicago, 1963, Pt. I, Ch. 31 on "the insufficiency of the human intellect and its having a limit at which it stops". Ch. 32: "In regard to matters that it is not in the nature of man to grasp, it is ... very harmful to occupy oneself with them." Ch. 72: "And just as in the body of man ... so are there in the world. ... Accordingly it behooves you to represent to yourself in this fashion the whole of this sphere as one living individual possessing a soul. ..." Ch. 73, the tenth premise, note to the reader: "We wish consequently to find something that would enable us to distinguish the things cognized intellectually from those imagined. For if the philosopher says, as he does: That which exists is my witness and by means of it we discern the necessary, the possible, and the impossible; the adherent of the Law says to him: The dispute between us is with regard to this point. For we claim that that which exists was made in virtue of will and was not a necessary consequence. Now if it was made in this fashion, it is admissible that it should be made in a different way, unless intellectual representation decides, as you think it decides, that something different from what exists at present is not admissible...."

The argument is repeated by Russell, see references in note 3 above.

[12] F. Bacon, *Principles and Origins According to the Fable of Cupid and Coelum etc.*, third paragraph (*Works, New Ed.*, London 1870, V, p. 465): "And certainly it is the prerogative of God alone, that when his nature is inquired by the senses, exclusion shall not end in affirmation."

See C. W. Lemmi, *Classical Deities in Bacon*, Baltimore 1933, pp. 50, 57, and 60. The note on p. 58 reads: "In other words, I think it probable that Bacon's *Thoughts on the Nature of Things* is an expansion of Comes's chapter of Cupid."

In *Novum Organum* (I, Aph. 75) Bacon speaks of systematic doubt as of "a calumny of nature herself" and (I, Aph. 129) of the arts and sciences as "that right over nature which belongs to [the human race] by divine bequest". His metaphor of speculations as chaining Nature, and his claim that we must woo Nature so that she reveals Her charms are systematic, and appear both in the Preface to his *Novum Organum* and early in that work (Book I, Aphs. 1, 3; also Aph. 102) and elsewhere.

This sentiment of Bacon's is deeply engrained in the western tradition. I shall quote only two passages illustrating it, on account of their interest.

Sir Charles Sherrington opens his classical *Man on His Nature*, Pelican 1940, thus: "As to Natural Theology and what we are to understand by it, more than one well-known statement offers us counsel. Bolingbroke, type in his way of eighteenth-century culture, wrote to Alexander Pope, the poet, 'What I understand by the first philosophy [metaphysics] is "natural theology", and I consider the constant contemplation of Nature, by which I mean the whole system of God's works as far as it lies open to us, as the common spring of all sciences, and of that', i.e. Natural Theology. There is, too, Lord Bacon's famous definition (*De Augmentis*, iii, 2), that 'spark of knowledge of God which may be had by the light of nature and the consideration of created things; and thus can be fairly held to be divine in respect of its object and natural in respect of its source of information'." (For the metaphors 'spark', 'light', etc., see Evelyn Underhill, *Mysticism*, 12th ed., London 1930, and G. G. Scholem, *op. cit.*)

It sounds funny to quote Sherrington quoting Bolingbroke quoting Bacon, but that is the stuff traditions are made of. The following is Basil Willey's comment (*op. cit.* p. 156–7) on Holbach's comment on Clarke; do I have to say not Clarke but Clarke's comment on Bacon? It is perhaps more important to stress that according to Willey, Holbach, in a manner characteristic of the whole 18th century, "declares that all our misfortunes are due to our neglecting and departing from Nature. ... Our errors *cannot* be 'natural', are not what Nature intended." And he goes on to expose Holbach's error as a typical "eighteenth century mental habit", namely "that of honouring Nature with a reverence which, in spite of professed atheism, is in fact religious or transfused from religion". Holbach, he says (p. 161) "treats Clarke rather as Marx afterwards treated Hegel; all that Clarke says of 'God', he assures us, may truly be said, and intelligibly said of 'Matter' or 'Nature' ...".

[13] Werner Heisenberg, 'Science as a Means of International Understanding', in his *Philosophic Problems of Nuclear Science* (transl. by F. C. Hayes, London 1952), reprinted in *Great Essays by Nobel Prize Winners* (ed. by C. Hamalian and E. L. Volpe), New York 1960. See also Heisenberg's 'The Role of Modern Physics in the Present Development of Human Thinking', in his *Physics and Philosophy*, New York 1958.

Heisenberg's metaphor of the way or highway to God may be an allusion to Max Brod's *Tycho Brahe's Way to God*: of course the word 'way', 'method', 'tao', frequently had religious as well as scientific-religious connotations, and so Brod's use, as well as Heisenberg's, is quite common.

[14] L. Wittgenstein, *Tractatus Logico-Philosophicus*, London 1922, p. 67. "4.022. The proposition *shows* its sense. The proposition *shows* how things stand, if it is true. And it *says*, that they do so stand." P. 79: "4.121. ... That which expresses *itself* in language, *we* cannot express by language. The proposition *shows* the logical form of reality. They exhibit it." No less! "4.1212. What *can* be shown *cannot* be said." What this last statement *can* mean without being *obviously* false I have *no idea*. No idea at all. P. 187: "6.522. There is indeed the inexpressible. This *shows* itself; it is the mystical." I do not know how *well* it shows itself, but I *do* agree.

See also M. Black, *A Companion to Wittgenstein's Tractatus*, Ithaca 1964, pp. 190–192: "The Notion of Showing. Wittgenstein uses the verb 'to show', or its cognates, very often ... forty occurrences. ... Unfortunately, this crucial concept is most elusive."

Black says that what *shows* itself is a symbol – and that the showing is immediate – 'in a flash'. This is not very illuminating.

See also L. Wittgenstein, *Remarks on the Foundations of Mathematics, op. cit.*, Pt. V, Section 9, p. 167: "Take a theme like that of Haydn's (St. Antony Chorale), take the part of one of Brahms's variations corresponding to the first part of the theme, and set the task of constructing the second part of the variation in the style of the first part. That is a problem of the same kind as mathematical problems are. If the solution is found, say as Brahms gives it, then there is no doubt; – that is the solution." And so on. This goes for rationality, though I think it can be sharply contrasted with the following quote from Husserl which is a marvellous open defense of a new, unknown, kind of rationality!

In the concluding summary of his 'The Crisis of European Man'. Husserl characterises the 'crises' as the "seeming collapse of rationalism". It is only a crisis of a false theory of rationality, he adds, based on 'naturalism' and 'objectivism', not of "the essence of rationalism itself" (E. Husserl, *Phenomenology and the Crisis of Philosophy*, Harper Torchbook 1965, p. 191). It never occurred to Husserl to doubt essentialism and thus justificationism and certitude, any more than to Wittgenstein. And so he could not but conclude that the choice is between "the ruin of Europe ... fallen into a barbarian hatred of spirit" and "a heroism of reason" (*ibid.*, p. 192). But at least he knew and confessed inability to offer a satisfactory theory of rationality. In other words, though he stuck to his concepts of rationality and of essence, he viewed them as open-textured, and stressed this openness beautifully.

Now, obviously, at least *prima facie*, the idea of rationality had better be open-textured, but this can hardly be said of the idea of essence. See the defence of the open-textured concept of rationality in Aron Gurwitsch, 'The Last Works of Edmund Husserl' in *Philosophy and Phenomenological Research* 17, No. 3 (1957), p. 396: "*Historical forms of rationality*, however, must be distinguished from the *idea of rationalism* or rationality, a Platonic idea which is specified and approximated in those historical forms. Surmounting a certain historical form of rationalism is one thing; abandoning the very idea of rationalism is quite another. ... *Philosophy and the idea of rationalism are one and the same. ...*"

For Bartley and for Lakatos see note 1 above.

[15] Sextus Empiricus, *Outlines of Pyrrhonism* (transl. by Rev. R. G. Bury, Loeb Classical Library, London and New York 1933) I, Ch. XII, "What is the End of Skepticism? ... the Skeptic's end is quietude in respect of matters of opinion and moderate feeling in respect of things unavoidable. ... The man who determines nothing ... neither shuns nor pursues anything eagerly; and, in consequence, he is unperturbed. ...

... The Skeptics were in hopes of gaining quietude by means of a decision ... and being unable to effect this they suspended judgment; and they found that quietude, as if by chance, followed upon their suspense. ... We do not suppose, however, that the Skeptic is wholly untroubled...."

As far as inner logic is concerned, one must consider the skeptical hypothesis – non-justificationism leads to ataraxia – empirically refuted. The Greek word 'ataraxia', here translated as 'quietude' and 'unperturbedness', is somewhat hard to render, since it is a quasi-religious term overloaded with nuance and overtone, and since modern experience and modern psychology enable us to distinguish with ease a variety of states of undisturbedness, from the insensitive yet most optimistic and the sensitive yet philosophical. It is easy, for example, to contrast the Zen-Buddhist matter-of-fact calm-through-cultivated-philosophic-indifference – particularly as understood and/or misunderstood in the West (see A. Koestler, *The Lotus and the Robot*, London 1960) – with the self-assured calm-through-cultivated-philosophic-optimism of early 19th-century English writings, whether literary (Jane Austen) or scientific (Sir John Leslie, John Dalton, Dr. Thomas Thomson). Perhaps the nearest modern variant of ataraxia is to be found in the 17th-century writings – due to imitation, I think, rather than either accident or inner logic. We have in the 17th century a skeptic and justificationist, Boyle and Spinoza, exhibiting it in a marvellous fashion. In central European late 19th century we find a certain degree of irritation accompanying all intellectual discourse, manifesting, it seems, an ideological opposition to ataraxia; so I read the atmosphere emanating from works of Kirchhoff and Boltzmann, Thomas Mann, and Freud. I should even say that a certain irritation at the idea of ataraxia is present in the works of Wittgenstein and of Popper alike – again, indifferently to the justificationism of the one and the skepticism of the other. Unless Popper will repudiate my calling him a skeptic, that is to say my identifying skepticism with non-justificationism. But even the refusal to identify skepticism with non-justificationism, to wit, the preference to ally skepticism with cynicism (in the modern deprecating sense, not in the ancient austere sense) and nihilism (alluded to in Heisenberg's text cited above) – even this refusal is but a profound distaste for ataraxia; which refusal, regrettable as it is, nonetheless refutes Sextus's hypothesis that skepticism leads to ataraxia; or, if you wish to put it in a different way, a non-justificationist's refusal (such as Popper's) to identify non-justificationism with skepticism refutes Sextus's thesis that non-justificationism is identical with skepticism.

Since the word 'skeptic' means searcher, it is a bit hard to see why the desire to have ataraxia seems to exclude the desire to search for the truth. There may be a different explanation for this. Possibly this is due to Kant's discussion of the difference between Hume's censorial skepticism and his own critical skepticism, which encourages the search for arguments even beyond their legitimate limit; see *Critique of Pure Reason*, Pt. II, Ch. I, Section 2, 'Impossibility of skeptical satisfaction of the Pure Reason that is in conflict with itself' (A760B788). Another possible line is expressed by Ludwig Edelstein (Crombie, *op. cit.*, p. 34). "A Platonic metaphor expresses the same thought in a different way. He who does not learn to work like a slave for the possession of the truth will never reach it" (*Republic* IV, 494D). This sentiment is expressed by Schopenhauer in his attack on Hegel, approvingly quoted by Popper (*Open Society*, II, Chapter 12), "Who can really believe that truth also will come to light, just as a by-product?"

The answer to this rhetoric question is, all those people who are able to maintain any measure of a critical attitude towards Protestant ethics. And, let me repeat, Protestant ethics so-called, is Cabbalistic-alchemist ritualistic apologetics. As a matter of fact, unpleasant to the labour pietists, truth ever so often does come to light as a by-product. Whatever is the proper method of science, be it the one described by Popper or not, clearly it is not the one consciously adopted by the fathers of the scientific revolution: if they have adopted it, they did so unintentionally – as the unintended consequence of their conscious and different intentions – and so the proper method, and the truths it had revealed, came quite as by-products.

It is amazing how similar in Baconian – 'Protestant', if you will – sentiment, are Popper and Wittgenstein. See Ludwig Wittgenstein, *Philosophische Bemerkungen* (ed. by R. Rhees), Oxford 1964, Foreword: "... I would like to say 'this book has been written for the greater glory of God', but nowadays this would be contemptible, i.e., it would not be understood aright. Namely, it has been written in good will and as far as it has not been written in good will but out of vanity etc., so far the author would like to have it condemmed. He cannot purge it further of these ingredients than he is himself clean of them." (My translation.) Clearly, according to this passage, vanity is a sinful intent that cannot possibly lead to the discovery of the truth. And, according to this passage, any motive other than "the greater glory of God" is vain or sinful: Of course, this does not mean that Wittgenstein – and his like – would oppose ataraxia, but ataraxia, so it seems, must only come at the very end of a very hard day's work: this is the ambivalent irritability about ataraxia here-and-now which I have alluded to above. I join Feyerabend in saying, a good intellectual debate over a glass of beer makes the beer taste better – even though beer is not exactly my cup of tea.

Thus, the question, does ataraxia help learning must be answered with, sometimes. It can only be reopened after developing a better theory of both peace of mind and work. There are varieties of peace of mind; and varieties of hard work; and of the virtues and defects of either. One might then work out a view of the impact of philosophy on states of mind – somewhat similarly to, but more critically than, Erik Erikson's study of William James. As he shows, James was depressed by determinism, and was finally able to overcome his depression by refuting determinism to his own satisfaction. (See Erikson's introduction, to *Emotional Problems of the Student* (ed. by G. B. Blaine Jr. and C. C. McArthur), Appleton-Century-Croft, New York, 1961). Or, in reverse, a state of mind may appear to influence philosophy. See John Stuart Mill's story in his *Autobiography*, where he blames his early utilitarianism for his depression and where he narrates how his emergence from the depression led him to a new version of his utilitarianism. (See, however, Ruth Borchard's *Mill, The Man*, London 1957.)

Finally, Bartley has drawn my attention to the possibility that not all ancient skeptics shared Sextus's view of ataraxia as a by-product, that for some of them possibly the end of all discourse should be ataraxia – thus making them more akin to Zen Buddhists then to Socrates. See John Owen, *Evenings with the Skeptics, or Free Discussion on Free Thinkers*, London 1881, "Arkesilaos might easily have taken his own test of, and ideas concerning, truth, as possessing not only subjective but an objective validity" (Vol. I, p. 307), and remarks on "the pursuit of ataraxia" (309), "the probability of Karneades is ... a compromise between dogmatists and absolute Skeptics" (318, 319), "the perpetual appeal Ataraxia [had for Sextus] in its accurate definition ... [is] opposed to a [truly skeptic] philosophy which makes non-definition the chief principle in its method" (338–9).

[16] R. Boyle, *Works* (ed. by Th. Birch), 1st ed., 1744, Vol. III, p. 432a ff.: "... the study of physick has one prerogative, (above divinity). ... I mean the certainty and clearness, and the resulting satisfactoriness of our knowledge of physical, in comparison with any we can have of theological, matters, whose being dark and uncertain, the nature of the things themselves, and the numerous controversies of differing sects about them, sufficiently manifest.

... Cartesius was [so] sensible of a dependence of physical demonstration upon metaphysical truths, that he would not allow any certainty not only to them, but even to geometrical demonstrations, until he had evinced that there is a God, and that he cannot deceive men, that make use of their faculties aright.

[But] when ... Descartes ... demonstrate[s] ..., the presumed physico-mathematical demonstration can produce in a wary mind but a moral certainty, and not the greatest ... that is possible to be attained...."

In the above paragraph the demonstration referred to pertains to comets, and comets were at the time the biggest trouble for Cartesians since the 1661 comet moved in the wrong direction along its orbit thus destroying the vortices. (See Laplace, *System of the World*, Book V.)

See also *op. cit.*, Vol. IV, p. 346a: "... And for the rule ... that there is always the same quantity of motion ... the proof he [Descartes] offers, being drawn from the immutability of God, seems very metaphysical, and not very cogent to me, who fear, that the properties and extent of the divine immutability are not so well known to us mortals, as to allow Cartesius to make it, in our case, an argument *apriori*."

B. Scharfstein stresses in his 'Descartes' Dream' (*Philosophical Forum*, Vol. I, 1968–69), the fact that Descartes thought active curiosity a sin and tried very hard to be a passive student; that, indeed, Descartes expressed this sentiment strongly on his death-bed. This is not merely a psychological factor, certainly not in the 17th century when receptivity was contrasted with (sinful) willfulness by Bacon (doctrine of idola, end of *Sylva*), Locke (*Conduct of the Understanding*), Descartes (*Discourse*) and Spinoza (*The Improvement of the Understanding*, and letter to Oldenburg cited above).

[17] The idea that development can be not from the little satisfactory to the much satisfactory (Bacon) but from the more unsatisfactory through the less unsatisfactory to the perfect, can be found in Plato's *Symposium* (211c), "starting from individual beauties, the quest for the universal beauty must find him ever mounting the heavenly ladder, stepping from rung to rung ... until at last he comes to know what beauty is ... and once you have seen it, you will never be seduced again" by (lesser) beauties of particular things, but stick to the highest "vision" of "the heavenly beauty" itself, i.e. the ideal of beauty, thus achieving knowledge, virtue, and perhaps even immortality.

The very same rejection of the ladder can be found in Aristotle, more didactically stated and with its edges trimmed: "the premisses of demonstrative knowledge must be true, primary, immediate, more knowable than, and prior to, the conclusion" where the premiss is a theory and the conclusion an observed fact (*Post. Anal.*, i, 2, 71b). It is quite clear that to Aristotle empirical experience is the ladder as far as demonstrable knowledge is concerned but the real basis as far as probable knowledge is concerned.

Whereas the empirical background to demonstrable knowledge is, according to Aristotle, removable – it is not a justification – as a part of the hierarchy of causal explanation it is no more removable than a corollary of a mathematical theorem is removable from mathematics: the first principles are the very causes of all facts; once we know them we have a causal hierarchy of causes and effects, each step being the effect of the higher step and the cause of the lower step, as well as being the conclusion from the higher step and the premiss to the lower step – except for the first principles which are causes or premisses alone and the (general) facts which are the effects or conclusion alone. Here is a ladder, or a pyramid, which was transmitted in the Renaissance by Sir Francis Bacon to the modern world, and reinforced by thinkers like August Comte. (See note 6 above.)

It is important, if confusion is to be avoided, to note that there are two ladders here. One is for justifying our theories and hence, as I am arguing, methodological; and this ladder can be thrown away once the goal – knowledge – has been achieved. The other ladder is imbedded in knowledge itself, and is, thus, epistemological; it cannot be removed. (See also the ladder in Boethius, *Consolatione*, I, 1.)

To return to Bacon, he uses the metaphor of removing the ladder, but only when he clearly deals with methodology, not with epistemology. When he insists that ancient knowledge was attained by induction, he worries about the absence of historical evidence for this. The evidence, he claims (*Novum Organum*, I, Aph. 125), has been removed – just in the manner in which builders remove scaffoldings and ladders out of sight.

Goethe reverted the metaphor: the scaffolding cannot usually be evidence as evidence is legitimately a part of science, and so he viewed hypotheses as scaffoldings. See his *Maximen und Reflexionen* (Goethe, *Gedenkausgabe der Werke, Briefe, und Gespräche*, Zürich 1949, Vol. 9, p. 653, para. 1222): "Hypotheses are scaffoldings which one puts up before building and which one tears down once the building is complete. They are indispensable for the worker: only one should not take the scaffolding for the building." (My translation.) The same idea is put more tersely in R. T. H. Laennec's *Traité de l'ausculation médiate et des maladies de poumons et du cœur*, Paris 1826, I, 280 (see E. H. Ackerknecht, *Medicine at the Paris Hospital, 1794–1848*, Baltimore 1967, p. 9) and also in Herschel's *Preliminary Discourse on the Study of Natural Philosophy*, London and Philadelphia 1830–31, Pt. II, Ch. 7, para. 216, p. 153: ..." to lay any great stress on hypotheses ... except inasmuch as they serve as a scaffold for the erection of general laws is to quite mistake the scaffold for the pile."

Let us not inquire as to the loss of the empirical data by the ancients, and overlook the questionability of Bacon's simile of movable ladders and scaffoldings here. Can an empiricist treat hypotheses as scaffoldings? (Not that Goethe was an empiricist; the above quoted empiricist maxim is followed by a few anti-empiricist ones. But Herschel was as much of an empiricist as any thinker, and his metaphor stuck.) Admittedly, it seems quite innocuous for an empiricist to say that hypotheses are mere scaffoldings, meaning, finally the basis for science is empirical fact not conjectures; but if epistemologically we remove empirical evidence too – as both Plato and Aristotle maintain – then where is the difference between apriorism and inductivism? We see here, again, how right Kant was to claim that the difference is subtle (see note 9 above).

One may defend Sir John Herschel by claiming that he was no Platonist-Aristotelian; he removes hypotheses as scaffoldings but never the empirical basis. But this is overshooting the target. Herschel insists on the certitude of Newtonian mechanics, but not on its empirical foundation: he quite permits *a priori* proof of it as much as *a posteriori* proofs. He does not really care, and so we cannot care more than he and insist on declaring him apriorist or inductivist.

Those who did insist on inductivism while rejecting apriorism, for example Ampère, had to be more subtle: for them both hypothesis and experience are ladders to be removed, but hypothesis must be removed first. This is not merely an order of chronology but also an order of priority, even of emotional priority. As G. E. M. de Ste. Croix suggests (Crombie, ed., *op. cit.*, p. 84), Plato's idea of removing the ladder presents not only apriorism, but also a characteristic aprioristic contempt towards experience, which Plato sometimes exhibited (in the *Republic* and later works). It is very interesting that Bacon quotes (*Advancement*, II: '*History of Nature Wrought Mechanical*') Plato's other discussion of the ladder, also leading to the first principles of beauty, not the *Symposium* (where neither contempt nor praise for facts is registered) but *Greater Hippias*, where Socrates pokes ironic fun at those who are impatiently contemptuous of lower forms of beauty. Bacon reads this to say, you can remove the ladder, but only after having used it. (Attitudes to empirical knowledge in Plato's works may be used as means of attacking the Socratic problem.)

There is also the question of the vulnerability of the ladder here. The empiricist as well as the apriorist view hypotheses as highly vulnerable and so, of course, something that must be sooner or later removed "out of sight" (Bacon). But whereas apriorists view evidence of the senses as vulnerable temporary means (Socrates of Plato's early dialogues, Bruno, Galileo, Descartes), empiricists view evidence of the senses as final even if removable (Aristotle, Bacon, Herschel, Whewell).

Thus, it is possible to view empiricism or inductivism as a double-secure system, where our hierarchy of final knowledge is based both on immediate intuitions and on empirical facts. Perhaps this represents Herschel's view better.

In my view, not only Plato and his followers, but, clearly Socrates too, felt that in principle apriorism should do.

This may explain why it was so hard to penetrate the classical arguments for scientific certitude and criticize them. For a discussion of the 20th-century background to E. A. Burtt's *Metaphysical Foundations of Modern Physical Science*, London 1925, see my 'Science in Flux', *Boston Studies*, III, p. 304.

It would be unfair to Burtt, however, to suggest that in our day and age his idea (there is no finality in science) is common property. Some writers fail to adhere to it even while advocating it. I shall give one example here: Heisenberg's 'Recent Changes in the Foundation of Exact Science' (*op. cit.*). Heisenberg attempts to criticize the view that there is finality in science. "Even Kant's philosophy, intended as a critique of premature dogmatizing in scientific concepts, could not prevent the torpescence of the scientific concept of the universe – it may even be said that it encouraged it" (p. 22). And Heisenberg blames this on Kant's apriorism. But Heisenberg's explanation may be false. How well Heisenberg himself – not an apriorist – has succeeded in preventing "premature dogmatizing and torpescence" may be surmised from the following sentence, two paragraphs later: "... modern physics has shown that the structure of classical physics – as that of modern physics – is complete in itself." What this exactly means I do not know, as I do not know how a theory can declare itself complete or incomplete, in itself or besides itself. But it sounds to me to be not exactly devoid of the attitude condemned in Kant. It sounds to me that Heisenberg's 'complete' means for modern physics what Kant's "premature dogmatizing" means for classical natural philosophy (see my 'Is Physics Complete?', *Synthese*, 1958). Indeed, one may ask, does Heisenberg disagree with Kant about the finality of classical physics? The answer is, no: "Columbus's discoveries were immaterial to the geography of the Mediterranean countries, and it would be quite wrong to claim that ... [he] had made obsolete the positive geographical knowledge of the day. It is equally wrong to speak today of a revolution in physics. Modern physics has changed nothing in the great classical disciplines ... only the conception of hitherto unexplored regions, formed prematurely ... has undergone a decisive transformation" (p. 18).

Now, the idea that Columbus's discoveries were immaterial to Mediterranean geography is such a folly, that Heisenberg himself has to modify it, replacing the word 'geography' with 'positive geography'. All that remains now to do is define 'positive'. 'Positive geography' evidently has nothing to do with views of the Mediterranean basin as the centre of the earth. Indeed the concept of positive geography, that is to say, of cartography of a refined form acceptable to Heisenberg's taste, did not exist in that day; nor were there at the time maps accurate enough to be viewed as positive; nor is positive cartography of the Mediterranean basin indifferent to the question of the curvature of the earth. At most Heisenberg could say that as a result of the Columbian revolution, the scientific revolution, etc., we can now say in retrospect that an idealized and improved version of pre-Columbian Mediterranean cartography is a very good approximation to post-Columbian cartography.

One may say that the idealization and correction of past theories have to be accepted anyway; and that then the difference between their being first approximation to, and being parts and parcels of, present day theories, is so negligible it may be ignored. First, this sounds dogmatic to me; second, when one insists on there being no correction at all, no change, and even in violent language ("it is equally wrong"), than it should be stressed that corrections were made, however small; third, the whole picture of science is rendered hyper-positivistic in order to render this idea plausible – which is a high cost for a very small return.

I cannot escape the impression that even Heisenberg feels uncomfortable about his own position and its all too seemingly dogmatic character. For, I think he proceeds from the above quoted passage to an attack on Kant's dogmatism as an expression of some unease. As if to draw attention to Kant's alleged even worse dogmatism. But Kant lived before Einstein, and so his error is easier to sympathise with than Heisenberg's.

[18] Robespierre's famous speech, his *Report on the Relations Between Religious and Moral Ideas and Republican Principles* of May 7, 1794, expresses all this very well. The art of government, he says, "has hitherto been the art of cheating and corrupting men, but ... ought to be that of enlightening and improving them". He proves the existence of God and the immortality of the soul thus. Man needs enlightenment and improvement. In order to achieve these man needs "more respect for himself and his fellowmen". In order to achieve these he needs faith in God and in the immortality of the soul. Finally, "I cannot see how Nature can have suggested to man fictions that were more useful than reality." This is fascinating, and a shrewd combination of a Cartesian and a Kantian mode of arguing. A Cartesian argument is from God's veracity and its form is, if an inquiry went properly yet the outcome was not the truth, then God would be a liar; which is absurd. A Kantian argument is transcendental and its form is, if such and such were not the case then knowledge would be impossible; but knowledge exists. Robespierre argues from the existence of knowledge and morality to the existence of religion and from the existence of religion in a proper manner to its truth – a combined Kantian-cum-Cartesian argument. This is not a small achievement for an allegedly verbose second-hand-ideas-spouting mere politician.

Since the above argument is rational, Robespierre continues, it is binding yet non-partisan: opponents to it are merely irrational. "You fanatics have nothing to hope from us. To recall men to the worship of the Supreme Being is to deal fanaticism a mortal blow. All follies fall to the ground before Reason; all fictions fade away in the light of truth. *Without compulsion, and without persecution, all sects are to be merged* in the universal religion of virtue." (Italics mine.) (J. M. Thompson, *Leaders of the French Revolution*, London 1929; New York 1967, pp. 236–7.)

In his second speech of July 26, 1794, he is even sharper. "I know but two parties, that of good citizens and that of the bad. ... There does exist a generous ambition ... an egoism of enlightened men. ... What then are we to do? ... to establish a single control ... and thus to crush all factions under the weight of national authority, and to build on their ruins the power of justice and freedom" (*ibid.*, pp. 139–141).

See also J. L. Talmon, *The Origins of Totalitarian Democracy*, London 1952, particularly the quote from Lemercier (p. 36) on the beneficial 'despotism of evidence', and on the "'natural and irresistible force of evidence' which rules out any arbitrary action on the part of the administration", and the quote from Dubois (p. 167): "Either there exists no demonstrable ethics at all, or there should exist only one – just as there exists only one geometry." (Translations mine.)

See also H. B. Acton, 'Prejudice', *Revue internationale de philosophie* **21** (1952);

J. W. N. Watkins, 'Milton's Vision of a Reformed England', *The Listener*, January 22, 1959; and F. A. Hayek, *The Constitution of Liberty*, London 1960, p. 527, n. 15 on Jefferson's opposition to academic freedom.

As to the sentiment prevalent today, Popper claims that the slogan 'Vox populi – vox dei', i.e. the idea that public opinion has already achieved a state of near perfection, is very widespread in western democracies, he even says it is the official ideology of western liberalism, and he wishes to combat it. See Popper's 'Public Opinion and Liberal Principles' in his *Conjectures and Refutations*. The same view, or a similar one, is criticized by Michael Oakeshott in his 'Rationalism in Politics', though from a very conservative viewpoint; M. Oakeshott, *Rationalism in Politics*, London 1962.

[19] De Finetti's view was foreshadowed by Wollaston, as reported by Faraday in 'Observations on Mental Education', alternatively, 'Observations on the Education of the Judgment', *Lectures on Education Delivered at the Royal Institution of Great Britain*, London 1854, reprint, 1855; also in *Modern Culture, etc.* (ed. by E. L. Youman), London 1867; also in *Science and Education: Lectures Delivered in the Royal Institution* (ed. by Sir E. Ray Lankester), London (1917); also in M. Faraday, *Experimental Researches in Chemistry and Physics*, London 1859.

For De Finetti's views, see *Studies in Subjective Probability* (ed. by H. E. Kyburg Jr. and H. E. Smokler), New York 1964, and bibliography there. See also H. E. Kyburg, 'Recent Work in Inductive Logic', *American Philosophical Quarterly* **1**, No. 4 (1964), and bibliography there.

The sentiment is well expressed in J. W. N. Watkins, 'Decision and Uncertainty' *Brit. J. Phil. Sci.* **6** (1955), which is a review-article of Shackle's *Expectation in Economics*. See his summary (p. 78): "Although there are ineradicable elements of uncertainty in human life, and although it is the most significant decisions whose outcomes tend to be the most uncertain, the classical theory of decision-taking, common to both the philosophical and the economic utilitarians of the 19th century, presupposed foreknowledge of the decision's outcome. In the 20th century foreknowledge was reduced to knowledge of the probabilities of the possible outcomes of a decision, in line with the tendency to substitute probability for certainty."

[20] S. E. Toulmin, 'Crucial Experiments: Priestley and Lavoisier', *J. Hist. Ideas* **18** (1957), reprinted in *Roots of Scientific Thought: A Cultural Perspective* (ed. by P. P. Wiener and A. Noland), New York 1957.

[21] Priestley's case is discussed in my *Towards An Historiography of Science*, Mouton, The Hague, 1963, '12. Priestley's Dissent', and in my 'Revolutions in Science, Occasional or Permanent?', *Organon* **3** (1966).

[22] The best description I know of all this is in F. L. Will, 'The Preferability of Probable Beliefs', *Journal of Philosophy* **62** (1965). I do not pretend to understand Will's discussion – nor do I consider it interesting – but I do think his conclusion (pp. 66–7) is striking. There is, he says, "no logical room for such questions" as why prefer probable beliefs. To say probably *A* often means, to say *A* is preferable; or else it "serves ... to express and appraise the grounds of possible assertions. ... And in some uses the specification of the strength of the grounds for a proposition may not, for somewhat special reasons, close the question of the credibility of that proposition. ... In some circumstances the appraisal ... may be made in such a way that the acceptability or credibility of the proposition ... [also depends] upon whether the person ... is, in fixing [!] his beliefs, acting within what may be referred to collectively as the institution of human knowledge." Whether Will has in mind *only* people who are contemptuous of science – who are at liberty to reject credible views and "what might be referred to

collectively as the institution of human knowledge" – I do not know. Interestingly, his description fits Einstein as well: "I live in that solitude which is painful in youth, but delicious in the years of maturity" ('Self Portrait', *Out of My Late Years*, London 1950, p. 5).

Does Will recommend alliance with "the institutions of human knowledge?" He does not say. It is not a matter upon which "the institution of human knowledge has a confirmed belief"; in the west that institution tolerates dissent and rebellion and the rejection of the most confirmed belief. See my 'The Confusion Between Physics and Metaphysics in the Standard Histories of Science', in *Proceedings of the Tenth International Congress for the History of Science, Ithaca, 1962*, Paris, 1964.

[23] The case of William James is rather complicated. On the one hand James was prone to assertions which are quite exasperating in their obvious unacceptability. On the other hand, James's desiderata were interesting though highly problematic. This may explain how come Russell wrote, in reply to the criticism that he had "caricatured pragmatism by saying that, according to it, truth is what pays"; merely "but this is a verbal quotation from William James" (Russell, 'Reply to Critics', in *The Philosophy of Bertrand Russell* (ed. by P. A. Schilpp), Evanston 1944, p. 731). James's desiderata were the defense of pluralism as well as of extreme empiricism, while being a reformist. See R. B. Perry, *The Thought and Character of William James*, Harper Torchbook 1964, chapters 24, 27, 30, 32, and especially the rather moving Conclusion.

[24] See Ellis' comment on *Novum Organum*, II, Aph. 36 (Experimentum Crucis), *Works*, new ed., p. 297, note 2: "Nothing shows better than an instance of this kind, the impossibility of reducing philosophical reasoning to a uniform method of exclusion. Bacon seems to recognize as the only true form of induction ... that ... which proceeds by exclusion. ... The argument depends on a wholly non-logical element, the conviction of the unity and harmony of nature."

For Bacon's pyramid of nature and science, see notes 7 and 17 above. The most explicit expressions of these sentiments concerning parallelism between mind and nature are to be found in Bacon's frankly mythological writings, such as his *Wisdom of The Ancients* (myths of Pan, Echo, and Proserpine), *Thoughts on the Nature of Things, De Augmentis Scientiarum*, Book II, Ch. XIII, as well as his preface to his *Great Instauration*, Preface and Conclusion of *Novum Organum*, *Parasceve*, etc. Those who wish to take Bacon's mythology lightly should consult Spedding's introductions to both *The Wisdom of the Ancients* and *The New Atlantis*, as well as C. W. Lemmi's *Classical Deities in Bacon*, Baltimore 1931. See also *De Augmentis* Book 7, Ch. III, where the parallel is more boldly stated as the topics are "the culture of the mind" and "moral knowledge" (*Works*, *op. cit.*, Vol. IV, p. 28). Moral knowledge, however, is declared in the opening of *Valerius Terminus*, *The Advancement*, and *De Augmentist*, to be the highest end of all learning. These are the most cabbalistic passages of Bacon (esp. *Valerius*, Ch. I), which adumbrate the intellectual love of God most clearly.

There is no doubt that one way or another all inductivist philosophers felt that God must have put some constraint on Nature so as to render Her comprehensible by inductive means (see quote from Bacon in note 12). It was J. M. Keynes who, in his *Treatise on Probability*, Cambridge 1921, Chapter 23, claimed that both Bacon and Mill were rather vague about it, and he, wishing to be clear and explicit, postulated his famous principle of limited variety. The principle is clearly enough stated by Ellis, and quoted in the beginning of this note.

The most subtle attitude towards this principle and similar ones is Kant's, need one say? On one hand, he says, any such principle is concerning reason and nature and so could only be judged from above both nature and reason – namely not by us. He even goes further and rejects violently any such principle even as the merest hypothesis. Transcendental hypotheses, as he calls these, "such as the appeal to a divine Author" (A773B801), are not explanatory, and lead the mind to unhealthy self-satisfaction (A772B800). "Order and purposiveness in nature must themselves be explained from natural grounds and according to natural laws; and the wildest hypotheses, if only they are physical, are here more tolerable than a hyperphysical [transcendental] hypothesis, such as an appeal to a divine Author, assumed simply in order that we may have an explanation."

On the other hand, Kant does not like Hume's 'censorship', and claims that transcendental hypotheses may be stated; that he always likes to read new books defending them. He adds that transcendental hypotheses are useful to combat contrary transcendental hypotheses – the result of the combat should be a draw. That even after the draw there is use – one may take such a hypothesis as an ideal, or as a regulative idea or principle, which should guide one's research (rather than elicit false satisfaction). What is unclear to me is whether the ideal is inter-subjective or private. It can be nothing else; if it were inter-subjective it would be apriori demonstrable, and if private they do not belong to the metaphysics of morals, namely, they are not rational guiding principles (of research). In a sense Kant says they are private (A782B810), in a sense he stresses the universality of the ideals or regulative ideas or principles of pure reason. I cannot follow his subtlety there. I can only say, surely he wants research, the intellectual love of God, to be a supreme universal principle, but he may perhaps claim that research is not necessarily bound to ideals. This is not satisfactory, but I leave it at that.

Much has been said in comment on Carl Becker's *Heavenly City of the Eighteenth Century Philosophers* and its claim that there is so much in common between the medieval and the 18th-century philosophers. Yet one might sum up the intellectually relevant similarities. First, the main difference between the Medieval thinkers and their heirs lay in the formers' sense of utter impotence. The medieval philosophers shared with the Renaissance philosophers the dream of the recapture of antiquity, but before Brunelleschi's success in constructing a dome in an ancient manner, followed by his disciples' success in sculpting, painting, and building, like the ancients, the general sense was that of impotence. Similarly, the idea of the hierarchy of knowledge and the knowability of the world, potent in the post-Baconian era, existed before. Indeed, the main message of Bacon, as he himself stresses, is a message of hope and of cajoling people to do research.

The second important difference is that the medieval writers are much more confused than the modern writers. This point is hard to divine from secondary sources, such as Crombie's in particular, because authors usually quote only what they think is clear or what they wish to clarify, or some such. Confusion in the Middle Ages much depends on slavery to diverse ancient authorities; but, as Galileo showed, the elimination of error is an arduous task, and of confusion even more so.

So much for the differences; as to the similarities, they lay in the confusions between Plato and Aristotle, between induction and deduction, between learning (methodology) and knowledge (epistemology), all as means for supporting the optimistic thesis of the knowability of the universe, to be found both in the Middle Ages and later. Yet, on the whole, in the 18th century this led to an ode to nature and to man's mastery of nature, whereas earlier it often sounded like a wail over a paradise lost.

(On the unity of science in the Middle Ages see M. DeWulf, *Philosophy and Civilization in the Middle Ages*, Princeton 1922, end of Ch. 4, vi and 5, i, ii; DeWulf is a Catholic apologist worse than Crombie, but he is still rather informative.)

And so, again, Crombie may score a point in viewing Grosseteste as a predecessor after a fashion to Bacon and to Galileo. But in a way I do not think he would like. In his second chapter he admits (p. 32 note) that the 12th-century philosophers received these ideas from Aristotle via Boethius; yet he calls them Platonists and complains of their mistrust of the senses – forgetting Kepler's and Galileo's mistrust of the senses; forgetting that Grosseteste himself does too (p. 73)!

And even Francis Bacon who is among the Moderns almost the only Quixotic believer in the senses, even he constantly and systematically refuses to side with Plato or Aristotle about the hierarchy of knowledge, speaking of 'axioms and definitions' in one breath, more systematically then Grosseteste and Roger Bacon, and even claiming explicitly that Plato is the only one who tried the inductive method, but only after corrupting science by mixing it with theology (*Novum Organum*, I, Aph. 105). (The widespread Ellis translation says, "Plato, who does indeed employ ... induction ... for the purpose of discussing definitions and ideas." The expression "of discussing" is a rendering of "executiendas"; in Khintchin's translation, Oxford 1855, it is rendered "of formation"; why not "of executing"? Ellis also translates "dialectics" as "logic". See my dissertation, University of London, 1956, unpublished.)

All this, of course, raises again the question of the difference between apriorism and inductivism, both in the Middle Ages and later. Now in the Middle Ages the hierarchy of causal theories was identified with Jacob's ladder, namely with divine illumination. During the late Renaissance, as the outcome of some measure of growth of both rationalism and optimism of sorts, two attitudes developed towards divine illumination. One was the neo-Platonist or light-mystic, i.e. Cabbalistic attitude run so optimistic as to expect illumination more or less here and now. The other, more academic, attitudes excluded divine illumination altogether and replaced it with the authority of the senses. And so, whereas in Grosseteste knowledge by divine illumination has no recourse to empirical basis but defective knowledge is empirical, in the late Renaissance academies both kinds of knowledge tend to merge and raise the serious problems *which are still with us*. And to think that it all starts with Aristotle's two systems – of certain and of probable knowledge!

[25] For the connection between Whewell and Duhem, see my 'Duhem Versus Galileo', *Brit. J. Phil. Sci.* **8** (1957), as well as my *Towards An Historiography of Science*, *op. cit.*, Section 10.

For Duhem's influence on Meyerson see Meyerson's preface to his *Identity and Reality* and references there.

See also my 'Sensationalism', *Mind*, 1966, concerning the difficulty involved in the view that there are uninterpreted data.

J. O. WISDOM

ON METHODS OF REFUTATION IN METAPHYSICS

A PROBLEM OF RATIONAL DISBELIEF

What I wish to open up is a sub-problem of the problem of rationality. Most of us who think of ourselves as rational men would like to be able to give reasonable reasons for the things we believe. What is much less explored is the opposite problem, regarding the things we do not believe, of being able to find reasonable reasons against the things we disbelieve. This problem arises because so many systems of philosophy are put on the shelf without being firmly refuted. Consider, for example, the system of Spinoza or of Berkeley or of Kant. It is the dominant outlook in the Anglo-American philosophical world to regard most of these philosophies as outdated and as saying very little that can be taken seriously. There may be justice in this view: quite possibly there are very few living issues in many of these philosophers. But one must take them seriously if for no other reason than that there is no standard refutation of them. But if you look back on attempts to discuss historical philosophies, it is extraordinary how little agreement there is to be found on what is wrong with them. We all know that there is no agreement about positive contributions to philosophy, but there is equally little agreement on what shall be rejected and how. It seems to me very curious that there should be no way of settling these matters. In view of this I have cast around for various methods by which metaphysical views, if they should be false, might be rejected. That, in short is my problem.

SYNTHETIC A PRIORI METAPHYSICS

If you cast your mind back over the great philosophies of the past, to consider how you might refute them, you are bound to consider the notion of a synthetic *a priori* proposition, upon which the whole Kantian philosophy was built; and the question must arise whether the notion of such propositions can be refuted.

A synthetic proposition is *a priori* if, though synthetic, it is necessarily true (or true independently of all experience).

Examples of supposedly synthetic *a priori* propositions may be taken from Kant himself. Thus causality was a category that applied to the world universally and necessarily. Kant also had some rather unfortunate but still rather instructive examples such as that two straight lines cannot enclose a space. For when he gives examples that we know are not true, this does not encourage us to place much confidence in the concept of the synthetic *a priori* proposition. Incidentally, I think it was a pity that he used the example of two straight lines not enclosing a space; if he had been bolder and said that one straight line cannot enclose a space it would have been more striking, seeing that in elliptic geometry one straight line can enclose a space.

To try to get some clear idea of what this very curious kind of proposition is, let us remember that a synthetic *a priori* proposition is not analytical, is not just a tautology. A tautology is a proposition that is true of all possible worlds. But although a synthetic *a priori* proposition is not tautological it is necessary. Thus it meant a proposition that is true, not of all logically possible worlds, but, I suggest, in all possible worlds of a certain sort, namely of all 'conceivable' worlds. The difference in scope may be shown like this. A tautology contains no information, but a synthetic *a priori* proposition does or is intended to do so. This, of course, is a stumbling block; practically no modern logician of any school could swallow the possibility of an *a priori* proposition that was other than analytical; and the synthetic *a priori* proposition is just this. It is worth turning this the other way round, even if it helps only a little, to see what would happen if you tried to deny a synthetic *a priori* proposition. The denial would be a proposition that is false, and necessarily false, without being self-contradictory. And this is very curious.

The notion of a synthetic *a priori* proposition is particularly important because it may well be central to most metaphysics. Although Kant introduced the idea of a synthetic *a priori* proposition only rather late in the history of philosophy, the idea, or its equivalent, can be detected much earlier. There are in particular two ancestors of the idea that I should like to mention. One is the doctrine of innate ideas in Descartes. It is to some extent obscure just what an innate idea really was in Descartes. Nonetheless certain features stand out. An innate idea is characterized as something that is clear and distinct and gives you necessary truth. But it also provided truth about reality, and hence was not simply tautologous; for innate ideas were not intended to be vacuous ideas that were uninformative about the world. So you get the same two characteristics in connection with innate ideas in Descartes as you get with the synthetic *a priori* in Kant. More interesting, perhaps, is the idea of an essence in Aristotle. Though he put it in terms of 'real definition', an essence was something that you contemplated in order to find out real knowledge about the world. If you wanted to find out about man, you would first of all ascertain what is the essence of man. The essence of man was said to be rationality and animality; and this would be your major premiss. Then, by deductive procedures of logic, you would arrive at other properties. Thus one could put Aristotle's view of scientific procedure like this: you do not have to build upon experience; you need impeccable logic, and a perfectly sound manner of ascertaining essences of things; given the essences, to be stated in the major premiss, apply impeccable logic, and you would derive necessary properties of things. In this way you learn about the universe (Popper, 1963). The point to note about this is that the properties of the essence were necessary (and therefore of course universal), but also provided information about the world. These two characteristics of necessity and information-content lead me to suppose that the old idea of an essence is something that can and must be the same as the idea of a synthetic *a priori* proposition, and expressible by means of such a proposition, and therefore that the Kantian doctrine of synthetic *a priori* propositions represents something that is well-nigh as old as philosophy.

My surmise is that really metaphysical metaphysics always turns on such propositions.

A THEORY OF THE SYNTHETIC A PRIORI

Be that as it may, we have to consider how we might handle these things. For this purpose, I would propose the following: that a synthetic *a priori* proposition has a universe of discourse, which so far from exhausting the entire universe, is limited to what is regarded as 'conceivable', and that this restriction is due to some limiting property that is taken to characterise the universe of discourse; such a property excludes a certain domain from discourse; moreover the excluded domain is kept outside our intellectual vision, because 'inconceivable'. An example I shall begin with comes from the Kantian distinction between phenomena and things-in-themselves. Kant's theory was that the only world that you can know is the world of experience. It is a short step to disregard the rest of the world so that the whole world comes to mean simply the world you know. The domain of things-in-themselves does not count in our experience, and becomes excluded from the universe of discourse. This world that you know is the world that is subject to the categories of causality and so on. Now if you make a restriction, excluding the domain of things-in-themselves, then you can do it in such a way that a category like causality will necessarily apply to the whole of the rest. If that is so, then the synthetic *a priori* proposition, to do with causality, is necessarily true of all phenomena or must in fact hold of all phenomena.

Now if we look more widely and try to see whether this idea has been employed elsewhere, I think we can find other examples of it. Consider Hegel, for instance. Though he is notoriously difficult to discuss, nonetheless one can fairly say that the Hegelian philosophy contains centrally the proposition that the real is rational. What does this tell us? That the real excludes the irrational; the irrational is not part of the real world. In that case, any proposition that is true of rational things holds for the whole world, and therefore has the look of being synthetic *a priori*, because it excludes the domain of the irrational, which may plausibly seem not to exist.

Take a different sort of example altogether. Supposing you are born color-blind, so that you can see only black and white. You might very well suppose that it was a necessary truth that colors are divisible into black and white. Or take a much more mundane example such as this: we may easily suppose that all rabbits are long-eared. If a new animal appeared, which might be a rabbit with short ears, a controversy could develop, perhaps among biologists, about whether this was a rabbit or not. Some might maintain it could not be a rabbit because it is synthetic *a priori* that rabbits are long-eared.

We may, if we wish, refer to the notion of the synthetic *a priori* proposition as the principle of 'synthetic *a priori* exclusion'. It is exemplified by the plausible exclusion of that other field of things-in-themselves we know nothing about.

What I am trying to bring out, by means of this device of analyzing the synthetic *a priori* proposition, is this. If such a proposition might by chance be true, then in particular it might so happen (though some of us do not believe it) that causation is a universal category, for it might so happen that everything in the universe without any exclusion is subject to laws of cause and effect. In that case, then this proposition would be factually true. Hence if one of these propositions might be true, then of course there could be no way by which you could ever refute them in principle. You might be able if you are lucky to refute one or two of the specified members, but you could not refute all of them in principle if any one of them could be contingently true.

Thus the first point I want to bring out about them is that any of them might be empirically true. This is part of the reason why they carry some air of plausibility. Next if you consider a supposed synthetic *a priori* proposition, it is always possible that somebody will be able to devise an area to which it does not apply. And if you can devise one you have a satisfactory refutation. For example, supposing you held that energy must always be a positive quantity, which might sound reasonable enough, and that this was synthetic *a priori.* Then you would not be receptive to the possibility of a counter-example. And yet, of course, Dirac did in fact postulate the existence of negative energy, leading to the discovery of the positron. This refutation of the idea that energy has to be essentially positive is, of course, highly imaginative. Moreover you find such a counter-example only when you are lucky. The feat of imagination, not to mention the luck required to find an area to which a synthetic *a priori* proposition fails to apply, readily explains why we regard such a possibility as inconceivable, and why the proposition seems necessary.

An interesting feature comes out here, that it is possible to broaden the scope of enquiries by deliberately trying to deny what appear to be universal necessary truths. An example that naturally comes to mind is the standard 'truth' of Euclidean geometry which I referred to before, namely that it was maintained that the Euclidean axioms were universally necessary and true. One can sit down deliberately, as in the end mathematicians did, to try to see whether one could find an alternative or find an area in which the 'truths' do not hold. In this connection, I should like to mention another example. A synthetic proposition that was denied *a priori,* was that of the spontaneous creation of matter out of nothing; but it has appeared in contemporary scientific cosmology, put forward by Kapp, Bondi, Gold, and Hoyle. This was a serviceable theory up to a point, in that it did some explanatory work, so it cannot be thrown out on the grounds that the idea does not make sense. If it could solve a problem, then the idea was good. A similar though stronger example concerns the spontaneous breakdown of radium atoms, which involves the idea of spontaneous destruction of matter. These theories violate the idea that matter cannot come out of nothing and cannot be destroyed. And most thinkers for thousands of years had held that these were necessary truths. Again, we can see that a synthetic *a priori* proposition can be denied effectively, but only when we can produce a testable theory that denies it.

Now to turn to synthetic *a priori* propositions in connection with metaphysics. If it is true that all metaphysical metaphysics centered on these things, then I suggest that we have a method of handling metaphysical questions. There has been an idea, of course, for a very long time, which Hume and logical positivists have had in common, that there exists a principle yielding a method of relegating all metaphysics to the flames. That is well known. And it is also known that it cannot be done: there is no principle of drawing the line or inserting a wedge between metaphysics that is nonsensical or false and science that is not.[1] Now if the foregoing analysis of synthetic *a priori* propositions is correct, we can see why this is so. The reason is that anyone might happen by luck to hit on a synthetic *a priori* proposition that happened to be true without excluding part of the universe from its scope. Hence a piece of metaphysics might by chance be true, and therefore would not be nonsensical. The question then arises, what should you do? I suggest that the procedure is to try to find the boundary line that renders it such a proposition, and then challenge the boundary – which means being able to invent something on the other side of the boundary.[2] This will usually prove too difficult because it may require that one should have a new scientific theory which one cannot invent on the spur of the moment.

As an example I would venture a possible solution to the old problem of natural necessity. This problem arises, as I have explained at length elsewhere (Wisdom, 1946), from the dillemma that the causal nexus must be either one of necessary connexion or one of *de facto* regularity; but necessary connexion (other than what is logically necessary) would seem to be a myth, while *de facto* regularity would allow anything to happen after anything and would therefore be inadequate to explain such orderliness as exists. In an infinitely extensible universe, however, it is 100% probable that an area in space and time, large to us but infinitely small in relation to the rest, exists characterised by orderliness as a chance effect. In other words we should have the impression of a law of orderliness, persisting and ubiquitous in our 'universe', which would in reality be but a random effect in an isolated pocket of the universe as a whole. For this to solve the problem of natural necessity, we should have to test the hypothesis that most of the universe is lawless or without order. But even the untested conjecture is enough to show that the causal nexus does not have to be either one of necessary connexion or one of *de facto* regularity throughout the entire universe.

The consequences of the foregoing theory of the nature of the synthetic *a priori* proposition are thus: any supposed example might be empirically true, hence such propositions are always plausible, and are never refutable in principle; on the other hand, individual counter-examples can sometimes be found, hence no such proposition can be necessarily true – a synthetic *a priori* candidate cannot be shown to be necessarily false or necessarily true. In particular the Humean or logical positivist general method of liquidating such propositions is impossible, though a method of modest challenge can be provided. The method of challenge, when no counter-example is actually found, leaves them unrefuted but shows that they are necessarily unprovable so that no conclusions that might be be based upon them are thereby reliable.

METAPHYSICS OF INADEQUATE SPECIFICATION

If metaphysical metaphysics is full-blooded metaphysics, there is also a milk-and-water metaphysics, which hinges not on the synthetic *a priori* proposition but on inadequate specification of basic notions and their relations. *Parturiunt montes*... The resulting cosmological vision may be described as an 'ontology of unspecified make-up'.

Such a position may be illustrated by Berkeley's cosmology. His problem was to explain the existence of sense-perceptions. We are not concerned with the very acute and subtle inferences that he made in the body of his work and the valid attacks he made on other philosophers, but only with his overall solution. His position was that a postulated entity, called God, with very few properties such as omnipotence and perfection specified, explains the existence of the perceptions that we have, by causing these perceptions in us or manifesting them to us as a language by which we read of God. This explanation is like an *ad hoc* hypothesis because the only evidence for it is the sensory experiences to be explained; there is no independent evidence for it. Further we are given practically no specification of the entity involved, and practically no specification of the machinery by which the effect is brought about, or by which the deduction is made, in short by which the explanation does its work.[3] This is what I am describing as inadequate specification.[4] The point about it is that there is no way of refuting such a doctrine because it is essentially too vague to be discussed. One cannot say that no claim at all is made because one or two specifications or properties of God are mentioned, and some process of causation also; but there is insufficient detail given to identify the object, and insufficient detail given about the chain of deduction by which the explanatory concept explains by a causal process our sense-perceptions. So there is only a very partial claim. I would summarize the position by saying that there are, in this Berkeleian solution, bits and pieces out of which a theory might be propounded, but which are only bits and pieces not really hewn into shape; you can hardly refute the position because you have hardly got a claim; and if you have a very mild claim, too weak to specify to any significant extent, then you have almost nothing to answer.

THE EMBEDDED ONTOLOGY OF SCIENCE

Let us turn to a method of refutation that relates to a different form of metaphysics altogether. It is well known that in Popperian methodology the aim of a scientist or experimentalist or methodologist is to consider the refutability of scientific theories (Popper, 1959). And the method dealt with by Popper in his first work was to find empirical refutations i.e., factual observations, that would refute a theory. Now Popper never suggested that refutation by observation was the only method of refutation; he was also concerned with criticism of theories in general; but it was refutation by observation that he developed a theory of. I noticed, however, that one can go further, and, in line with his broader idea of criticism, can devise a different method of refutation for certain statements to which the method of refutation by observation does not apply (Wisdom, 1963b).

For example in all classical physics, it is an implication that energy can exist at all levels, i.e., there is a continuity in the values energy can take (although this is nowhere said in a textbook of classical physics from the dawn of man till now). There is nothing to rule it out. But if you were to ask yourself, as a physicist, how you would test this, you would be defeated: there is no observation that you could make that would go against it; for if you found some break in observations of energy-values, the reply would always be open that we have only to look just a bit more to find an energy-value to fill the gap. So if in the nature of things you cannot get an observation that could refute this particular statement, is it therefore metaphysics? Popper is not specially concerned about such a statement, because it is a consequence of and part of a testable theory. But in itself, it is not refutable by observation. Is it, then a piece of metaphysics that forms part of science?

The method I have put forward for handling such a statement is this. Begin by finding another theory that refutes it. And such is to hand in quantum physics. Quantum physics is first of all a testable theory, secondly it has been tested, and thirdly it has been corroborated. Now if you have a theory that is testable, tested, and corroborated, and that theory has a logical consequence, you have to accept that consequence; and if it is incompatible with the former theory, the former one has to be rejected. In particular, quantum physics, which is testable, tested, and corroborated, has a consequence that energy exists only at discrete levels (possesses only eigen-values). This contradicts the consequence about continuous energy-levels I have drawn from classical physics; therefore quantum physics refutes this statement from classical physics. The statement is not, Popperwise, observation-refutable; it is theory-refutable.

Now of course there is more to be said on the matter. This is not an absolute refutation, because after all we might have to give up quantum physics. In which case the refutation would go. But this is really not very detrimental to the purpose in hand, for our purpose is not to find absolute refutations. Even such apparently absolute things as contradictions may be found to be wrongly ascribed, because we misconstrue a theory or its symbolism may have been too vague. Again, even an observation-refutation of a theory is not absolute, because we may find that the observation was at fault (Popper, 1959, 104). It is true that refutation by means of observation is in a slightly stronger position than refutation by means of a theory; but that does not produce a difference of principle. The point is that, so long as we do accept quantum physics, in the rational sense that it is tested, testable, and corroborated, that is to say, so long as it has not been refuted, then we have to accept its logical consequences, and that means that we have to deny something that is incompatible with its logical consequences (in this case we have to deny the classical proposition that energy exists with continuous values).

But are we dealing with metaphysics? Certainly to say that energy exists at all levels (or only has discrete values) is to say something about the 'furniture of the earth'. We are therefore dealing with what I will call the 'embedded ontology of science'. Other examples, however, which seem more metaphysical, may help better.

Consider the idea of absolute space in Newtonian mechanics. This idea is no part of the empirical content of Newtonian theory in the sense that whenever you carry out any calculation or try to prove any theorem in the system you never write down the hypothesis that space is absolute. So in this sense there is a piece of metaphysical framework, or even metaphysical warp and weft, different from the empirical theory. Nonetheless the idea of absolute space in Newtonian mechanics is refuted by the theory of general relativity; for it is a consequence of this theory that space is non-absolute. And here you have a perfectly good metaphysical hypothesis, namely that of absolute space, refuted, not by an observation, but by a theory. And the doctrine of absolute space is surely metaphysical in some sense. And it is perhaps the sort of metaphysics that scientists should indulge in, the sort that is and probably must be indispensible to science.

I would like to mention one or two bizarre examples of ideas that can be refuted by this method. One might suppose that things like values were outside the sphere of refutability altogether. But consider preclassical physics, before Galileo. In the days of Aristotelian physics, or of the theory of impetus, it was common doctrine accepted by all that the earth was not only the center of the universe, but was the center of the universe because the dignity of man required it to be so. And another idea was that the heavenly bodies had to be round, because that was the perfect shape. Now these value judgments, so far as I know, actually entered into the body of the theory, i.e. you could not carry out the deductions or explanations in Aristotelian physics or in the physics of impetus, without making use of these value-judgments. If this is so, then the refutation of these systems of physics would carry refutations of these value-judgments as well, because the refuting theories entail that the earth is not the centre and that orbits are not necessarily circular.

Now let me take one or two other examples of problems from the history of philosophy that can be handled in a similar manner. These is a very interesting question about whether matter is infinitely divisible or not. I think it is very reasonable on the part of mankind to wonder about such a question. Unfortunately philosophers find it less interesting than they used to, and it has become unfashionable. What sort of question it may be is hard to characterize. It looks like the sort of thing people might have called the philosophy of nature; it differs from methodology in that it is not a second-order subject; and it is not a subject of scientific inquiry. It is considered by the philosopher in his arm-chair directly concerning himself with the real world. And there are other questions of the same sort, e.g. whether space is infinite or not. It happens that philosophers have not got very far with such questions; but I think we can now understand why. Philosophers have produced logical antimonies about them and have got nowhere with resolving these. But an interesting fact is that there are hypothetical answers to some of these questions to be got from modern physical theory. General relativity gives an answer to one of them and quantum mechanics gives an answer to another. Thus one answer would be that matter is not infinitely divisible, because in quantum mechanics matter must have a certain minimum size or ceases to be matter (in terms of the ordinary conception). And in the case of the infinity *versus* finitude of space, general relativity requires that space-time should be finite (although you meet no boundary no matter how far you move about in space-time). These two answers are consequences of scientific theories, though the consequences themselves are not directly checkable. So, in these cases of philosophical investigations of nature, philosophers have had to wait for answers depending on results of science.

THE ROLE OF METAPHYSICS IN SCIENCE

All of these examples differ from empirical theories in not being observation-refutable, but they have all played a part in science and have all been woven into some scientific theory or other. Such an example I call the 'embedded ontology', of a scientific theory. Some of the examples have also been discussed extensively by philosophers *in vacuo*; thus the absoluteness of space and the infinite divisibility of matter have been discussed in isolation from any scientific theory. Whichever way they are handled, they resemble other forms of metaphysics; and I think that there is always some metaphysics of this kind, or 'embedded ontology', forming the warp and weft of a scientific theory.

We need not consider the relationship of embedded ontology to the milk-and-water ontology of unspecific make-up, but we do need to consider its relation to the full-blooded metaphysics of the synthetic *a priori*. The same propositions may occur in both. And we have seen that some supposed synthetic *a priori* propositions can on occassion be refuted by a scientific theory. So there is no essential difference between the two classes. They differ mainly about the context in which they are entertained and as regards putative method of justification: they may be entertained in the context of an empirical theory or not; if they are their justification hangs on the testability, actual test, and corroboration of the empirical theory, and if not they usually lean on some synthetic *a priori* foundation. They may, however, belong to a deeper level of abstraction.

If it is true that empirical scientific theories always have associated with them, or always have a warp and weft of, an embedded ontology, or even if they only sometimes do, we are faced with the questions about the role of metaphysics in science.

It becomes evident that it was not wrong for philosophers to wonder about these ontological problems, if they conducted their speculations in the context of science. But this is what Plato, Aristotle, Descartes, Leibniz, and Kant usually did; full-blooded metaphysics, divorced from science tended to be, historically speaking, a minority deviation by the lesser figures.

We may demarcate an area of problems in which philosophers and scientists can, and indeed should, speculate, although they cannot solve these problems by speculating; they will have to wait, without any guarantee that waiting will yield results, until some new scientific theory is able to provide an empirical answer, or rather an answer in terms of an empirical or testable theorem. So part of what I wish to unfold about the role of metaphysics is that it is vitally important, though assured of short-run failure, to conjure with problems that science cannot yet tackle: at best this may foster progress with empirical scientific theory; at worst it will fail but may be stimulating.

APPENDIX

There are subsidiary methods that can be used for refuting isolated cases in metaphysics. They are of limited interest, because they do not apply widely.

(i) The first is the method of self-contradiction. One might suppose that one could run self-contradiction to earth in the work of great philosophers and great systems, but if this were the case, then surely these contradictions would have appeared long ago. Very clever critics have been examining the great systems of philosophy for generations and centuries and almost no agreed contradictions have turned up. Now and then the accusation of self-contradiction is made, but it is not a very common accusation. An example is to be found, I think, in Leibniz. He held that monads were wholly and completely independent of each other; secondly, God pre-established a harmony of corresponding events between different monads; and thirdly, God was a monad. From which three propositions, it follows that God interferred with independent monads. A rarity is to be found in Locke's account of an abstract general triangle which was said to be isoceles, scalene, right-angled, equiangular, and all and none of these at once.

(ii) The second is empirical refutation by observation. It might come as a surprise that this method can be used at all. Full-blooded metaphysics would not stoop to anything connected with observation. None the less, the realistic theory of perception, which involves that sense-data are part of the surface of the object you are looking at, is commonly refuted by pressing the eyeball and getting a double image, which shows that these images cannot both be part of the surface of the physical object. Here is an observation that is generally taken to constitute a refutation. I would be hard put to it to provide another example in philosophy.

(iii) There is what I will call the fallacy of Repeating the Question. An example will bring out what is meant. There is an old essentialistic version of the theory of inductive reasoning, which is that an inductive generalization is known to be true because it connects the essence of the subject with the essence of the predicate. You do not find much about this in modern works, mainly because this kind of doctrine has not been fashionable or even interesting to most philosophers for a great many years. The last person I know of to have written about it in a logic book was Joseph who was an Aristotelian. The idea was that the essence of a swan would be connected with the essence white, and therefore you could trust the generalization, 'All swans are white', because of the intrinsic connection of the essences, swaniness and whiteness. But in order to achieve this end, one must begin with the observation of a swan that is white, and then pass from the observation of the swan to the essence of a swan, and finally in order to achieve the desired result, one would have to have the inductive generalization holding good that the observation of the swan was universally and necessarily connected with the essence of the swan. Which is the problem of induction again. So it seems to me that what happens, in this alleged solution of the problem of induction, is to raise the question once more, exactly the same question, at a slightly different point. And this seems to me to be as invalid as the classical fallacies, such as that of begging the question, and I call it the fallacy of *repeating the question.*

This kind of fallacy is of a little more interest, because it is a little more widespread, than the first two. Another example concerns the controversy between idealism and materialism. Presumably subjective idealism arose because there was a belief that there could be no relation between two utterly unlike entities such as mind and matter. Therefore it had to be supposed that matter was not really matter but was mentalistic in some way. So the dualistic difficulty was overcome by a monistic reduction treating the two entities as essentially of the same sort. Let us say that matter equals a kind of disguised mind. Then for the idealist it becomes conceivable that this mentalistic matter should be relatable to mind. But you have then on your hands the problem of understanding the relation between matter and its disguise which would seem to raise the same problem all over again. And you can apply the same criticism to materialism if you treat the mind as kind of a thin matter.

We may formulate the fallacy, or the refutation, as follows:

Consider a metaphysical position (M) designed to meet a problem-situation containing a difficulty or a question Q. If M necessarily leads to Q, then in answering the question, M raises the same question again. Then M is untenable: M repeats the question.

(iv) Let us take another example, which rests upon what may be called the fallacy of *begging the denial.* In the doctrine of occasionalism as usually interpreted, it was regarded as inconceivable that mind and matter should be related because the two entities were so unlike. In particular they could not be related by the relation of interaction, i.e. there could be no causal relation between mind and matter or mind and body. The device was therefore used that an intermediary occurred consisting of the agency of God. According to the doctrine, on the occasion when I will to raise my arm, God ensures that my arm actually rises. It does not take a great amount of perspicacity to see that God, i.e. another mental entity, is now acting on my physical body. Therefore there is a relationship between a mind and my body, which was presumed inconceivable. So the occasionalist solution would seem to presuppose the relation that it actually denied.

We may formulate the fallacy, or refutation, as follows:

Consider a metaphysical position, M, designed to meet a problem-situation ostensibly containing a factor, F, that is regarded as inconceivable. If M contains

F, M assumes the possibility of a factor presumed to be impossible. Then *M* is untenable: *M* begs the denial.

Examples of (iii) and (iv) are forms of solution depending on the idea of a *tertium quid*. So there is a certain field of application of this sort of fallacy.

BIBLIOGRAPHY

Church, A. (1949), Review of A. J. Ayer's *Language, Truth and Logic*, in *J. Symb. Log.* **14**, 53.

Kapp, R. O. (1960), *Towards a Unified Cosmology*, London, Ch. 2.

Popper, K. R. (1959), *The Logic of Scientific Discovery*, London.

Popper, K. R. (1963), *The Open Society and its Enemies*, Vol. II, London, Ch. 11, pp. 9–12.

Wisdom, J. O. (1946), *Causation and the Foundations of Science*, Paris.

Wisdom, J. O. (1953), *The Unconscious Origin of Berkeley's Philosophy*, London, p. 28.

Wisdom, J. O. (1963a), 'Metamorphoses of the Verifiability Theory of Meaning', *Mind* **72**, 335.

Wisdom, J. O. (1963b), 'The Refutability of 'Irrefutable' Laws', *Brit. J. Philos. Sc.* **13**, 303.

REFERENCES

[1] Church (1946) torpedoed the main formulation of the principle of verifiability by showing that this in fact admitted all metaphysical statements as verifiable. Wisdom (1963a) tried to show that the matter was not one of formulation, but that any principle driving such a wedge is equivalent to a synthetic *a priori* proposition.

[2] This is the same in spirit (though put forward here for a different purpose) as the methodological principle of dropping restrictions, introduced by Kapp (1960), which he used to derive a number of results in physics.

[3] This was the only criticism of his philosophy I gave in my book on Berkeley (Wisdom, 1953); so far as I am aware, it had not been made before.

[4] Locke's theory that substance, which can hardly be said to be specified at all when described as an 'unknown somewhat', provides a base for our perceptions; it is structurally just the same as Berkeley's. Other parallels are to be found in Occasionalism and Kant's theory of the thing-in-itself.

SYNTHESE LIBRARY

Monographs on Epistemology, Logic, Methodology,
Philosophy of Science, Sociology of Science and of Knowledge, and on the
Mathematical Methods of Social and Behavioral Sciences

p.t.o.

*EVERT W. BETH and JEAN PIAGET, *Mathematical Epistemology and Psychology*. 1966, XXII + 326 pp. Dfl. 54.—

‡GUIDO KÜNG, *Ontology and the Logistic Analysis of Language. An Enquiry into the Contemporary Views on Universals*. 1967, XI + 210 pp. Dfl. 34.—

‡ROBERT S. COHEN and MARX W. WARTOFSKY (eds.), *Boston Studies in the Philosophy of Science*. Volume III: *Proceedings of the Boston Colloquium for the Philosophy of Science 1964/1966*. 1967, XLIX + 489 pp. Dfl. 65.—

‡C. D. BROAD, *Induction, Probability, and Causation. Selected Papers*. 1968, XI + 296 pp. Dfl. 48.—

‡GÜNTHER PATZIG, *Aristotle's Theory of the Syllogism. A Logica-Philological Study of Book A of the Prior Analytics*. 1968, XVII + 215 pp. Dfl. 45.—

‡NICHOLAS RESCHER, *Topics in Philosophical Logic*. 1968, XIV + 347 pp. Dfl. 62.—

‡ROBERT S. COHEN and MARX W. WARTOFSKY (eds.), *Boston Studies in the Philosophy of Science*. Volume V: *Proceedings of the Boston Colloquium for the Philosophy of Science 1966/1968*. 1969, VIII + 482 pp. Dfl. 58.—

Sole Distributors in the U.S.A. and Canada:

*GORDON & BREACH, INC., 150 Fifth Avenue, New York, N.Y. 10011
‡HUMANITIES PRESS, INC., 303 Park Avenue South, New York, N.Y. 10010

GPSR Compliance

The European Union's (EU) General Product Safety Regulation (GPSR) is a set of rules that requires consumer products to be safe and our obligations to ensure this.

If you have any concerns about our products, you can contact us on ProductSafety@springernature.com

In case Publisher is established outside the EU, the EU authorized representative is:

Springer Nature Customer Service Center GmbH
Europaplatz 3
69115 Heidelberg, Germany

Batch number: 10815974

Printed by Printforce, the Netherlands